D1229450

Statistics for Biology and Health

Series Editors
K. Dietz, M. Gail, K. Krickeberg, J. Samet, A. Tsiatis

Statistics for Biology and Health

SURVIVAL ANALYSIS
Techniques for Censored and Truncated Data

Second Edition

John P. Klein
Medical College of Wisconsin

Melvin L. Moeschberger
The Ohio State University Medical Center

With 97 Illustrations

Springer

John P. Klein
Division of Biostatistics
Medical College of Wisconsin
Milwaukee, WI 53226
USA

Melvin L. Moeschberger
School of Public Health
Division of Epidemiology and Biometrics
The Ohio State University Medical Center
USA

Series Editors
K. Dietz
Institut für Medizinisch Biometric
Universität Tübingen
Westbahnhofstrasse 55
D-72070 Tübingen
Germany

M. Gail
National Cancer Institute
Rockville, MD 20892
USA

K. Krickeberg
Le Chatelet
F-63270 Manglieu
France

J. Samet
School of Public Health
Department of Epidemiology
Johns Hopkins University
615 Wolfe St.
Baltimore, MD 21205-2103
USA

A. Tsiatis
Department of Statistics
North Carolina State University
Raleigh, NC 27695
USA

Library of Congress Cataloging-in-Publication Data
Klein, John P., 1950–
    Survival analysis : techniques for censored and truncated data / John P. Klein, Melvin
L. Moeschberger.—2nd ed.
        p. cm. — (Statistics for biology and health)
    Includes bibliographical references and index.

    1. Survival analysis (Biometry)   I. Moeschberger, Melvin L.
II. Title.   II. Series.
R853.S7 K535   2003
    610'.7'27—dc21   2002026667

ISBN 978-1-4419-2985-3   e-ISBN 978-0-387-21645-4

9 8 7 6 5 4 3   Corrected third printing, 2005

springeronline.com

Preface

The second edition contains some new material as well as solutions to the odd-numbered revised exercises. New material consists of a discussion of summary statistics for competing risks probabilities in Chapter 2 and the estimation process for these probabilities in Chapter 4. A new section on tests of the equality of survival curves at a fixed point in time is added in Chapter 7. In Chapter 8 an expanded discussion is presented on how to code covariates and a new section on discretizing a continuous covariate is added. A new section on Lin and Ying's additive hazards regression model is presented in Chapter 10. We now proceed to a general discussion of the usefulness of this book incorporating the new material with that of the first edition.

A problem frequently faced by applied statisticians is the analysis of time to event data. Examples of such data arise in diverse fields such as medicine, biology, public health, epidemiology, engineering, economics and demography. While the statistical tools we shall present are applicable to all these disciplines our focus is on applications of the techniques to biology and medicine. Here interest is, for example, on analyzing data on the time to death from a certain cause, duration of response to treatment, time to recurrence of a disease, time to development of a disease, or simply time to death.

The analysis of survival experiments is complicated by issues of censoring, where an individual's life length is known to occur only in a certain period of time, and by truncation, where individuals enter the study only if they survive a sufficient length of time or individuals are included in the study only if the event has occurred by a given date. The use of counting process methodology has, in recent years, allowed for substantial advances in the statistical theory to account for censoring and truncation in survival experiments. The book by Andersen et al. (1993) provides an excellent survey of the mathematics of this theory. In this book we shall attempt to make these complex methods more accessible to applied researchers without an advanced mathematical background by presenting the essence of the statistical methods and illustrating these results in an applied framework. Our emphasis is on applying these techniques, as well as classical techniques not based on the counting process theory, to data rather than on the theoretical development of these tools. Practical suggestions for implementing the various methods are set off in a series of practical notes at the end of each section. Technical details of the derivation of these techniques (which are helpful to the understanding of concepts, though not essential to using the methods of this book) are sketched in a series of theoretical notes at the end of each section or are separated into their own sections. Some more advanced topics, for which some additional mathematical sophistication is needed for their understanding or for which standard software is not available, are given in separate chapters or sections. These notes and advanced topics can be skipped without a loss of continuity.

We envision two complementary uses for this book. The first is as a reference book for investigators who find the need to analyze censored or truncated life time data. The second use is as a textbook for a graduate level course in survival analysis. The minimum prerequisite for such course is a traditional course in statistical methodology. The material included in this book comes from our experience in teaching such a course for master's level biostatistics students at The Ohio State University and at the Medical College of Wisconsin, as well as from our experience in consulting with investigators from The Ohio State University, The University of Missouri, The Medical College of Wisconsin, The Oak Ridge National Laboratory, The National Center for Toxicological Research, and The International Bone Marrow Transplant Registry.

The book is divided into thirteen chapters that can be grouped into five major themes. The first theme introduces the reader to basic concepts and terminology. It consists of the first three chapters which deal with examples of typical data sets one may encounter in biomedical applications of this methodology, a discussion of the basic parameters to which inference is to be made, and a detailed discussion of censoring and truncation. New to the second edition is Section 2.7 that presents a discussion of summary statistics for competing risks probabilities. Section 3.6 gives a brief introduction to counting processes, and is included for those individuals with a minimal background in this area who wish to have a conceptual understanding of this methodology. This section can be omitted without jeopardizing the reader's understanding of later sections of the book.

The second major theme is the estimation of summary survival statistics based on censored and/or truncated data. Chapter 4 discusses estimation of the survival function, the cumulative hazard rate, and measures of centrality such as the median and the mean. The construction of pointwise confidence intervals and confidence bands is presented. Here we focus on right censored as well as left truncated survival data since this type of data is most frequently encountered in applications. New to the second edition is a section dealing with estimation of competing risks probabilities. In Chapter 5 the estimation schemes are extended to other types of survival data. Here methods for double and interval censoring; right truncation; and grouped data are presented. Chapter 6 presents some additional selected topics in univariate estimation, including the construction of smoothed estimators of the hazard function, methods for adjusting survival estimates for a known standard mortality and Bayesian survival methods.

The third theme is hypothesis testing. Chapter 7 presents one-, two-, and more than two-sample tests based on comparing the integrated difference between the observed and expected hazard rate. These tests include the log rank test and the generalized Wilcoxon test. Tests for trend and stratified tests are also discussed. Also discussed are Renyi tests which are based on sequential evaluation of these test statistics and have greater power to detect crossing hazard rates. This chapter also presents some other censored data analogs of classical tests such as the Cramer–Von Mises test, the t test and median tests are presented. New to this second edition is a section on tests of the equality of survival curves at a fixed point in time.

The fourth theme, and perhaps the one most applicable to applied work, is regression analysis for censored and/or truncated data. Chapter 8 presents a detailed discussion of the proportional hazards model used most commonly in medical applications. New sections in this second edition include an expanded discussion of how to code covariates and a section on discretizing a continuous covariate. Recent advances in the methodology that allows for this model to be applied to left truncated data, provides the investigator with new regression diagnostics, suggests improved point and interval estimates of the predicted survival function, and makes more accessible techniques for handling time-dependent covariates (including tests of the proportionality assumption) and the synthesis of intermediate events in an analysis are discussed in Chapter 9.

Chapter 10 presents recent work on the nonparametric additive hazard regression model of Aalen (1989) and a new section on Lin and Ying's (1994) additive hazards regression models. One of these models model may be the model of choice in situations where the proportional hazards model or a suitable modification of it is not applicable. Chapter 11 discusses a variety of residual plots one can make to check the fit of the Cox proportional hazards regression models. Chapter 12 discusses parametric models for the regression problem. Models presented include those available in most standard computer packages. Techniques for assessing the fit of these parametric models are also discussed.

The final theme is multivariate models for survival data. In Chapter 13, tests for association between event times, adjusted for covariates, are given. An introduction to estimation in a frailty or random effect model is presented. An alternative approach to adjusting for association between some individuals based on an analysis of an independent working model is also discussed.

There should be ample material in this book for a one or two semester course for graduate students. A basic one semester or one quarter course would cover the following sections:

Chapter 2

Chapter 3, Sections 1–5

Chapter 4

Chapter 7, Sections 1–6, 8

Chapter 8

Chapter 9, Sections 1–4

Chapter 11

Chapter 12

In such a course the outlines of theoretical development of the techniques, in the theoretical notes, would be omitted. Depending on the length of the course and the interest of the instructor, these details could be added if the material in section 3.6 were covered or additional topics from the remaining chapters could be added to this skeleton outline. Applied exercises are provided at the end of the chapters. Solutions to odd numbered exercises are new to the second edition. The data used in the examples and in most of the exercises is available from us at our Web site which is accessible through the Springer Web site at http://www.springer-ny.com or http://www.biostat.mcw.edu/homepgs/klein/book.html.

Milwaukee, Wisconsin   John P. Klein
Columbus, Ohio   Melvin L. Moeschberger

Contents

        Right-Censored and Left-Truncated Data   91

        4.1   Introduction   91

        4.2   Estimators of the Survival and Cumulative Hazard
               Functions for Right-Censored Data   92

        4.3   Pointwise Confidence Intervals for the Survival Function   104

        4.4   Confidence Bands for the Survival Function   109

        4.5   Point and Interval Estimates of the Mean and
               Median Survival Time   117

        4.6   Estimators of the Survival Function for Left-Truncated
               and Right-Censored Data   123

        4.7   Summary Curves for Competing Risks   127

        4.8   Exercises   133

Chapter 5 — Estimation of Basic Quantities for
        Other Sampling Schemes   139

        5.1   Introduction   139

        5.2   Estimation of the Survival Function for Left, Double,
               and Interval Censoring   140

        5.3   Estimation of the Survival Function for
               Right-Truncated Data   149

        5.4   Estimation of Survival in the Cohort Life Table   152

        5.5   Exercises   158

Chapter 6 — Topics in Univariate Estimation   165

        6.1   Introduction   165

        6.2   Estimating the Hazard Function   166

        6.3   Estimation of Excess Mortality   177

        6.4   Bayesian Nonparametric Methods   187

        6.5   Exercises   198

1
Examples of Survival Data

1.1   Introduction

The problem of analyzing time to event data arises in a number of applied fields, such as medicine, biology, public health, epidemiology, engineering, economics, and demography. Although the statistical tools we shall present are applicable to all these disciplines, our focus is on applying the techniques to biology and medicine. In this chapter, we present some examples drawn from these fields that are used throughout the text to illustrate the statistical techniques we shall describe.

A common feature of these data sets is they contain either *censored* or *truncated observations*. Censored data arises when an individual's life length is known to occur only in a certain period of time. Possible censoring schemes are *right censoring*, where all that is known is that the individual is still alive at a given time, *left censoring* when all that is known is that the individual has experienced the event of interest prior to the start of the study, or *interval censoring*, where the only information is that the event occurs within some interval. Truncation schemes are *left truncation*, where only individuals who survive a sufficient time are included in the sample and *right truncation*, where only individuals who have experienced the event by a specified time are included in the sample. The issues of censoring and truncation are defined more carefully in Chapter 3.

1.2   Remission Duration from a Clinical Trial for Acute Leukemia

Freireich et al. (1963) report the results of a clinical trial of a drug 6-mercaptopurine (6-MP) versus a placebo in 42 children with acute leukemia. The trial was conducted at 11 American hospitals. Patients were selected who had a complete or partial remission of their leukemia induced by treatment with the drug prednisone. (A complete or partial remission means that either most or all signs of disease had disappeared from the bone marrow.) The trial was conducted by matching pairs of patients at a given hospital by remission status (complete or partial) and randomizing within the pair to either a 6-MP or placebo maintenance therapy. Patients were followed until their leukemia returned (relapse) or until the end of the study (in months). The data is reported in Table 1.1.

TABLE 1.1

Remission duration of 6-MP versus placebo in children with acute leukemia

Pair	Remission Status at Randomization	Time to Relapse for Placebo Patients	Time to Relapse for 6-MP Patients
1	Partial Remission	1	10
2	Complete Remission	22	7
3	Complete Remission	3	32^+
4	Complete Remission	12	23
5	Complete Remission	8	22
6	Partial Remission	17	6
7	Complete Remission	2	16
8	Complete Remission	11	34^+
9	Complete Remission	8	32^+
10	Complete Remission	12	25^+
11	Complete Remission	2	11^+
12	Partial Remission	5	20^+
13	Complete Remission	4	19^+
14	Complete Remission	15	6
15	Complete Remission	8	17^+
16	Partial Remission	23	35^+
17	Partial Remission	5	6
18	Complete Remission	11	13
19	Complete Remission	4	9^+
20	Complete Remission	1	6^+
21	Complete Remission	8	10^+

$^+$Censored observation

This data set is used in Chapter 4 to illustrate the calculation of the estimated probability of survival using the product-limit estimator, the calculation of the Nelson-Aalen estimator of the cumulative hazard function, and the calculation of the mean survival time, along with their standard errors. It is further used in section 6.4 to estimate the survival function using Bayesian approaches. Matched pairs tests for differences in treatment efficacy are performed using the stratified log rank test in section 7.5 and the stratified proportional hazards model in section 9.3.

1.3 Bone Marrow Transplantation for Leukemia

Bone marrow transplants are a standard treatment for acute leukemia. Recovery following bone marrow transplantation is a complex process. Prognosis for recovery may depend on risk factors known at the time of transplantation, such as patient and/or donor age and sex, the stage of initial disease, the time from diagnosis to transplantation, etc. The final prognosis may change as the patient's posttransplantation history develops with the occurrence of events at random times during the recovery process, such as development of acute or chronic graft-versus-host disease (GVHD), return of the platelet count to normal levels, return of granulocytes to normal levels, or development of infections. Transplantation can be considered a failure when a patient's leukemia returns (relapse) or when he or she dies while in remission (treatment related death).

Figure 1.1 shows a simplified diagram of a patient's recovery process based on two intermediate events that may occur in the recovery process. These intermediate events are the possible development of acute GVHD that typically occurs within the first 100 days following transplantation and the recovery of the platelet count to a self-sustaining level $\geq 40 \times 10^9/l$ (called platelet recovery in the sequel). Immediately following transplantation, patients have depressed platelet counts and are free of acute GVHD. At some point, they may develop acute GVHD or have their platelets recover at which time their prognosis (probabilities of treatment related death or relapse at some future time) may change. These events may occur in any order, or a patient may die or relapse without any of these events occurring. Patients may, then, experience the other event, which again modifies their prognosis, or they may die or relapse.

To illustrate this process we consider a multicenter trial of patients prepared for transplantation with a radiation-free conditioning regimen.

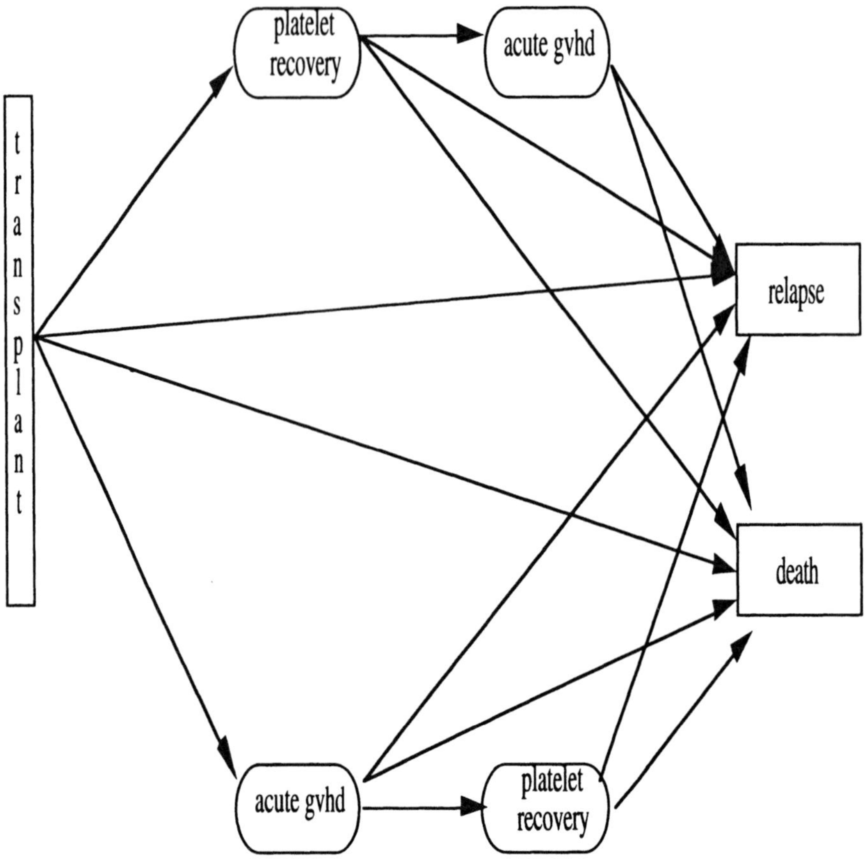

Figure 1.1   *Recovery Process from a Bone Marrow Transplant*

Details of the study are found in Copelan et al. (1991). The preparative regimen used in this study of allogeneic marrow transplants for patients with acute myeloctic leukemia (AML) and acute lymphoblastic leukemia (ALL) was a combination of 16 mg/kg of oral Busulfan (BU) and 120 mg/kg of intravenous cyclophosphamide (Cy). A total of 137 patients (99 AML, 38 ALL) were treated at one of four hospitals: 76 at The Ohio State University Hospitals (OSU) in Columbus; 21 at Hahnemann University (HU) in Philadelphia; 23 at St. Vincent's Hospital (SVH) in Sydney Australia; and 17 at Alfred Hospital (AH) in Melbourne. The study consists of transplants conducted at these institutions from March 1, 1984, to June 30, 1989. The maximum follow-up was 7 years. There were 42 patients who relapsed and 41 who died while in remission. Twenty-six patients had an episode of acute GVHD, and 17 patients either relapsed or died in remission without their platelets returning to normal levels.

Several potential risk factors were measured at the time of transplantation. For each disease, patients were grouped into risk categories based on their status at the time of transplantation. These categories were as follows: ALL (38 patients), AML low-risk first remission (54 patients), and AML high-risk second remission or untreated first relapse (15 patients) or second or greater relapse or never in remission (30 patients). Other risk factors measured at the time of transplantation included recipient and donor gender (80 and 88 males respectively), recipient and donor cytomegalovirus immune status (CMV) status (68 and 58 positive, respectively), recipient and donor age (ranges 7–52 and 2–56, respectively), waiting time from diagnosis to transplantation (range 0.8–87.2 months, mean 19.7 months), and, for AML patients, their French-American-British (FAB) classification based on standard morphological criteria. AML patients with an FAB classification of M4 or M5 (45/99 patients) were considered to have a possible elevated risk of relapse or treatment-related death. Finally, patients at the two hospitals in Australia (SVH and AH) were given a graft-versus-host prophylactic combining methotrexate (MTX) with cyclosporine and possibly methylprednisolone. Patients at the other hospitals were not given methotrexate but rather a combination of cyclosporine and methylprednisolone. The data is presented in Table D.1 of Appendix D.

This data set is used throughout the book to illustrate the methods presented. In Chapter 4, it is used to illustrate the product-limit estimator of the survival function and the Nelson–Aalen estimator of the cumulative hazard rate of treatment failure. Based on these statistics, pointwise confidence intervals and confidence bands for the survival function are constructed. The data is also used to illustrate point and interval estimation of summary survival parameters, such as the mean and median time to treatment failure in this chapter.

This data set is also used in Chapter 4 to illustrate summary probabilities for competing risks. The competing risks, where the occurrence of one event precludes the occurrence of the other event, in this example, are relapse and death.

In section 6.2, the data set is used to illustrate the construction of estimates of the hazard rate. These estimates are based on smoothing the crude estimates of the hazard rate obtained from the jumps of the Nelson–Aalen estimator found in Chapter 4 using a weighted average of these estimates in a small interval about the time of interest. The weights are chosen using a kernel weighting function.

In Chapter 7, this data is used to illustrate tests for the equality of K survival curves. Both stratified and unstratified tests are discussed.

In Chapter 8, the data is used to illustrate tests of the equality of K hazard rates adjusted for possible fixed-time confounders. A proportional hazards model is used to make this adjustment. Model building for this problem is illustrated. In Chapter 9, the models found in Chapter 8 are further refined to include covariates, whose values change over time, and to allow for stratified regression models. In Chapter 11, regression diagnostics for these models are presented.

1.4   Times to Infection of Kidney Dialysis Patients

In a study (Nahman et al., 1992) designed to assess the time to first exit-site infection (in months) in patients with renal insufficiency, 43 patients utilized a surgically placed catheter (Group 1), and 76 patients utilized a percutaneous placement of their catheter (Group 2). Cutaneous exit-site infection was defined as a painful cutaneous exit site and positive cultures, or peritonitis, defined as a presence of clinical symptoms, elevated peritoneal dialytic fluid, elevated white blood cell count (100 white blood cells/μl with >50% neutrophils), and positive peritoneal dialytic fluid cultures. The data appears in Table 1.2.

TABLE 1.2
Times to infection (in months) of kidney dialysis patients with different catheter-ization procedures

---

Surgically Placed Catheter

Infection Times: 1.5, 3.5, 4.5, 4.5, 5.5, 8.5, 8.5, 9.5, 10.5, 11.5, 15.5, 16.5, 18.5, 23.5, 26.5
Censored Observations: 2.5, 2.5, 3.5, 3.5, 3.5, 4.5, 5.5, 6.5, 6.5, 7.5, 7.5, 7.5, 7.5, 8.5, 9.5, 10.5, 11.5, 12.5, 12.5, 13.5, 14.5, 14.5, 21.5, 21.5, 22.5, 22.5, 25.5, 27.5

Percutaneous Placed Catheter

Infection Times: 0.5, 0.5, 0.5, 0.5, 0.5, 0.5, 2.5, 2.5, 3.5, 6.5, 15.5
Censored Observations: 0.5, 0.5, 0.5, 0.5, 0.5, 0.5, 0.5, 0.5, 0.5, 0.5, 1.5, 1.5, 1.5, 1.5, 2.5, 2.5, 2.5, 2.5, 2.5, 3.5, 3.5, 3.5, 3.5, 3.5, 4.5, 4.5, 4.5, 5.5, 5.5, 5.5, 5.5, 5.5, 6.5, 7.5, 7.5, 7.5, 8.5, 8.5, 8.5, 9.5, 9.5, 10.5, 10.5, 10.5, 11.5, 11.5, 12.5, 12.5, 12.5, 12.5, 14.5, 14.5, 16.5, 16.5, 18.5, 19.5, 19.5, 19.5, 20.5, 22.5, 24.5, 25.5, 26.5, 26.5, 28.5

The data is used in section 7.3 to illustrate how the inference about the equality of two survival curves, based on a two-sample weighted, log-rank test, depends on the choice of the weight function. In section 7.7, it is used to illustrate the two-sample Cramer–von Mises test for censored data. In the context of the proportional hazards model, this data is used in Chapter 8 to illustrate the different methods of constructing the partial likelihoods and the subsequent testing of equality of the survival curves when there are ties present. Testing for proportional hazards is illustrated in section 9.2. The test reveals that a proportional hazards assumption for this data is not correct. A model with a time-varying, covariate effect is more appropriate, and in that section the optimal cutoff for "early" and "late" covariate effect on survival is found.

1.5   Times to Death for a Breast-Cancer Trial

In a study (Sedmak et al., 1989) designed to determine if female breast cancer patients, originally classified as lymph node negative by standard light microscopy (SLM), could be more accurately classified by immunohistochemical (IH) examination of their lymph nodes with an anticytokeratin monoclonal antibody cocktail, identical sections of lymph nodes were sequentially examined by SLM and IH. The significance of this study is that 16% of patients with negative axillary lymph nodes, by standard pathological examination, develop recurrent disease within 10 years. Forty-five female breast-cancer patients with negative axillary lymph nodes and a minimum 10-year follow-up were selected from The Ohio State University Hospitals Cancer Registry. Of the 45 patients, 9 were immunoperoxidase positive, and the remaining 36 remained negative. Survival times (in months) for both groups of patients are given in Table 1.3 ($^+$ denotes a censored observation).

TABLE 1.3

Times to death (in months) for breast cancer patients with different immunohistochemical responses

---

Immunoperoxidase Negative: 19, 25, 30, 34, 37, 46, 47, 51, 56, 57, 61, 66, 67, 74, 78, 86, 122^+, 123^+, 130^+, 130^+, 133^+, 134^+, 136^+, 141^+, 143^+ ,148^+, 151^+, 152^+,153^+,154^+, 156^+, 162^+, 164^+, 165^+, 182^+,189^+,
Immunoperoxidase Positive: 22, 23, 38, 42, 73, 77, 89, 115, 144^+

$^+$Censored observation

This data is used to show the construction of the likelihood function and in calculating a two-sample test based on proportional hazards with no ties with right-censored data in Chapter 8. It is also used in Chapter 10 to illustrate the least-squares estimation methodology in the context of the additive hazards model. In that chapter, we also used this data to illustrate estimation for an additive model with constant excess risk over time.

1.6   Times to Infection for Burn Patients

In a study (Ichida et al., 1993) to evaluate a protocol change in disinfectant practices in a large midwestern university medical center, 154 patient records and charts were reviewed. Infection of a burn wound is a common complication resulting in extended hospital stays and in the death of severely burned patients. Control of infection remains a prominent component of burn management. The purpose of this study is to compare a routine bathing care method (initial surface decontamination with 10% povidone-iodine followed with regular bathing with Dial soap) with a body-cleansing method using 4% chlorhexidine gluconate. Medical records of patients treated during the 18-month study period provided information on burn wound infections and other medical information. The time until staphylococcus infection was recorded (in days), along with an indicator variable—whether or not an infection had occurred. Other fixed covariates recorded were gender (22% female), race (88% white), severity of the burn as measured by percentage of total surface area of body burned (average of 24.7% range 2–95%), burn site (45% on head, 23% on buttocks, 84% on trunk, 41% on upper legs, 31% on lower legs, or 29% in respiratory tract), and type of burn (6% chemical, 12% scald, 7% electric, or 75% flame). Two time-dependent covariates were recorded, namely, time to excision and time to prophylactic antibiotic treatment administered, along with the two corresponding indicator variables, namely, whether the patient's wound had been excised (64%) and whether the patient had been treated with an antibiotic sometime during the course of the study (41%). Eighty-four patients were in the group which received the new bathing solution, chlorhexidine, and 70 patients served as the historical control group which received the routine bathing care, povidone-iodine. The data is available on the authors' web site and is used in the exercises.

1.7   Death Times of Kidney Transplant Patients

Data on the time to death of 863 kidney transplant patients is available on the authors' web site. All patients had their transplant performed at The Ohio State University Transplant Center during the period 1982–1992. The maximum follow-up time for this study was 9.47 years. Patients were censored if they moved from Columbus (lost to follow-up) or if they were alive on June 30, 1992. In the sample, there were 432 white males, 92 black males, 280 white females, and 59 black females. Patient ages at transplant ranged from 9.5 months to 74.5 years with a mean age of 42.8 years. Seventy-three (16.9%) of the white males, 14 (15.2%) of the black males, 39 (13.9%) of the white females and 14 (23.7%) of the black females died prior to the end of the study.

In Chapter 6, the problem of estimating the hazard rate, using a kernel smoothing procedure, is discussed. In particular, the effect of changing the bandwidth and the choice of a kernel are considered. In Chapter 8 this data is also used to illustrate methods for discretizing a continuous covariate.

1.8 Death Times of Male Laryngeal Cancer Patients

Kardaun (1983) reports data on 90 males diagnosed with cancer of the larynx during the period 1970–1978 at a Dutch hospital. Times recorded are the intervals (in years) between first treatment and either death or the end of the study (January 1, 1983). Also recorded are the patient's age at the time of diagnosis, the year of diagnosis, and the stage of the patient's cancer. The four stages of disease in the study were based on the T.N.M. (primary tumor (T), nodal involvement (N) and distant metastasis (M) grading) classification used by the American Joint Committee for Cancer Staging (1972). The four groups are Stage I, $T_1N_0M_0$ with 33 patients; Stage II, $T_2N_0M_0$ with 17 patients; Stage III, $T_3N_0M_0$ and $T_xN_1M_0$, with 27 patients; $x = 1$, 2, or 3; and Stage IV, all other TNM combinations except TIS with 13 patients. The stages are ordered from least serious to most serious. The data is available on the authors' web site.

In section 7.4, the data is used to illustrate a test for trend to confirm the hypothesis that the higher the stage the greater the chance of dying. In Chapter 8, a global test for the effect of stage on survival is performed in the context of the proportional hazards model, and local tests are illustrated, after an adjustment for the patient's age. An analysis of variance (ANOVA) table is presented to summarize the effects of stage and age on survival. Contrasts are used to test the hypothesis that linear combinations of stage effects are zero. The construction of confidence intervals for different linear combinations of the stage effects is illustrated. The concept of an interaction in a proportional hazards regression model is illustrated through a stage by age interaction factor. The survival curve is estimated for each stage based on the Cox proportional hazards model.

This data is also used in Chapter 10 to illustrate estimation methodology in the additive hazards model. In Chapter 12, this data set is used to illustrate the fit of parametric models, using the accelerated failure-time model. The goodness of fit of these models is also discussed. The log logistic model is used in section 12.5 to illustrate using deviance residuals.

1.9 Autologous and Allogeneic Bone Marrow Transplants

The data in Table 1.4 is a sample of 101 patients with advanced acute myelogenous leukemia reported to the International Bone Marrow Transplant Registry. Fifty-one of these patients had received an autologous (auto) bone marrow transplant in which, after high doses of chemotherapy, their own marrow was reinfused to replace their destroyed immune system. Fifty patients had an allogeneic (allo) bone marrow transplant where marrow from an HLA (Histocompatibility Leukocyte Antigen)-matched sibling was used to replenish their immune systems.

An important question in bone marrow transplantation is the comparison of the effectiveness of these two methods of transplant as mea-

TABLE 1.4

Leukemia free-survival times (in months) for Autologous and Allogeneic Transplants

The leukemia-free survival times for the 50 allo transplant patients were 0.030, 0.493, 0.855, 1.184, 1.283, 1.480, 1.776, 2.138, 2.500, 2.763, 2.993, 3.224, 3.421, 4.178, 4.441^+, 5.691, 5.855^+, 6.941^+, 6.941, 7.993^+, 8.882, 8.882, 9.145^+, 11.480, 11.513, 12.105^+, 12.796, 12.993^+, 13.849^+, 16.612^+, 17.138^+, 20.066, 20.329^+, 22.368^+, 26.776^+, 28.717^+, 28.717^+, 32.928^+, 33.783^+, 34.211^+, 34.770^+, 39.539^+, 41.118^+, 45.033^+, 46.053^+, 46.941^+, 48.289^+, 57.401^+, 58.322^+, 60.625^+;

and, for the 51 auto patients, 0.658, 0.822, 1.414, 2.500, 3.322, 3.816, 4.737, 4.836^+, 4.934, 5.033, 5.757, 5.855, 5.987, 6.151, 6.217, 6.447^+, 8.651, 8.717, 9.441^+, 10.329, 11.480, 12.007, 12.007^+, 12.237, 12.401^+, 13.059^+, 14.474^+, 15.000^+, 15.461, 15.757, 16.480, 16.711, 17.204^+, 17.237, 17.303^+, 17.664^+, 18.092, 18.092^+, 18.750^+, 20.625^+, 23.158, 27.730^+, 31.184^+, 32.434^+, 35.921^+, 42.237^+, 44.638^+, 46.480^+, 47.467^+, 48.322^+, 56.086.

As usual, $^+$ denotes a censored observation.

sured by the length of patients' leukemia-free survival, the length of time they are alive, and how long they remain free of disease after their transplants. In Chapter 7, this comparison is made using a weighted log-rank test, and a censored data version of the median test and the t-test.

This data is used in Chapter 11 to illustrate graphical methods for checking model assumptions following a proportional hazards regression analysis. In section 11.3, the martingale residuals are used to check overall model fit. In section 11.4, score residuals are used to check the proportional hazards assumption on disease-free survival for type of transplant. In section 11.5, the use of deviance residuals is illustrated for checking for outliers and, in section 11.6, the influence of individual observations is examined graphically.

In Chapter 12, this data set is used to illustrate the fit of parametric models using the accelerated failure-time model. The goodness of fit of these models is also discussed. Diagnostic plots for checking the fit of a parametric regression model using this data set are illustrated in section 12.5.

1.10   Bone Marrow Transplants for Hodgkin's and Non-Hodgkin's Lymphoma

The data in Table 1.5 was collected on 43 bone marrow transplant patients at The Ohio State University Bone Marrow Transplant Unit. Details of this study can be found in Avalos et al. (1993). All patients had either Hodgkin's disease (HOD) or non-Hodgkin's lymphoma (NHL) and were given either an allogeneic (Allo) transplant from an HLA match sibling donor or an autogeneic (Auto) transplant; i.e., their own marrow was cleansed and returned to them after a high dose of chemotherapy. Also included are two possible explanatory variables, Karnofsky score at transplant and the waiting time in months from diagnosis to transplant. Of interest is a test of the null hypothesis of no difference in the leukemia-free survival rate between patients given an Allo or Auto transplant, adjusting for the patient's disease state. This test, which requires stratification of the patient's disease, is presented in section 7.5. We also use this data in section 11.3 to illustrate how the martingale residual can be used to determine the functional form of a covariate. The data, in Table 1.5, consists of the time on study for each patient, T_i, and the event indicator $\delta_i = 1$ if dead or relapsed; 0 otherwise; and two covariates Z_1, the pretransplant Karnofsky score and Z_2, the waiting time to transplant.

TABLE 1.5
Times to death or relapse (in days) for patients with bone marrow transplants for Hodgkin's and non-Hodgkin's lymphoma

Allo NHL				Auto NHL				Allo HOD				Auto HOD			
T_i	δ_i	Z_1	Z_2	T_i	δ_i	Z_1	Z_2	T_i	δ_i	Z_1	Z_2	T_i	δ_i	Z_1	Z_2
28	1	90	24	42	1	80	19	2	1	20	34	30	1	90	73
32	1	30	7	53	1	90	17	4	1	50	28	36	1	80	61
49	1	40	8	57	1	30	9	72	1	80	59	41	1	70	34
84	1	60	10	63	1	60	13	77	1	60	102	52	1	60	18
357	1	70	42	81	1	50	12	79	1	70	71	62	1	90	40
933	0	90	9	140	1	100	11					108	1	70	65
1078	0	100	16	176	1	80	38					132	1	60	17
1183	0	90	16	210	0	90	16					180	0	100	61
1560	0	80	20	252	1	90	21					307	0	100	24
2114	0	80	27	476	0	90	24					406	0	100	48
2144	0	90	5	524	1	90	39					446	0	100	52
				1037	0	90	84					484	0	90	84
												748	0	90	171
												1290	0	90	20
												1345	0	80	98

TABLE 1.6
Death times (in weeks) of patients with cancer of the tongue

Aneuploid Tumors:
Death Times: 1, 3, 3, 4, 10, 13, 13, 16, 16, 24, 26, 27, 28, 30, 30, 32, 41, 51, 65, 67, 70, 72, 73, 77, 91, 93, 96, 100, 104, 157, 167
Censored Observations: 61, 74, 79, 80, 81, 87, 87, 88, 89, 93, 97, 101, 104, 108, 109, 120, 131, 150, 231, 240, 400
Diploid Tumors:
Death Times: 1, 3, 4, 5, 5, 8, 12, 13, 18, 23, 26, 27, 30, 42, 56, 62, 69, 104, 104, 112, 129, 181
Censored Observations: 8, 67, 76, 104, 176, 231

1.11   Times to Death for Patients with Cancer of the Tongue

A study was conducted on the effects of ploidy on the prognosis of patients with cancers of the mouth. Patients were selected who had a paraffin-embedded sample of the cancerous tissue taken at the time of surgery. Follow-up survival data was obtained on each patient. The tissue samples were examined using a flow cytometer to determine if the tumor had an aneuploid (abnormal) or diploid (normal) DNA profile using a technique discussed in Sickle–Santanello et al. (1988). The data in Table 1.6 is on patients with cancer of the tongue. Times are in weeks.

The data is used in exercises.

1.12 Times to Reinfection for Patients with Sexually Transmitted Diseases

A major problem in certain subpopulations is the occurrence of sexually transmitted diseases (STD). Even if one ignores the lethal effects of the Acquired Immune Deficiency Syndrome (AIDS), other sexually transmitted diseases still have a significant impact on the morbidity of the community. Two of these sexually transmitted diseases are the focus of this investigation: gonorrhea and chlamydia. These diseases are of special interest because they are often asymptomatic in the female, and, if left untreated, can lead to complications including sterility.

Both of these diseases can be easily prevented and effectively treated. Therefore, it is a mystery why the incidence of these diseases remain high in several subpopulations. One theory is that a core group of individuals experience reinfections, thereby, serving as a natural reservoir of the disease organism and transferring the disease to uninfected individuals.

The purpose of this study is to identify those factors which are related to time until reinfection by either gonorrhea or chlamydia, given an initial infection of gonorrhea or chlamydia. A sample of 877 individuals, with an initial diagnosis of gonorrhea or chlamydia were followed for reinfection. In addition to the primary outcome variable just stated, an indicator variable which indicates whether a reinfection occurred was recorded. Demographic variables recorded were race (33% white, 67% black), marital status (7% divorced/separated, 3% married and 90% single), age of patient at initial infection (average age is 20.6 years with a range of 13–48 years), years of schooling (11.4 years with a range of 6–18 years), and type of initial infection (16% gonorrhea, 45% chlamydia and 39% both gonorrhea and chlamydia). Behavioral factors recorded at the examination, when the initial infection was diagnosed, were number of partners in the last 30 days (average is 1.27 with a range of 0–19), oral sex within past 12 months (33%), rectal sex within past 12 months (6%), and condom use (6% always, 58% sometimes, and 36%

never). Symptoms noticed at time of initial infection were presence of abdominal pain (14%), sign of discharge (46%), sign of dysuria (13%), sign of itch (19%), sign of lesion (3%), sign of rash (3%), and sign of lymph involvement (1%). If the factors related to a greater risk of reinfection can be identified, then, interventions could be targeted to those individuals who are at greatest risk for reinfection. This, in turn, should reduce the size of the core group and, thereby, reduce the incidence of the diseases. The data for this study is available on our web site.

This data is used in the exercises.

1.13   Time to Hospitalized Pneumonia in Young Children

Data gathered from 3,470 annual personal interviews conducted for the National Longitudinal Survey of Youth (NLSY, 1995) from 1979 through 1986 was used to study whether the mother's feeding choice (breast feeding vs. never breast fed) protected the infant against hospitalized pneumonia in the first year of life. Information obtained about the child included whether it had a normal birthweight, as defined by weighing at least 5.5 pounds (36%), race (56% white, 28% black, and 16% other), number of siblings (range 0–6), and age at which the child was hospitalized for pneumonia, along with an indicator variable as to whether the child was hospitalized. Demographic characteristics of the mother, such as age (average is 21.64 years with a range of 14–29 years), years of schooling (average of 11.4 years), region of the country (15% Northeast, 25% North central, 40% South, and 20% West), poverty (92%), and urban environment (76%). Health behavior measures during pregnancy, such as alcohol use (36%) and cigarette use (34%), were also recorded. The data for this study is available on our web site.

This data is used in the exercises.

1.14   Times to Weaning of Breast-Fed Newborns

The National Longitudinal Survey of Youth is a stratified random sample which was begun in 1979. Youths, aged 14 to 21 in 1979, have been interviewed yearly through 1988. Beginning in 1983, females in the survey were asked about any pregnancies that have occurred since they were last interviewed (pregnancies before 1983 were also documented). Questions regarding breast feeding are included in the questionnaire.

This data set consists of the information from 927 first-born children to mothers who chose to breast feed their children and who have complete information for all the variables of interest. The sample was restricted to children born after 1978 and whose gestation age was between 20 and 45 weeks. The year of birth restriction was included in an attempt to eliminate recall problems.

The response variable in the data set is duration of breast feeding in weeks, followed by an indicator of whether the breast feeding was completed (i.e., the infant is weaned). Explanatory variables for breast-feeding duration include race of mother (1 if white, 2 if black, 3 if other); poverty status indicator (1 if mother in poverty); smoking status of mother (1 if smoking at birth of child); alcohol-drinking status of mother (1 if drinking at birth of child); age of mother at child's birth, year of child's birth, education of mother (in years); and lack of prenatal care status (1 if mother sought prenatal care after third month or never sought prenatal care, 0 if mother sought prenatal care in first three months of pregnancy). The complete data for this study is available on our web site.

This data is used in section 5.4 to illustrate the construction of the cohort life table. In Chapter 8, it is used to show how to build a model where predicting the outcome is the main purpose, i.e., interest is in finding factors which contribute to the distribution of the time to weaning.

1.15   Death Times of Psychiatric Patients

Woolson (1981) has reported survival data on 26 psychiatric inpatients admitted to the University of Iowa hospitals during the years 1935–1948. This sample is part of a larger study of psychiatric inpatients discussed by Tsuang and Woolson (1977). Data for each patient consists of age at first admission to the hospital, sex, number of years of follow-up (years from admission to death or censoring) and patient status at the follow-up time. The data is given in Table 1.7. In section 6.3, the estimate of the relative mortality function and cumulative excess mortality of these patients, compared to the standard mortality rates of residents of Iowa in 1959, is considered. In section 7.2, this data is used to illustrate one-sample hypothesis tests. Here, a comparison of the survival experience of these 26 patients is made to the standard mortality of residents of Iowa to determine if psychiatric patients tend to have shorter lifetimes. It is used in Chapter 9 to illustrate left truncation in the context of proportional hazards models.

TABLE 1.7
Survival data for psychiatric inpatients

Gender	Age at Admission	Time of Follow-up
Female	51	1
Female	58	1
Female	55	2
Female	28	22
Male	21	30^+
Male	19	28
Female	25	32
Female	48	11
Female	47	14
Female	25	36^+
Female	31	31^+
Male	24	33^+
Male	25	33^+
Female	30	37^+
Female	33	35^+
Male	36	25
Male	30	31^+
Male	41	22
Female	43	26
Female	45	24
Female	35	35^+
Male	29	34^+
Male	35	30^+
Male	32	35
Female	36	40
Male	32	39^+

$^+$Censored observation

1.16 Death Times of Elderly Residents of a Retirement Community

Channing House is a retirement center located in Palo Alto, California. Data on ages at death of 462 individuals (97 males and 365 females) who were in residence during the period January 1964 to July 1975 has been reported by Hyde (1980). A distinctive feature of these individuals was that all were covered by a health care program provided by the center which allowed for easy access to medical care without any additional financial burden to the resident. The age in months when members of the community died or left the center and the age when individuals entered the community is available on the authors' web site.

The life lengths in this data set are *left truncated* because an individual must survive to a sufficient age to enter the retirement community. Individuals who die at an early age are excluded from the study. Ignoring this left truncation leads to the problem of length-biased sampling. The concept of left truncation and the bias induced into the estimation process by ignoring it is discussed in section 3.4.

This data will be used in section 4.6 to illustrate how one estimates the conditional survival function for left-truncated data. The data is used in section 7.3 to illustrate the comparison of two samples (male and female), when there is left truncation and right censoring employing the log-rank test, and in Chapter 9 employing the Cox proportional hazards model.

1.17   Time to First Use of Marijuana

Turnbull and Weiss (1978) report part of a study conducted at the Stanford-Palo Alto Peer Counseling Program (see Hamburg et al. [1975] for details of the study). In this study, 191 California high school boys were asked, "When did you first use marijuana?" The answers were the exact ages (uncensored observations); "I never used it," which are right-censored observations at the boys' current ages; or "I have used it but can not recall just when the first time was," which is a left-censored observation (see section 3.3). Notice that a left-censored observation

TABLE 1.8
Marijuana use in high school boys

Age	Number of Exact Observations	Number Who Have Yet to Smoke Marijuana	Number Who Have Started Smoking at an Earlier Age
10	4	0	0
11	12	0	0
12	19	2	0
13	24	15	1
14	20	24	2
15	13	18	3
16	3	14	2
17	1	6	3
18	0	0	1
>18	4	0	0

tells us only that the event has occurred prior to the boy's current age. The data is in Table 1.8.

This data is used in section 5.2 to illustrate the calculation of the survival function for both left- and right-censored data, commonly referred to as doubly censored data.

1.18   Time to Cosmetic Deterioration of Breast Cancer Patients

Beadle et al. (1984a and b) report a retrospective study carried out to compare the cosmetic effects of radiotherapy alone versus radiotherapy and adjuvant chemotherapy on women with early breast cancer. The use of an excision biopsy, followed by radiation therapy, has been suggested as an alternative to mastectomy. This therapy preserves the breast and, hence, has the benefit of an improved cosmetic effect. The use of adjuvant chemotherapy is often indicated to prevent recurrence of the cancer, but there is some clinical evidence that it enhances the effects of radiation on normal tissue, thus, offsetting the cosmetic benefit of this procedure.

To compare the two treatment regimes, a retrospective study of 46 radiation only and 48 radiation plus chemotherapy patients was made. Patients were observed initially every 4–6 months, but, as their recovery progressed, the interval between visits lengthened. At each visit, the clinician recorded a measure of breast retraction on a three-point scale (none, moderate, severe). The event of interest was the time to first

TABLE 1.9

Time to cosmetic deterioration (in months) in breast cancer patients with two treatment regimens

---

Radiotherapy only: (0, 7]; (0, 8]; (0, 5]; (4, 11]; (5, 12]; (5, 11]; (6, 10]; (7, 16]; (7, 14]; (11, 15]; (11, 18]; ≥15; ≥17; (17, 25]; (17, 25]; ≥18; (19, 35]; (18, 26]; ≥22; ≥24; ≥24; (25, 37]; (26, 40]; (27, 34]; ≥32; ≥33; ≥34; (36, 44]; (36, 48]; ≥36; ≥36; (37, 44]; ≥37; ≥37; ≥37; ≥38; ≥40; ≥45; ≥46; ≥46; ≥46; ≥46; ≥46; ≥46; ≥46.
Radiotherapy and Chemotherapy: (0, 22]; (0, 5]; (4, 9]; (4, 8]; (5, 8]; (8, 12]; (8, 21]; (10, 35]; (10, 17]; (11, 13]; ≥11; (11, 17]; ≥11; (11, 20]; (12, 20]; ≥13; (13, 39]; ≥13; ≥13; (14, 17]; (14, 19]; (15, 22]; (16, 24]; (16, 20]; (16, 24]; (16, 60]; (17, 27]; (17, 23]; (17, 26]; (18, 25]; (18, 24]; (19, 32]; ≥21; (22, 32]; ≥23; (24, 31]; (24, 30]; (30, 34]; (30, 36]; ≥31; ≥32; (33, 40]; ≥34; ≥34; ≥35; (35, 39]; (44, 48]; ≥48.

(a, b]—interval in which deterioration took place.

appearance of moderate or severe breast retraction. Due to the fact that patients were observed only at these random times, the exact time of breast retraction is known only to fall in the interval between visits. This type of data is call *interval-censored data* (see section 3.3).

The data for the two groups is shown in Table 1.9. The data consists of the interval, in months, in which deterioration occurred or the last time a patient was seen without deterioration having yet occurred (right-censored observations). This data is used in section 5.2 to illustrate the computation of an estimate of the survival function based on interval-censored data.

1.19 Time to AIDS

Lagakos et al. (1988) report data on the infection and induction times for 258 adults and 37 children who were infected with the AIDS virus and developed AIDS by June 30, 1986. The data consists of the time in years, measured from April 1, 1978, when adults were infected by the virus from a contaminated blood transfusion, and the waiting time to development of AIDS, measured from the date of infection. For the pediatric population, children were infected in utero or at birth, and the infection time is the number of years from April 1, 1978 to birth. The data is in Table 1.10.

In this sampling scheme, only individuals who have developed AIDS prior to the end of the study period are included in the study. Infected individuals who have yet to develop AIDS are not included in the sample. This type of data is called *right-truncated* data (see section 3.4). Estimation of the survival function for this data with right-truncated data is discussed in section 5.3.

TABLE 1.10

Induction times (in years) for AIDS in adults and children

Infection Time	Adult Induction Time	Child Induction Time
0.00	5	
0.25	6.75	
0.75	5, 5, 7.25	
1.00	4.25, 5.75, 6.25, 6.5	5.5
1.25	4, 4.25, 4.75, 5.75	
1.50	2.75, 3.75, 5, 5.5, 6.5	2.25
1.75	2.75, 3, 5.25, 5.25	
2.00	2.25, 3, 4, 4.5, 4.75, 5, 5.25, 5.25, 5.5, 5.5, 6	
2.25	3, 5.5	3
2.50	2.25, 2.25, 2.25, 2.25, 2.5, 2.75, 3, 3.25, 3.25, 4, 4, 4	
2.75	1.25, 1.5, 2.5, 3, 3, 3.25, 3.75, 4.5, 4.5, 5, 5, 5.25, 5.25, 5.25, 5.25, 5.25	1
3.00	2, 3.25, 3.5, 3.75, 4, 4, 4.25, 4.25, 4.25, 4.75, 4.75, 4.75, 5	1.75
3.25	1.25, 1.75, 2, 2, 2.75, 3, 3, 3.5, 3.5, 4.25, 4.5	
3.50	1.25, 2.25, 2.25, 2.5, 2.75, 2.75, 3, 3.25, 3.5, 3.5, 4, 4, 4.25, 4.5, 4.5	0.75
3.75	1.25, 1.75, 1.75, 2, 2.75, 3, 3, 3, 4, 4.25, 4.25	0.75, 1, 2.75, 3, 3.5, 4.25
4.00	1, 1.5, 1.5, 2, 2.25, 2.75, 3.5, 3.75, 3.75, 4	1
4.25	1.25, 1.5, 1.5, 2, 2, 2, 2.25, 2.5, 2.5, 2.5, 3, 3.5, 3.5	1.75
4.50	1, 1.5, 1.5, 1.5, 1.75, 2.25, 2.25, 2.5, 2.5, 2.5, 2.5, 2.75, 2.75, 2.75, 2.75, 3, 3, 3, 3.25, 3.25	3.25
4.75	1, 1.5, 1.5, 1.5, 1.75, 1.75, 2, 2.25, 2.75, 3, 3, 3.25, 3.25, 3.25, 3.25, 3.25, 3.25	1, 2.25
5.00	0.5, 1.5, 1.5, 1.75, 2, 2.25, 2.25, 2.25, 2.5, 2.5, 3, 3, 3	0.5, 0.75, 1.5, 2.5
5.25	0.25, 0.25, 0.75, 0.75, 0.75, 1, 1, 1.25, 1.25, 1.5, 1.5, 1.5, 1.5, 2.25, 2.25, 2.5, 2.5, 2.75	0.25, 1, 1.5
5.50	1, 1, 1, , 1.25, 1.25, 1.75, 2, 2.25, 2.25, 2.5	.5, 1.5, 2.5
5.75	0.25, 0.75, 1, 1.5, 1.5, 1.5, 2, 2, 2.25	1.75
600	0.5, 0.75, 0.75, 0.75, 1, 1, 1, 1.25, 1.25, 1.5, 1.5, 1.75, 1.75, 1.75, 2	0.5, 1.25
6.25	0.75, 1, 1.25, 1.75, 1.75	0.5, 1.25
6.50	0.25, 0.25, 0.75, 1, 1.25, 1.5	0.75
6.75	0.75, 0.75, 0.75, 1, 1.25, 1.25, 1.25	0.5, 0.75
7.00	0.75	0.75
7.25	0.25	0.25

2
Basic Quantities and Models

2.1 Introduction

In this chapter we consider the basic parameters used in modeling survival data. We shall define these quantities and show how they are interrelated in sections 2.2–2.4. In section 2.5 some common parametric models are discussed. The important application of regression to survival analysis is covered in section 2.6, where both parametric and semiparametric models are presented. Models for competing risks are discussed in section 2.7.

Let X be the time until some specified event. This event may be death, the appearance of a tumor, the development of some disease, recurrence of a disease, equipment breakdown, cessation of breast feeding, and so forth. Furthermore, the event may be a good event, such as remission after some treatment, conception, cessation of smoking, and so forth. More precisely, in this chapter, X is a nonnegative random variable from a homogeneous population. Four functions characterize the distribution of X, namely, the *survival function*, which is the probability of an individual surviving to time x; the *hazard rate (function)*, sometimes termed *risk function*, which is the chance an individual of age x experiences the event in the next instant in time; the *probability density (or probability mass) function*, which is the unconditional probability of the event's occurring at time x; and the *mean residual life* at time x, which is the mean time to the event of interest, given the event has not occurred at x. If we know any one of these four functions, then the other three can be uniquely determined. In practice, these four functions, along with another useful quantity, the *cumulative hazard function*, are used to illustrate different aspects of the distribution of X. In the competing risk context, the *cause-specific hazard rate*, which is the rate at which subjects who have yet to experience any of the competing risks are experiencing the ith competing cause of failure, is often used. This quantity and other competing risk quantities are discussed in detail in section 2.7. In Chapters 4–6, we shall see how these functions are estimated and how inferences are drawn about the survival (or failure) distribution.

2.2   The Survival Function

The basic quantity employed to describe time-to-event phenomena is the survival function, the probability of an individual surviving beyond time x (experiencing the event after time x). It is defined as

$$S(x) = Pr(X > x). \tag{2.2.1}$$

In the context of equipment or manufactured item failures, $S(x)$ is referred to as the reliability function. If X is a continuous random variable, then, $S(x)$ is a continuous, strictly decreasing function.

When X is a continuous random variable, the survival function is the complement of the cumulative distribution function, that is, $S(x) = 1 - F(x)$, where $F(x) = Pr(X \le x)$. Also, the survival function is the integral of the probability density function, $f(x)$, that is,

$$S(x) = Pr(X > x) = \int_x^\infty f(t)\, dt. \tag{2.2.2}$$

Thus,

$$f(x) = -\frac{dS(x)}{dx}.$$

Note that $f(x)\, dx$ may be thought of as the "approximate" probability that the event will occur at time x and that $f(x)$ is a nonnegative function with the area under $f(x)$ being equal to one.

EXAMPLE 2.1   The survival function for the Weibull distribution, discussed in more detail in section 2.5, is $S(x) = \exp(-\lambda x^\alpha)$, $\lambda > 0$, $\alpha > 0$. The exponential distribution is a special case of the Weibull distribution when $\alpha = 1$. Survival curves with a common median of 6.93 are exhibited in Figure 2.1 for $\lambda = 0.26328$, $\alpha = 0.5$; $\lambda = 0.1$, $\alpha = 1$; and $\lambda = 0.00208$, $\alpha = 3$.

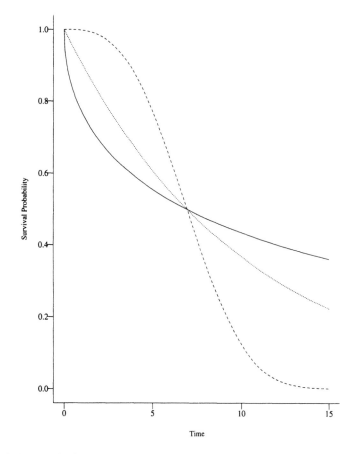

Figure 2.1 *Weibull Survival functions for $\alpha = 0.5$, $\lambda = 0.26328$ (————);*
$\alpha = 1.0$, $\lambda = 0.1$ ($\cdots\cdots$); $\alpha = 3.0$, $\lambda = 0.00208$ (------).

Many types of survival curves can be shown but the important point to note is that they all have the same basic properties. They are monotone, nonincreasing functions equal to one at zero and zero as the time approaches infinity. Their rate of decline, of course, varies according to the risk of experiencing the event at time x but it is difficult to determine the essence of a failure pattern by simply looking at the survival curve. Nevertheless, this quantity continues to be a popular description of survival in the applied literature and can be very useful in comparing two or more mortality patterns. Next, we present one more survival curve, which will be discussed at greater length in the next section.

EXAMPLE 2.2   The U.S. Department of Health and Human Services publishes yearly survival curves for all causes of mortality in the United States and each of the fifty states by race and sex in their Vital Statistics of the United

TABLE 2.1

Survival Functions of U.S. Population By Race and Sex in 1989

Age	White Male	White Female	Black Male	Black Female	Age	White Male	White Female	Black Male	Black Female
0	1.00000	1.00000	1.00000	1.00000	43	0.93771	0.97016	0.85917	0.93361
1	0.99092	0.99285	0.97996	0.98283	44	0.93477	0.96862	0.85163	0.92998
2	0.99024	0.99232	0.97881	0.98193	45	0.93161	0.96694	0.84377	0.92612
3	0.98975	0.99192	0.97792	0.98119	46	0.92820	0.96511	0.83559	0.92202
4	0.98937	0.99160	0.97722	0.98059	47	0.92450	0.96311	0.82707	0.91765
5	0.98905	0.99134	0.97664	0.98011	48	0.92050	0.96091	0.81814	0.91300
6	0.98877	0.99111	0.97615	0.97972	49	0.91617	0.95847	0.80871	0.90804
7	0.98850	0.99091	0.97571	0.97941	50	0.91148	0.95575	0.79870	0.90275
8	0.98825	0.99073	0.97532	0.97915	51	0.90639	0.95273	0.78808	0.89709
9	0.98802	0.99056	0.97499	0.97892	52	0.90086	0.94938	0.77685	0.89103
10	0.98782	0.99041	0.97472	0.97870	53	0.89480	0.94568	0.76503	0.88453
11	0.98765	0.99028	0.97449	0.97847	54	0.88810	0.94161	0.75268	0.87754
12	0.98748	0.99015	0.97425	0.97823	55	0.88068	0.93713	0.73983	0.87000
13	0.98724	0.98999	0.97392	0.97796	56	0.87250	0.93222	0.72649	0.86190
14	0.98686	0.98977	0.97339	0.97767	57	0.86352	0.92684	0.71262	0.85321
15	0.98628	0.98948	0.97258	0.97735	58	0.85370	0.92096	0.69817	0.84381
16	0.98547	0.98909	0.97145	0.97699	59	0.84299	0.91455	0.68308	0.83358
17	0.98445	0.98862	0.97002	0.97658	60	0.83135	0.90756	0.66730	0.82243
18	0.98326	0.98809	0.96829	0.97612	61	0.81873	0.89995	0.65083	0.81029
19	0.98197	0.98755	0.96628	0.97559	62	0.80511	0.89169	0.63368	0.79719
20	0.98063	0.98703	0.96403	0.97498	63	0.79052	0.88275	0.61584	0.78323
21	0.97924	0.98654	0.96151	0.97429	64	0.77501	0.87312	0.59732	0.76858
22	0.97780	0.98607	0.95873	0.97352	65	0.75860	0.86278	0.57813	0.75330
23	0.97633	0.98561	0.95575	0.97267	66	0.74131	0.85169	0.55829	0.73748
24	0.97483	0.98514	0.95267	0.97174	67	0.72309	0.83980	0.53783	0.72104
25	0.97332	0.98466	0.94954	0.97074	68	0.70383	0.82702	0.51679	0.70393
26	0.97181	0.98416	0.94639	0.96967	69	0.68339	0.81324	0.49520	0.68604
27	0.97029	0.98365	0.94319	0.96852	70	0.66166	0.79839	0.47312	0.66730
28	0.96876	0.98312	0.93989	0.96728	71	0.63865	0.78420	0.45058	0.64769
29	0.96719	0.98257	0.93642	0.96594	72	0.61441	0.76522	0.42765	0.62723
30	0.96557	0.98199	0.93273	0.96448	73	0.58897	0.74682	0.40442	0.60591
31	0.96390	0.98138	0.92881	0.96289	74	0.56238	0.72716	0.38100	0.58375
32	0.96217	0.98073	0.92466	0.96118	75	0.53470	0.70619	0.35749	0.56074
33	0.96038	0.98005	0.92024	0.95934	76	0.50601	0.68387	0.33397	0.53689
34	0.95852	0.97933	0.91551	0.95740	77	0.47641	0.66014	0.31050	0.51219
35	0.95659	0.97858	0.91044	0.95336	78	0.44604	0.63494	0.28713	0.48663
36	0.95457	0.97779	0.90501	0.95321	79	0.41503	0.60822	0.26391	0.46020
37	0.95245	0.97696	0.89922	0.95095	80	0.38355	0.57991	0.24091	0.43291
38	0.95024	0.97607	0.89312	0.94855	81	0.35178	0.54997	0.21819	0.40475
39	0.94794	0.97510	0.88677	0.94598	82	0.31991	0.51835	0.19583	0.37573
40	0.94555	0.97404	0.88021	0.94321	83	0.28816	0.48502	0.17392	0.34588
41	0.94307	0.97287	0.87344	0.94023	84	0.25677	0.44993	0.15257	0.31522
42	0.94047	0.97158	0.86643	0.93703	85	0.22599	0.41306	0.13191	0.28378

States Series. In Table 2.1, we present the overall survival probabilities for males and females, by race, taken from the 1990 report (U.S. Department of Health and Human Services, 1990). Figure 2.2 shows the survival curves and allows a visual comparison of the curves. We can see that white females have the best survival probability, white males and black females are comparable in their survival probabilities, and black males have the worst survival.

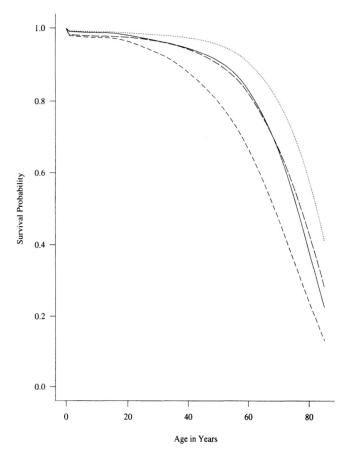

Figure 2.2 *Survival Functions for all cause mortality for the US population in 1989. White males (—————); white females (·······); black males (------); black females (——— ———).*

When X is a discrete, random variable, different techniques are required. Discrete, random variables in survival analyses arise due to rounding off measurements, grouping of failure times into intervals, or when lifetimes refer to an integral number of units. Suppose that X can take on values x_j, $j = 1, 2, \ldots$ with probability mass function (p.m.f.) $p(x_j) = Pr(X = x_j)$, $j = 1, 2, \ldots$, where $x_1 < x_2 < \cdots$.

The survival function for a discrete random variable X is given by

$$S(x) = Pr(X > x) = \sum_{x_j > x} p(x_j). \qquad (2.2.3)$$

EXAMPLE 2.3   Consider, for pedagogical purposes, the lifetime X, which has the p.m.f. $p(x_j) = Pr(X = j) = 1/3$, $j = 1, 2, 3$, a simple discrete uniform distribution. The corresponding survival function, plotted in Figure 2.3, is expressed by

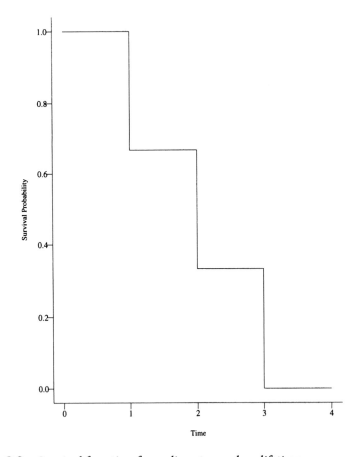

Figure 2.3   *Survival function for a discrete random lifetime*

$$S(x) = Pr(X > x) = \sum_{x_j > x} p(x_j) = \begin{cases} 1 & \text{if } 0 \le x < 1, \\ 2/3 & \text{if } 1 \le x < 2, \\ 1/3 & \text{if } 2 \le x < 3, \\ 0 & \text{if } x \ge 3. \end{cases}$$

Note that, when X is discrete, the survival function is a nonincreasing step function.

2.3   The Hazard Function

A basic quantity, fundamental in survival analysis, is the hazard function. This function is also known as the conditional failure rate in reliability, the force of mortality in demography, the intensity function in stochastic processes, the age-specific failure rate in epidemiology, the inverse of the Mill's ratio in economics, or simply as the hazard rate. The hazard rate is defined by

$$h(x) = \lim_{\Delta x \to 0} \frac{P[x \le X < x + \Delta x \mid X \ge x]}{\Delta x}. \qquad (2.3.1)$$

If X is a continuous random variable, then,

$$h(x) = f(x)/S(x) = -d \ln[S(x)]/dx. \qquad (2.3.2)$$

A related quantity is the cumulative hazard function $H(x)$, defined by

$$H(x) = \int_0^x h(u) \, du = -\ln[S(x)]. \qquad (2.3.3)$$

Thus, for continuous lifetimes,

$$S(x) = \exp[-H(x)] = \exp\left[-\int_0^x h(u) \, du\right]. \qquad (2.3.4)$$

From (2.3.1), one can see that $h(x)\Delta x$ may be viewed as the "approximate" probability of an individual of age x experiencing the event in the next instant. This function is particularly useful in determining the appropriate failure distributions utilizing qualitative information about the mechanism of failure and for describing the way in which the chance of experiencing the event changes with time. There are many general shapes for the hazard rate. The only restriction on $h(x)$ is that it be nonnegative, i.e., $h(x) \ge 0$.

Some generic types of hazard rates are plotted in Figure 2.4. For example, one may believe that the hazard rate for the occurrence of a particular event is increasing, decreasing, constant, bathtub-shaped,

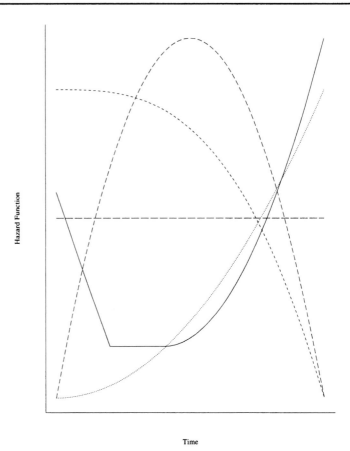

Figure 2.4 *Shapes of hazard functions. Constant hazard (————); increasing hazard (------); decreasing hazard (- - - - - -); bathtub shaped (————); humpshaped (————).*

hump-shaped, or possessing some other characteristic which describes the failure mechanism.

Models with increasing hazard rates may arise when there is natural aging or wear. Decreasing hazard functions are much less common but find occasional use when there is a very early likelihood of failure, such as in certain types of electronic devices or in patients experiencing certain types of transplants. Most often, a bathtub-shaped hazard is appropriate in populations followed from birth. Similarly, some manufactured equipment may experience early failure due to faulty parts, followed by a constant hazard rate which, in the later stages of equipment life, increases. Most population mortality data follow this type of hazard function where, during an early period, deaths result, primarily, from infant diseases, after which the death rate stabilizes, followed by an increasing hazard rate due to the natural aging process. Finally, if the hazard rate is increasing early and eventually begins declining, then, the hazard is termed hump-shaped. This type of hazard rate is often used in modeling survival after successful surgery where there is an initial increase in risk due to infection, hemorrhaging, or other complications just after the procedure, followed by a steady decline in risk as the patient recovers. Specific distributions which give rise to these different types of failure rates are presented in section 2.5.

EXAMPLE 2.1   *(continued)* One particular distribution, which is flexible enough to accommodate increasing ($\alpha > 1$), decreasing ($\alpha < 1$), or constant hazard rates ($\alpha = 1$), is the Weibull distribution introduced in Example 2.1. Hazard rates, $h(x) = \alpha\lambda x^{\alpha-1}$, are plotted for the same values of the parameters used in Figure 2.1, namely, $\lambda = 0.26328$, $\alpha = 0.5$; $\lambda = 0.1$, $\alpha = 1$; and $\lambda = 0.00208$, $\alpha = 3$ in Figure 2.5. One can see

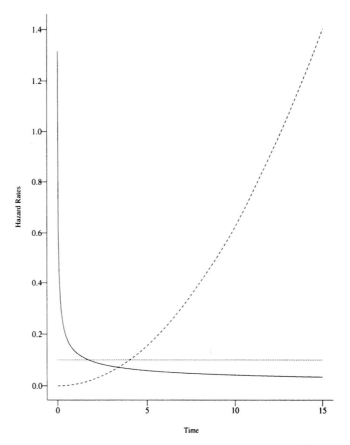

Figure 2.5 *Weibull hazard functions for $\alpha = 0.5$, $\lambda = 0.26328$ (————); $\alpha = 1.0$, $\lambda = 0.1$ (------); $\alpha = 3.0$, $\lambda = 0.00208$ (— — —).*

that, though the three survival functions have the same basic shape, the hazard functions are dramatically different.

An example of a bathtub-shaped hazard rate is presented in the following example.

EXAMPLE 2.2   *(continued)* The 1989 U.S. mortality hazard rates, by sex and race, are presented in Figure 2.6. One can see the decreasing hazard rates early in all four groups, followed, approximately, by, a constant hazard rate, eventually leading to an increasing hazard rate starting at different times for each group.

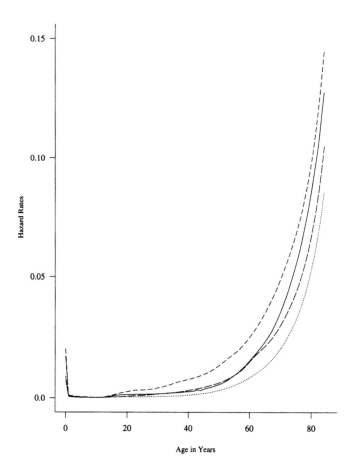

Figure 2.6   *Hazard functions for all cause mortality for the US population in 1989. White males (————); white females (······); black males (- - - - -); black females (————).*

When X is a discrete random variable, the hazard function is given by

$$h(x_j) = Pr(X = x_j \mid X \ge x_j) = \frac{p(x_j)}{S(x_{j-1})}, \quad j = 1, 2, \ldots \quad (2.3.5)$$

where $S(x_0) = 1$. Because $p(x_j) = S(x_{j-1}) - S(x_j)$, in conjunction with (2.3.5), $h(x_j) = 1 - S(x_j)/S(x_{j-1})$, $j = 1, 2, \ldots$.

Note that the survival function may be written as the product of conditional survival probabilities

$$S(x) = \prod_{x_j \le x} S(x_j)/S(x_{j-1}). \quad (2.3.6)$$

Thus, the survival function is related to the hazard function by

$$S(x) = \prod_{x_j \le x} [1 - h(x_j)]. \quad (2.3.7)$$

EXAMPLE 2.3 *(continued)* Let us reconsider the discrete random variable X in Example 2.3 with $p(x_j) = Pr(X = j) = 1/3$, $j = 1, 2, 3$. The hazard function may be obtained by direct application of (2.3.5). This leads to

$$h(x_j) = \begin{array}{l} 1/3, \text{ for } j = 1, \\ 1/2, \text{ for } j = 2, \\ 1, \text{ for } j = 3, \text{ and} \\ 0, \text{ elsewhere}. \end{array}$$

Note that the hazard rate is zero for a discrete random variable except at points where a failure could occur.

Practical Notes

1. Though the three survival functions in Figure 2.1 have the same basic shape, one can see that the three hazard functions shown in Figure 2.5 are dramatically different. In fact, the hazard function is usually more informative about the underlying mechanism of failure than the survival function. For this reason, consideration of the hazard function may be the dominant method for summarizing survival data.

2. The relationship between some function of the cumulative hazard function and some function of time has been exploited to develop hazard papers (Nelson, 1982), which will give the researcher an intuitive impression as to the desirability of the fit of specific models. For example, if X has a Weibull distribution, as in Example 2.1, then

its cumulative hazard rate is $H(x) = \lambda x^\alpha$, so a plot of $\ln H(x)$ versus $\ln x$ is a straight line with slope α and y intercept $\ln \lambda$. Using a nonparametric estimator of $H(x)$, developed in Chapter 4, this relationship can be exploited to provide a graphical check of the goodness of fit of the Weibull model to data (see section 12.5 for details and examples).

Theoretical Notes

1. For discrete lifetimes, we shall define the cumulative hazard function by

$$H(x) = \sum_{x_j \leq x} h(x_j). \qquad (2.3.8)$$

Notice that the relationship $S(x) = \exp\{-H(x)\}$ for this definition no longer holds true. Some authors (Cox and Oakes, 1984) prefer to define the cumulative hazard for discrete lifetimes as

$$H(x) = -\sum_{x_j \leq x} \ln[1 - h(x_j)], \qquad (2.3.9)$$

because the relationship for continuous lifetimes $S(x) = \exp[-H(x)]$ will be preserved for discrete lifetimes. If the $h(x_j)$ are small, (2.3.8) will be an approximation of (2.3.9). We prefer the use of (2.3.8) because it is directly estimable from a sample of censored or truncated lifetimes and the estimator has very desirable statistical properties. This estimator is discussed in Chapter 4.

2. For continuous lifetimes, the failure distribution is said to have an increasing failure-rate (IFR) property, if the hazard function $h(x)$ is nondecreasing for $x \geq 0$, and an increasing failure rate on the average (IFRA) if the ratio of the cumulative hazard function to time $H(x)/x$ is nondecreasing for $x > 0$.

3. For continuous lifetimes, the failure distribution is said to have a decreasing failure-rate (DFR) property if the hazard function $h(x)$ is nonincreasing for $x \geq 0$.

2.4   The Mean Residual Life Function and Median Life

The fourth basic parameter of interest in survival analyses is the *mean residual life* at time x. For individuals of age x, this parameter measures their expected remaining lifetime. It is defined as $\mathrm{mrl}(x) = E(X - x \mid X > x)$. It can be shown (see Theoretical Note 1) that the mean residual life is the area under the survival curve to the right of x divided by $S(x)$. Note that the mean life, $\mu = \mathrm{mrl}(0)$, is the total area under the survival curve.

For a continuous random variable,

$$\mathrm{mrl}(x) = \frac{\int_x^\infty (t - x) f(t) dt}{S(x)} = \frac{\int_x^\infty S(t) \, dt}{S(x)} \qquad (2.4.1)$$

and

$$\mu = E(X) = \int_0^\infty t f(t) \, dt = \int_0^\infty S(t) \, dt. \qquad (2.4.2)$$

Also the variance of X is related to the survival function by

$$\mathrm{Var}(X) = 2 \int_0^\infty t S(t) dt - \left[\int_0^\infty S(t) \, dt \right]^2. \qquad (2.4.3)$$

The pth quantile (also referred to as the $100p$th percentile) of the distribution of X is the smallest x_p so that

$$S(x_p) \le 1 - p, \text{ i.e., } x_p = \inf\{t : S(t) \le 1 - p\}. \qquad (2.4.4)$$

If X is a continuous random variable, then the pth quantile is found by solving the equation $S(x_p) = 1 - p$. The median lifetime is the 50th percentile $x_{0.5}$ of the distribution of X. It follows that the median lifetime for a continuous random variable X is the value $x_{0.5}$ so that

$$S(x_{0.5}) = 0.5. \qquad (2.4.5)$$

EXAMPLE 2.4   The mean and median lifetimes for an exponential life distribution are $1/\lambda$ and $(\ln 2)/\lambda$ as determined from equations (2.4.2) and (2.4.5), respectively. Furthermore, the mean residual life for an exponential distribution is also $1/\lambda$ from equation (2.4.1). Distributions with this property are said to exhibit lack of memory. The exponential distribution is the unique continuous distribution possessing this characteristic.

EXAMPLE 2.1   (continued) For the Weibull distribution the $100p$th percentile is found by solving the equation $1 - p = \exp\{-\lambda x_p^\alpha\}$ so that $x_p = \{-\ln[1 - p]/\lambda\}^{1/\alpha}$.

EXAMPLE 2.2   (continued) The median and other percentiles for the population mortality distribution of black men may be determined graphically by using

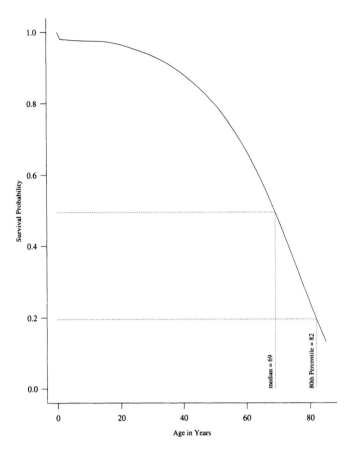

Figure 2.7 *Determination of the median lifetime and 80th percentile of lifetimes for black men in the US population in 1989*

the survival function plot depicted in Figure 2.2. First, find the appropriate survival probability, and, then, interpolate to the appropriate time. Determination of the median and 80th percentile, as illustrated in Figure 2.7, give values of about 69 and 82 years, respectively. More accurate values can be found by linear interpolation in Table 2.1. We see that $S(68) = 0.51679 > 0.5$ and $S(69) = 0.49520 < 0.5$, so the median lies between 68 and 69 years. By linear interpolation,

$$x_{0.5} = 68 + \frac{S(68) - 0.5}{S(68) - S(69)} = 68.78 \text{ years.}$$

Similar calculations yield $x_{0.8} = 81.81$ years.

Theoretical Notes

1. For a continuous random variable X,

$$E(X - x \mid X > x) = \frac{\int_x^\infty (t - x) f(t) dt}{S(x)}.$$

   We integrate by parts to establish equation (2.4.1) using the fact that $f(t) dt = -dS(t)$, so that $E(X - x \mid X > x) S(x) = -(t - x) S(t) \mid_x^\infty + \int_x^\infty S(t) dt$. The first term on the right-hand side of the equation is 0 because $S(\infty)$ is 0. For a discrete, random variable, the result that the mean residual life is related to the area under the survival curve is obtained by using a partial summation formula.

2. Interrelationships between the various quantities discussed earlier, for a continuous lifetime X, may be summarized as

$$S(x) = \int_x^\infty f(t) dt$$

$$= \exp\left[-\int_0^x h(u) du\right]$$

$$= \exp[-H(x)]$$

$$= \frac{\mathrm{mrl}(0)}{\mathrm{mrl}(x)} \exp\left[-\int_0^x \frac{du}{\mathrm{mrl}(u)}\right].$$

$$f(x) = -\frac{d}{dx} S(x)$$

$$= h(x) S(x)$$

$$= \left(\frac{d}{dx} \mathrm{mrl}(x) + 1\right) \left(\frac{\mathrm{mrl}(0)}{\mathrm{mrl}(x)^2}\right) \exp\left[-\int_0^x \frac{du}{\mathrm{mrl}(u)}\right]$$

$$h(x) = -\frac{d}{dx} ln[S(x)]$$

$$= \frac{f(x)}{S(x)}$$

$$= \left(\frac{d}{dx} \mathrm{mrl}(x) + 1\right) / \mathrm{mrl}(x)$$

$$\mathrm{mrl}(x) = \frac{\int_x^\infty S(u) du}{S(x)}$$

$$= \frac{\int_x^\infty (u - x) f(u) du}{S(x)}.$$

3. Interrelationships between the various quantities discussed earlier, for discrete lifetimes X, may be summarized as

$$S(x) = \sum_{x_j > x} p(x_j)$$

$$= \prod_{x_j \leq x} [1 - h(x_j)],$$

$$p(x_j) = S(x_{j-1}) - S(x_j) = h(x_j)S(x_{j-1}), \quad j = 1, 2, \ldots,$$

$$h(x_j) = \frac{p(x_j)}{S(x_{j-1})},$$

$$\text{mrl}(x) = \frac{(x_{i+1} - x)S(x_i) + \sum_{j \geq i+1}(x_{j+1} - x_j)S(x_j)}{S(x)},$$

for $x_i \leq x < x_{i+1}$.

4. If X is a positive random variable with a hazard rate $h(t)$, which is a sum of a continuous function $h_c(t)$ and a discrete function which has mass $h_d(x_j)$ at times $0 \leq x_1 \leq x_2 \leq \cdots$, then the survival function is related to the hazard rate by the so called "product integral" of $[1 - h(t)]dt$ defined as follows:

$$S(x) = \prod_{x_j \leq x}[1 - h_d(x_j)]\exp\left[-\int_0^x h_c(t)dt\right].$$

5. Sometimes (particularly, when the distribution is highly skewed), the median is preferred to the mean, in which case, the quantity median residual lifetime at time x, mdrl(x), is preferred to the mean residual lifetime at time x, mrl(x), as defined in (2.4.1). The median residual lifetime at time x is defined to be the median of the conditional distribution of $X - x \,|\, X > x$ and is determined using (2.4.4) except that the conditional distribution is used. It is the length of the interval from x to the time where one-half of the individuals alive at time x will still be alive. Note that the mdrl(0) is simply the median of the unconditional distribution.

2.5   Common Parametric Models for Survival Data

Although nonparametric or semiparametric models will be used extensively, though not exclusively, in this book, it is appropriate and necessary to discuss the more widely used parametric models. These models are chosen, not only because of their popularity among researchers who analyze survival data, but also because they offer insight into the nature of the various parameters and functions discussed in previous sections, particularly, the hazard rate. Some of the important models discussed include the exponential, Weibull, gamma, log normal, log logistic, normal, exponential power, Gompertz, inverse Gaussian, Pareto, and the generalized gamma distributions. Their survival functions, hazard rates, density functions, and expected lifetimes are summarized in Table 2.2.

First, because of its historical significance, mathematical simplicity, and important properties, we shall discuss the *exponential* distribution. Its survival function is $S(x) = \exp[-\lambda x], \lambda > 0, x \geq 0$. The density function is $f(x) = \lambda \exp[-\lambda x]$, and it is characterized by a constant hazard function $h(x) = \lambda$.

The exponential distribution has the following properties. The first, referred to as the lack of memory property, is given by

$$P(X \geq x + z \mid X \geq x) = P(X \geq z), \qquad (2.5.1)$$

which allows for its mathematical tractability but also reduces its applicability to many realistic applied situations. Because of this distributional property, it follows that $E(X - x \mid X > x) = E(X) = 1/\lambda$; that is, the mean residual life is constant. Because the time until the future occurrence of an event does not depend upon past history, this property is sometimes called the "no-aging" property or the "old as good as new" property. This property is also reflected in the exponential distribution's constant hazard rate. Here, the conditional probability of failure at any time t, given that the event has not occurred prior to time t, does not depend upon t. Although the exponential distribution has been historically very popular, its constant hazard rate appears too restrictive in both health and industrial applications.

The mean and standard deviation of the distribution are $1/\lambda$ (thus, the coefficient of variation is unity) and the pth quantile is $x_p = -\ln(1 - p)/\lambda$. Because the exponential distribution is a special case of both the Weibull and gamma distributions, considered in subsequent paragraphs, other properties will be implicit in the discussion of those distributions.

Though not the first to suggest the use of this next distribution, Rosen and Rammler (1933) used it to describe the "laws governing the fineness of powdered coal," and Weibull (1939, 1951) proposed the same distribution, to which his name later became affixed, for describing the life length of materials. Its survival function is $S(x) = \exp[-\lambda x^\alpha]$, for $x > 0$. Here $\lambda > 0$ is a scale parameter, and $\alpha > 0$ is a shape parameter. The two-parameter Weibull was previously introduced in Example 2.1. The exponential distribution is a special case when $\alpha = 1$. Figure 2.1, already presented, exhibits a variety of Weibull survival functions. Its

TABLE 2.2

Hazard Rates, Survival Functions, Probability Density Functions, and Expected Lifetimes for Some Common Parametric Distributions

Distribution	Hazard Rate $h(x)$	Survival Function $S(x)$	Probability Density Function $f(x)$	Mean $E(X)$
Exponential $\lambda > 0, x \geq 0$	λ	$\exp[-\lambda x]$	$\lambda \exp(-\lambda x)$	$\dfrac{1}{\lambda}$
Weibull $\alpha, \lambda > 0,$ $x \geq 0$	$\alpha \lambda x^{\alpha-1}$	$\exp[-\lambda x^\alpha]$	$\alpha \lambda x^{\alpha-1} \exp(-\lambda x^\alpha)$	$\dfrac{\Gamma(1 + 1/\alpha)}{\lambda^{1/\alpha}}$
Gamma $\beta, \lambda > 0,$ $x \geq 0$	$\dfrac{f(x)}{S(x)}$	$1 - I(\lambda x, \beta)^*$	$\dfrac{\lambda^\beta x^{\beta-1} \exp(-\lambda x)}{\Gamma(\beta)}$	$\dfrac{\beta}{\lambda}$
Log normal $\sigma > 0, x \geq 0$	$\dfrac{f(x)}{S(x)}$	$1 - \Phi\left[\dfrac{\ln x - \mu}{\sigma}\right]$	$\dfrac{\exp\left[-\frac{1}{2}\left(\frac{\ln x - \mu}{\sigma}\right)^2\right]}{x(2\pi)^{1/2}\sigma}$	$\exp(\mu + 0.5\sigma^2)$
Log logistic $\alpha, \lambda > 0, x \geq 0$	$\dfrac{\alpha x^{\alpha-1}\lambda}{1 + \lambda x^\alpha}$	$\dfrac{1}{1 + \lambda x^\alpha}$	$\dfrac{\alpha x^{\alpha-1}\lambda}{[1 + \lambda x^\alpha]^2}$	$\dfrac{\pi \operatorname{Csc}(\pi/\alpha)}{\alpha \lambda^{1/\alpha}}$ if $\alpha > 1$
Normal $\sigma > 0,$ $-\infty < x < \infty$	$\dfrac{f(x)}{S(x)}$	$1 - \Phi\left[\dfrac{x - \mu}{\sigma}\right]$	$\dfrac{\exp\left[-\frac{1}{2}\left(\frac{x - \mu}{\sigma}\right)^2\right]}{(2\pi)^{1/2}\sigma}$	μ
Exponential power $\alpha, \lambda > 0, x \geq 0$	$\alpha \lambda^\alpha x^{\alpha-1} \exp\{[\lambda x]^\alpha\}$	$\exp\{1 - \exp[(\lambda x)^\alpha]\}$	$\alpha e \lambda^\alpha x^{\alpha-1} \exp[(\lambda x)^\alpha] - \exp\{\exp[(\lambda x)^\alpha]\}$	$\int_0^\infty S(x)dx$
Gompertz $\theta, \alpha > 0, x \geq 0$	$\theta e^{\alpha x}$	$\exp\left[\dfrac{\theta}{\alpha}(1 - e^{\alpha x})\right]$	$\theta e^{\alpha x} \exp\left[\dfrac{\theta}{\alpha}(1 - e^{\alpha x})\right]$	$\int_0^\infty S(x)dx$
Inverse Gaussian $\lambda \geq 0, x \geq 0$	$\dfrac{f(x)}{S(x)}$	$\Phi\left[\left(\frac{\lambda}{x}\right)^{1/2}\left(1 - \frac{x}{\mu}\right)\right] - e^{2\lambda/\mu}\Phi\left\{-\left[\frac{\lambda}{x}\right]^{1/2}\left(1 + \frac{x}{\mu}\right)\right\}$	$\left(\dfrac{\lambda}{2\pi x^3}\right)^{1/2}\exp\left[\dfrac{\lambda(x-\mu^2)}{2\mu^2 x}\right]$	μ
Pareto $\theta > 0, \lambda > 0$ $x \geq \lambda$	$\dfrac{\theta}{x}$	$\dfrac{\lambda^\theta}{x^\theta}$	$\dfrac{\theta \lambda^\theta}{x^{\theta+1}}$	$\dfrac{\theta\lambda}{\theta - 1}$ if $\theta > 1$
Generalized gamma $\lambda > 0, \alpha > 0,$ $\beta > 0, x \geq 0$	$\dfrac{f(x)}{S(x)}$	$1 - I[\lambda x^\alpha, \beta]$	$\dfrac{\alpha \lambda^\beta x^{\alpha\beta-1}\exp(-\lambda x^\alpha)}{\Gamma(\beta)}$	$\int_0^\infty S(x)dx$

$^* I(t, \beta) = \int_0^t u^{\beta-1}\exp(-u)\,du / \Gamma(\beta)$.

$$\int (\alpha) \quad or \quad \int (\alpha)$$

hazard function has the fairly flexible form

$$h(x) = \lambda \alpha x^{\alpha - 1}. \tag{2.5.2}$$

One can see from Figure 2.5 that the Weibull distribution is flexible enough to accommodate increasing ($\alpha > 1$), decreasing ($\alpha < 1$), or constant hazard rates ($\alpha = 1$). This fact, coupled with the model's relatively simple survival, hazard, and probability density functions, have made it a very popular parametric model. It is apparent that the shape of the Weibull distribution depends upon the value of α, thus, the reason for referring to this parameter as the "shape" parameter.

The rth moment of the Weibull distribution is $[\Gamma(1 + r/\alpha)]\lambda^{-r/\alpha}$. The mean and variance are $[\Gamma(1 + 1/\alpha)]\lambda^{-1/\alpha}$ and $\{\Gamma(1 + 2/\alpha) - [\Gamma(1 + 1/\alpha)]^2\}\lambda^{-2/\alpha}$, respectively, where $\Gamma[\alpha] = \int_0^\infty u^{\alpha-1}e^{-u}du$ is the well-known gamma function. $\Gamma[\alpha] = (\alpha - 1)!$ when α is an integer and is tabulated in Beyer (1968) when α is not an integer. The pth quantile of the Weibull distribution is expressed by

$$x_p = \{-[\ln(1 - p)]/\lambda\}^{1/\alpha}.$$

It is sometimes useful to work with the logarithm of the lifetimes. If we take $Y = \ln X$, where X follows a Weibull distribution, then, Y has the density function

$$\alpha \exp\{\alpha[y - (-(\ln\lambda)/\alpha)] - \exp\{\alpha[y - (-(\ln\lambda)/\alpha)]\}\}, -\infty < y < \infty. \tag{2.5.3}$$

Writing the model in a general linear model format, $Y = \mu + \sigma E$, where $\mu = (-\ln\lambda)/\alpha, \sigma = \alpha^{-1}$ and E has the standard extreme value distribution with density function

$$\exp(w - e^w), -\infty < w < \infty. \tag{2.5.4}$$

A random variable (more familiar to the traditional linear model audience) X is said to follow the *log normal* distribution if its logarithm $Y = \ln X$, follows the normal distribution. For time-to-event data, this distribution has been popularized because of its relationship to the *normal* distribution (a distribution which we assume is commonly known from elementary statistics courses and whose hazard rate, survival function, density function and mean are reported in Table 2.2 for completeness) and because some authors have observed that the log normal distribution approximates survival times or ages at the onset of certain diseases (Feinleib, 1960 and Horner, 1987).

Like the normal distribution, the log normal distribution is completely specified by two parameters μ and σ, the mean and variance of Y. Its density function is expressed by

$$f(x) = \frac{\exp\left[-\frac{1}{2}\left(\frac{\ln x - \mu}{\sigma}\right)^2\right]}{x(2\pi)^{1/2}\sigma} = \phi\left(\frac{\ln x - u}{\sigma}\right)/x \tag{2.5.5}$$

and its survival function is given by

$$S(x) = 1 - \Phi \left[\frac{\ln x - \mu}{\sigma} \right], \qquad (2.5.6)$$

where $\Phi(\phi)$ is the cumulative distribution function (density function) of a standard normal variable.

The hazard rate of the log normal is hump-shaped, that is, its value at 0 is zero, and it increases to a maximum and, then, decreases to 0 as x approaches infinity (see Figure 2.8). This model has been criticized as a lifetime distribution because the hazard function is decreasing for large x which seems implausible in many situations. The model may fit certain cases where large values of x are not of interest.

For the log normal distribution the mean lifetime is given by $\exp(\mu + \sigma^2/2)$ and the variance by $[\exp(\sigma^2) - 1] \exp(2\mu + \sigma^2)$. The pth percentile,

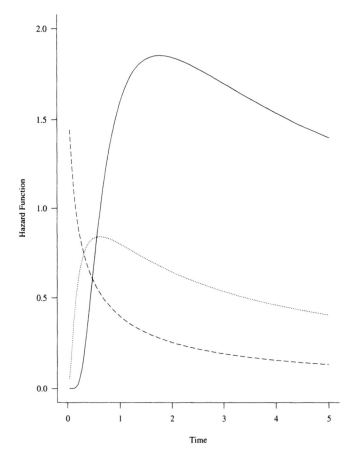

Figure 2.8 *Log normal hazard rates.* $\mu = 0, \sigma = 0.5$ (————); $\mu = 0, \sigma = 0.1$ (------); $\mu = 0, \sigma = 2.0$ (— — —)

x_p is expressed as $\exp(\mu + \sigma z_p)$, where z_p is the pth percentile of a standard normal distribution.

A variable X is said to follow the *log logistic* distribution if its logarithm $Y = \ln X$ follows the logistic distribution, a distribution closely resembling the normal distribution, but the survival function is mathematically more tractable. The density function for Y is expressed by

$$\frac{\exp(\frac{y-\mu}{\sigma})}{\sigma[1 + \exp(\frac{y-\mu}{\sigma})]^2}, \quad -\infty < y < \infty, \tag{2.5.7}$$

where μ and σ^2 are, respectively, the mean and scale parameter of Y. Again, we can cast this distribution in the linear model format by taking $Y = \mu + \sigma W$, where W is the standardized logistic distribution with $\mu = 0$ and $\sigma = 1$.

The hazard rate and survival function, respectively, for the log logistic distribution may be written as relatively simple expressions:

$$h(x) = \frac{\alpha \lambda x^{\alpha-1}}{1 + \lambda x^\alpha}, \tag{2.5.8}$$

and

$$S(x) = \frac{1}{1 + \lambda x^\alpha}, \tag{2.5.9}$$

where $\alpha = 1/\sigma > 0$ and $\lambda = \exp(-\mu/\sigma)$.

The numerator of the hazard function is the same as the Weibull hazard, but the denominator causes the hazard to take on the following characteristics: monotone decreasing for $\alpha \leq 1$. For $\alpha > 1$, the hazard rate increases initially to a maximum at time $[(\alpha - 1)/\lambda]^{1/\alpha}$ and then decreases to zero as time approaches infinity, as shown in Figure 2.9. The mean and variance of X are given by $E[X] = \pi \csc(\pi/\alpha)/(\alpha\lambda^{1/\alpha})$, if $\alpha > 1$, and $\text{Var}(X) = 2\pi\csc(2\pi/\alpha)/(\alpha\lambda^{2/\alpha}) - E[X]^2$, if $\alpha > 2$. The pth percentile is $x_p = \{p/[\lambda(1-p)]\}^{1/\alpha}$.

This distribution is similar to the Weibull and exponential models because of the simple expressions for $h(x)$ and $S(x)$ above. Its hazard rate is similar to the log normal, except in the extreme tail of the distribution, but its advantage is its simpler hazard function $h(x)$ and survival function $S(x)$.

The *gamma* distribution has properties similar to the Weibull distribution, although it is not as mathematically tractable. Its density function is given by

$$f(x) = \lambda^\beta x^{\beta-1} \exp(-\lambda x)/\Gamma(\beta), \tag{2.5.10}$$

where $\lambda > 0$, $\beta > 0$, $x \geq 0$, and $\Gamma(\beta)$ is the gamma function. For reasons similar to those of the Weibull distribution, λ is a scale parameter and β is called the shape parameter. This distribution, like the Weibull, includes the exponential as a special case ($\beta = 1$), approaches

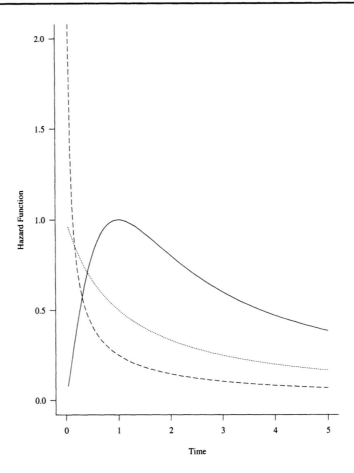

Figure 2.9 *Log logistic hazard rates.* $\lambda = 1, \sigma = 0.5$ (————); $\lambda = 1, \sigma = 1.0$ (······); $\lambda = 1, \sigma = 2.0$ (------)

a normal distribution as $\beta \to \infty$, and gives the chi-square distribution with ν degrees of freedom when $\nu = 2\beta$ (β, an integer) and $\lambda = 1/2$. The mean and variance of the gamma distribution are β/λ and β/λ^2, respectively.

The hazard function for the gamma distribution is monotone increasing for $\beta > 1$, with $h(0) = 0$ and $h(x) \to \lambda$ as $x \to \infty$, and monotone decreasing for $\beta < 1$, with $h(0) = \infty$ and $h(x) \to \lambda$ as $x \to \infty$. When $\beta > 1$, the mode is at $x = (\beta - 1)/\lambda$. A plot of the gamma hazard function is presented in Figure 2.10.

The survival function of the gamma distribution is expressed as

$$S(x) = \left[\int_x^\infty \lambda(\lambda t)^{\beta-1} \exp(-\lambda t) dt \right] / \Gamma(\beta) \qquad (2.5.11)$$

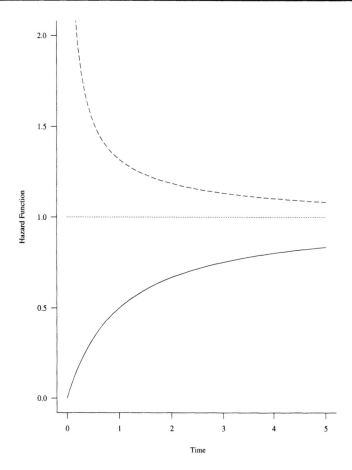

Figure 2.10   *Gamma hazard rates.* $\lambda = 1$, $\beta = 2.0$ (————); $\lambda = 1$, $\beta = 1.0$
(········); $\lambda = 1$, $\beta = 0.5$ (— — —)

$$= 1 - \left[\int_0^{\lambda x} u^{\beta-1} \exp(-u)\, du \right] / \Gamma(\beta)$$

$$= 1 - I(\lambda x, \beta),$$

where I is the incomplete gamma function.

For $\beta = n$, an integer, we obtain the Erlangian distribution whose survival function and hazard function, respectively, calculated from (2.2.2) and (2.3.2) simplify to

$$S(x) = \exp(-\lambda x) \sum_{k=0}^{n-1} (\lambda x)^k / k! \qquad (2.5.12)$$

and

$$b(x) = \lambda(\lambda x)^{n-1} \left[(n-1)! \sum_{k=0}^{n-1} (\lambda x)^k / k! \right]^{-1}.$$

Practical Notes

1. A relationship for the exponential distribution is $H(x) = -\ln S(x) = \lambda x$. This provides an empirical check for an exponential fit to data by plotting $H(x)$ vs x. The resulting plot should be a straight line through the origin with slope λ.

2. An empirical check of the Weibull distribution is accomplished by plotting $\ln[H(x)]$ vs. $\ln(x)$ (utilizing the relationship in the continuation of Example 2.1). The plot should result in a straight line with slope α and y intercept $\ln (\lambda)$. Later, in Chapter 12, we shall use this technique to give crude estimates of the parameters.

3. The *generalized gamma* distribution introduces an additional parameter α allowing additional flexibility in selecting a hazard function. This model has density function

$$f(x) = \frac{\alpha \lambda^\beta x^{\alpha\beta-1} \exp\{-\lambda x^\alpha\}}{\Gamma(\beta)} \tag{2.5.13}$$

and survival function

$$S(x) = 1 - I(\lambda x^\alpha, \beta).$$

This distribution reduces to the exponential when $\alpha = \beta = 1$, to the Weibull when $\beta = 1$, to the gamma when $\alpha = 1$, and approaches the log normal as $\beta \rightarrow \infty$. It is a useful distribution for model checking.

4. Occasionally, the event of interest may not occur until a threshold time ϕ is attained, in which case, $S(x) < 1$ only for $x > \phi$. In reliability theory, ϕ is called the "guarantee time." For example, in this instance, the Weibull survival function may be modified as follows:

$$S(x) = \exp[-\lambda(x - \phi)^\alpha], \lambda > 0, \alpha > 0 \text{ and } x > \phi.$$

Similar modifications may be made to the other distributions discussed in this section to accommodate the notion of a threshold parameter.

5. A model that has a hazard function capable of being bathtub-shaped, i.e., decreasing initially and, then, increasing as time increases, is the *exponential power* distribution with $\alpha < 1$ (Smith-Bain, 1975).

6. A distribution with a rich history in describing mortality curves is one introduced by Gompertz (1825) and later modified by Makeham (1860) by adding a constant to the hazard function (see Chiang, 1968, pp. 61–62). Again, the hazard function, survival function, density function, and mean of the *Gompertz* distribution are summarized in Table 2.2.

7. Other distributions which have received some attention in the literature are the *inverse Gaussian* and the *Pareto* distributions. These distributions are tabulated in Table 2.2 along with their survival, hazard, and probability density functions.

Theoretical Notes

1. The exponential distribution is summarized, with references, by Galambos (1982), Galambos and Kotz (1978), and Johnson and Kotz (1970). It is known to have been studied as early as the nineteenth century by Clausius (1858) in connection with the kinetic theory of gases. More recently, in studies of manufactured items (Davis, 1952; Epstein and Sobel, 1954; Epstein, 1958) and, to a lesser extent, in health studies (Feigl and Zelen, 1965; Sheps, 1966), the exponential distribution has historically been used in describing time to failure. As has been already noted, its constant hazard rate and lack of memory property greatly limit its applicability to modern survival analyses.
2. The Weibull distribution has been widely used in both industrial and biomedical applications. Lieblein and Zelen (1956), Berretoni (1964), and Nelson (1972) used it to describe the life length of ball bearings, electron tubes, manufactured items, and electrical insulation, respectively. Pike (1966) and Peto and Lee (1973) have given a theoretical motivation for its consideration in representing time to appearance of tumor or until death in animals which were subjected to carcinogenic insults over time (the multi-hit theory). Lee and Thompson (1974) argued, in a similar vein, that, within the class of proportional hazard rate distributions, the Weibull appears to be the most appropriate choice in describing lifetimes. Other authors (Lee and O'Neill, 1971; Doll, 1971) claim that the Weibull model fits data involving time to appearance of tumors in animals and humans quite well.
3. The Weibull distribution is also called the first asymptotic distribution of extreme values (see Gumbel, 1958, who popularized its use). The Weibull distribution arises as the limiting distribution of the minimum of a sample from a continuous distribution. For this reason, the Weibull distribution has been suggested as the appropriate distribution in certain circumstances.

2.6   Regression Models for Survival Data

Until this point, we have dealt exclusively with modeling the survival experience of a homogeneous population. However, a problem frequently encountered in analyzing survival data is that of adjusting the survival function to account for concomitant information (sometimes referred to as covariates, explanatory variables or independent variables). Populations which exhibit such heterogeneity are prevalent whether the study involves a clinical trial, a cohort study, or an observational study.

Consider a failure time $X > 0$, as has been discussed in the previous sections, and a vector $\mathbf{Z}^t = (Z_1, \ldots, Z_p)$ of explanatory variables associated with the failure time X. $\mathbf{Z}^t$ may include quantitative variables (such as blood pressure, temperature, age, and weight), qualitative variables (such as gender, race, treatment, and disease status) and/or time-dependent variables, in which case $\mathbf{Z}^t(x) = [Z_1(x), \ldots, Z_p(x)]$. Typical time-dependent variables include whether some intermediate event has or has not occurred by time x, the amount of time which has passed since the same intermediate event, serial measurements of covariates taken since a treatment commenced or special covariates created to test the validity of given model. Previously, we have stressed the importance of modeling the survival function, hazard function, or some other parameter associated with the failure-time distribution. Often a matter of greater interest is to ascertain the relationship between the failure time X and one or more of the explanatory variables. This would be the case if one were comparing the survival functions for two or more treatments, wanting to determine the prognosis of a patient presenting with various characteristics, or identifying pertinent risk factors for a particular disease, controlling for relevant confounders.

Two approaches to the modeling of covariate effects on survival have become popular in the statistical literature. The first approach is analogous to the classical linear regression approach. In this approach, the natural logarithm of the survival time $Y = \ln(X)$ is modeled. This is the natural transformation made in linear models to convert positive variables to observations on the entire real line. A linear model is assumed for Y, namely,

$$Y = \mu + \boldsymbol{\gamma}^t \mathbf{Z} + \sigma W, \qquad (2.6.1)$$

where $\boldsymbol{\gamma}^t = (\gamma_1, \ldots, \gamma_p)$ is a vector of regression coefficients and W is the error distribution. Common choices for the error distribution include the standard normal distribution which yields a log normal regression model, the extreme value distribution (2.5.4), which yields a Weibull regression model, or a logistic distribution (2.5.7), which yields a log logistic regression model. Estimation of regression coefficients, which is discussed in detail in Chapter 12, is performed using maximum likelihood methods and is readily available in most statistical packages.

This model is called the accelerated failure-time model. To see why this is so, let $S_o(x)$ denote the survival function of $X = e^Y$ when $\mathbf{Z}$ is zero, that is, $S_o(x)$ is the survival function of $\exp(\mu + \sigma W)$.

Now,

$$Pr[X > x \mid \mathbf{Z}] = Pr[Y > \ln x \mid \mathbf{Z}]$$
$$= Pr[\mu + \sigma W > \ln x - \boldsymbol{\gamma}'\mathbf{Z} \mid \mathbf{Z}]$$
$$= Pr[e^{\mu + \sigma W} > x \exp(-\boldsymbol{\gamma}'\mathbf{Z}) \mid \mathbf{Z}]$$
$$= S_o[x \exp(-\boldsymbol{\gamma}'\mathbf{Z})].$$

Notice that the effect of the explanatory variables in the original time scale is to change the time scale by a factor $\exp(-\boldsymbol{\gamma}'\mathbf{Z})$. Depending on the sign of $\boldsymbol{\gamma}'\mathbf{Z}$, the time is either accelerated by a constant factor or degraded by a constant factor. Note that the hazard rate of an individual with a covariate value $\mathbf{Z}$ for this class of models is related to a baseline hazard rate h_o by

$$h(x \mid \mathbf{Z}) = h_o[x \exp(-\boldsymbol{\gamma}'\mathbf{Z})] \exp(-\boldsymbol{\gamma}'\mathbf{Z}). \qquad (2.6.2)$$

EXAMPLE 2.5   Suppose that the survival time X follows a Weibull distribution with parameters λ and α. Recall that in section 2.5 we saw that the natural logarithm of X, $Y = \ln(X)$, can be written as a linear model, $Y = \mu + \sigma W$, where $\mu = (-\ln(\lambda)/\alpha)$, $\sigma = \alpha^{-1}$, and W has a standard extreme value distribution with density function $f(w) = \exp\{w - e^w\}$, $-\infty < w < \infty$. Suppose that we also have a set of $p - 1$ covariates, $\{Z_2, \ldots, Z_p\}$ which can explain some of the patient to patient variability observed for the lifetimes under study. We shall define the covariate $Z_1 = 1$ to allow for an intercept term in our log linear model and $\mathbf{Z}^t = (Z_1, \ldots, Z_p)$. Let $\boldsymbol{\gamma}' = (\gamma_1, \ldots, \gamma_p)$ be a vector of regression coefficients. The natural log linear model for Y is given by

$$Y = \boldsymbol{\gamma}'\mathbf{Z} + \sigma W.$$

With this model, the survival function for Y is expressed as

$$S_Y(y \mid \mathbf{Z}) = \exp\left[-\exp\left(\frac{y - \boldsymbol{\gamma}'\mathbf{Z}}{\sigma}\right)\right].$$

On the original time scale the survival function for X is given by

$$S_X(x \mid Z) = \exp\left[-x^{1/\sigma} \exp\left(\frac{-\boldsymbol{\gamma}'\mathbf{Z}}{\sigma}\right)\right] = \exp\{-[x \exp(-\boldsymbol{\gamma}'\mathbf{Z})]^\alpha\}$$
$$= S_o(x \exp\{-\boldsymbol{\gamma}'\mathbf{Z}\}),$$

where $S_o(x) = \exp(-x^\alpha)$ is a Weibull survival function.

Although the accelerated failure-time model provides a direct extension of the classical linear model's construction for explanatory variables for conventional data, for survival data, its use is restricted by the error distributions one can assume. As we have seen earlier in this chapter, the easiest survival parameter to model is the hazard rate which tells us how quickly individuals of a certain age are experiencing the event of interest. The major approach to modeling the effects of covariates on survival is to model the conditional hazard rate as a function of the covariates. Two general classes of models have been used to relate covariate effects to survival, the family of multiplicative hazard models and the family of additive hazard rate models.

For the family of multiplicative hazard rate models the conditional hazard rate of an individual with covariate vector $\mathbf{z}$ is a product of a baseline hazard rate $h_o(x)$ and a non-negative function of the covariates, $c(\boldsymbol{\beta}'\mathbf{z})$, that is,

$$h(x \mid \mathbf{z}) = h_o(x)c(\boldsymbol{\beta}'\mathbf{z}). \tag{2.6.3}$$

In applications of the model, $h_o(x)$ may have a specified parametric form or it may be left as an arbitrary nonnegative function. Any nonnegative function can be used for the link function $c(\ )$. Most applications use the Cox (1972) model with $c(\boldsymbol{\beta}'\mathbf{z}) = \exp(\boldsymbol{\beta}'\mathbf{z})$ which is chosen for its simplicity and for the fact it is positive for any value of $\boldsymbol{\beta}'\mathbf{z}$.

A key feature of multiplicative hazards models is that, when all the covariates are fixed at time 0, the hazard rates of two individuals with distinct values of $\mathbf{z}$ are proportional. To see this consider two individuals with covariate values $\mathbf{z}_1$ and $\mathbf{z}_2$. We have

$$\frac{h(x \mid \mathbf{z}_1)}{h(x \mid \mathbf{z}_2)} = \frac{h_0(x)c(\boldsymbol{\beta}'\mathbf{z}_1)}{h_0(x)c(\boldsymbol{\beta}'\mathbf{z}_2)} = \frac{c(\boldsymbol{\beta}'\mathbf{z}_1)}{c(\boldsymbol{\beta}'\mathbf{z}_2)},$$

which is a constant independent of time.

Using (2.6.3), we see that the conditional survival function of an individual with covariate vector $\mathbf{z}$ can be expressed in terms of a baseline survival function $S_o(x)$ as

$$S(x \mid \mathbf{z}) = S_o(x)^{c(\boldsymbol{\beta}'\mathbf{z})}. \tag{2.6.4}$$

This relationship is also found in nonparametric statistics and is called a "Lehmann Alternative."

Multiplicative hazard models are used for modeling relative survival in section 6.3 and form the basis for modeling covariate effects in Chapters 8 and 9.

EXAMPLE 2.5   *(continued)* The multiplicative hazard model for the Weibull distribution with baseline hazard rate $h_o(x) = \alpha\lambda x^{\alpha-1}$ is $h(x \mid \mathbf{z}) = \alpha\lambda x^{\alpha-1}c(\boldsymbol{\beta}'\mathbf{z})$. When the Cox model is used for the link function, $h(x \mid z) = \alpha\lambda x^{\alpha-1}\exp(\boldsymbol{\beta}'\mathbf{z})$. Here the conditional survival function is given by $S(x \mid \mathbf{z}) = \exp[-\lambda x^\alpha]^{\exp[\boldsymbol{\beta}'\mathbf{z}]} = \exp[-\lambda x^\alpha \exp(\boldsymbol{\beta}'\mathbf{z})] = \exp[-\lambda (x \exp[\boldsymbol{\beta}'\mathbf{z}/\alpha])^\alpha]$, which is of the form of an accelerated failure-time model (2.6.2). The Weibull is the only continuous distribution which has the property of being both an accelerated failure-time model and a multiplicative hazards model.

A second class of models for the hazard rate is the family of additive hazard rate models. Here, we model the conditional hazard function by

$$h(x \mid \mathbf{z}) = h_o(x) + \sum_{j=1}^{p} z_j(x)\beta_j(x). \qquad (2.6.5)$$

The regression coefficients for these models are functions of time so that the effect of a given covariate on survival is allowed to vary over time. The p regression functions may be positive or negative, but their values are constrained because (2.6.5) must be positive.

Estimation for additive models is typically made by nonparametric (weighted) least-squares methods. Additive models are used in section 6.3 to model excess mortality and, in Chapter 10, to model regression effects.

Practical Notes

1. From Theoretical Note 1 of section 2.4,

$$S(x \mid \mathbf{z}) = \exp\left[-\int_0^x h(t \mid \mathbf{z})dt\right] \qquad (2.6.6)$$

and, in conjunction with (2.6.4),

$$S(x \mid \mathbf{z}) = \exp\left[-\int_0^x h_o(t) \exp[\boldsymbol{\beta}'\mathbf{z}]dt\right]$$

$$= \left\{\exp\left[-\int_0^x h_o(t)dt\right]\right\}^{\exp[\boldsymbol{\beta}'\mathbf{z}]}$$

$$= [S_o(x)]^{\exp[\boldsymbol{\beta}'\mathbf{z}]}$$

which implies that

$$\ln[-\ln S(x \mid \mathbf{z})] = \boldsymbol{\beta}'\mathbf{z} + \ln[-\ln S_o(x)]. \qquad (2.6.7)$$

So the logarithms of the negative logarithm of the survival functions of X, given different regressor variables $\mathbf{z}_i$, are parallel. This relationship will serve as a check on the proportional hazards assumption discussed further in Chapter 11.

2.7   Models for Competing Risks

In the previous sections of this chapter we have examined parameters which can be used to describe the failure time, T, of a randomly selected individual. Here T may be the time to death (see, for example, sections 1.5, 1.7, 1.8, 1.11), the time to treatment failure (see, for example, sections 1.9, 1.10), time to infection (see sections 1.6, 1.12), time to weaning (section 1.14), etc.

In some medical experiments we have the problem of competing risks. Here each subject may fail due to one of K ($K \geq 2$) causes, called competing risks. An example of competing risks is found in section 1.3. Here the competing risks for treatment failure are relapse and death in remission. Occurrence of one of these events precludes us from observing the other event on this patient. Another classical example of competing risks is cause-specific mortality, such as death from heart disease, death from cancer, death from other causes, etc.

To discuss parameters for the competing-risks problem we shall formulate the model in terms of a latent failure time approach. Other formulations, as discussed in Kalbfleisch and Prentice (1980), give similar representations. Here we let $X_i, i = 1, \ldots, K$ be the potential unobservable time to occurrence of the ith competing risk. What we observe for each patient is the time at which the subject fails from any cause, $T = \text{Min}(X_1, \ldots, X_p)$ and an indicator δ which tells which of the K risks caused the patient to fail, that is, $\delta = i$ if $T = X_i$.

The basic competing risks parameter is the *cause-specific hazard rate for risk i* defined by

$$
b_i(t) = \lim_{\Delta t \to 0} \frac{P[t \leq T < t + \Delta t, \delta = i \mid T \geq t]}{\Delta t} \tag{2.7.1}
$$

$$
= \lim_{\Delta t \to 0} \frac{P[t \leq X_i < t + \Delta t, \delta = i \mid X_j \geq t, j = 1, \ldots, K]}{\Delta t}
$$

Here $b_i(t)$ tells us the rate at which subjects who have yet to experience any of the competing risks are experiencing the ith competing cause of failure. The overall hazard rate of the time to failure, T, given by (2.3.1) is the sum of these K cause-specific hazard rates; that is

$$
b_T(t) = \sum_{i=1}^{K} b_i(t).
$$

The cause-specific hazard rate can be derived from the joint survival function of the K competing risks. Let $S(t_1, \ldots, t_K) = \text{Pr}[X_1 >$

$t_1, \ldots, X_K > t_K]$. The cause specific hazard rate is given by

$$h_i(t) = \frac{-\partial S(t_1, \ldots, t_K)/\partial t_i\big|_{t_1 = \cdots t_K = t}}{S(t, \ldots, t)} \tag{2.7.2}$$

EXAMPLE 2.6   Suppose that we have K competing risks and that the potential failure times are independent with survival functions $S_i(t)$ for $i = 1, 2, \ldots, K$. Then the joint survival function is $S(t_1, \ldots, t_K) = \prod_{i=1}^{K} S_i(t_i)$, and by (2.7.2) we have

$$h_i(t) = \frac{-\partial \prod_{j=1}^{K} S_j(t_j)/\partial t_i\big|_{t_1 = \cdots t_K = t}}{\prod_{j=1}^{K} S_j(t)} = \frac{-\partial S_i(t_i)/\partial t_i\big|_{t_i = t}}{S_i(t)},$$

which is precisely the hazard rate of X_i.

Example 2.6 shows that for independent competing risks the marginal and cause-specific hazard rates are identical. This need not be the case when the risks are dependent as we see in the following example.

EXAMPLE 2.7   Suppose we have two competing risks and the joint survival function is $S(t_1, t_2) = [1 + \theta(\lambda_1 t_1 + \lambda_2 t_2)]^{-1/\theta}, \theta \geq 0, \lambda_1, \lambda_2 \geq 0$. Here the two potential failure times are correlated with a Kendall's τ of $(\theta/(\theta + 2))$ (see section 13.3 for a discussion and derivation of this model). By (2.7.2) we have

$$h_i(t) = \frac{-\partial [1 + \theta(\lambda_1 t_1 + \lambda_2 t_2)]^{-1/\theta}/\partial t_i\big|_{t_1 = t_2 = t}}{1 + \theta t(\lambda_1 + \lambda_2)]^{-1/\theta}}$$

$$= \frac{\lambda_i}{1 + \theta t(\lambda_1 + \lambda_2)}, i = 1, 2.$$

Here the survival function of the time to failure, $T = \min(X_1, X_2)$ is $S(t, t) = S_T(t) = [1 + \theta t(\lambda_1 + \lambda_2)]^{-1/\theta}$ and its hazard rate is $(\lambda_1 + \lambda_2)/[1 + \theta t(\lambda_1 + \lambda_2)]$. Note that the marginal survival function for X_1 is given by $S(t_1, 0) = [1 + \theta t \lambda_1]^{-1/\theta}$ and the marginal hazard rate is, from (2.3.2), $\lambda_1/(1 + \theta \lambda_1 t)$, which is not the same as the crude hazard rate.

In competing-risks modeling we often need to make some assumptions about the dependence structure between the potential failure times. Given that we can only observe the failure time and cause and not the potential failure times these assumptions are not testable with only competing risks data. This is called the *identifiability dilemma*.

We can see the problem clearly by careful examination of Example 2.7. Suppose we had two independent competing risks with hazard rates $\lambda_1/[1 + \theta t(\lambda_1 + \lambda_2)]$ and $\lambda_2/[1 + \theta t(\lambda_1 + \lambda_2)]$, respectively. By Example 2.6 the cause-specific hazard rates and the marginal hazard rates are identical when we have independent competing risks. So the crude hazard rates for this set of independent competing risks are identical to the set of dependent competing risks in Example 2.7. This means that given what we actually see, (T, δ), we can never distinguish a pair of dependent competing risks from a pair of independent competing risks.

In competing risks problems we are often interested not in the hazard rate but rather in some probability which summarizes our knowledge about the likelihood of the occurrence of a particular competing risk. Three probabilities are computed, each with their own interpretation. These are the *crude*, *net*, and *partial crude* probabilities. The crude probability is the probability of death from a particular cause in the real world where all other risks are acting on the individual. For example, if the competing risk is death from heart disease, then an example of a crude probability is the chance a man will die from heart disease prior to age 50. The net probability is the probability of death in a hypothetical world where the specific risk is the only risk acting on the population. In the potential failure time model this is a marginal probability for the specified risk. For example, a net probability is the chance that a man will die from heart disease in the counterfactual world where men can only die from heart disease. Partial crude probabilities are the probability of death in a hypothetical world where some risks of death have been eliminated. For example, a partial crude probability would be the chance a man dies from heart disease in a world where cancer has been cured.

Crude probabilities are typically expressed by the cause-specific sub-distribution function. This function, also known as the *cumulative incidence function*, is defined as $F_i(t) = P[T \leq t, \delta = i]$. The cumulative incidence function can be computed directly from the joint density function of the potential failure times or it can be computed from the cause specific hazard rates. That is,

$$F_i(t) = P[T \leq t, \delta = i] = \int_0^t h_i(u) \exp\{-H_T(u)\} \, du. \qquad (2.7.3)$$

Here $H_T(t) = \sum_{j=1}^K \int_0^t h_j(u) \, du$ is the cumulative hazard rate of T. Note that the value of $F_i(t)$ depends on the rate at which all the competing risks occur, not simply on the rate at which the specific cause of interest is occurring. Also, since $h_i(t)$ can be estimated directly from the observed data, $F_i(t)$ is directly estimable without making any assumptions about the joint distribution of the potential failure times (see section 4.7). $F_i(t)$ is not a true distribution function since $F_i(\infty) = P[\delta = i]$. It has the property that it is non-decreasing with $F_i(0) = 0$ and $F_i(\infty) < 1$. Such a function is called a "sub-distribution" function.

The net survival function, $S_i(t)$, is the marginal survival function found from the joint survival function by taking $t_j = 0$ for all $j \neq i$. When the competing risks are independent then the net survival function is related to the crude probabilities by

$$S_i(t) = \exp\left\{ -\int_0^t \frac{dF_i(u)}{S_T(u)} \right\}.$$

This relationship is used in Chapter 4 to allow us to estimate probabilities when there is a single independent competing risk which is regarded as random censoring (see section 3.2 for a discussion of random censoring).

When the risks are dependent, Peterson (1976) shows that net survival probabilities can be bounded by the crude probabilities. He shows that

$$S_T(t) \leq S_i(t) \leq 1 - F_i(t).$$

The lower (upper) bounds correspond to perfect positive (negative) correlation between the risks. These bounds may be quite wide in practice. Klein and Moeschberger (1988) and Zheng and Klein (1994) show that these bounds can be tightened by assuming a family of dependence structures for the joint distribution of the competing risks.

For partial crude probabilities we let $\mathbf{J}$ be the set of causes that an individual can fail from and $\mathbf{J}^C$ the set of causes which are eliminated from consideration. Let $T^J = \min(X_i, i \in \mathbf{J})$ then we can define the partial crude sub-distribution function by $F_i^J(t) = \Pr[T^J \leq t, \delta = i], i \in \mathbf{J}$. Here the ith partial crude probability is the chance of dying from cause i in a hypothetical patient who can only experience one of the causes of death in the set $\mathbf{J}$. One can also define a partial crude hazard rate by

$$\lambda_i^J(t) = \frac{-\partial S(t_1, \ldots, t_K)/\partial t_i\big|_{t_j=t, t_j \in \mathbf{J}, t_j=0, t_j \in \mathbf{J}^C}}{S(t_1, \ldots, t_p)\big|_{t_j=t, t_j \in \mathbf{J}, t_j=0, t_j \in \mathbf{J}^C}}. \tag{2.7.4}$$

As in the case of the crude partial incidence function we can express the partial crude sub-distribution function as

$$F_i^J(t) = \Pr[T^J \leq t, \delta = i] = \int_0^t \lambda_i^J(x) \exp\left\{ -\sum_{j \in \mathbf{J}} \int_0^t \lambda_j^J(u)\, du \right\} dx. \tag{2.7.5}$$

When the risks are independent then the partial crude hazard rate can be expressed in terms of the crude probabilities as

$$\lambda_i^J(t) = \frac{dF_i(t)/dt}{S_T(t)}. \tag{2.7.6}$$

EXAMPLE 2.8   Suppose we have three independent exponential competing risks with hazard rates $\lambda_1, \lambda_2, \lambda_3$, respectively. In this case, as seen in Example 2.6, the net and crude hazard rates for the first competing risk are equal to λ_1. The hazard rate of T is $h_T(t) = \lambda_1 + \lambda_2 + \lambda_3$. Equation (2.7.3), the crude sub-distribution function for the first competing risk is

$$F_1(t) = \int_0^t \lambda_1 \exp(-u(\lambda_1 + \lambda_2 + \lambda_3)) \, du$$

$$= \frac{\lambda_1}{\lambda_1 + \lambda_2 + \lambda_3}\{1 - \exp\{-t(\lambda_1 + \lambda_2 + \lambda_3)\}\}.$$

Note that the crude probability of death from cause 1 in the interval $[0, t]$ is not the same as the net (marginal) probability of death in this interval given by $1 - \exp\{-\lambda_1 t\}$. Also $F_i(\infty) = \lambda_1/(\lambda_1 + \lambda_2 + \lambda_3)$, which is the probability that the first competing risk occurs first. If we consider a hypothetical world where only the first two competing risks are operating ($\mathbf{J} = \{1, 2\}$), the partial crude hazard rates are $\lambda_i^J(t) = \lambda_i, i = 1, 2$, and the partial crude sub-distribution function is given by

$$F_1^J(t) = \int_0^t \lambda_1 \exp(-u(\lambda_1 + \lambda_2)) \, du = \frac{\lambda_1}{\lambda_1 + \lambda_2}\{1 - \exp\{-t(\lambda_1 + \lambda_2)\}\}.$$

EXAMPLE 2.7   *(continued)*   Suppose we have two competing risks with joint survival function $S(t_1, t_2) = [1 + \theta(\lambda_1 t_1 + \lambda_2 t_2)]^{-1/\theta}, \theta \geq 0, \lambda_1, \lambda_2 \geq 0$. Here the crude hazard rates are given by $\lambda_i/[1 + \theta t(\lambda_1 + \lambda_2)]$, for $i = 1, 2$. The cause-specific cumulative incidence function for the ith risk is

$$F_i(t) = \int_0^t \frac{\lambda_i}{[1 + \theta x(\lambda_1 + \lambda_2)]} \exp\left\{-\int_0^x \frac{\lambda_1 + \lambda_2}{[1 + \theta u(\lambda_1 + \lambda_2)]} \, du\right\} dx$$

$$= \frac{\lambda_i}{\lambda_1 + \lambda_2}\left\{1 - [1 + \theta t(\lambda_1 + \lambda_2)]^{-1/\theta}\right\}.$$

In Figure 2.11 we plot the cumulative incidence function and the net probability for cause 1 when $\lambda_1 = 1, \lambda_2 = 2$, and $\theta = 2$. Here we

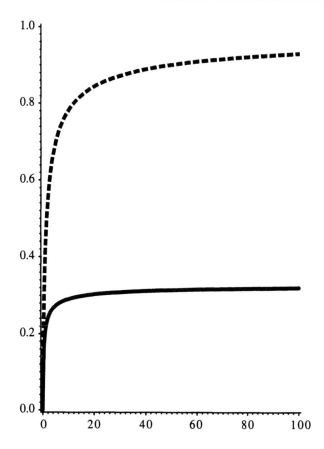

Figure 2.11 *Cumulative incidence function (solid line) and net probability for the first competing risk in Example 2.7.*

see clearly that the cumulative incidence function levels off at one-third the probability that the first competing risk fails first. Also we see quite clearly that the crude probability is always less than the net probability.

Practical Notes

1. Competing risk theory has an intriguing history going back to a memoir read in 1760 by Daniel Bernoulli before the French Academy of Sciences and published in 1765. It was motivated by a controversy on the merits of smallpox inoculation. Using Halley's Breslau life table of 1693, Bernoulli constructed a hypothetical lifetable, which reflected the mortality structure at different ages if smallpox was eliminated. A key assumption was, as Bernoulli recognized, that the hypothetical lifetimes of individuals saved from smallpox were independent of lifetimes associated with the other causes of death. Bernoulli's question "What would be the effect on mortality if the occurrence of one

or more causes of death were changed?" and the untestable assumption of independence of causes of death are still very much with us today.

2. For simplicity, we shall only assume one competing risk, whose event time will be denoted by Y (although all results may be generalized to many competing risks). In the competing-risks framework, as we have seen, we can only observe $T = $ minimum (X, Y) and $\delta = I(X < Y)$, an indicator function which indicates whether or not the main event of interest has occurred. The early observation by Cox (1959, 1962) that there was a difficulty in the interpretation of bivariate data in the competing risk context was elucidated and clarified by later authors. Berman (1963) showed explicitly that the distribution of (T, δ) determined that of X, if X and Y are assumed to be independent. Tsiatis (1975) proved a nonidentifiability theorem which concluded that a dependent-risk model is indistinguishable from some independent risk model and that any analysis of such data should include a careful analysis of biological circumstances. Peterson (1976) argued that serious errors can be made in estimating the survival function in the competing risk problem because one can never know from the data whether X and Y are independent or not.

3. Heckman and Honore (1989) show, under certain regularity conditions, for both proportional hazards and accelerated failure time models that if there is an explanatory covariate, $\mathbf{Z}$, whose support is the entire real line then the joint distribution of (X, Y) is identifiable from $(T, \delta, \mathbf{Z})$. Slud (1992), in a slightly different vein, shows how the marginal distribution of the survival time X can be nonparametrically identifiable when only the data $(T, \delta, \mathbf{Z})$ are observed, where $\mathbf{Z}$ is an observed covariate such that the competing risk event time, Y, and $\mathbf{Z}$ are conditionally independent given X.

Theoretical Notes

1. Slud and Rubinstein (1983) have obtained tighter bounds on $S(x)$ than the Peterson bounds described earlier, in this framework, by utilizing some additional information. Their method requires the investigator to bound the function

$$\rho(t) = \frac{\{[f_i(t)/q_i(t)] - 1\}}{\{[S_i(t)/S_T(t)] - 1\}}$$

where

$$f_i(t) = -\frac{dS_i(t)}{dt},$$

and

$$q_i(t) = \frac{d}{dt}F_i(t).$$

Knowledge of the function $\rho(t)$ and the observable information, (T, δ), is sufficient to determine uniquely the marginal distribution of X. The resulting estimators $\hat{S}_\rho(x)$ are decreasing functions of $\rho(\cdot)$. These resulting bounds are obtained by the investigator's specification of two functions, $\rho_i(t)[\rho_1(t) < \rho_2(t)]$, so that if the true $\rho(t)$ function is in the interval $[\rho_1(t) < \rho_2(t)]$, for all t, then $\hat{S}\rho_2(t) \le S(t) \le \hat{S}\rho_1(t)$.

2. Pepe (1991) and Pepe and Mori (1993) interpret the cumulative incidence function as a "marginal probability." Note that this function is not a true marginal distribution as discussed earlier but rather is the chance that the event of interest will occur prior to time t in a system where an individual is exposed to both risks. Pepe and Mori suggest as an alternative to the cumulative incidence function the "conditional probability" of X, defined by

$$P(\{X \le t, X < Y\} \mid \{Y < t, X > Y\}^c) = \frac{F_i(t)}{F_i^c(t)},$$

which they interpret as the probability of X's occurring in $[0, t)$, given nonoccurrence of Y in $[0, t)$, where F^c denotes the complement of F.

2.8 Exercises

2.1   The lifetime of light bulbs follows an exponential distribution with a hazard rate of 0.001 failures per hour of use.

(a) Find the mean lifetime of a randomly selected light bulb.

(b) Find the median lifetime of a randomly selected light bulb.

(c) What is the probability a light bulb will still function after 2,000 hours of use?

2.2   The time in days to development of a tumor for rats exposed to a carcinogen follows a Weibull distribution with $\alpha = 2$ and $\lambda = 0.001$.

(a) What is the probability a rat will be tumor free at 30 days? 45 days? 60 days?

(b) What is the mean time to tumor? (Hint $\Gamma(0.5) = \sqrt{\pi}$.)

(c) Find the hazard rate of the time to tumor appearance at 30 days, 45 days, and 60 days.

(d) Find the median time to tumor.

2.3   The time to death (in days) following a kidney transplant follows a log logistic distribution with $\alpha = 1.5$ and $\lambda = 0.01$.

(a) Find the 50, 100, and 150 day survival probabilities for kidney transplantation in patients.

(b) Find the median time to death following a kidney transplant.

(c) Show that the hazard rate is initially increasing and, then, decreasing over time. Find the time at which the hazard rate changes from increasing to decreasing.

(d) Find the mean time to death.

2.4   A model for lifetimes, with a bathtub-shaped hazard rate, is the exponential power distribution with survival function $S(x) = \exp\{1 - \exp[(\lambda x)^\alpha]\}$.

(a) If $\alpha = 0.5$, show that the hazard rate has a bathtub shape and find the time at which the hazard rate changes from decreasing to increasing.

(b) If $\alpha = 2$, show that the hazard rate of x is monotone increasing.

2.5   The time to death (in days) after an autologous bone marrow transplant, follows a log normal distribution with $\mu = 3.177$ and $\sigma = 2.084$. Find

(a) the mean and median times to death;

(b) the probability an individual survives 100, 200, and 300 days following a transplant; and

(c) plot the hazard rate of the time to death and interpret the shape of this function.

2.6   The Gompertz distribution is commonly used by biologists who believe that an exponential hazard rate should occur in nature. Suppose that the time to death in months for a mouse exposed to a high dose of radiation follows a Gompertz distribution with $\theta = 0.01$ and $\alpha = 0.25$. Find

(a) the probability that a randomly chosen mouse will live at least one year,

(b) the probability that a randomly chosen mouse will die within the first six months, and

(c) the median time to death.

2.7   The time to death, in months, for a species of rats follows a gamma distribution with $\beta = 3$ and $\lambda = 0.2$. Find

(a) the probability that a rat will survive beyond age 18 months,

(b) the probability that a rat will die in its first year of life, and

(c) the mean lifetime for this species of rats.

2.8   The battery life of an internal pacemaker, in years, follows a Pareto distribution with $\theta = 4$ and $\lambda = 5$.

(a) What is the probability the battery will survive for at least 10 years?

(b) What is the mean time to battery failure?

(c) If the battery is scheduled to be replaced at the time t_o, at which 99% of all batteries have yet to fail (that is, at t_o so that $Pr(X > t_o) = .99$), find t_o.

2.9   The time to relapse, in months, for patients on two treatments for lung cancer is compared using the following log normal regression model:

$$Y = \text{Ln}(X) = 2 + 0.5Z + 2W,$$

where W has a standard normal distribution and $Z = 1$ if treatment A and 0 if treatment B.

(a) Compare the survival probabilities of the two treatments at 1, 2, and 5 years.

(b) Repeat the calculations if W has a standard logistic distribution. Compare your results with part (a).

2.10   A model used in the construction of life tables is a piecewise, constant hazard rate model. Here the time axis is divided into k intervals, $[\tau_{i-1}, \tau_i)$, $i = 1, \ldots, k$, with $\tau_o = 0$ and $\tau_k = \infty$. The hazard rate on the ith interval is a constant value, θ_i; that is

$$h(x) = \begin{cases} \theta_1 & 0 \le x < \tau_1 \\ \theta_2 & \tau_1 \le x < \tau_2 \\ \vdots & \\ \theta_{k-1} & \tau_{k-2} \le x < \tau_{k-1} \\ \theta_k & x \ge \tau_{k-1} \end{cases}.$$

(a) Find the survival function for this model.

(b) Find the mean residual-life function.

(c) Find the median residual-life function.

2.11   In some applications, a third parameter, called a guarantee time, is included in the models discussed in this chapter. This parameter ϕ is the smallest time at which a failure could occur. The survival function of the three-parameter Weibull distribution is given by

$$S(x) = \begin{cases} 1 & \text{if } x < \phi \\ \exp[-\lambda(x - \phi)^\alpha] & \text{if } x \ge \phi. \end{cases}$$

(a) Find the hazard rate and the density function of the three- parameter Weibull distribution.

(b) Suppose that the survival time X follows a three-parameter Weibull distribution with $\alpha = 1$, $\lambda = 0.0075$ and $\phi = 100$. Find the mean and median lifetimes.

2.12   Let X have a uniform distribution on the interval 0 to θ with density function

$$f(x) = \begin{array}{l} 1/\theta, \text{ for } 0 \le x \le \theta \\ 0, \text{ otherwise}. \end{array}$$

(a) Find the survival function of X.

(b) Find the hazard rate of X.

(c) Find the mean residual-life function.

2.13   Suppose that X has a geometric distribution with probability mass function

$$p(x) = p(1 - p)^{x-1}, x = 1, 2, \ldots$$

(a) Find the survival function of X. (Hint: Recall that for $0 < \theta < 1$, $\sum_{j=k}^{\infty} \theta^j = \theta^k/(1 - \theta)$).

(b) Find the hazard rate of X. Compare this rate to the hazard rate of an exponential distribution.

2.14   Suppose that a given individual in a population has a survival time which is exponential with a hazard rate θ. Each individual's hazard rate θ is potentially different and is sampled from a gamma distribution with density function

$$f(\theta) = \frac{\lambda^\beta \theta^{\beta-1} e^{-\lambda\theta}}{\Gamma(\beta)}$$

Let X be the life length of a randomly chosen member of this population.

(a) Find the survival function of X.

(Hint: Find $S(x) = E_\theta[e^{-\theta x}]$.)

(b) Find the hazard rate of X. What is the shape of the hazard rate?

2.15   Suppose that the hazard rate of X is a linear function $h(x) = \alpha + \beta x$, with α and $\beta > 0$. Find the survival function and density function of x.

2.16   Given a covariate Z, suppose that the log survival time Y follows a linear model with a logistic error distribution, that is,

$$Y = \ln(X) = \mu + \beta Z + \sigma W \text{ where the pdf of } W \text{ is given by}$$

$$f(w) = \frac{e^w}{(1 + e^w)^2}, -\infty < w < \infty.$$

(a) For an individual with covariate Z, find the conditional survival function of the survival time X, given Z, namely, $S(x \mid Z)$.

(b) The odds that an individual will die prior to time x is expressed by $[1 - S(x \mid Z)]/S(x \mid Z)$. Compute the odds of death prior to time x for this model.

(c) Consider two individuals with different covariate values. Show that, for any time x, the ratio of their odds of death is independent of x. The log logistic regression model is the only model with this property.

2.17 Suppose that the mean residual life of a continuous survival time X is given by $\mathrm{MRL}(x) = x + 10$.

(a) Find the mean of X.

(b) Find $h(x)$.

(c) Find $S(x)$.

2.18 Let X have a uniform distribution on 0 to 100 days with probability density function

$$f(x) = 1/100 \text{ for } 0 < x < 100,$$

$$= 0, \text{ elsewhere}.$$

(a) Find the survival function at 25, 50, and 75 days.

(b) Find the mean residual lifetime at 25, 50, and 75 days.

(c) Find the median residual lifetime at 25, 50, and 75 days.

2.19 Suppose that the joint survival function of the latent failure times for two competing risks, X and Y, is

$$S(x, y) = (1 - x)(1 - y)(1 + .5xy), \quad 0 < x < 1, \quad 0 < y < 1.$$

(a) Find the marginal survival function for X.

(b) Find the cumulative incidence of X.

2.20 Let X and Y be two competing risks with joint survival function

$$S(x, y) = \exp\{-x - y - .5xy\}, 0 < x, y.$$

(a) Find the marginal cumulative distribution function of X.

(b) Find the cumulative incidence function of X.

3
Censoring and Truncation

3.1 Introduction

Time-to-event data present themselves in different ways which create special problems in analyzing such data. One peculiar feature, often present in time-to-event data, is known as censoring, which, broadly speaking, occurs when some lifetimes are known to have occurred only within certain intervals. The remainder of the lifetimes are known exactly. There are various categories of censoring, such as right censoring, left censoring, and interval censoring. Right censoring will be discussed in section 3.2. Left or interval censoring will be discussed in section 3.3. To deal adequately with censoring in the analysis, we must consider the design which was employed to obtain the survival data. There are several types of censoring schemes within both left and right censoring. Each type will lead to a different likelihood function which will be the basis for the inference. As we shall see in section 3.5, though the likelihood function is unique for each type of censoring, there is a common approach to be used in constructing it.

A second feature which may be present in some survival studies is that of *truncation*, discussed in section 3.4. Left truncation occurs when subjects enter a study at a particular age (not necessarily the origin for the event of interest) and are followed from this *delayed entry time* until the event occurs or until the subject is censored. Right truncation occurs when only individuals who have experienced the event of interest are observable. The main impact on the analysis, when data are truncated, is that the investigator must use a conditional distribution in constructing the likelihood, as shown in section 3.5, or employ a statistical method which uses a selective risk set to be explained in more detail in Chapter 4.

Sections 3.5 and 3.6 present an overview of some theoretical results needed to perform modern survival analysis. Section 3.5 shows the construction of likelihoods for censored and truncated data. These likelihoods are the basis of inference techniques for parametric models and, suitably modified, as partial likelihoods for semiparametric models. Section 3.6 gives a brief introduction to the theory of counting processes. This very general theory is used to develop most nonparametric techniques for censored and truncated data and is the basis for developing the statistical properties of both parametric and nonparametric methods in survival analysis.

3.2   Right Censoring

First, we will consider *Type I censoring* where the event is observed only if it occurs prior to some prespecified time. These censoring times may vary from individual to individual. A typical animal study or clinical trial starts with a fixed number of animals or patients to which a treatment (or treatments) is (are) applied. Because of time or cost considerations, the investigator will terminate the study or report the results before all subjects realize their events. In this instance, if there are no accidental losses or subject withdrawals, all censored observations have times equal to the length of the study period.

Generally, it is our convention that random variables are denoted by upper case letters and fixed quantities or realizations of random variables are denoted by lower case letters. With censoring, this convention will obviously present some difficulties in notation because, as we shall see, some censoring times are fixed and some are random. At the risk of causing some confusion we will stick to upper case letters for censoring times. The reader will be expected to determine from the context whether the censoring time is random or fixed.

In right censoring, it is convenient to use the following notation. For a specific individual under study, we assume that there is a lifetime X and a fixed censoring time, C_r (C_r for "right" censoring time). The X's are assumed to be independent and identically distributed with probability density function $f(x)$ and survival function $S(x)$. The exact lifetime X of an individual will be known if, and only if, X is less than or equal to C_r. If X is greater than C_r, the individual is a survivor, and his or her event time is censored at C_r. The data from this experiment can be conveniently represented by pairs of random variables (T, δ), where δ indicates whether the lifetime X corresponds to an event ($\delta = 1$) or is censored ($\delta = 0$), and T is equal to X, if the lifetime is observed, and to C_r if it is censored, i.e., $T = \min(X, C_r)$.

EXAMPLE 3.1   Consider a large scale animal experiment conducted at the National Center for Toxicological Research (NCTR) in which mice were fed a particular dose of a carcinogen. The goal of the experiment was to assess the effect of the carcinogen on survival. Toward this end, mice were followed from the beginning of the experiment until death or until a prespecified censoring time was reached, when all those still alive were sacrificed (censored). This example is illustrated in Figure 3.1.

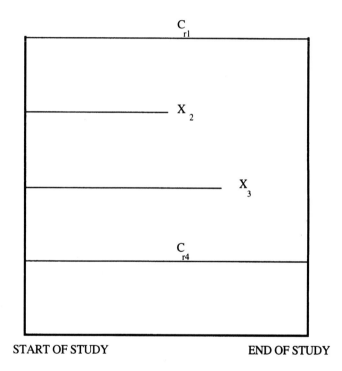

Figure 3.1   *Example of Type I censoring*

When animals have different, fixed-sacrifice (censoring) times, this form of Type I censoring is called *progressive Type I censoring*. An advantage of this censoring scheme is that the sacrificed animals give information on the natural history of nonlethal diseases. This type of censoring is illustrated in the following example.

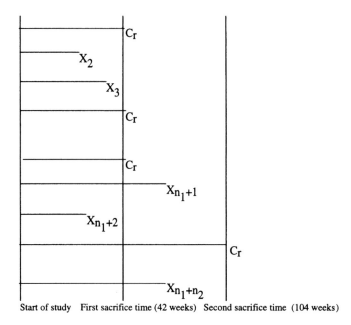

Figure 3.2   *Type I censoring with two different sacrifice times*

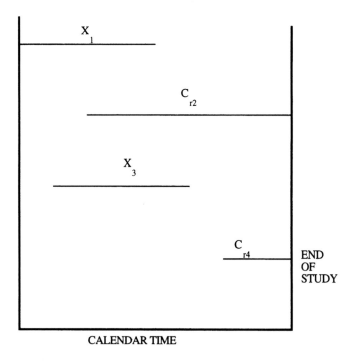

Figure 3.3   *Generalized Type I censoring when each individual has a different starting time*

EXAMPLE 3.2 Consider a mouse study where, for each sex, 200 mice were randomly divided into four dose-level groups and each mouse was followed until death or until a prespecified sacrifice time (42 or 104 weeks) was reached (see Figure 3.2 for a schematic of this trial for one gender and one dose level). The two sacrifice times were chosen to reduce the cost of maintaining the animals while allowing for limited information on the survival experience of longer lived mice.

Another instance, which gives rise to Type I censoring, is when individuals enter the study at different times and the terminal point of the study is predetermined by the investigator, so that the censoring times are known when an individual is entered into the study. In such studies (see Figure 3.3 for a hypothetical study with only four subjects), individuals have their own specific, fixed, censoring time. This form of censoring has been termed *generalized Type I censoring* (cf. David and Moeschberger, 1978). A convenient representation of such data is to shift each individual's starting time to 0 as depicted in Figure 3.4. Another method for representing such data is the Lexis diagram (Keiding, 1990). Here calendar time is on the horizontal axis, and life length is represented by a 45° line. The time an individual spends on study is represented by the height of the ray on the vertical axis. Figure 3.5 shows a Lexis diagram for the generalized Type I censoring scheme depicted in Figure 3.4. Here patients 1 and 3 experience the event of interest prior to the end of the study and are exact observations with $\delta = 1$. Patients 2 and 4, who experience the event after the end of the study, are only known to be alive at the end of the study and are censored observations ($\delta = 0$). Examples of studies with generalized Type I censoring are the breast-cancer trial in section 1.5, the acute leukemia trial in section 1.2, the study of psychiatric patients in section 1.15, and the study of weaning of newborns in section 1.14.

A second type of right censoring is *Type II censoring* in which the study continues until the failure of the first r individuals, where r is some predetermined integer ($r < n$). Experiments involving Type II censoring are often used in testing of equipment life. Here, all items are put on test at the same time, and the test is terminated when r of the n items have failed. Such an experiment may save time and money because it could take a very long time for all items to fail. It is also true that the statistical treatment of Type II censored data is simpler because the data consists of the r smallest lifetimes in a random sample of n lifetimes, so that the theory of order statistics is directly applicable to determining the likelihood and any inferential technique employed. Here, it should be noted that r the number of failures and $n - r$ the

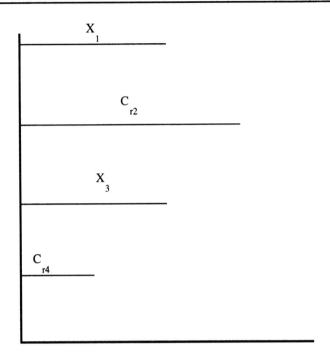

TIME ON STUDY

Figure 3.4 *Generalized Type I censoring for the four individuals in Figure 3.3 with each individual's starting time backed up to 0. $T_1 = X_1$ (death time for first individual) ($\delta_1 = 1$); $T_2 = C_{r2}$ (right censored time for second individual) ($\delta_2 = 0$); $T_3 = X_3$ (death time for third individual) ($\delta_3 = 1$); $T_4 = C_{r4}$ (right censored time for fourth individual) ($\delta_4 = 0$).*

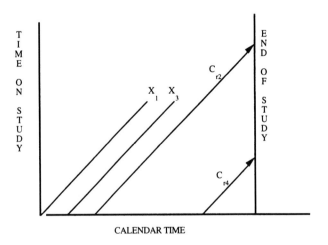

Figure 3.5 *Lexis diagram for generalized Type I censoring in Figure 3.3*

number of censored observations are fixed integers and the censoring time $T_{(r)}$, the rth ordered lifetime is random.

A generalization of Type II censoring, similar to the generalization of Type I censoring with different sacrifice times, is *progressive Type II censoring*. Here the first r_1 failures (an integer chosen prior to the start of the study) in a sample of n items (or animals) are noted and recorded. Then $n_1 - r_1$ of the remaining $n - r_1$ unfailed items (or animals) are removed (or sacrificed) from the experiment, leaving $n - n_1$ items (or animals) on study. When the next r_2 items (another integer chosen prior to the start of the study) fail, $n_2 - r_2$ of the unfailed items are removed (or animals sacrificed). This process continues until some predecided series of repetitions is completed. Again, r_i and n_i ($i = 1, 2$) are fixed integers and the two censoring times, $T_{(r_1)}$ and $T_{(n_1+r_2)}$, are random.

A third type of right censoring is *competing risks* censoring. A special case of competing risks censoring is random censoring. This type of censoring arises when we are interested in estimation of the marginal distribution of some event but some individuals under study may experience some competing event which causes them to be removed from the study. In such cases, the event of interest is not observable for those who experience the competing event and these subjects are random right censored at that time. As shown in section 2.7, in the competing risk framework, to be able to identify the marginal distribution from competing risks data we need the event time and censoring times to be independent of each other. This relationship cannot be determined from the data alone. Typical examples of where the random censoring times may be thought to be independent of the main event time of interest are accidental deaths, migration of human populations, and so forth.

Whenever we encounter competing risks it is important to determine precisely what quantity we wish to estimate. We need to decide if we want to estimate a marginal (net), crude, or partial crude probability as discussed in section 2.7. If we wish to estimate a marginal probability, which is the chance of the event's occurring in a world where all other risks cannot occur, the other competing risks are random observations. Here we need an assumption of independence between the time to the event of interest and the competing events to make a meaningful inference. Techniques for estimation in this framework are discussed in sections 4.1–4.6. When interest centers on estimation of crude probabilities (that is, the probability of the event in the real world where a person can fail from any of the competing causes), then each competing risk is modeled by a cumulative incidence curve (see section 4.7) and no independence assumption is needed. For partial crude probabilities (that is, the probability of the event's occurring in a world where only a subset of competing risks are possible causes of failure) some of the competing risks are treated as random censored observations (those to be eliminated) and others are modeled by a cumulative incidence curve. In this case we require that those causes treated as random censored observations need to be independent of the other causes to obtain consistent estimates of the desired probabilities.

In many studies, the censoring scheme is a combination of random and Type I censoring. In such studies, some patients are randomly censored when, for example, they move from the study location for reasons unrelated to the event of interest, whereas others are Type I censored when the fixed study period ends.

Theoretical Note

1. In Type I progressive censoring, the sacrifice times are fixed (predetermined prior to the start of the study), whereas, in Type II progressive censoring, the sacrifice times are random times at which a predetermined number of deaths has occurred. This distinction is extremely important in constructing the likelihood function in section 3.5. An advantage of either type of censoring scheme is that the sacrificed animals give information on the natural history of nonlethal diseases.

3.3   Left or Interval Censoring

A lifetime X associated with a specific individual in a study is considered to be *left censored* if it is less than a censoring time C_l (C_l for "left" censoring time), that is, the event of interest has already occurred for the individual before that person is observed in the study at time C_l. For such individuals, we know that they have experienced the event sometime before time C_l, but their exact event time is unknown. The exact lifetime X will be known if, and only if, X is greater than or equal to C_l. The data from a left-censored sampling scheme can be represented by pairs of random variables (T, ε), as in the previous section, where T is equal to X if the lifetime is observed and ε indicates whether the exact lifetime X is observed ($\varepsilon = 1$) or not ($\varepsilon = 0$). Note that, for left censoring as contrasted with right censoring, $T = \max(X, C_l)$.

EXAMPLE 3.3   In a study to determine the distribution of the time until first marijuana use among high school boys in California, discussed in section 1.17, the question was asked, When did you first use marijuana?" One of the responses was "I have used it but can not recall just when the first time was." A boy who chose this response is indicating that the event had occurred prior to the boy's age at interview but the exact age at which he started using marijuana is unknown. This is an example of a left-censored event time.

EXAMPLE 3.4   In early childhood learning centers, interest often focuses upon testing children to determine when a child learns to accomplish certain specified tasks. The age at which a child learns the task would be considered the time-to-event. Often, some children can already perform the task when they start in the study. Such event times are considered left censored.

Often, if left censoring occurs in a study, right censoring may also occur, and the lifetimes are considered *doubly censored* (cf. Turnbull, 1974). Again, the data can be represented by a pair of variables (T, δ), where $T = \max[\min(X, C_r), C_l]$ is the on study time; δ is 1 if T is a death time, 0 if T is a right-censored time, and -1 if T is a left-censored time. Here C_l is the time before which some individuals experience the event and C_r is the time after which some individuals experience the event. X will be known exactly if it is less than or equal to C_r and greater than or equal to C_l.

EXAMPLE 3.3   *(continued)* An additional possible response to the question "When did you first use marijuana?" was "I never used it" which indicates a right-censored observation. In the study described in section 1.17, both left-censored observations and right-censored observations were present, in addition to knowing the exact age of first use of marijuana (uncensored observations) for some boys. Thus, this is a doubly censored sampling scheme.

EXAMPLE 3.4   *(continued)* Some children undergoing testing, as described in Example 3.4, may not learn the task during the entire study period, in which case such children would be right-censored. Coupled with the left-censored observations discussed earlier, this sample would also contain doubly censored data.

A more general type of censoring occurs when the lifetime is only known to occur within an interval. Such *interval censoring* occurs when patients in a clinical trial or longitudinal study have periodic follow-up and the patient's event time is only known to fall in an interval $(L_i, R_i]$ (L for left endpoint and R for right endpoint of the censoring interval). This type of censoring may also occur in industrial experiments where there is periodic inspection for proper functioning of equipment items. Animal tumorigenicity experiments may also have this characteristic.

EXAMPLE 3.5   In the Framingham Heart Study, the ages at which subjects first developed coronary heart disease (CHD) are usually known exactly. However, the ages of first occurrence of the subcategory angina pectoris may be known only to be between two clinical examinations, approximately two years apart (Odell et al., 1992). Such observations would be interval-censored.

EXAMPLE 3.6   In section 1.18, the data from a retrospective study to compare the cosmetic effects of radiotherapy alone versus radiotherapy and adjuvant chemotherapy on women with early breast cancer are reported. Patients were observed initially every 4–6 months but, as their recovery progressed, the interval between visits lengthened. The event of interest was the first appearance of moderate or severe breast retraction, a cosmetic deterioration of the breast. The exact time of retraction was known to fall only in the interval between visits (interval-censored) or after the last time the patient was seen (right-censored).

In view of the last two examples, it is apparent that any combination of left, right, or interval censoring may occur in a study. Of course, interval censoring is a generalization of left and right censoring because, when the left end point is 0 and the right end point is C_l we have left censoring and, when the left end point is C_r and the right end point is infinite, we have right censoring.

The main impact on the analysis, when data are truncated, is that the investigator must use a conditional distribution in constructing the likelihood, as shown in section 3.5, or employ a statistical method which uses a selective risk set, explained in more detail in section 4.6.

3.4   Truncation

A second feature of many survival studies, sometimes confused with censoring, is *truncation*. Truncation of survival data occurs when only those individuals whose event time lies within a certain observational window (Y_L, Y_R) are observed. An individual whose event time is not in this interval is not observed and no information on this subject is available to the investigator. This is in contrast to censoring where there is at least partial information on each subject. Because we are only aware of individuals with event times in the observational window, the inference for truncated data is restricted to conditional estimation.

When Y_R is infinite then we have *left truncation*. Here we only observe those individuals whose event time X exceeds the truncation time Y_L. That is we observe X if and only if $Y_L < X$. A common example of left truncation is the problem of estimating the distribution of the diameters of microscopic particles. The only particles big enough to be seen based on the resolution of the microscope are observed and smaller particles do not come to the attention of the investigator. In survival studies the truncation event may be exposure to some disease, diagnosis of a disease, entry into a retirement home, occurrence of some intermediate event such as graft-versus-host disease after a bone marrow transplantation, etc. In this type of truncation any subjects who experience the event of interest prior to the truncation time are not observed. The truncation time is often called a *delayed entry time* since we only observe subjects from this time until they die or are censored. Note that, as opposed to left censoring where we have partial information on individuals who experience the event of interest prior to age at entry, for left truncation these individuals were never considered for inclusion into the study.

EXAMPLE 3.7   In section 1.16, a survival study of residents of the Channing House retirement center located in California is described. Ages at death (in months) are recorded, as well as ages at which individuals entered the retirement community (the truncation event). Since an individual must survive to a sufficient age to enter the retirement center, all individuals who died earlier will not enter the center and thus are out of the investigator's cognizance; i.e., such individuals have no chance to be in the study and are considered left truncated. A survival analysis of this data set needs to account for this feature.

Right truncation occurs when Y_L is equal to zero. That is, we observe the survival time X only when $X \leq Y_R$. Right truncation arises, for example, in estimating the distribution of stars from the earth in that stars too far away are not visible and are right truncated. A second example of a right-truncated sample is a mortality study based on death records. Right-truncated data is particularly relevant to studies of AIDS.

EXAMPLE 3.8   Consider the AIDS study described in section 1.19. Here cases of patients with transfusion-induced AIDS were sampled. Retrospective determination of the transfusion times were used to estimate the waiting time from infection at transfusion to clinical onset of AIDS. The registry was sampled on June 30, 1986, so only those whose waiting time from transfusion to AIDS was less than the time from transfusion to June 30, 1986, were available for observation. Patients transfused prior to June 30, 1986, who developed AIDS after June 30, 1986, were not observed and are right truncated.

The main impact on the analysis when data are truncated is that the investigator must use a conditional distribution in constructing the likelihood, as shown in section 3.5, or employ a statistical method which uses a selective risk set, which will be explained in more detail in section 4.6.

3.5 Likelihood Construction for Censored and Truncated Data

As stated previously, the design of survival experiments involving censoring and truncation needs to be carefully considered when constructing likelihood functions. A critical assumption is that the lifetimes and censoring times are independent. If they are not independent, then specialized techniques must be invoked. In constructing a likelihood function for censored or truncated data we need to consider carefully what information each observation gives us. An observation corresponding to an exact event time provides information on the probability that the event's occurring at this time, which is approximately equal to the density function of X at this time. For a right-censored observation all we know is that the event time is larger than this time, so the information is the survival function evaluated at the on study time. Similarly for a left-censored observation, all we know is that the event has already occurred, so the contribution to the likelihood is the cumulative distribution function evaluated at the on study time. Finally, for interval-censored data we know only that the event occurred within the interval, so the information is the probability that the event time is in this interval. For truncated data these probabilities are replaced by the appropriate conditional probabilities.

More specifically, the likelihoods for various types of censoring schemes may all be written by incorporating the following components:

exact lifetimes	- $f(x)$
right-censored observations	- $S(C_r)$
left-censored observations	- $1 - S(C_l)$
interval-censored observations	- $[S(L) - S(R)]$
left-truncated observations	- $f(x)/S(Y_L)$
right-truncated observations	- $f(x)/[1 - S(Y_R)]$
interval-truncated observations	- $f(x)/[S(Y_L) - S(Y_R)]$

The likelihood function may be constructed by putting together the component parts as

$$L \propto \prod_{i\varepsilon D} f(x_i) \prod_{i\varepsilon R} S(C_r) \prod_{i\varepsilon L}(1 - S(C_l)) \prod_{i\varepsilon I}[S(L_i) - S(R_i)], \qquad (3.5.1)$$

where D is the set of death times, R the set of right-censored observations, L the set of left-censored observations, and I the set of interval-censored observations. For left-truncated data, with truncation interval (Y_{Li}, Y_{Ri}) independent from the jth death time, we replace $f(x_i)$ by

$f(x_i)/[S(Y_{Li}) - S(Y_{Ri})]$ and $S(C_i)$ by $S(C_i)/[S(Y_{Li}) - S(Y_{Ri})]$ in (3.5.1).

For right-truncated data, only deaths are observed, so that the likelihood is of the form

$$L \propto \prod_i f(Y_i)/[1 - S(Y_i)].$$

If each individual has a different failure distribution, as might be the case when regression techniques are used,

$$L = \prod_{i \in D} f_i(x_i) \prod_{i \in R} S_i(C_r) \prod_{i \in L} [1 - S_i(C_l)] \prod_{i \in I} [S_i(L_i) - S_i(R_i)]. \qquad (3.5.2)$$

We will proceed with explicit details in constructing the likelihood function for various types of censoring and show how they all basically lead to equation (3.5.1).

Data from experiments involving right censoring can be conveniently represented by pairs of random variables (T, δ), where δ indicates whether the lifetime X is observed ($\delta = 1$) or not ($\delta = 0$), and T is equal to X if the lifetime is observed and to C_r if it is right-censored, i.e., $T = \min(X, C_r)$.

Details of constructing the likelihood function for *Type I censoring* are as follows. For $\delta = 0$, it can be seen that

$$Pr[T, \delta = 0] = Pr[T = C_r \mid \delta = 0] Pr[\delta = 0] = Pr(\delta = 0)$$

$$= Pr(X > C_r) = S(C_r).$$

Also, for $\delta = 1$,

$$Pr(T, \delta = 1) = Pr(T = X \mid \delta = 1) Pr(\delta = 1),$$

$$= Pr(X = T \mid X \leq C_r) Pr(X \leq C_r)$$

$$= \left[\frac{f(t)}{1 - S(C_r)} \right] [1 - S(C_r)] = f(t).$$

These expressions can be combined into the single expression

$$Pr(t, \delta) = [f(t)]^\delta [S(t)]^{1-\delta}.$$

If we have a random sample of pairs (T_i, δ_i), $i = 1, \ldots, n$, the likelihood function is

$$L = \prod_{i=1}^{n} Pr[t_i, \delta_i] = \prod_{i=1}^{n} [f(t_i)]^{\delta_i} [S(t_i)]^{1-\delta_i} \qquad (3.5.3)$$

which is of the same form as (3.5.1). Because we can write $f(t_i) = h(t_i)S(t_i)$ we can write this likelihood as

$$L = \prod_{i=1}^{n}[h(t_i)]^{\delta_i} \exp[-H(t_i)]$$

EXAMPLE 3.9   Assume $f(x) = \lambda e^{-\lambda x}$.
Then, the likelihood function is

$$L_I = \prod_{i=1}^{n}(\lambda e^{-\lambda t_i})^{\delta_i} \exp[-\lambda t_i(1 - \delta_i)] \qquad (3.5.4)$$

$$= \lambda^r \exp[-\lambda S_T],$$

where $r = \sum \delta_i$ is the observed number of events and S_T is the total time on test for all n individuals under study.

EXAMPLE 3.10   A simple random censoring process encountered frequently is one in which each subject has a lifetime X and a censoring time C_r, X and C_r being independent random variables with the usual notation for the probability density and survival function of X as in Type I censoring and the p.d.f. and survival function of C_r denoted by $g(c_r)$ and $G(c_r)$, respectively. Furthermore, let $T = \min(X, C_r)$ and δ indicates whether the lifetime X is censored ($\delta = 0$) or not ($\delta = 1$). The data from a sample of n subjects consist of the pairs (t_i, δ_i), $i = 1, \ldots, n$. The density function of this pair may be obtained from the joint density function of X and C_r, $f(x, c_r)$, as

$$Pr(T_i = t, \delta = 0) = Pr(C_{r,i} = t, X_i > C_{r,i})$$

$$= \frac{d}{dt}\int_0^t \int_v^\infty f(u, v)du\, dv. \qquad (3.5.5)$$

When X and C_r are independent with marginal densities f and g, respectively, (3.5.5) becomes

$$= \frac{d}{dt}\int_0^t \int_v^\infty f(u)g(v)du\, dv$$

$$= \frac{d}{dt}\int_0^t S(v)g(v)dv$$

$$= S(t)g(t)$$

and, similarly,

$$Pr(T_i = t, \delta = 1) = Pr(X_i = t, X_i < C_{r,i}) = f(t)G(t).$$

So,

$$L = \prod_{i=1}^{n} [f(t_i)G(t_i)]^{\delta_i} [g(t_i)S(t_i)]^{1-\delta_i}$$

$$= \left\{ \prod_{i=1}^{n} G(t_i)^{\delta_i} g(t_i)^{1-\delta_i} \right\} \left\{ \prod_{i=1}^{n} f(t_i)^{\delta_i} S(t_i)^{1-\delta_i} \right\}.$$

If the distribution of the censoring times, as alluded to earlier, does not depend upon the parameters of interest, then, the first term will be a constant with respect to the parameters of interest and the likelihood function takes the form of (3.5.1)

$$L \propto \prod_{i=1}^{n} [f(t_i)]^{\delta_i} [S(t_i)]^{1-\delta_i}. \tag{3.5.6}$$

Practical Notes

1. The likelihoods constructed in this section are used primarily for analyzing parametric models, as discussed in Chapter 12. They also serve as a basis for determining the partial likelihoods used in the semiparametric regression methods discussed in Chapters 8 and 9.
2. Even though, in most applications of analyzing survival data, the likelihoods constructed in this section will not be explicitly used, the rationale underlying their construction has value in understanding the contribution of the individual data components depicted in (3.5.1).

Theoretical Notes

1. For Type II censoring, the data consist of the rth smallest lifetimes $X_{(1)} \le X_{(2)} \le \cdots \le X_{(r)}$ out of a random sample of n lifetimes $X_1, \ldots, X_n$ from the assumed life distribution. Assuming $X_1, \ldots, X_n$ are i.i.d. and have a continuous distribution with p.d.f. $f(x)$ and survival function $S(x)$, it follows that the joint p.d.f. of $X_{(1)}, \ldots, X_{(r)}$ is (cf. David, 1981)

$$L_{\text{II},1} = \frac{n!}{(n-r)!} \left[\prod_{i=1}^{r} f(x_{(i)}) \right] [S(x_{(r)})]^{n-r}. \tag{3.5.7}$$

2. For simplicity, in the progressive Type II censoring case, assume that the censoring (or serial sacrifice) has just two repetitions. Here we observe the r_1 ordered failures $X_{(1)} \leq X_{(2)} \leq \cdots \leq X_{(r_1)}$, then, n_1 items are removed from the study and sacrificed. Of the remaining $(n - r_1 - n_1)$ items we observe the next r_2 ordered failures $X^*_{(1)} \leq X*_{(2)} \leq \cdots \leq X^*_{(r_2)}$ after which the study stops with the remaining $n - n_1 - r_1 - r_2$ items being censored at $X^*_{r_2}$. The likelihood for this type of data may be written as

$$Pr(X_{(1)}, \ldots, X_{(r_1)}, X^*_{(1)}, \ldots, X^*_{(r_2)},)$$

$$= P_1(X_{(1)}, \ldots, X_{(r_1)}) P_2(X^*_{(1)}, \ldots, X^*_{(r_2)} \mid X_{(1)}, \ldots, X_{(r_1)}).$$

By equation (3.5.7), the first term above becomes

$$\frac{n!}{(n - r_1)!} \prod_{i=1}^{r_1} f(t_{(i)})[S(t_{(r_1)})]^{n-r_1}$$

and, by a theorem in order statistics (David, 1981), the second term above becomes

$$\frac{(n - r_1 - n_1)!}{(n - r_1 - n_1 - r_2)!} \prod_{i=1}^{r_2} f^*(x^*_{(i)})[S^*(x^*_{(r_2)})]^{n-r_1-n_1-r_2}$$

where $f^*(x) = \frac{f(x)}{S(x_{r_1})}$, $x \geq x_{(r_1)}$ is the truncated p.d.f. and $S^*(x) = \frac{S(x)}{S(x_{r_1})}$, $x \geq x_{(r_1)}$ is the truncated survival function so that

$$L_{II,2} = \frac{n!(n - r_1 - n_1)!}{(n - r_1)!(n - r_1 - n_1 - r_2)!} \prod_{i=1}^{r_1} f(x_{(i)})[S(t_{(r_1)})]^{n-r_1}$$

$$x \frac{\prod_{i=1}^{r_2} f(t^*_{(i)})}{[S(t_{(r_1)})]^{r_2}} \left[\frac{S(t^*_{(r_2)})}{S(t_{(r_1)})} \right]^{n-r_1-n_1-r_2}$$

so that

$$L_{II,2} \propto \prod_{i=1}^{r_1} f(x_{(i)})[S(x_{(r_1)})]^{n_1} \prod_{i=1}^{r_2} f(x^*_{(i)})[S(x^*_{(r_2)})]^{n-r_1-n_1-r_2}$$

which, again, can be written in the form of (3.5.1).

3. For random censoring, when X and C_r are not independent, the likelihood given by (3.5.6) is not correct. If the joint survival function of X and C_r is $S(x, c)$, then, the likelihood is of the form

$$L_{III} \propto \prod_{i=1}^{n} \{[-\partial S(x, t_i)/\partial x]_{x=t_i}\}^{\delta_i} \{[-\partial S(t_i, c)/\partial c]_{c=t_i}\}^{1-\delta_i},$$

which may be appreciably different from (3.5.6).

3.6   Counting Processes

In the previous section, we discussed the construction of classical like-lihoods for censored and truncated data. These likelihoods can be used to develop some of the methods described in the remainder of this book. An alternative approach to developing inference procedures for censored and truncated data is by using counting process methodology. This approach was first developed by Aalen (1975) who combined ele-ments of stochastic integration, continuous time martingale theory and counting process theory into a methodology which quite easily allows for development of inference techniques for survival quantities based on censored and truncated data. These methods allow relatively simple development of the large sample properties of such statistics. Although complete exposition of this theory is beyond the scope of this book, we will give a brief survey in this section. For a more rigorous survey of this area, the reader is referred to books by Andersen et al. (1993) and Fleming and Harrington (1991).

We start by defining a counting process $N(t), t \geq 0$, as a stochastic process with the properties that $N(0)$ is zero; $N(t) < \infty$, with probability one; and the sample paths of $N(t)$ are right-continuous and piecewise constant with jumps of size $+1$. Given a right-censored sample, the processes, $N_i(t) = I[T_i \leq t, \delta_i = 1]$, which are zero until individual i dies and then jumps to one, are counting processes. The process $N(t) = \sum_{i=1}^{n} N_i(t) = \sum_{t_i \leq t} \delta_i$ is also a counting process. This process simply counts the number of deaths in the sample at or prior to time t.

The counting process gives us information about when events occur. In addition to knowing this information, we have additional information on the study subjects at a time t. For right censored data, this informa-tion at time t includes knowledge of who has been censored prior to time t and who died at or prior to time t. In some problems, our in-formation may include values for a set of fixed time covariates, such as age, sex, treatment at time 0 and possibly the values of time-dependent covariates, at all times prior to t. This accumulated knowledge about what has happened to patients up to time t is called the *history* or *fil-tration* of the counting process at time t and is denoted by $\mathbf{F}_t$. As time progresses, we learn more and more about the sample so that a natural requirement is that $\mathbf{F}_s \subset \mathbf{F}_t$ for $s \leq t$. In the case of right-censored data, the history at time t, $\mathbf{F}_t$, consists of knowledge of the pairs (T_i, δ_i) provided $T_i \leq t$ and the knowledge that $T_i > t$ for those individuals still under study at time t. We shall denote the history at an instant just prior to time t by $\mathbf{F}_{t-}$. The history $\{\mathbf{F}_t, t \geq 0\}$ for a given problem depends on the observer of the counting process.

For right-censored data, if the death times X_i and censoring times C_i are independent, then, the chance of an event at time t, given the history just prior to t, is given by

$$Pr[t \leq T_i \leq t + dt, \delta_i = 1 | F_{t-}] \tag{3.6.1}$$

$$= \begin{cases} Pr[t \leq X_i \leq t + dt, C_i > t + dt_i | X_i \geq t, C_i \geq t] = h(t)dt & \text{if } T_i \geq t, \\ 0 & \text{if } T_i < t \end{cases}$$

For a given counting process, we define $dN(t)$ to be the change in the process $N(t)$ over a short time interval $[t, t + dt)$. That is $dN(t) = N[(t + dt)^-] - N(t^-)$ (Here t^- is a time just prior to t). In the right-censored data example (assuming no ties), $dN(t)$ is one if a death occurred at t or 0, otherwise. If we define the process $Y(t)$ as the number of individuals with a study time $T_i \geq t$, then, using (3.6.1),

$$E[dN(t)|\mathbf{F}_{t-}] = E[\text{Number of observations with}$$

$$t \leq X_i \leq t + dt, C_i > t + dt_i \mid \mathbf{F}_{t-}]$$

$$= Y(t)h(t)dt. \tag{3.6.2}$$

The process $\lambda(t) = Y(t)h(t)$ is called the *intensity process* of the counting process. $\lambda(t)$ is itself a stochastic process that depends on the information contained in the history process, $\mathbf{F}_t$ through $Y(t)$.

The stochastic process $Y(t)$ is the process which provides us with the number of individuals at risk at a given time and, along with $N(t)$, is a fundamental quantity in the methods presented in the sequel. Notice that, if we had left truncated data and right-censored data, the intensity process would be the same as in (3.6.2) with the obvious modification to $Y(t)$ as the number of individuals with a truncation time less than t still at risk at time t.

We define the process $\Lambda(t)$ by $\int_0^t \lambda(s)ds, t \geq 0$. This process, called the *cumulative intensity process*, has the property that $E[N(t)|\mathbf{F}_{t-}] = E[\Lambda(t) \mid \mathbf{F}_{t-}] = \Lambda(t)$. The last equality follows because, once we know the history just prior to t, the value of $Y(t)$ is fixed and, hence, $\Lambda(t)$ is nonrandom. The stochastic process $M(t) = N(t) - \Lambda(t)$ is called *the counting process martingale*. This process has the property that increments of this process have an expected value, given the strict past, $\mathbf{F}_{t-}$, that are zero. To see this,

$$E(dM(t) \mid \mathbf{F}_{t-}) = E[dN(t) - d\Lambda(t) \mid \mathbf{F}_{t-}]$$

$$= E[dN(t) \mid \mathbf{F}_{t-}] - E[\lambda(t)dt \mid \mathbf{F}_{t-}]$$

$$= 0.$$

The last inequality follows because $\lambda(t)$ has a fixed value, given $\mathbf{F}_{t-}$.

A stochastic process with the property that its expected value at time t, given its history at time $s < t$, is equal to its value at time s is called a *martingale*, that is, $M(t)$ is a martingale if

$$E[M(t) \mid \mathbf{F}_s] = M(s), \text{for all } s < t. \tag{3.6.3}$$

To see that this basic definition is equivalent to having $E[dM(t) \mid \mathbf{F}_{t^-}] = 0$ for all t, note that, if $E[dM(t) \mid \mathbf{F}_{t^-}] = 0$, then,

$$E(M(t) \mid \mathbf{F}_s) - M(s) = E[M(t) - M(s) \mid \mathbf{F}_s]$$

$$= E\left[\int_s^t dM(u) \mid \mathbf{F}_s\right]$$

$$= \int_s^t E[E[dM(u) \mid \mathbf{F}_{u^-}] \mid \mathbf{F}_s]$$

$$= 0.$$

Thus the counting process martingale is indeed a martingale.

The counting process martingale, $M(t) = N(t) - \Lambda(t)$ is made up of two parts. The first is the process $N(t)$, which is a nondecreasing step function. The second part $\Lambda(t)$ is a smooth process which is predictable in that its value at time t is fixed just prior to time t. This random function is called a *compensator* of the counting process. The martingale can be considered as mean zero noise which arises when we subtract the smoothly varying compensator from the counting process

To illustrate these concepts, a sample of 100 observations was generated from an exponential population with hazard rate $h_X(t) = 0.2$. Censoring times were generated from an independent exponential distribution with hazard rate $h_C(t) = 0.05$. Figure 3.6 shows the processes $N(t)$ and the compensator of $N(t)$, $\Lambda(t) = \int_0^t h(u)Y(u)du$, of a single sample drawn from these distributions. Note that $N(t)$ is an increasing step function with jumps at the observed death times, $Y(t)$ is a decreasing step function with steps of size one at each death or censoring time, and $\Lambda(t)$ is an increasing continuous function that is quite close to $N(t)$.

Figure 3.7 depicts the values of $M(t)$ for 10 samples generated from this population. The sample in Figure 3.6 is the solid line on this figure. We can see in this figure that the sample paths of $M(t)$ look like a sample of random, mean 0, noise.

An additional quantity needed in this theory is the notion of the *predictable variation process* of $M(t)$, denoted by $\langle M \rangle(t)$. This quantity is defined as the compensator of the process $M^2(t)$. Although $M(t)$ reflects the noise left after subtracting the compensator, $M^2(t)$ tends to increase with time. Here, $\langle M \rangle(t)$ is the systematic part of this increase and is the predictable process needed to be subtracted from $M^2(t)$ to produce a martingale. The name, predictable variation process, comes from the fact that, for a martingale $M(t)$, $\mathrm{var}(dM(t) \mid \mathbf{F}_{t^-}) = d\langle M \rangle(t)$. To see this, recall that, by definition, $E[dM(t)] = 0$. Now,

$$dM^2(t) = M[(t + dt)^-]^2 - M(t^-)^2$$

$$= [M(t^-) + dM(t)]^2 - M(t^-)^2$$

$$= [dM(t)]^2 + 2M(t^-)dM(t).$$

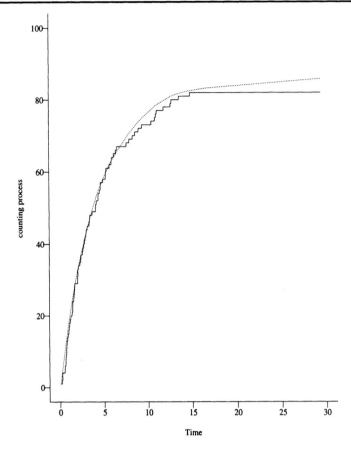

Figure 3.6 *Example of a counting process, N(t) (solid line) and its compensator, Λ(t), (dashed line) for a sample of 100 individuals*

So,

$$\text{Var}[dM(t) \mid \mathbf{F}_{t^-}] = E[(dM(t))^2 \mid \mathbf{F}_{t^-}]$$

$$= E[(dM^2(t)) \mid \mathbf{F}_{t^-}] - 2E[M(t^-)dM(t) \mid \mathbf{F}_{t^-}]$$

$$= d\langle M \rangle(t) - 2M(t^-)E[dM(t) \mid \mathbf{F}_{t^-}] = d\langle M \rangle(t)$$

because once $\mathbf{F}_{t^-}$ is known, $M(t^-)$ is a fixed quantity and $E[dM(t) \mid \mathbf{F}_{t^-}] = 0$.

To find $\text{Var}[dM(t) \mid \mathbf{F}_{t^-}]$ recall that $dN(t)$ is a zero-one random variable with a probability, given the history, of $\lambda(t)$ of having a jump of size one at time t. The variance of such a random variable is $\lambda(t)[1 - \lambda(t)]$. If there are no ties in the censored data case, $\lambda(t)^2$ is close to zero so that $\text{Var}[dM(t) \mid \mathbf{F}_{t^-}] \cong \lambda(t) = Y(t)h(t)$. In this case, notice that the conditional mean and variance of the counting process $N(t)$ are the same and one can show that locally, conditional on the past history, the counting

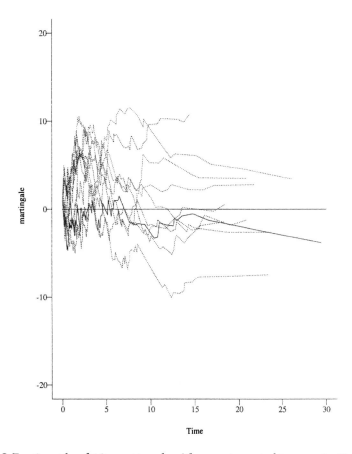

Figure 3.7 *Sample of 10 martingales. The compensated process in Figure 3.6 is the solid line.*

process behaves like a Poisson process with rate $\lambda(t)$. When there are ties in the data, the Bernoulli variance is used. Of course, in either case, these variances are conditional variances in that they depend on the history at time t^- through $Y(t)$. In many applications these conditional variances serve as our estimator of the variance of $dM(t)$.

Many of the statistics in later sections are *stochastic integrals* of the basic martingale discussed above. Here, we let $K(t)$ be a *predictable process*. That is $K(t)$ is a stochastic process whose value is known, given the history just prior to time t, $\mathbf{F}_{t^-}$. An example of a predictable process is the process $Y(t)$. Over the interval 0 to t, the stochastic integral of such a process, with respect to a martingale, is denoted by $\int_0^t K(u)dM(u)$. Such stochastic integrals have the property that they themselves are martingales as a function of t and their predictable variation process can be found from the predictable variation process of the original martingale by

$$\left\langle \int_0^t K(u)dM(u) \right\rangle = \int_0^t K(u)^2 d\langle M\rangle(u). \qquad (3.6.4)$$

To illustrate how these tools can be used to derive nonparametric estimators of parameters of interest, we shall derive a nonparametric estimator of the cumulative hazard rate $H(t)$ based on right-censored data, the so-called Nelson–Aalen estimator. Recall that we can write $dN(t) = Y(t)h(t)dt + dM(t)$. If $Y(t)$ is nonzero, then,

$$\frac{dN(t)}{Y(t)} = h(t)dt + \frac{dM(t)}{Y(t)}. \qquad (3.6.5)$$

If $dM(t)$ is noise, then, so is $dM(t)/Y(t)$. Because the value of $Y(t)$ is fixed just prior to time t,

$$E\left[\frac{dM(t)}{Y(t)} \mid \mathbf{F}_{t^-}\right] = \frac{E[dM(t) \mid \mathbf{F}_{t^-}]}{Y(t)} = 0.$$

Also, the conditional variance of the noise can be found as

$$\mathrm{Var}\left[\frac{dM(t)}{Y(t)} \mid \mathbf{F}_{t^-}\right] = \frac{\mathrm{Var}[dM(t) \mid \mathbf{F}_{t^-})]}{Y(t)^2} = \frac{d\langle M\rangle(t)}{Y(t)^2}.$$

If we let $J(t)$ be the indicator of whether $Y(t)$ is positive and we define $0/0 = 0$, then, integrating both sides of equation (3.6.5),

$$\int_0^t \frac{J(u)}{Y(u)} dN(u) = \int_0^t J(u)h(u)du + \int_0^t \frac{J(u)}{Y(u)} dM(u).$$

The integral $\int_0^t \frac{J(u)}{Y(u)} dN(u) = \hat{H}(t)$ is the Nelson–Aalen estimator of $H(t)$. The integral, $W(t) = \int_0^t \frac{J(u)}{Y(u)} dM(u)$, is the stochastic integral of the predictable process $\frac{J(u)}{Y(u)}$ with respect to a martingale and, hence, is also a martingale. Again, we can think of this integral as random noise or the statistical uncertainty in our estimate. The random quantity $H^*(t) = \int_0^t J(u)h(u)du$, for right-censored data is equal to $H(t)$ in the range where we have data, and, ignoring the statistical uncertainty in $W(t)$, the statistic $\hat{H}(t)$ is a nonparametric estimator of the random quantity $H^*(t)$.

Because $W(t) = \hat{H}(t) - H^*(t)$ is a martingale, $E[\hat{H}(t)] = E[H^*(t)]$. Note that $H^*(t)$ is a random quantity and its expectation is not, in general, equal to $H(t)$. The predictable variation process of $W(t)$ is found quite simply, using (3.6.4), as

$$\langle W\rangle(t) = \int_0^t \left[\frac{J(u)}{Y(u)}\right]^2 d\langle M\rangle(u) = \int_0^t \left[\frac{J(u)}{Y(u)}\right]^2 Y(u)h(u)du$$

$$= \int_0^t \left[\frac{J(u)}{Y(u)}\right] h(u)du.$$

A final strength of the counting process approach is the martingale central limit theorem. Recall that $Y(t)/n$ and $N(t)/n$ are sample averages and that, for a large sample, the random variation in both should be small. For large n, suppose that $Y(t)/n$ is close to a deterministic function $y(t)$. Let $Z^{(n)}(t) = \sqrt{n}\,W(t) = \sqrt{n}[\hat{H}(t) - H^*(t)]$. This process is almost equal to $\sqrt{n}[\hat{H}(t) - H(t)]$, because for large samples $H^*(t)$ is very close to $H(t)$. Given the history, the conditional variance of the jumps in $Z^{(n)}(t)$ are found to converge to $h(t)/y(t)$. To see this

$$\operatorname{Var}[dZ^{(n)}(t) \mid \mathbf{F}_{t^-}] = n\operatorname{Var}[dW(t) \mid \mathbf{F}_{t^-}]$$

$$= n\operatorname{Var}[\frac{dM(t)}{Y(t)} \mid \mathbf{F}_{t^-}]$$

$$= n\frac{d\langle M(t)\rangle}{Y(t)^2}$$

$$= n\frac{\lambda(t)dt}{Y(t)^2}$$

$$= n\frac{Y(t)h(t)dt}{Y(t)^2} = \frac{h(t)dt}{Y(t)/n}\,,$$

which converges to $h(t)dt/y(t)$ for large samples. Also, for large samples, $Z^{(n)}$ will have many jumps, but all of these jumps will be small and of order $1/\sqrt{n}$.

The above heuristics tell us that, for large samples, $Z^{(n)}$ has almost continuous sample paths and a predictable variation process very close to

$$\langle Z^{(n)}\rangle \approx \int_0^t \frac{h(u)du}{y(u)}\,. \tag{3.6.6}$$

It turns out that there is one and only one limiting process, $Z^{(\infty)}$ which is a martingale with continuous sample paths and a deterministic predictable variation $\langle Z^{(\infty)}\rangle$ exactly equal to (3.6.6). This limiting process has independent increments and normally distributed finite-dimensional distributions. A process has independent increments if, for any set of nonoverlapping intervals (t_{i-1}, t_i), $i = 1, \ldots, k$ the random variables $Z^{(\infty)}(t_i) - Z^{(\infty)}(t_{i-1})$ are independent. The limiting process has normally distributed finite-dimensional distributions if the joint distribution of $[Z^{(\infty)}(t_1), \ldots, Z^{(\infty)}(t_k)]$ is multivariate normal for any value of k. For the process $Z^{(\infty)}$, $[Z^{(\infty)}(t_1), \ldots, Z^{(\infty)}(t_k)]$ has a k-variate normal distribution with mean 0 and a covariance matrix with entries

$$\operatorname{cov}[Z^{(\infty)}(t), Z^{(\infty)}(s)] = \int_0^{\min(s,t)} \frac{h(u)du}{y(u)}\,.$$

This basic convergence allows us to find confidence intervals for the cumulative hazard rate at a fixed time because $\sqrt{n}[\hat{H}(t) - H^*(t)]$ will have an approximate normal distribution with mean 0 and variance

$$\sigma[Z^{(\infty)}] = \int_0^t \frac{b(u)\,du}{y(u)}.$$

An estimate of the variance can be obtained from

$$n \int_0^t \frac{dN(u)}{Y(u)^2}$$

because we can estimate $y(t)$ by $Y(t)/n$ and $b(t)$ by $dN(t)/Y(t)$. The fact that, as a process, $Z^{(n)}$ is approximated by a continuous process with normal margins also allows us to compute confidence bands for the cumulative hazard rate (see section 4.4).

To estimate the survival function, recall that, for a continuous random variable, $S(t) = \exp[-H(t)]$ and, for a discrete, random variable, $S(t) = \prod_{s=0}^t [1 - d\hat{H}(s)]$. Here, we say that $S(t)$ is the *product integral* of $1 - d\hat{H}(t)$. To obtain an estimator of the survival function, we take the product integral of $1 - d\hat{H}(t)$ to obtain

$$\hat{S}(t) = \prod_{s=0}^t [1 - d\hat{H}(t)] = \prod_{s=0}^t \left[1 - \frac{dN(s)}{Y(s)}\right].$$

This is the Kaplan–Meier estimator (see section 4.2) which is a step function with steps at the death times where $dN(t) > 0$. It turns out that $\hat{S}(t)/S(t) - 1$ is a stochastic integral with respect to the basic martingale M and is also a martingale. Thus confidence intervals and confidence bands for the survival function can be found using the martingale central limit theorem discussed above (see sections 4.3 and 4.4).

Counting processes methods can be used to construct likelihoods for survival data in a natural way. To derive a likelihood function based on $N(t)$ consider a separate counting process, $N_j(t)$, for each individual in the study. Given the history up to time t, $dN_j(t)$ has an approximate Bernoulli distribution with probability $\lambda_j(t)dt$ of having $dN_j(t) = 1$. The contribution to the likelihood at a given time is, then, proportional to

$$\lambda_j(t)^{dN_j(t)}[1 - \lambda_j(t)dt]^{1 - dN_j(t)}.$$

Integrating this quantity over the range $[0, \tau]$ gives a contribution to the likelihood of

$$\lambda_j(t)^{dN_j(t)} \exp\left[-\int_0^\tau \lambda_j(u)\,du\right].$$

The full likelihood for all n observations based on information up to time τ is, then, proportional to

$$L = \left[\prod_{j=1}^n \lambda_j(t)^{dN_j(t)}\right] \exp\left[-\sum_{j=1}^n \int_0^\tau \lambda_j(u)\,du\right].$$

For right-censored data, where $\lambda_j(t) = Y_j(t)h(t)$, with $Y_j(t) = 1$ if $t \leq t_j, 0$ if $t > t_j$, so

$$L \propto \left[\prod_{j=1}^{n} h(t_j)^{\delta_j} \right] \exp\left(- \sum_{j=1}^{n} H(t_j) \right),$$

which is exactly the same form as (3.5.1). This heuristic argument is precisely stated in Chapter 2 of Andersen et al. (1993).

The counting process techniques illustrated in this section can be used to derive a wide variety of statistical techniques for censored and truncated survival data. They are particularly useful in developing nonparametric statistical methods. In particular, they are the basis of the univariate estimators of the survival function and hazard rate discussed in Chapter 4, the smoothed estimator of the hazard rate and the models for excess and relative mortality discussed in Chapter 6, most of the k-sample nonparametric tests discussed in Chapter 7, and the regression methods discussed in Chapters 8, 9, and 10. A check of the martingale property is used to test model assumptions for regression models, as discussed in Chapter 11. Most of the statistics developed in the sequel can be shown to be stochastic integrals of some martingale, so large sample properties of the statistics can be found by using the predictable variation process and the martingale central limit theorem. In the theoretical notes, we shall point out where these methods can be used and provide references to the theoretical development of the methods. The books by Andersen et al. (1993) or Fleming and Harrington (1991) provide a sound reference for these methods.

3.7 Exercises

3.1 Describe, in detail, the types of censoring which are present in the following studies.

(a) The example dealing with remission duration in a clinical trial for acute leukemia described in section 1.2.

(b) The example studying the time to death for breast cancer patients described in section 1.5.

3.2 A large number of disease-free individuals were enrolled in a study beginning January 1, 1970, and were followed for 30 years to assess the age at which they developed breast cancer. Individuals had clinical exams every 3 years after enrollment. For four selected individuals described below, discuss in detail, the types of censoring and truncation that are represented.

(a) A healthy individual, enrolled in the study at age 30, never developed breast cancer during the study.

(b) A healthy individual, enrolled in the study at age 40, was diagnosed with breast cancer at the fifth exam after enrollment (i.e., the disease started sometime between 12 and 15 years after enrollment).

(c) A healthy individual, enrolled in the study at age 50, died from a cause unrelated to the disease (i.e., not diagnosed with breast cancer at any time during the study) at age 61.

(d) An individual, enrolled in the study at age 42, moved away from the community at age 55 and was never diagnosed with breast cancer during the period of observation.

(e) Confining your attention to the four individuals described above, write down the likelihood for this portion of the study.

3.3   An investigator, performing an animal study designed to evaluate the effects of vegetable and vegetable-fiber diets on mammary carcinogenesis risk, randomly assigned female Sprague-Dawley rats to five dietary groups (control diet, control diet plus vegetable mixture, 1; control diet plus vegetable mixture, 2; control diet plus vegetable-fiber mixture, 1; and control diet plus vegetable-fiber mixture, 2). Mammary tumors were induced by a single oral dose (5 mg dissolved in 1.0 ml. corn oil) of 7,12-dimethylbenz(α)anthracene (DMBA) administered by intragastric intubation, i.e., the starting point for this study is when DMBA was given.

Starting 6 weeks after DMBA administration, each rat was examined once weekly for 14 weeks (post DMBA administration) and the time (in days) until onset of the first palpable tumor was recorded. We wish to make an inference about the marginal distribution of the time until a tumor is detected. Describe, in detail, the types of censoring that are represented by the following rats.

(a) A rat who had a palpable tumor at the first examination at 6 weeks after intubation with DMBA.

(b) A rat that survived the study without having any tumors.

(c) A rat which did not have a tumor at week 12 but which had a tumor at week 13 after intubation with DMBA.

(d) A rat which died (without tumor present and death was unrelated to the occurrence of cancer) at day 37 after intubation with DMBA.

(e) Confining our attention to the four rats described above, write down the likelihood for this portion of the study.

3.4   In section 1.2, a clinical trial for acute leukemia is discussed. In this trial, the event of interest is the time from treatment to leukemia relapse. Using the data for the 6-MP group and assuming that the time to relapse distribution is exponential with hazard rate λ, construct the likelihood function. Using this likelihood function, find the maximum likelihood estimator of λ by finding the value of λ which maximizes this likelihood.

3.5 Suppose that the time to death has a log logistic distribution with parameters λ and α. Based on the following left-censored sample, construct the likelihood function.

DATA: 0.5, 1, 0.75, 0.25-, 1.25-, where - denotes a left- censored observation.

3.6 The following data consists of the times to relapse and the times to death following relapse of 10 bone marrow transplant patients. In the sample patients 4 and 6 were alive in relapse at the end of the study and patients 7–10 were alive, free of relapse at the end of the study. Suppose the time to relapse had an exponential distribution with hazard rate λ and the time to death in relapse had a Weibull distribution with parameters θ and α.

Patient	Relapse Time (months)	Death Time (months)
1	5	11
2	8	12
3	12	15
4	24	33+
5	32	45
6	17	28+
7	16+	16+
8	17+	17+
9	19+	19+
10	30+	30+

+ Censored observation

(a) Construct the likelihood for the relapse rate λ.

(b) Construct a likelihood for the parameters θ and α.

(c) Suppose we were only allowed to observe a patients death time if the patient relapsed. Construct the likelihood for θ and α based on this truncated sample, and compare it to the results in (b).

3.7 To estimate the distribution of the ages at which postmenopausal woman develop breast cancer, a sample of eight 50-year-old women were given yearly mammograms for a period of 10 years. At each exam, the presence or absence of a tumor was recorded. In the study, no tumors were detected by the women by self-examination between the scheduled yearly exams, so all that is known about the onset time of breast cancer is that it occurs between examinations. For four of the eight women, breast cancer was not detected during the 10 year study period. The age at onset of breast cancer for the eight subjects was in the following intervals:

$$(55, 56], (58, 59], (52, 53], (59, 60], \geq 60, \geq 60, \geq 60, \geq 60.$$

(a) What type of censoring or truncation is represented in this sample?

(b) Assuming that the age at which breast cancer develops follows a Weibull distribution with parameters λ and α, construct the likelihood function.

3.8   Suppose that the time to death X has an exponential distribution with hazard rate λ and that the right-censoring time C is exponential with hazard rate θ. Let $T = \min(X, C)$ and $\delta = 1$ if $X \leq C; 0$, if $X > C$. Assume that X and C are independent.

(a) Find $P(\delta = 1)$

(b) Find the distribution of T.

(c) Show that δ and T are independent.

(d) Let $(T_1, \delta_1), \ldots, (T_n, \delta_n)$ be a random sample from this model. Show that the maximum likelihood estimator of λ is $\sum_{i=1}^{n} \delta_i / \sum_{i=1}^{n} T_i$. Use parts a–c to find the mean and variance of $\hat{\lambda}$.

3.9   An example of a counting process is a Poisson process $N(t)$ with rate λ. Such a process is defined by the following three properties:

(a) $N(0) = 0$ with probability 1.

(b) $N(t) - N(s)$ has a Poisson distribution with parameter $\lambda(t - s)$ for any $0 \leq s \leq t$.

(c) $N(t)$ has independent increments, that is, for $0 \leq t_1 < t_2 < t_3 < t_4$, $N(t_2) - N(t_1)$ is independent of $N(t_4) - N(t_3)$.

Let $\mathbf{F}_s$ be the σ-algebra defined by $N(s)$. Define the process $M(t) = N(t) - \lambda t$.

i. Show that $E|M(t)| < \infty$.

ii. Show that $E[M(t) \mid N(s)] = M(s)$ for $s < t$, and conclude that $M(t)$ is a martingale and that λt is the compensator of $N(t)$. (Hint: Write $M(t) = M(t) - M(s) + M(s)$.)

4
Nonparametric Estimation of Basic Quantities for Right-Censored and Left-Truncated Data

4.1 Introduction

In this chapter we shall examine techniques for drawing an inference about the distribution of the time to some event X, based on a sample of right-censored survival data. A typical data point consists of a time on study and an indicator of whether this time is an event time or a censoring time for each of the n individuals in the study. We assume throughout this chapter that the potential censoring time is unrelated to the potential event time. The methods are appropriate for Type I, Type II, progressive or random censoring discussed in section 3.2.

To allow for possible ties in the data, suppose that the events occur at D distinct times $t_1 < t_2 < \cdots < t_D$, and that at time t_i there are d_i events (sometimes simply referred to as deaths). Let Y_i be the number of individuals who are at risk at time t_i. Note that Y_i is a count of the number of individuals with a time on study of t_i or more (i.e., the number of individuals who are alive at t_i or experience the event of interest at t_i). The quantity d_i/Y_i provides an estimate of the conditional probability that an individual who survives to just prior to time t_i experiences the event at time t_i. As we shall see, this is the basic quantity from which we will construct estimators of the survival function and the cumulative hazard rate.

Basic estimators of the survival function $S(t)$, the cumulative hazard function $H(t)$, and their standard errors based on right-censored data are discussed in section 4.2. In section 4.3, confidence intervals for $S(t)$ and $H(t)$ for a fixed value of t are presented, and section 4.4 presents confidence bands which provide a specified coverage probability for a range of times. Section 4.5 discusses inference for the mean time to event and for percentiles of X based on right-censored data. The final section shows how the estimators developed for right-censored data can be extended to left-truncated data. Estimating for other censoring and truncating schemes is considered in Chapter 5.

4.2   Estimators of the Survival and Cumulative Hazard Functions for Right-Censored Data

The standard estimator of the survival function, proposed by Kaplan and Meier (1958), is called the Product-Limit estimator. This estimator is defined as follows for all values of t in the range where there is data:

$$\hat{S}(t) = \begin{cases} 1 & \text{if } t < t_1, \\ \prod_{t_i \le t}[1 - \frac{d_i}{Y_i}], & \text{if } t_1 \le t \end{cases} \quad . \tag{4.2.1}$$

For values of t beyond the largest observation time this estimator is not well defined (see Practical Notes 2 and 3 for suggestions as to solutions to this problem).

The Product-Limit estimator is a step function with jumps at the observed event times. The size of these jumps depends not only on the number of events observed at each event time t_i, but also on the pattern of the censored observations prior to t_i.

The variance of the Product-Limit estimator is estimated by Greenwood's formula:

$$\hat{V}[\hat{S}(t)] = \hat{S}(t)^2 \sum_{t_i \le t} \frac{d_i}{Y_i(Y_i - d_i)}. \tag{4.2.2}$$

The standard error of the Product-Limit estimator is given by $\{\hat{V}[\hat{S}(t)]\}^{1/2}$.

EXAMPLE 4.1   We consider the data in section 1.2 on the time to relapse of patients in a clinical trial of 6-MP against a placebo. We shall consider only the 6-MP patients. The calculations needed to construct the Product-Limit estimator and its estimated variance are in Table 4.1A. The Product-Limit estimator, found in Table 4.1B, is a step function. Figure 4.1A shows a plot of this estimated survival function. Note that the survival curve is defined only up to 35 weeks, the largest of the observation times.

TABLE 4.1A

Construction of the Product-Limit Estimator and its Estimated Variance for the 6-MP Group

Time t_i	Number of events d_i	Number at risk Y_i	Product-Limit Estimator $\hat{S}(t) = \prod_{t_i \le t}[1 - \frac{d_i}{Y_i}]$	$\sum_{t_i \le t} \frac{d_i}{Y_i(Y_i - d_i)}$	$\hat{S}(t)^2 \sum_{t_i \le t} \frac{d_i}{Y_i(Y_i - d_i)}.$
6	3	21	$[1 - \frac{3}{21}] = 0.857$	$\frac{3}{21 \times 18} = 0.0079$	$0.857^2 \times 0.0079 = 0.0058$
7	1	17	$[0.857](1 - \frac{1}{17}) = 0.807$	$0.0079 + \frac{1}{17 \times 16} = 0.0116$	$0.807^2 \times 0.0116 = 0.0076$
10	1	15	$[0.807](1 - \frac{1}{15}) = 0.753$	$0.0116 + \frac{1}{15 \times 14} = 0.0164$	$0.753^2 \times 0.0164 = 0.0093$
13	1	12	$[0.753](1 - \frac{1}{12}) = 0.690$	$0.0164 + \frac{1}{12 \times 11} = 0.0240$	$0.690^2 \times 0.0240 = 0.0114$
16	1	11	$[0.690](1 - \frac{1}{11}) = 0.628$	$0.0240 + \frac{1}{11 \times 10} = 0.0330$	$0.628^2 \times 0.0330 = 0.0130$
22	1	7	$[0.628](1 - \frac{1}{7}) = 0.538$	$0.0330 + \frac{1}{7 \times 6} = 0.0569$	$0.538^2 \times 0.0569 = 0.0164$
23	1	6	$[0.538](1 - \frac{1}{6}) = 0.448$	$0.0569 + \frac{1}{6 \times 5} = 0.0902$	$0.448^2 \times 0.0902 = 0.0181$

TABLE 4.1B

The Product-Limit Estimator and Its Estimated Standard Error for the 6-MP Group

Time on Study (t)	$\hat{S}(t)$	Standard Error
$0 \le t < 6$	1.000	0.000
$6 \le t < 7$	0.857	0.076
$7 \le t < 10$	0.807	0.087
$10 \le t < 13$	0.753	0.096
$13 \le t < 16$	0.690	0.107
$16 \le t < 22$	0.628	0.114
$22 \le t < 23$	0.538	0.128
$23 \le t < 35$	0.448	0.135

The Product-Limit estimator provides an efficient means of estimating the survival function for right-censored data. It can also be used to estimate the cumulative hazard function $H(t) = -\ln[S(t)]$. The estimator is $\hat{H}(t) = -\ln[\hat{S}(t)]$. An alternate estimator of the cumulative hazard rate, which has better small-sample-size performance than the estimator based on the Product-Limit estimator, was first suggested by Nelson (1972) in a reliability context. The estimator was rediscovered by Aalen (1978b) who derived the estimator using modern counting process techniques (see section 3.6 for a sketch of this derivation). This estimator, which shall be referred to as the Nelson–Aalen estimator of the cumulative hazard, is defined up to the largest observed time on study as follows:

$$\tilde{H}(t) = \begin{cases} 0, & \text{if } t \leq t_1, \\ \sum_{t_i \leq t} \frac{d_i}{Y_i}, & \text{if } t_1 \leq t. \end{cases} \qquad (4.2.3)$$

The estimated variance of the Nelson–Aalen estimator is due to Aalen (1978b) and is given by

$$\sigma_H^2(t) = \sum_{t_i \leq t} \frac{d_i}{Y_i^2}. \qquad (4.2.4)$$

Based on the Nelson–Aalen estimator of the cumulative hazard rate (4.2.3), an alternate estimator of the survival function is given by $\tilde{S}(t) = \exp[-\tilde{H}(t)]$.

The Nelson–Aalen estimator has two primary uses in analyzing data. The first is in selecting between parametric models for the time to

TABLE 4.2

Construction of the Nelson–Aalen Estimator and its Estimated Variance for the 6-MP Group

Time t	$\tilde{H}(t) = \sum_{t_i \leq t} \frac{d_i}{Y_i}$	$\sigma_H^2 = \sum_{t_i \leq t} \frac{d_i}{Y_i^2}$	Standard Error
$0 \leq t < 6$	0	0	0
$6 \leq t < 7$	$\frac{3}{21} = 0.1428$	$\frac{3}{21^2} = 0.0068$	0.0825
$7 \leq t < 10$	$0.1428 + \frac{1}{17} = 0.2017$	$0.0068 + \frac{1}{17^2} = 0.0103$	0.1015
$10 \leq t < 13$	$0.2017 + \frac{1}{15} = 0.2683$	$0.0103 + \frac{1}{15^2} = 0.0147$	0.1212
$13 \leq t < 16$	$0.2683 + \frac{1}{12} = 0.3517$	$0.0147 + \frac{1}{12^2} = 0.0217$	0.1473
$16 \leq t < 22$	$0.3517 + \frac{1}{11} = 0.4426$	$0.0217 + \frac{1}{11^2} = 0.0299$	0.1729
$22 \leq t < 23$	$0.4426 + \frac{1}{7} = 0.5854$	$0.0299 + \frac{1}{7^2} = 0.0503$	0.2243
$23 \leq t < 35$	$0.5854 + \frac{1}{6} = 0.7521$	$0.0503 + \frac{1}{6^2} = 0.0781$	0.2795

event. Here, one plots the Nelson–Aalen estimator on special paper so that, if a given parametric model fits the data, the resulting graph will be approximately linear. For example, a plot of $\tilde{H}(t)$ versus t will be approximately linear if the exponential distribution, with hazard rate λ, fits the data. The use of the Nelson–Aalen estimators in model identification is discussed further in Chapter 12.

A second use of the Nelson–Aalen estimator is in providing crude estimates of the hazard rate $h(t)$. These estimates are the slope of the Nelson–Aalen estimator. Better estimates of the hazard rate are obtained by smoothing the jump sizes of the Nelson–Aalen estimator with a parametric kernel (see Chapter 6).

EXAMPLE 4.1   *(continued)* The construction of the Nelson–Aalen estimator of the cumulative hazard and its estimated variance for the 6-MP group is given in Table 4.2.

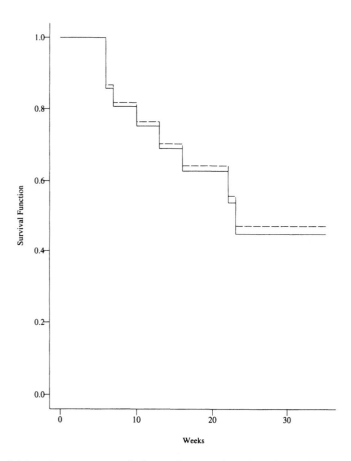

Figure 4.1A *Comparison of the Nelson–Aalen (------) and Product-Limit (————) estimates of the survival function for the 6-MP group.*

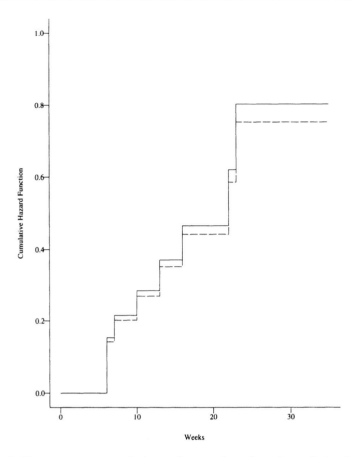

Figure 4.1B *Comparison of the Nelson–Aalen (------) and Product-Limit (————) estimates of the cumulative hazard rate for the 6-MP group.*

Figure 4.1A shows the two estimates of the survival function for the 6-MP data and Figure 4.1B the two estimates of the cumulative hazard rate. Note that all estimates are step functions with jumps at the observed deaths.

EXAMPLE 4.2   To illustrate the use of the Product-Limit estimator and the Nelson–Aalen estimator in providing summary information about survival, consider the data on the efficiency of a bone marrow transplant in acute leukemia. Using the data reported in section 1.3, we shall focus on the disease-free survival probabilities for ALL, AML low risk and AML high risk patients. An individual is said to be disease-free at a given time after transplant if that individual is alive without the recurrence of leukemia. The event indicator for disease-free survival is $\delta_3 = 1$ if the individual has died or has relapsed ($\delta_3 = \max(\delta_1, \delta_2)$ in Table D.1 of Appendix D).

The days on study for a patient is the smaller of their relapse or death time.

Table 4.3 shows the calculations needed for constructing the estimated survival function and hazard rate for the ALL group. Similar calculations for the two AML groups are left as an exercise.

Figure 4.2 shows a plot of the estimated disease-free survival curves (4.2.1) for the three groups. In this figure, first notice that the curves end at different points, because the largest times on study are different for the three groups (2081 days for ALL, 2569 for AML low risk, and 2640 for AML high risk). Secondly, the figure suggests that AML low risk patients have the best and AML high risk patients the least favorable prognosis. The three year disease-free survival probabilities are 0.3531 ($SE = 0.0793$) for the ALL group; 0.5470 ($SE = 0.0691$) for the AML

TABLE 4.3

Estimators of the Survival Function and Cumulative Hazard Rate for ALL Patients

			Product-Limit Estimator		Nelson–Aalen Estimator	
t_i	d_i	Y_i	$\hat{S}(t_i)$	$\sqrt{\hat{V}[\hat{S}(t_i)]}$	$\tilde{H}(t_i)$	$\sigma_H(t_i)$
1	1	38	0.9737	0.0260	0.0263	0.0263
55	1	37	0.9474	0.0362	0.0533	0.0377
74	1	36	0.9211	0.0437	0.0811	0.0468
86	1	35	0.8947	0.0498	0.1097	0.0549
104	1	34	0.8684	0.0548	0.1391	0.0623
107	1	33	0.8421	0.0592	0.1694	0.0692
109	1	32	0.8158	0.0629	0.2007	0.0760
110	1	31	0.7895	0.0661	0.2329	0.0825
122	2	30	0.7368	0.0714	0.2996	0.0950
129	1	28	0.7105	0.0736	0.3353	0.1015
172	1	27	0.6842	0.0754	0.3723	0.1081
192	1	26	0.6579	0.0770	0.4108	0.1147
194	1	25	0.6316	0.0783	0.4508	0.1215
230	1	23	0.6041	0.0795	0.4943	0.1290
276	1	22	0.5767	0.0805	0.5397	0.1368
332	1	21	0.5492	0.0812	0.5873	0.1449
383	1	20	0.5217	0.0817	0.6373	0.1532
418	1	19	0.4943	0.0819	0.6900	0.1620
466	1	18	0.4668	0.0818	0.7455	0.1713
487	1	17	0.4394	0.0815	0.8044	0.1811
526	1	16	0.4119	0.0809	0.8669	0.1916
609	1	14	0.3825	0.0803	0.9383	0.2045
662	1	13	0.3531	0.0793	1.0152	0.2185
2081	0	1	0.3531	0.0793	1.0152	0.2185

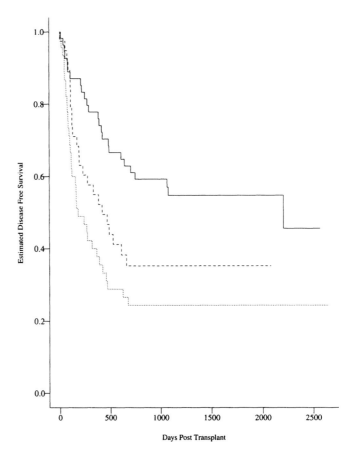

Figure 4.2   *Estimated disease free survival for the 137 bone marrow transplant patients. AML-Low risk (————); AML-High risk (------); ALL (— — —)*

low risk group; and 0.2444 (*SE* = 0.0641) for the AML high risk group. Whether these apparent differences are statistically significant will be addressed in later sections.

Figure 4.3 is a plot of the estimated cumulative hazard rates (4.2.3) for the three disease groups. Again, this plot shows that AML high risk patients have the highest combined relapse and death rate, whereas AML low risk patients have the smallest rate. For each disease group, the cumulative hazard rates appear to be approximately linear in the first two years, suggesting that the hazard rate is approximately constant. A crude estimate of these constant hazard rates is the slopes of the Nelson–Aalen estimators. These estimates give a rate of about 0.04 events per month for ALL patients, 0.02 events per month for AML low risk patients, and 0.06 events per month for AML high risk patients.

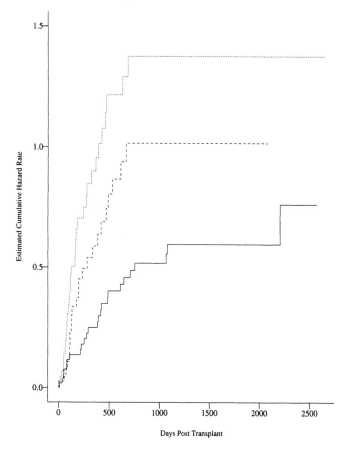

Figure 4.3 *Estimated cumulative hazard rate for the 137 bone marrow transplant patients. AML-Low risk (————); AML-High risk (------); ALL (— — —)*

Practical Notes

1. Both the Nelson–Aalen and Product-Limit estimator are based on an assumption of noninformative censoring which means that knowledge of a censoring time for an individual provides no further information about this person's likelihood of survival at a future time had the individual continued on the study. This assumption would be violated, for example, if patients with poor prognosis were routinely censored. When this assumption is violated, both estimators are estimating the wrong function and the investigator can be appreciably misled. See Klein and Moeschberger (1984) for details.

2. The Kaplan–Meier estimator of the survival function is well defined for all time points less than the largest observed study time t_{max}. If the largest study time corresponds to a death time, then, the estimated

survival curve is zero beyond this point. If the largest time point is censored, the value of $S(t)$ beyond this point is undetermined because we do not know when this last survivor would have died if the survivor had not been censored. Several nonparametric suggestions have been made to account for this ambiguity. Efron (1967) suggests estimating $\hat{S}(t)$ by 0 beyond the largest study time t_{max}. This corresponds to assuming that the survivor with the largest time on study would have died immediately after the survivor's censoring time and leads to an estimator which is negatively biased. Gill (1980) suggests estimating $\hat{S}(t)$ by $\hat{S}(t_{max})$ for $t > t_{max}$, which corresponds to assuming this individual would die at ∞ and leads to an estimator which is positively biased. Although both estimators have the same large-sample properties and converge to the true survival function for large samples, a study of small-sample properties of the two estimators by Klein(1991) shows that Gill's version of the Kaplan–Meier is preferred.

3. The two nonparametric techniques for estimation beyond t_{max} correspond to the two most extreme situations one may encounter. Brown, Hollander, and Kowar (1974) suggest completing the tail by an exponential curve picked to give the same value of $S(t_{max})$. The estimated survival function for $t > t_{max}$ is given by $\hat{S}(t) = \exp\{t \ln[\hat{S}(t_{max})]/t_{max}\}$. For the data in Example 4.2, this method yields estimates of $\hat{S}(t) = \exp(-0.0005t)$ for $t > 2081$ days for the ALL Group; $\hat{S}(t) = \exp(-0.00035t)$ for $t > 2569$ days for the AML low risk group; and $\hat{S}(t) = \exp(-0.000053t)$ for $t > 2640$ for the AML high risk group. Based on these estimates, Figure 4.4 shows a comparison of the disease-free survival of three-risk groups for the first eight years after transplant. Moeschberger and Klein (1985) have extended these techniques to allow using the more flexible Weibull distribution to complete the tail of the Product-Limit estimator.

4. An alternative estimator of the variance of $\hat{S}(t)$, due to Aalen and Johansen (1978) is given by

$$\tilde{V}[\hat{S}(t)] = \hat{S}(t)^2 \sum_{t_i \leq t} \frac{d_i}{Y_i^2}. \qquad (4.2.5)$$

Both this estimator and Greenwood's estimator tend to underestimate the true variance of the Kaplan–Meier estimator for small to moderate samples. On average, Greenwood's estimator tends to come closest to the true variance and has a smaller variance except when Y_i is very small (see Klein, 1991).

5. An alternate estimator of the variance of $\tilde{H}(t)$ is found in Klein (1991). This estimator is given as

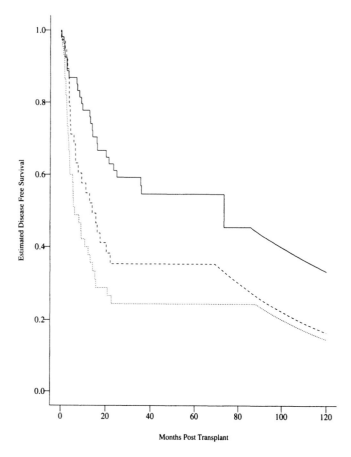

Figure 4.4 *Estimated disease free survival for the 137 bone marrow transplant patients using the Brown-Hollander-Kowar tail estimate. AML-low risk (————); AML-high risk (------); ALL (— — —)*

$$\hat{V}[\tilde{H}(t)] = \sum_{t_i \le t} \frac{(Y_i - d_i)d_i}{Y_i^3} . \tag{4.2.6}$$

This estimator tends to be too small, whereas the estimator (4.2.4) tends to be too large. The estimator (4.2.4) has a smaller bias than (4.2.6) and is preferred.

6. An estimator of the variability of the Nelson–Aalen estimator of the survival function $\hat{S}(t)$ is found by substituting $\tilde{S}(t)$ for $\hat{S}(t)$ in either Eq. 4.2.2 or 4.2.5.

7. When there is no censoring, the Product-Limit estimator reduces to the empirical survival function.

8. The statistical packages SAS, STATA, SPSS, and S-Plus provide procedures for computing the Product-Limit estimator and the estimated

cumulative hazard rate based on this statistic. S-Plus also provides the Nelson–Aalen estimator of the survival function and an estimate of its variance using (4.2.5).

Theoretical Notes

1. The Product-Limit estimator was first constructed by using a *reduced-sample* approach. In this approach, note that, because events are only observed at the times t_i, $S(t)$ should be a step function with jumps only at these times, there being no information on events occurring at other times. We will estimate $S(t)$ by a discrete distribution with mass at the time points $t_1, t_2, \ldots, t_D$. We can estimate the $Pr[T > t_i \mid T \geq t_i]$ as the fraction of individuals who are at risk at time t_i but who do not die at this time, that is

$$\hat{Pr}[T > t_i \mid T \geq t_i] = \frac{Y_i - d_i}{Y_i}, \text{for } i = 1, 2, \ldots, D.$$

To estimate $S(t_i)$, recall that

$$S(t_i) = \frac{S(t_i)}{S(t_{i-1})} \frac{S(t_{i-1})}{S(t_{i-2})} \cdots \frac{S(t_2)}{S(t_1)} \frac{S(t_1)}{S(0)} S(0)$$

$$= P[T > t_i \mid T \geq t_i]P[T > t_{i-1} \mid T \geq t_{i-1}] \cdots$$

$$P[T > t_2 \mid T \geq t_1]P[T > t_1 \mid T \geq t_1],$$

because $S(0) = 1$ and, for a discrete distribution, $S(t_{i-1}) = Pr[T > t_{i-1}] = Pr[T \geq t_i]$. Simplifying this expression yields the Product-Limit estimator (4.2.1).

2. *Redistribute to the right algorithm.* This derivation is best explained by an example. Suppose we have ten individuals on study with the following times on study (+ denotes a censored observation): 3, 4, 5+ ,6, 6+, 8+, 11, 14, 15, 16+. We start the algorithm by assigning mass $1/n$ to each observation (the estimator we would have if there was no censoring). Now, start at the far left and take the mass at each censored observation and redistribute it equally to each observation greater than this value. (Here censored observations tied with events are treated as being just to the right of the event.) This process is repeated until the largest observation is reached. The survival curve obtained from this final set of probabilities is the Kaplan–Meier estimate. If the largest observation is censored, the mass can be either left at that point, so that the Kaplan-Meier estimator drops to zero, or redistributed to $+\infty$, so that the curve is constant beyond this value.

3. *Self Consistency.* If we had no censored observations, the estimator of the survival function at a time t is the proportion of observations

Data Points	Step 0	Step 1	Step 2	Step 3	$S(t)$
3	$\frac{1}{10}$	0.100	0.100	0.100	0.900
4	$\frac{1}{10}$	0.100	0.100	0.100	0.800
5+	$\frac{1}{10}$	0.000	0.000	0.000	0.800
6	$\frac{1}{10}$	$\frac{1}{10} + \left(\frac{1}{7}\right)\frac{1}{10} = 0.114$	0.114	0.114	0.686
6+	$\frac{1}{10}$	$\frac{1}{10} + \left(\frac{1}{7}\right)\frac{1}{10} = 0.114$	0.000	0.000	0.686
8+	$\frac{1}{10}$	$\frac{1}{10} + \left(\frac{1}{7}\right)\frac{1}{10} = 0.114$	$0.114 + \frac{1}{5}0.114 = 0.137$	0.000	0.686
11	$\frac{1}{10}$	$\frac{1}{10} + \left(\frac{1}{7}\right)\frac{1}{10} = 0.114$	$0.114 + \frac{1}{5}0.114 = 0.137$	$0.137 + \frac{1}{4}0.137 = 0.171$	0.515
14	$\frac{1}{10}$	$\frac{1}{10} + \left(\frac{1}{7}\right)\frac{1}{10} = 0.114$	$0.114 + \frac{1}{5}0.114 = 0.137$	$0.137 + \frac{1}{4}0.137 = 0.171$	0.343
15	$\frac{1}{10}$	$\frac{1}{10} + \left(\frac{1}{7}\right)\frac{1}{10} = 0.114$	$0.114 + \frac{1}{5}0.114 = 0.137$	$0.137 + \frac{1}{4}0.137 = 0.171$	0.171
16+	$\frac{1}{10}$	$\frac{1}{10} + \left(\frac{1}{7}\right)\frac{1}{10} = 0.114$	$0.114 + \frac{1}{5}0.114 = 0.137$	$0.137 + \frac{1}{4}0.137 = 0.171$	0.000*

*Efron's Estimator

which are larger than t, that is,

$$\hat{S}(t) = \frac{1}{n}\sum_{i=1}^{n}\phi(X_i)$$

where

$$\phi(y) = 1 \text{ if } y > t; 0, \text{ if } y \le t.$$

For right-censored data, we want to construct our estimator in a similar manner by redefining the scoring function ϕ. Let $T_1, T_2, \ldots, T_n$ be the observed times on study. If T_i is a death time, we know with certainty whether T_i is smaller or greater than t. If T_i is a censored time greater than or equal to t, then, we know that the true death time must be larger than t because it is larger than T_i for this individual. For a censored observation less than t, we do not know if the corresponding death time is greater than t because it could fall between T_i and t. If we knew $S(t)$, we could estimate the probability of this censored observation being larger than t by $S(t)/S(T_i)$. Using these revised scores, we will call an estimator $\hat{S}(t)$ a self-consistent estimator of S if

$$\hat{S}(t) = \frac{1}{n}\left[\sum_{T_i > t} \phi(T_i) + \sum_{\delta_i = 0, T_i \leq t} \frac{\hat{S}(t)}{\hat{S}(T_i)}\right]. \qquad (4.2.7)$$

To find $\hat{S}(t)$ from this formula one starts with any estimate of S and substitutes this in the right hand side of (4.2.7) to get an updated estimate of S. This new estimate of $\hat{S}(t)$ is, then, used in the next step to obtain a revised estimate. This procedure continues until convergence. Efron (1967) shows that the final estimate of S is exactly the Product-Limit estimator for t less than the largest observed time.

4. Both the Product-Limit estimator and the Nelson–Aalen estimator can be derived using the theory of counting processes. Details of this construction can be found in Andersen et al. (1993) or Fleming and Harrington (1991). An outline of this approach is found in section 3.6.

5. Under certain regularity conditions, one can show that the Nelson–Aalen estimator and the Product-Limit estimator are nonparametric maximum likelihood estimators.

6. Both the Product-Limit and Nelson–Aalen estimators of either the survival function or the cumulative hazard rate are consistent. The statistics are asymptotically equivalent.

7. The Nelson–Aalen estimator of the cumulative hazard rate is the first term in a Taylor series expansion of minus the logarithm of the Product-Limit estimator.

8. Small-sample-size properties of the Product-Limit estimator have been studied by Guerts (1987) and Wellner (1985). Small sample size properties of the variance estimators for the Product-Limit estimator and the Nelson–Aalen estimator are found in Klein (1991).

9. Under suitable regularity conditions, both the Nelson–Aalen and Product-Limit estimators converge weakly to Gaussian processes. This fact means that for fixed t, the estimators have an approximate normal distribution.

4.3 Pointwise Confidence Intervals for the Survival Function

The Product-Limit estimator provides a summary estimate of the mortality experience of a given population. The corresponding standard error provides some limited information about the precision of the estimate. In this section, we shall use these estimators to provide confidence intervals for the survival function at a fixed time t_o. The intervals are constructed to assure, with a given confidence level $1 - \alpha$ that the true value of the survival function, at a predetermined time t_o, falls in the interval we shall construct.

Before introducing the confidence intervals, we need some additional notation. Let $\sigma_S^2(t) = \hat{V}[\hat{S}(t)]/\hat{S}^2(t)$. Note that $\sigma_S^2(t)$ is the sum in Greenwood's formula (4.2.2).

The most commonly used $100 \times (1 - \alpha)\%$ confidence interval for the survival function at time t_o, termed the linear confidence interval, is defined by

$$\hat{S}(t_o) - Z_{1-\alpha/2}\sigma_S(t_o)\hat{S}(t_o), \hat{S}(t_o) + Z_{1-\alpha/2}\sigma_S(t_o)\hat{S}(t_o) \qquad (4.3.1)$$

Here $Z_{1-\alpha/2}$ is the $1 - \alpha/2$ percentile of a standard normal distribution. This is the confidence interval routinely constructed by most statistical packages.

Better confidence intervals can be constructed by first transforming $\hat{S}(t_o)$. These improved estimators were proposed by Borgan and Liestøl (1990). The first transformation suggested is a log transformation (see Theoretical Note 4) of the cumulative hazard rate. The $100 \times (1 - \alpha)\%$ log-transformed confidence interval for the survival function at t_o is given by

$$[\hat{S}(t_o)^{1/\theta}, \hat{S}(t_o)^{\theta}], \text{ where } \theta = \exp\left\{\frac{Z_{1-\alpha/2}\sigma_S(t_o)}{\ln[\hat{S}(t_o)]}\right\}. \qquad (4.3.2)$$

Note that this interval is not symmetric about the estimate of the survival function.

The second transformation is an arcsine-square root transformation of the survival function which yields the following $100 \times (1 - \alpha)\%$ confidence interval for the survival function:

$$\sin^2\left\{\max\left[0, \arcsin(\hat{S}(t_o)^{1/2}) - 0.5Z_{1-\alpha/2}\sigma_S(t_o)\left(\frac{\hat{S}(t_o)}{1 - \hat{S}(t_o)}\right)^{1/2}\right]\right\}$$

$$\leq S(t_o) \leq \qquad (4.3.3)$$

$$\sin^2\left\{\min\left[\frac{\pi}{2}, \arcsin(\hat{S}(t_o)^{1/2}) + 0.5Z_{1-\alpha/2}\sigma_S(t_o)\left(\frac{\hat{S}(t_o)}{1 - \hat{S}(t_o)}\right)^{1/2}\right]\right\}.$$

EXAMPLE 4.2   *(continued)* To illustrate these confidence intervals, we shall use the estimated disease-free survival function and cumulative hazard rate for ALL patients in Table 4.3. Note that at 1 year (365 days) the estimated survival function S(365) was found to be 0.5492 with an estimated variance of 0.0812^2. Thus, $\sigma_S^2(365) = (0.0812/0.5492)^2 = 0.1479^2$. A 95% linear confidence interval for the survival function at year one is $0.5492 \pm 1.96 \times 0.1479 \times 0.5492 = (0.3900, 0.7084)$.

To find the 95% log-transformed confidence interval for the one year survival function, we find that $\theta = \exp[\frac{1.96 \times 0.1479}{\ln(0.5492)}] = 0.6165$, so that the interval is $(0.54921^{1/0.6165}, 0.5492^{0.6165}) = (0.3783, 0.6911)$.

The 95% arcsine-square root transformation confidence interval for the one year survival function is

$$\sin^2 \left\{ \max \left[0, \arcsin(0.5492^{1/2}) - 0.5 \times 1.96 \times 0.1479 \times \left(\frac{0.5492}{1 - 0.5492} \right)^{1/2} \right] \right\}$$

to

$$\sin^2 \left\{ \min \left[\frac{\pi}{2}, \arcsin(0.5492^{1/2}) + 0.5 \times 1.96 \times 0.1479 \times \left(\frac{0.5492}{1 - 0.5492} \right)^{1/2} \right] \right\}$$

$$= (0.3903, 0.7032).$$

Table 4.4 shows the three possible 95% confidence intervals that can be constructed for the disease-free survival function for each of the three risk groups presented in Figures 4.2. We can see that AML high risk patients have a smaller chance of surviving beyond one year than the AML low risk patients.

TABLE 4.4

95% Confidence Intervals for Disease-Free Survival One Year After Transplant

	ALL	AML low risk	AML high risk
$\hat{S}(365)$	0.5492	0.7778	0.3778
$\hat{V}[\hat{S}(365)]$	0.0812^2	0.0566^2	0.0723^2
$\sigma_s(365)$	0.1479	0.0728	0.1914
Linear confidence interval for $S(365)$	0.3900, 0.7084	0.6669, 0.8887	0.2361, 0.5195
Log-transformed confidence interval for $S(365)$	0.3783, 0.6911	0.6419, 0.8672	0.2391, 0.5158
Arcsine square-root confidence interval for $S(365)$	0.3903, 0.7032	0.6583, 0.8776	0.2433, 0.5227

Practical Notes

1. Bie et al. (1987) have presented $100(1 - \alpha)$% pointwise confidence intervals for the cumulative hazard function. Similar to the confi-

dence intervals constructed for the survival function, there are three possible intervals, which correspond to the three transformations of the cumulative hazard function. The intervals are

Linear:

$$\tilde{H}(t_o) - Z_{1-\alpha/2}\sigma_H(t_o), \ \tilde{H}(t_o) + Z_{1-\alpha/2}\sigma_H(t_o). \quad (4.3.4)$$

Log-Transformed

$$[\tilde{H}(t_o)/\phi, \ \phi\tilde{H}(t_o)] \text{ where } \phi = \exp[\frac{Z_{1-\alpha/2}\sigma_H(t_o)}{\tilde{H}(t_o)}]. \quad (4.3.5)$$

Arcsine-Square Root Transformed

$$-2\ln\left\{\sin\left[\min\left(\frac{\pi}{2}, \arcsin[\exp\{-\tilde{H}(t_o)/2\}]\right.\right.\right.$$
$$\left.\left.\left. +0.5Z_{1-\alpha/2}\sigma_H(t_o)\{\exp\{\tilde{H}(t_o)\} - 1\}^{-1/2}\right)\right]\right\}$$
$$\leq H(t_o) \leq \quad\quad\quad (4.3.6)$$
$$-2\ln\{\sin[\max(0, \arcsin[\exp\{-\tilde{H}(t_o)/2\}]$$
$$-0.5Z_{1-\alpha/2}\sigma_H(t_o)\{\exp\tilde{H}(t_o)\} - 1\}^{-1/2})]\}$$

Using the data in Example 4.2, we have the following 95% confidence intervals for the cumulative hazard rate at one year after transplant:

	ALL	AML low risk	AML high risk
Linear confidence interval for H(365)	0.3034, 0.8713	0.1076, 0.3898	0.5875, 1.3221
Log-transformed confidence interval for H(365)	0.3622, 0.9524	0.1410, 0.4385	0.6499, 1.4028
Arcsin square root confidence interval for H(365)	0.3451, 0.9217	0.1293, 0.4136	0.6366, 1.3850

2. Borgan and Liestøl (1990) have shown that both the log-transformed and arcsine-square root transformed confidence intervals for S perform better than the usual linear confidence interval. Both give about the correct coverage probability for a 95% interval for samples as small as 25 with as much as 50% censoring except in the extreme right-hand tail where there will be little data. The sample size needed for the standard linear confidence interval to have the correct coverage probability is much larger. For very small samples, the arcsine-square root interval tends to be a bit conservative in that the actual

coverage probability is a bit greater than $(1 - \alpha)$, whereas, for the log-transformed interval, the coverage probability is a bit smaller than $(1 - \alpha)$. The coverage probability for the linear interval in these cases is much smaller than $(1 - \alpha)$. Similar observations were made by Bie et al. (1987) for the corresponding interval estimates of the cumulative hazard rate. For very large samples, the three methods are equivalent.

3. Alternative confidence intervals for the cumulative hazard rate can be found by taking (minus) the natural logarithm of the confidence intervals constructed for the survival function. Similarly the exponential of (minus) the confidence limits for the cumulative hazard yields a confidence interval for the survival function.

4. Both the log-transformed and arcsine-square root transformed confidence intervals, unlike the linear interval, are not symmetric about the point estimator of the survival function or cumulative hazard rate. This is appropriate for small samples where the point estimators are biased and the distribution of the estimators is skewed.

5. The confidence intervals constructed in this section are valid only at a single point t_o. A common incorrect use of these intervals is to plot them for all values of t and interpret the curves obtained as a confidence band, that is, these curves are interpreted as having, for example, 95% confidence that the *entire* survival function lies within the band. The bands obtained this way are too narrow to make this inference. The proper bands are discussed in the following section.

6. Confidence intervals for the survival function are available in the S-Plus routine surv.fit. The intervals can be constructed using either the linear or the log-transformed method.

Theoretical Notes

1. Construction of the linear confidence intervals follows directly from the asymptotic normality of the Product-Limit or Nelson–Aalen estimators.
2. The log-transformed interval was first suggested by Kalbfleisch and Prentice (1980).
3. The arcsine-square root transformed interval was first suggested by Nair (1984).
4. The "log"-transformed confidence interval is based on first finding a confidence interval for the log of the cumulative hazard function. This is sometimes called a log-log transformed interval since the cumulative hazard function is the negative log of the survival function.

4.4   Confidence Bands for the Survival Function

In section 4.3, pointwise confidence intervals for the survival function were presented. These intervals are valid for a single fixed time at which the inference is to be made. In some applications it is of interest to find upper and lower confidence bands which guarantee, with a given confidence level, that the survival function falls within the band for all t in some interval, that is, we wish to find two random functions $L(t)$ and $U(t)$, so that $1 - \alpha = Pr[L(t) \leq S(t) \leq U(t)$, for all $t_L \leq t \leq t_U]$. We call such a $[L(t), U(t)]$ a $(1 - \alpha) \times 100\%$ confidence band for $S(t)$.

We shall present two approaches to constructing confidence bands for $S(t)$. The first approach, originally suggested by Nair (1984), provides confidence bounds which are proportional to the pointwise confidence intervals discussed in section 4.3. These bands are called the equal probability or *EP* bands. To implement these bands we pick $t_L < t_U$ so that t_L is greater than or equal to the smallest observed event time and t_U is less than or equal to the largest observed event time. To construct confidence bands for $S(t)$, based on a sample of size n, define

$$a_L = \frac{n\sigma_S^2(t_L)}{1 + n\sigma_S^2(t_L)} \qquad (4.4.1)$$

and

$$a_U = \frac{n\sigma_S^2(t_U)}{1 + n\sigma_S^2(t_U)}.$$

The construction of the EP confidence bands requires that $0 < a_L < a_U < 1$.

To construct a $100(1 - \alpha)\%$ confidence band for $S(t)$ over the range $[t_L, t_U]$, we, first, find a confidence coefficient, $c_\alpha(a_L, a_U)$ from Table C.3 in Appendix C. As in the case of $100(1 - \alpha)\%$ pointwise confidence intervals at a fixed time, there are three possible forms for the confidence bands. The three bands are the linear bands, the log-transformed bands, and the arcsine-square root transformed bands expressed as follows:

Linear:

$$\hat{S}(t) - c_\alpha(a_L, a_U)\sigma_S(t)\hat{S}(t), \ \hat{S}(t) + c_\alpha(a_L, a_U)\sigma_S(t)\hat{S}(t). \qquad (4.4.2)$$

Log-Transformed:

$$(\hat{S}(t)^{1/\theta}, \hat{S}(t)^\theta),$$

$$\text{where } \theta = \exp\left[\frac{c_\alpha(a_L, a_U)\sigma_S(t)}{\ln[\hat{S}(t)]}\right]. \qquad (4.4.3)$$

Arcsine-Square Root Transformed:

$$\sin^2\{\max[0, \arcsin\{\hat{S}(t)^{1/2}\} - 0.5c_\alpha(a_L, a_U)\sigma_S(t)[\hat{S}(t)/(1 - \hat{S}(t))]^{1/2}]\}$$

$$\leq S(t) \leq \qquad\qquad\qquad (4.4.4)$$

$$\sin^2\left\{\min\left[\frac{\pi}{2}, \arcsin\{\hat{S}(t)^{1/2}\} + 0.5c_\alpha(a_L, a_U)\sigma_S(t)[\hat{S}(t)/(1 - \hat{S}(t))]^{1/2}\right]\right\}.$$

EXAMPLE 4.2   *(continued)* To illustrate these confidence intervals, we shall use the estimated disease-free survival function for ALL patients in Table 4.3. We construct confidence bands for $S(t)$ over the range $100 \leq t \leq 600$ days. Here, we have $\sigma_S^2(100) = \sigma_S^2(86) = 0.0498^2/0.8947^2 = 0.0031$ and $\sigma_S^2(600) = \sigma_S^2(526) = 0.0809^2/0.4119^2 = 0.0386$. From 4.4.1 we find $a_L = 38(0.0031)/[1 + 38(0.0031)] = 0.1$ and $a_U = 38(0.0386)/[1 + 38(0.0386)] = 0.6$. For a 95% confidence band, we find, from Table C.3 in Appendix C, that $c_{05}(0.1, 0.6) = 2.8826$.

Table 4.5 shows the three 95% confidence bands for the survival function based on the EP method. Note that the calculation of the entries in this table is identical to the calculations performed in section 4.3 for the 95% pointwise confidence intervals at day 365 with the exception that the Z coefficient, 1.96 is replaced by the appropriate value from Table C.3 of Appendix C.

An alternate set of confidence bounds has been suggested by Hall and Wellner (1980). These bands are not proportional to the pointwise confidence bounds. For these bounds, a lower limit, t_L, of zero is allowed. To construct a $100 \times (1 - \alpha)\%$ confidence band for $S(t)$ over the region $[t_L, t_U]$, we find the appropriate confidence coefficient $k_\alpha(a_L, a_U)$, from Table C.4 of Appendix C. Again, there are three possible forms for the confidence bands. These are the linear bands, the log-transformed bands and the arcsine-transformed bands. These $100 \times (1 - \alpha)\%$ confidence bands are expressed as

Linear:

$$\hat{S}(t) - \frac{k_\alpha(a_L, a_U)[1 + n\sigma_S^2(t)]}{n^{1/2}}\hat{S}(t), \quad \hat{S}(t) + \frac{k_\alpha(a_L, a_U)[1 + n\sigma_S^2(t)]}{n^{1/2}}\hat{S}(t).$$

$$(4.4.5)$$

Log-Transformed:

$$[\hat{S}(t)^{1/\theta}, \hat{S}(t)^\theta], \qquad\qquad\qquad (4.4.6)$$

$$\text{where } \theta = \exp\left\{\frac{k_\alpha(a_L, a_U)[1 + n\sigma_S^2(t)]}{n^{1/2}\ln[\hat{S}(t)]}\right\}.$$

TABLE 4.5

95% EP Confidence Bands for the Disease Free Survival Function

t_i	$\hat{S}(t_i)$	$\sqrt{\hat{V}[\hat{S}(t_i)]}$	σ_S^2	Linear		Log-Transformed		Arcsine-Transformed	
100	0.8947	0.0498	0.0031	0.7511	1.0000	0.6246	0.9740	0.7139	0.9907
104	0.8684	0.0548	0.0040	0.7104	1.0000	0.5992	0.9619	0.6766	0.9812
107	0.8421	0.0592	0.0049	0.6715	1.0000	0.5719	0.9485	0.6408	0.9698
109	0.8158	0.0629	0.0059	0.6345	0.9971	0.5452	0.9339	0.6071	0.9567
110	0.7895	0.0661	0.0070	0.5990	0.9800	0.5188	0.9184	0.5748	0.9421
122	0.7368	0.0714	0.0094	0.5310	0.9426	0.4666	0.8848	0.5130	0.9098
129	0.7105	0.0736	0.0107	0.4983	0.9227	0.4410	0.8670	0.4834	0.8924
172	0.6842	0.0754	0.0121	0.4669	0.9015	0.4162	0.8485	0.4549	0.8739
192	0.6579	0.0770	0.0137	0.4359	0.8799	0.3917	0.8294	0.4270	0.8549
194	0.6316	0.0783	0.0154	0.4059	0.8573	0.3678	0.8097	0.3999	0.8350
230	0.6041	0.0795	0.0173	0.3749	0.8333	0.3431	0.7886	0.3720	0.8137
276	0.5767	0.0805	0.0195	0.3447	0.8087	0.3187	0.7672	0.3448	0.7920
332	0.5492	0.0812	0.0219	0.3151	0.7833	0.2951	0.7451	0.3183	0.7694
383	0.5217	0.0817	0.0245	0.2862	0.7572	0.2719	0.7224	0.2925	0.7462
418	0.4943	0.0819	0.0275	0.2582	0.7304	0.2496	0.6993	0.2675	0.7223
468	0.4668	0.0818	0.0307	0.2310	0.7026	0.2280	0.6753	0.2433	0.6976
487	0.4394	0.0815	0.0344	0.2045	0.6743	0.2069	0.6510	0.2198	0.6723
526	0.4119	0.0809	0.0386	0.1787	0.6451	0.1865	0.6259	0.1970	0.6462
600	0.4119	0.0809	0.0386	0.1787	0.6451	0.1865	0.6259	0.1970	0.6462

TABLE 4.6

95% Hall–Wellner Confidence Bands for the Disease-Free Survival Function

t_i	$\hat{S}(t_i)$	σ_S^2	Linear		Log-Transformed		Arcsine-square root Transformed	
100	0.8947	0.0031	0.6804	1.0000	0.3837	0.9872	0.6050	1.0000
104	0.8684	0.0040	0.6541	1.0000	0.4445	0.9757	0.5966	0.9971
107	0.8421	0.0049	0.6277	1.0000	0.4696	0.9617	0.5824	0.9869
109	0.8158	0.0059	0.6015	1.0000	0.4771	0.9455	0.5652	0.9723
110	0.7895	0.0070	0.5752	1.0000	0.4747	0.9278	0.5459	0.9550
122	0.7368	0.0094	0.5225	0.9511	0.4532	0.8888	0.5034	0.9152
129	0.7105	0.0107	0.4961	0.9249	0.4377	0.8682	0.4810	0.8939
172	0.6842	0.0121	0.4699	0.8985	0.4205	0.8468	0.4582	0.8718
192	0.6579	0.0137	0.4435	0.8723	0.4018	0.8251	0.4349	0.8492
194	0.6316	0.0154	0.4172	0.8460	0.3822	0.8029	0.4114	0.8262
230	0.6041	0.0173	0.3894	0.8188	0.3606	0.7796	0.3864	0.8021
276	0.5767	0.0195	0.3616	0.7918	0.3383	0.7561	0.3612	0.7779
332	0.5492	0.0219	0.3337	0.7647	0.3156	0.7324	0.3359	0.7535
383	0.5217	0.0245	0.3057	0.7377	0.2925	0.7087	0.3104	0.7290
418	0.4943	0.0275	0.2779	0.7107	0.2694	0.6849	0.2851	0.7046
468	0.4668	0.0307	0.2500	0.6836	0.2462	0.6609	0.2599	0.6799
487	0.4394	0.0344	0.2221	0.6567	0.2230	0.6372	0.2347	0.6555
526	0.4119	0.0386	0.1942	0.6296	0.2000	0.6133	0.2097	0.6310
600	0.4119	0.0386	0.1942	0.6296	0.2000	0.6133	0.2097	0.6310

Arcsine-Square Root Transformation:

$$\sin^2\{\max[0, \arcsin\{\hat{S}(t)^{1/2}\} - 0.5\frac{k_\alpha(a_L, a_U)[1 + n\sigma_S^2(t)]}{n^{1/2}}[(\hat{S}(t)/(1 - \hat{S}(t))]^{1/2}\}$$

$$\leq S(t) \leq \qquad\qquad\qquad\qquad\qquad\qquad\qquad\qquad (4.4.7)$$

$$\sin^2\left\{\min\left[\frac{\pi}{2}, \arcsin\{\hat{S}(t)^{1/2}\} + 0.5\frac{k_\alpha(a_L, a_U)\{1 + n\sigma_S^2(t)\}}{n^{1/2}}[\hat{S}(t)/(1 - \hat{S}(t))]^{1/2}\right]\right\}.$$

EXAMPLE 4.2   *(continued)*   To illustrate the Hall-Wellner confidence bands, again, we consider the disease-free survival estimates for $S(t)$ obtained from the 38 ALL patients in Table 4.3. As for the EP bands, we construct 95%

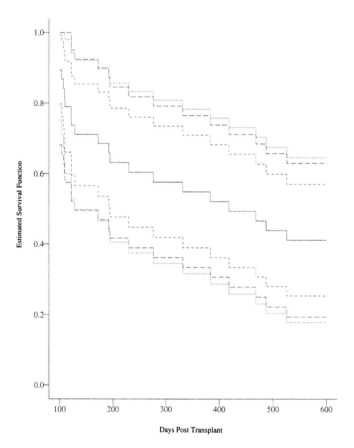

Days Post Transplant

Figure 4.5   *Comparison of 95% pointwise confidence interval, EP confidence band and Hall-Wellner confidence band for the disease free survival function based on the untransformed survival functions for ALL patients. Estimated Survival (————); Pointwise confidence interval (— — —); EP confidence band (------); Hall-Wellner band (—— ——)*

confidence bands for $S(t)$ in the range $100 \leq t \leq 600$. The required confidence coefficient, from Table C.4 of Appendix C, is $k_{05}(0.1, 0.6) = 1.3211$. Table 4.6 shows the Hall-Wellner 95% confidence bands based on the three transformations.

Figures 4.5–4.7 show the 95% confidence bands for the disease-free survival function based on either the EP or Hall–Wellner bands for the three transformations. Also included are the 95% pointwise confidence intervals obtained from the results of section 4.3. These figures show that the Hall–Wellner bands are wider for small t and shorter for large t. Both bands are wider than the curves one obtains by using pointwise confidence intervals.

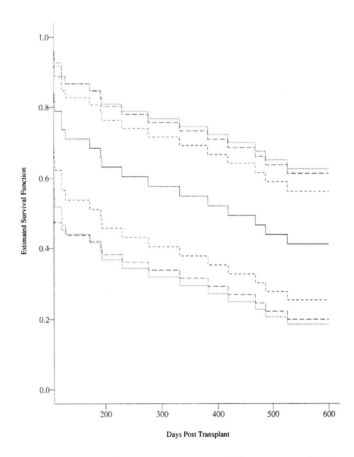

Figure 4.6 *Comparison of 95% pointwise confidence interval, EP confidence band and Hall-Wellner confidence band for the disease free survival function found using the log transformation for ALL patients. Estimated Survival (————); Pointwise confidence interval (— — —); EP confidence band (------); Hall-Wellner band (——— ———)*

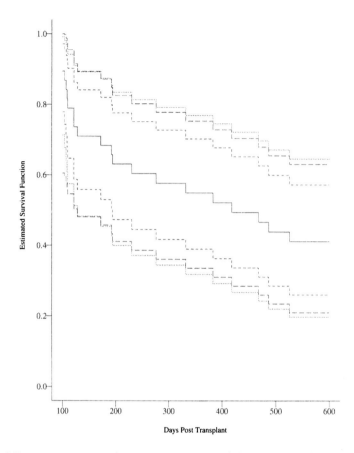

Figure 4.7 *Comparison of 95% pointwise confidence interval, EP confidence band and Hall-Wellner confidence band for the disease free survival function found using the arc sine transformation for ALL patients. Estimated Survival (————); Pointwise confidence interval (— — —); EP confidence band (------); Hall-Wellner band (—— ——)*

Figure 4.8 shows the 95% EP arcsine-square root transformed confidence bands for the three disease categories over the range of 100 to 600 days.

Practical Notes

1. Confidence bands for the cumulative hazard rate can also be constructed by either the EP or Hall–Wellner method. To construct these

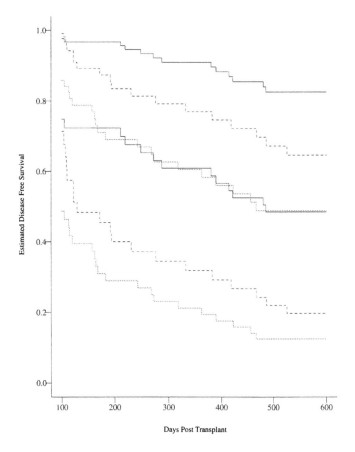

Figure 4.8   *EP confidence bands for the disease free survival function based on the arc sine transformation for bone marrow transplant patients. AML-Low risk (————); AML-High risk (------); ALL (— — —)*

bands, we first compute

$$a_{\mathrm{L}} = \frac{n\sigma_H^2(t_{\mathrm{L}})}{1 + n\sigma_H^2(t_{\mathrm{L}})} \qquad (4.4.8)$$

and

$$a_{\mathrm{U}} = \frac{n\sigma_H^2(t_{\mathrm{U}})}{1 + n\sigma_H^2(t_{\mathrm{U}})}.$$

The EP confidence bands, which are valid over the range $t_{\mathrm{L}} \leq t \leq t_{\mathrm{U}}$, with $0 < a_{\mathrm{L}} < a_{\mathrm{U}} < 1$, are found by substituting for $Z_{1-\alpha/2}$ in (4.3.4)–(4.3.6) the appropriate confidence coefficient $c_\alpha(a_{\mathrm{L}}, a_{\mathrm{U}})$ from Table C.3 of Appendix C.

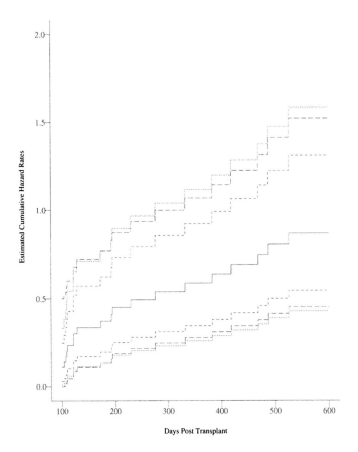

Figure 4.9 *Comparison of 95% pointwise confidence interval, EP confidence band and Hall-Wellner confidence band for the cumulative hazard function found using the arc sine transformation for ALL patients. Estimated Survival (————); Pointwise confidence interval (— — —); EP confidence band (------); Hall-Wellner band (—— ——)*

The Hall–Wellner confidence bands for the cumulative hazard rate are found by substituting $\frac{k_\alpha(a_L, a_U)[1 + n\sigma_H^2(t)]}{n^{1/2}}$ for $Z_{1-\alpha/2}\sigma_H(t)$ in (4.3.4)–(4.3.6).

Figure 4.9 shows the 95% arcsine-square root transformed EP and Hall–Wellner confidence bands and the 95% pointwise confidence interval for the cumulative hazard rate of the ALL patients over the interval 100 to 600 days.

2. For the EP bounds for the survival function, Borgan and Liestøl (1990) have shown that the linear confidence band given by formula (4.4.2) performs very poorly when the sample size is small (< 200). The coverage probability for this bound is considerably smaller than the target level. For both the log- and arcsine-square root transformed

bands, the coverage probability is approximately correct for smaller sample sizes. Both seem to give reasonable results for samples with as few as 20 observed events. The arcsine-square root transformed band seems to perform a bit better than the log-transformed interval and is recommended. Similar properties hold for the confidence bands for the cumulative hazard rate as discussed in Bie et al. (1987).

3. For the Hall–Wellner bounds, Borgan and Liestøl (1990) show that all three bands for $S(t)$ perform reasonably well for samples with as few as 20 observed events. For $H(t)$, Bie et al. (1987) show that the performance of the linear bands is poor for small samples, whereas the two transformed bands perform well for relatively small samples.

4. For the confidence bands for $H(t)$, linear EP confidence bands tend to have an upper band which is a bit too low, whereas the log-transformed lower band is too high for small t and the upper band too low for large t. For the EP arcsine-square root band, the majority of the errors occur when the upper boundary is too low. For the HW bands, the majority of the errors occur in the midrange of t.

Theoretical Notes

1. These bounds are based on weak convergence of the Product-Limit estimator or the Nelson–Aalen estimator to a mean zero Gaussian process. The EP bounds are based on the transformation $q(x) = [x(1 - x)]^{-1/2}$ of the standardized estimator, whereas for the Hall–Wellner bounds, no transformation of this process is made.

2. The critical values found in Table C.3 of Appendix C are the upper αth fractile of the random variable $U = \sup\{| W°(x)[x(1 - x)]^{-1/2} |, a_L \leq x \leq a_U\}$, where $W°$ is a standard Brownian bridge (see Nair, 1984). Miller and Siegmund (1982) show that, for large d, $Pr[U \geq d] \cong 4\phi(d)/d + \phi(d)(d - d^{-1}) \log[\frac{a_U(1-a_L)}{a_L(1-a_U)}]$, where $\phi()$ is the standard normal density function.

3. The critical values for the Hall–Wellner bands (Table C.4 of Appendix C) are the upper αth fractile of a Brownian bridge, computed from results in Chung (1986).

4.5 Point and Interval Estimates of the Mean and Median Survival Time

The Product-Limit estimator provides an estimator of the survival function $S(t)$. In section 2.4, we saw that other summary measures of an individual's survival experience, such as the mean or median time to the event, are functions of the survival function. Nonparametric estimates of these quantities can be obtained in a straightforward manner by substituting the Product-Limit estimator for the unknown survival function in the appropriate formula.

In section 2.4, it was shown that the mean time to the event μ is given by $\mu = \int_0^\infty S(t)dt$. A natural estimator of μ is obtained by substituting $\hat{S}(t)$ for $S(t)$ in this expression. This estimator is appropriate only when the largest observation corresponds to a death because in other cases, the Product-Limit estimator is not defined beyond the largest observation. Several solutions to this problem are available. First, one can use Efron's tail correction to the Product-Limit estimator (see Practical Note 2 of section 4.2) which changes the largest observed time to a death if it was a censored observation. An estimate of the mean restricted to the interval 0 to t_{max} is made. A second solution is to estimate the mean restricted to some preassigned interval $[0, \tau]$, where τ is chosen by the investigator to be the longest possible time to which anyone could survive. For either case, the estimated mean restricted to the interval $[0, \tau]$, with τ either the longest observed time or preassigned by the investigator, is given by

$$\hat{\mu}_\tau = \int_0^\tau \hat{S}(t)dt. \tag{4.5.1}$$

The variance of this estimator is

$$\hat{V}[\hat{\mu}_\tau] = \sum_{i=1}^{D}\left[\int_{t_i}^{\tau}\hat{S}(t)dt\right]^2 \frac{d_i}{Y_i(Y_i - d_i)} \tag{4.5.2}$$

A $100(1 - \alpha)$% confidence interval for the mean is expressed by

$$\hat{\mu}_\tau \pm Z_{1-\alpha/2}\sqrt{\hat{V}[\hat{\mu}_\tau]}. \tag{4.5.3}$$

EXAMPLE 4.1   *(continued)* Consider estimating the mean survival time for the 6-MP patients based on the Product-Limit estimator presented in Table 4.1. Because the largest observation is censored, an estimate of the mean restricted to 35 weeks will be constructed. The following integrals are needed as intermediate calculations in estimating the variance of our estimate and serve as a convenient bookkeeping method for constructing the estimate of the mean:

$$\int_{23}^{35} \hat{S}(t)dt = 0.448(35 - 23) = 5.376;$$

$$\int_{22}^{35} \hat{S}(t)dt = 5.376 + 0.538(23 - 22) = 5.914;$$

$$\int_{16}^{35} \hat{S}(t)dt = 5.914 + 0.628(22 - 16) = 9.682;$$

$$\int_{13}^{35} \hat{S}(t)dt = 9.682 + 0.690(16 - 13) = 11.752;$$

$$\int_{10}^{35} \hat{S}(t)dt = 11.752 + 0.753(13 - 10) = 14.011;$$

$$\int_{7}^{35} \hat{S}(t)dt = 14.011 + 0.807(10 - 7) = 16.429;$$

$$\int_{6}^{35} \hat{S}(t)dt = 16.429 + 0.857(7 - 6) = 17.286;$$

$$\int_{0}^{35} \hat{S}(t)dt = 17.286 + 1.0(6 - 0) = 23.286.$$

Thus, $\hat{\mu}_{35} = 23.286$ weeks, and

$$\hat{V}[\hat{\mu}_{35}] = \frac{3 \times 17.286^2}{21 \times 18} + \frac{16.429^2}{17 \times 16} + \frac{14.011^2}{15 \times 14} + \frac{11.752^2}{12 \times 11} + \frac{9.682^2}{11 \times 10}$$
$$+ \frac{5.914^2}{7 \times 6} + \frac{5.376^2}{6 \times 5} = 7.993.$$

The standard error of the estimated mean time to relapse is $7.993^{1/2} = 2.827$ weeks.

EXAMPLE 4.2 (*continued*) Using Efron's tail correction, the estimated mean disease-free survival time for ALL patients is $\hat{\mu}_{2081} = 899.28$ days with a standard error of 150.34 days. A 95% confidence interval for the mean disease-free survival time for ALL patients is $899.28 \pm 1.96(150.34) = (606.61, 1193.95)$ days. Similar calculations for the AML low risk group yields an estimated mean disease-free survival time of $\hat{\mu}_{2569} = 1548.84$ days with a standard error of 150.62 days (95% confidence interval: (1253.62, 1844.07) days.) For the AML high- risk group, $\hat{\mu}_{2640} = 792.31$ days with a standard error of 158.25 days (95% confidence interval: (482.15, 1102.5) days).

Comparison of the duration of the mean disease-free survival time for the three disease categories is complicated by the differences in the largest study times between the groups. To make comparisons which adjust for these differences, the estimated mean, restricted to the interval 0 to 2081 days, is computed for each group. Here, we find the following estimates:

Disease Group	Mean Restricted to 2081 days	Standard Error	95% Confidence Interval
ALL	899.3 days	150.3 days	606.6–1193.9 days
AML low risk	1315.2 days	118.8 days	1082.4–1548.0 days
AML high risk	655.67 days	122.9 days	414.8–896.5 days

Again, these results suggest that AML high risk patients have a lower survival rate than AML low risk patients, whereas ALL patients may be comparable with either of the two AML risk groups.

The Product-Limit estimator can also be used to provide estimates of quantiles of the distribution of the time-to-event distribution. Recall that the pth quantile of a random variable X with survival function $S(x)$, is defined by $x_p = \inf\{t : S(t) \le 1 - p\}$, that is, x_p is the smallest time at which the survival function is less than or equal to $1 - p$. When $p = 1/2$, x_p is the median time to the event of interest. To estimate x_p, we find the smallest time $\hat{x}_p$ for which the Product-Limit estimator is less than or equal to $1 - p$. That is, $\hat{x}_p = \inf\{t : \hat{S}(t) \le 1 - p\}$. In practice, the standard error of $\hat{x}_p$ is difficult to compute because it requires an estimate of the density function of X at $\hat{x}_p$ (see Practical Note 3 below). Brookmeyer and Crowley (1982a) have constructed confidence intervals for $\hat{x}_p$, based on a modification of the confidence interval construction for $S(t)$ discussed in section 4.3, which do not require estimating the density function. A $100(1 - \alpha)\%$ confidence interval for x_p, based on the linear confidence interval, is the set of all time points t which satisfy the following condition:

$$- Z_{1-\alpha/2} \le \frac{\hat{S}(t) - (1 - p)}{\hat{V}^{1/2}[\hat{S}(t)]} \le Z_{1-\alpha/2}. \qquad (4.5.4)$$

The $100(1 - \alpha)\%$ confidence interval for x_p based on the log-transformed interval is the set of all points t which satisfy the condition:

$$- Z_{1-\alpha/2} \le \frac{[\ln\{-\ln[\hat{S}(t)]\} - \ln\{-\ln[1 - p]\}][\hat{S}(t)\ln[\hat{S}(t)]]}{\hat{V}^{1/2}[\hat{S}(t)]} \le Z_{1-\alpha/2}. \qquad (4.5.5)$$

The $100(1 - \alpha)\%$ confidence interval for x_p based on the arcsine-square root transformation is given by

$$- Z_{1-\alpha/2} \le \frac{2\{\arcsine[\sqrt{\hat{S}(t)}] - \arcsine[\sqrt{(1 - p)}]\}[\hat{S}(t)(1 - \hat{S}(t))]^{1/2}}{\hat{V}^{1/2}[\hat{S}(t)]} \le Z_{1-\alpha/2}. \qquad (4.5.6)$$

EXAMPLE 4.2   *(continued)* We shall estimate the median disease-free survival time for the ALL group. From Table 4.3 we see that $\hat{S}(383) = 0.5217 > 0.5$

TABLE 4.7

Construction of a 95% Confidence Interval for the Median

t_i	$\hat{S}(t_i)$	$\sqrt{\hat{V}[\hat{S}(t_i)]}$	Linear (4.5.4)	Log (4.5.5)	Arcsine (4.5.6)
1	0.9737	0.0260	18.242	3.258	7.674
55	0.9474	0.0362	12.350	3.607	6.829
74	0.9211	0.0437	9.625	3.691	6.172
86	0.8947	0.0498	7.929	3.657	5.609
104	0.8684	0.0548	6.719	3.557	5.107
107	0.8421	0.0592	5.783	3.412	4.645
109	0.8158	0.0629	5.022	3.236	4.214
110	0.7895	0.0661	4.377	3.036	3.806
122	0.7368	0.0714	3.316	2.582	3.042
129	0.7105	0.0736	2.862	2.334	2.679
172	0.6842	0.0754	2.443	2.074	2.326
192	0.6579	0.0770	2.052	1.804	1.981
194	0.6316	0.0783	1.681	1.524	1.642
230	0.6041	0.0795	1.309	1.220	1.290
276	0.5767	0.0805	0.952	0.909	0.945
332	0.5492	0.0812	0.606	0.590	0.604
383	0.5217	0.0817	0.266	0.263	0.266
418	0.4943	0.0819	−0.070	−0.070	−0.070
468	0.4668	0.0818	−0.406	−0.411	−0.405
487	0.4394	0.0815	−0.744	−0.759	−0.741
526	0.4119	0.0809	−1.090	−1.114	−1.078
609	0.3825	0.0803	−1.464	−1.497	−1.437
662	0.3531	0.0793	−1.853	−1.886	−1.798
2081	0.3531	0.0793	−1.853	−1.886	−1.798

and $\hat{S}(418) = 0.4943 \leq 0.5$, so $\hat{x}_{0.5} = 418$ days. To construct 95% confidence intervals for the median, we complete Table 4.7. To illustrate the calculations which enter into construction of this Table, consider the first row. Here the entry in the fourth column is the middle term in (4.5.4), namely, $(0.9737 − 0.5)/0.0260 = 18.242$. The entry in the fifth column is the middle term in (4.5.5), namely,

$$([\ln(−\ln(0.9737)) − \ln(−\ln(0.5))]\{0.9737\ln[0.9737]\}/0.0260) = 3.258,$$

and the entry in the last column is the middle term in (4.5.6), namely, $2[\arcsine(\sqrt{0.9737}) − \arcsine(\sqrt{0.5})][0.9737(1 − 0.9737)]^{1/2}/0.0260 = 7.674$. To find the linear 95% confidence interval, we find all those values of t_i which have a value, in column four between −1.96 and 1.96. Thus the 95% linear confidence interval for $x_{0.5}$ is $x_{0.05} > 194$ days. The upper limit of this interval is undetermined because (4.5.4) never drops below -1.96 due to the heavy censoring. Based on the log transformation, a 95% confidence interval for x_p is $x_{0.05} > 192$ days. The interval based on the arcsine-transformed interval is $x_{0.05} > 194$ days.

Similar calculations for the two groups of AML patients show that the median disease-free survival time, for the low risk group, is 2204 days and, for the high risk group, is 183 days. For the low risk group, the lower end points of the 95% confidence intervals for the median disease-free survival time are 704 days, based on the linear approximation and 641 days based on either the log or arcsine transformation. For the high risk group, the 95% confidence intervals for the median are $(115, 363)$ days for the linear and arcsine-square root transformed intervals and $(113, 363)$, based on the log-transformed interval.

Practical Notes

1. If there is no censoring, then, the estimator of the mean time to death reduces to the sample mean. In addition, if there are no ties, then the estimated variance of the mean estimate reduces to the sample variance divided by n.

2. Alternate estimators of the mean survival time can be found by finding the area under one of the tail-corrected Product-Limit estimators discussed in Practical Note 2 of section 4.2.

3. An estimator of the large sample variance of the estimator of the pth percentile is given by $\hat{V}[\hat{x}_p] = \frac{\hat{V}[\hat{S}(x_p)]}{\hat{f}(x_p)^2}$, where $\hat{f}(x_p)$ is an estimate of the density function at the pth percentile. A crude estimate of $\hat{f}(t)$ is $\frac{\hat{S}(t-b)-\hat{S}(t+b)}{2b}$ based on a uniform kernel density estimate. Here, b is some small number.

4. Most major statistical packages provide an estimate of the mean lifetime. When the largest observation is censored, one must carefully examine the range over which the mean is computed.

Theoretical Notes

1. The asymptotic properties of the estimators of the mean and pth quantile follow directly from the weak convergence of the Product-Limit estimator. Details can be found in Andersen et al. (1993).

2. Details of constructing the confidence interval for median survival are found in Brookmeyer and Crowley (1982a) who also present a Monte Carlo study of the performance of the linear interval.

4.6 Estimators of the Survival Function for Left-Truncated and Right-Censored Data

The estimators and confidence intervals presented in sections 4.2–4.5 were based on right-censored samples. In this section, we shall show how these statistics can be modified to handle left-truncated and right-censored data. Here, we have associated, with the jth individual, a random age L_j at which he/she enters the study and a time T_j at which he/she either dies or is censored. As in the case of right-censored data, define $t_1 < t_2 < \cdots < t_D$ as the distinct death times and let d_i be the number of individuals who experience the event of interest at time t_i. The remaining quantity needed to compute the statistics in the previous sections is the number of individuals who are at risk of experiencing the event of interest at time t_i, namely Y_i. For right-censored data, this quantity was the number of individuals on study at time 0 with a study time of at least t_i. For left-truncated data, we redefine Y_i as the number of individuals who entered the study prior to time t_i and who have a study time of at least t_i, that is, Y_i is the number of individuals with $L_j \le t_i \le T_j$.

Using Y_i as redefined for left-truncated data, all of the estimation procedures defined in sections 4.2–4.4 are now applicable. However, one must take care in interpreting these statistics. For example, the Product-Limit estimator of the survival function at a time t is now an estimator of the probability of survival beyond t, conditional on survival to the smallest of the entry times L, $Pr[X > t \mid X \ge L] = S(t)/S(L)$. Similarly the Nelson–Aalen statistic estimates the integral of the hazard rate over the interval L to t. Note that the slope of the Nelson–Aalen estimator still provides an estimator of the unconditional hazard rate.

Some care in directly applying these estimators is needed. For left-truncated data, it is possible for the number at risk to be quite small for small values of t_i. If, for some t_i, Y_i and d_i are equal, then, the Product-Limit estimator will be zero for all t beyond this point, even though we are observing survivors and deaths beyond this point. In such cases, it is common to estimate the survival function conditional on survival to a time where this will not happen by considering only those death times beyond this point. This is illustrated in the following example.

EXAMPLE 4.3   To illustrate how the statistics developed in the previous sections can be applied to left-truncated data, consider the Channing House data described in section 1.16. The data is found on our web site. Here the truncation times are the ages, in months, at which individuals

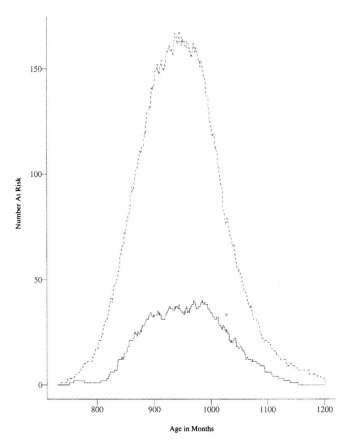

Figure 4.10   *Number at risk as a function of age for the 97 males (————)
and the 365 females (-----) in the Channing house data set*

entered the community. We shall focus on estimating the conditional
survival function.

Figure 4.10 shows the number of individuals at risk as a function of
the age at which individuals die for both males and females. Note that
the number at risk initially increases as more individuals enter into the
study cohort and that this number decreases for later ages as individuals
die or are censored.

Consider the data on males. Here the risk set is empty until 751
months when one individual enters the risk set. At 759 months, a second
individual enters the risk set. These two individuals die at 777 and 781
months. A third individual enters the risk set at 782 months. Computing
the Product-Limit estimator of $S(t)$ directly by (4.2.1) based on this
data would yield an estimate of $\hat{S}(t) = 1$ for $t < 777$, $\hat{S}(t) = 1/2$
for $777 \leq t < 781$, and $\hat{S}(t) = 0$ for $t \geq 781$. This estimate has little

meaning since the majority of the males in the study clearly survive beyond 781 months.

Rather than estimating the unconditional survival function, we estimate the conditional probability of surviving beyond age t, given survival to age a. We estimate $S_a(t) = Pr[X > t \mid X \geq a]$ by considering only those deaths that occur after age a, that is,

$$\hat{S}_a(t) = \prod_{a \leq t_i \leq t} \left[1 - \frac{d_i}{Y_i} \right], t \geq a. \qquad (4.6.1)$$

Similarly for Greenwood's formula (4.2.2) or for the Nelson–Aalen estimator (4.2.3), only deaths beyond a are considered.

Figure 4.11 shows the estimated probability of surviving beyond age t, given survival to 68 or 80 years for both males and females.

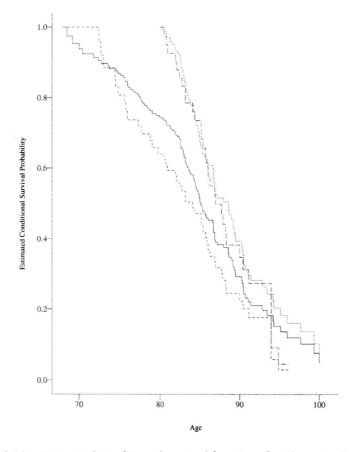

Figure 4.11   *Estimated conditional survival functions for Channing house residents. 68 year old females (————); 80 year old females (------); 68 year old males (———); 80 year old males (————).*

As in the unconditional Product-Limit estimator, the estimates are computed only over the range where $Y_i > 0$. These estimates could be extended beyond this time by the methods discussed in Practical Note 3 of section 4.2.

Practical Notes

1. A key assumption needed for making an inference with left-truncated data is the notion of 'independent truncation', that is, the Product-Limit estimator is a maximum likelihood estimator of the survival function if

$$\frac{Pr[X = x \mid L = l, X > L]}{Pr[X \geq x \mid L = l, X > L]} = \frac{Pr[X = x]}{Pr[X \geq x]} = h(x),$$

the hazard rate of X. Tsai (1990) provides a formal test of this hypothesis which is valid under independent censoring. See Keiding (1992) for further discussion of this point and further examples.

2. When the data is left truncated and right censored, the Product-Limit estimator may have a relatively large variance for small t, where the risk sets are small, and also for large t. This early instability in the point estimator of the survival function may propagate throughout the entire curve. Lai and Ying (1991) suggest a solution to this problem by a slight modification of the Product-Limit estimators where deaths are ignored when the risk set is small. Their estimator is given by

$$\tilde{S}(t) = \prod_{t_i \leq t} \left\{ 1 - \frac{d_i}{Y_i} I[Y_i \geq cn^\alpha] \right\},$$

where I is the indicator of the set A, n is the sample size, and $c > 0$, $0 < \alpha < 1$ are constants. This estimator is asymptotically equivalent to the usual product limit estimator.

Theoretical Note

1. The derivation of the Product-Limit estimator and the Nelson–Aalen estimator follows directly from the theory of counting processes as presented in section 3.6 with the modified definition of $Y(t)$ as discussed in Practical Note 2 of that section.

4.7   Summary Curves for Competing Risks

The summary survival curves presented in sections 4.2–4.6 are based on the assumption that the event and censoring times are independent. In the case of competing risks data, as discussed in section 2.7, this untestable assumption may be suspect. In this section we present three techniques for summarizing competing risks data.

To help in understanding the difference between the three estimators and their interpretation, consider the bone marrow transplant study discussed in section 1.3. In earlier sections of this chapter we considered estimation of the survival function for the time to treatment failure. Recall that treatment failure is defined as death in remission or relapse, whichever comes first. Here death in remission and relapse are competing risks and we are interested in summary curves that tell us how the likelihood of these events develops over time. Occurrence of one of the events precludes occurrence of the other event.

The first estimator which is commonly used is the complement of the Kaplan-Meier estimator. Here occurrences of the other event are treated as censored observations. For example, the estimated probability of relapsing before time t is one minus the Kaplan-Meier estimator of relapse obtained by treating occurrences of relapse as events and occurrences of death before relapse as censored observations. This estimator is an attempt to estimate the probability of relapsing before time t. It can be interpreted as the probability of relapse by time t if the risk of non-relapse death was removed. It is the probability of relapse in a hypothetical world where it is impossible for patients to die in remission. Reversing the roles of death in remission and relapse yields the treatment-related mortality or death in remission probability. Here this is an estimate of death in the world where relapse is not possible. These are rarely the probabilities of clinical interest and we cannot recommend the use of this estimator.

The second estimator is the cumulative incidence function. This estimator is constructed as follows. Let $t_1 < t_2 < \cdots < t_K$ be the distinct times where one of the competing risks occurs. At time t_i let Y_i be the number of subjects at risk, r_i be the number of subjects with an occurrence of the event of interest at this time, and d_i be the number of subjects with an occurrence of any of the other events of interest at this time. Note that $d_i + r_i$ is the number of subjects with an occurrence of any one of the competing risks at this time. Independent random censoring due to a patient being lost to follow-up is not counted here as one of the competing risks and affects only the value of Y_i. The cumulative incidence function is defined by

$$\text{CI}(t) = \begin{cases} 0 & \text{if } t \leq t_1 \\ \displaystyle\sum_{t_i \leq t} \left\{ \prod_{j=1}^{i-1} 1 - \frac{[d_j + r_j]}{Y_j} \right\} \frac{r_i}{Y_i} & \text{if } t_1 \leq t \end{cases} \qquad (4.7.1)$$

Note that for $t \geq t_1$ the cumulative incidence function is

$$\text{CI}(t) = \sum_{t_i \leq t} \hat{S}(t_i-) \frac{r_i}{Y_i}$$

where $\hat{S}(t_i-)$ is the Kaplan-Meier estimator, evaluated at just before t_i, obtained by treating any one of the competing risks as an event. The cumulative incidence function estimates the probability that the event of interest occurs before time t and that it occurs before any of the competing causes of failure. It is the estimate of the probability of the event of interest in the real world where a subject may fail from any of the competing causes of failure. For example, the relapse cumulative incidence is the chance a patient will have relapsed in the interval 0 to t in the real world where they may die in remission. The treatment related mortality cumulative incidence is the chance of death before relapse in the real world. Note that the sum of the cumulative incidences for all the competing risks is $1 - \hat{S}(t)$, which in the bone marrow transplant example is the complement of the treatment failure Kaplan-Meier estimate found in section 4.2.

The variance of the cumulative incidence is estimated by

$$V[\text{CI}(t)] = \sum_{t_i \leq t} \hat{S}(t_i)^2 \left\{ [\text{CI}(t) - \text{CI}(t_i)]^2 \frac{r_i + d_i}{Y_i^2} \right.$$

$$\left. + [1 - 2(\text{CI}(t) - \text{CI}(t_i))] \frac{r_i}{Y_i^2} \right\}. \qquad (4.7.2)$$

Confidence pointwise $(1 - \alpha)$ 100% confidence intervals for the cumulative incidence are given by $\text{CI}(t) \pm Z_{1-\alpha/2} V[\text{CI}(t)]^{1/2}$.

The third probability used to summarize competing risks data is the conditional probability function for the competing risk. For a particular risk, K, let $\text{CI}_K(t)$ and $\text{CI}_{K^c}(t)$ be the cumulative incidence functions for risk K and for all other risks lumped together, respectively. Then the conditional probability function is defined by

$$\text{CP}_K(t) = \frac{\text{CI}_K(t)}{1 - \text{CI}_{K^c}(t)}. \qquad (4.7.3)$$

The variance of this statistic is estimated by

$$V[\text{CP}_K(t)] = \frac{\hat{S}(t-)^2}{\{1 - \text{CI}_{K^c}(t)\}^4} \sum_{t_i \leq t} \frac{[1 - \text{CI}_{K^c}(t_i)]^2 r_i + \text{CI}_K(t_i)^2 d_i}{Y_i^2}. \qquad (4.7.4)$$

The conditional probability is an estimate of the conditional probability of event K's occurring by t given that none of the other causes have occurred by t. In the bone marrow transplantation example the conditional probability of relapse is the probability of relapsing before time t given the patient is not dead from other causes prior to t. It is the probability of relapsing among survivors who have not died from non-relapse-related toxicities.

To understand these probabilities better, consider a hypothetical bone marrow transplant experiment involving 100 patients. Suppose that there is no independent censoring and at one year after transplant 10 patients have relapsed and 30 patients have died in remission. When there is no censoring the cumulative incidence reduces to the cumulative number of events of the given type divided by n so the relapse cumulative incidence is 10/100 and the death in remission cumulative incidence is 30/100. The death in remission incidence is clearly interpreted as the proportion of patients who died in complete remission

TABLE 4.8

Estimates of Relapse and Death in Remission (TRM) for ALL Patients

t_i	d_i	r_i	Y_i	TRM 1-KME	Relapse 1-KME	TRM CI	Relapse CI	TRM CP	Relapse CP
1	1	0	38	0.0263	0.0000	0.0263	0.0000	0.0263	0.0000
55	0	1	37	0.0263	0.0270	0.0263	0.0263	0.0270	0.0270
74	0	1	36	0.0263	0.0541	0.0263	0.0526	0.0278	0.0541
86	1	0	35	0.0541	0.0541	0.0526	0.0526	0.0556	0.0556
104	0	1	34	0.0541	0.0819	0.0526	0.0789	0.0571	0.0833
107	1	0	33	0.0828	0.0819	0.0789	0.0789	0.0857	0.0857
109	0	1	32	0.0828	0.1106	0.0789	0.1053	0.0882	0.1143
110	0	1	31	0.0828	0.1393	0.0789	0.1316	0.0909	0.1429
122	1	1	30	0.1134	0.1680	0.1053	0.1579	0.1250	0.1765
129	0	1	28	0.1134	0.1977	0.1053	0.1842	0.1290	0.2059
172	1	0	27	0.1462	0.1977	0.1316	0.1842	0.1613	0.2121
192	0	1	26	0.1462	0.2285	0.1316	0.2105	0.1667	0.2424
194	1	0	25	0.1804	0.2285	0.1579	0.2105	0.2000	0.2500
230	0	1	23	0.1804	0.2621	0.1579	0.2380	0.2072	0.2826
276	1	0	22	0.2176	0.2621	0.1854	0.2380	0.2432	0.2921
332	0	1	21	0.2549	0.2621	0.2128	0.2380	0.2793	0.3023
383	0	1	20	0.2549	0.2990	0.2128	0.2654	0.2897	0.3372
418	1	0	19	0.2941	0.2990	0.2403	0.2654	0.3271	0.3494
466	1	0	18	0.3333	0.2990	0.2677	0.2654	0.3645	0.3625
487	1	0	17	0.3725	0.2990	0.2952	0.2654	0.4019	0.3766
526	1	0	16	0.4117	0.2990	0.3227	0.2654	0.4393	0.3919
609	0	1	14	0.4117	0.3490	0.3227	0.2949	0.4576	0.4353
662	0	1	13	0.4117	0.3991	0.3227	0.3243	0.4775	0.4788

prior to one year. The conditional probabilities estimates are 10/70 for relapse and 30/90 for death in remission. Here the death in remission probability is estimated by the number who die in remission divided by the number who could have died in remission which is the number at risk at one year who have yet to relapse. The complement of the Kaplan-Meier estimate depends on the pattern of occurrences of deaths and relapses. If all deaths occur before the first relapse then the relapse probability is 10/70 while if all the relapses occurred before the first death we get an estimate of 10/100. Any value between these two extremes is possible. Clearly this estimate has no meaningful interpretation.

EXAMPLE 4.2   *(continued)* We consider the data on the 38 patients with ALL given a transplant and examine the three probabilities for relapse and for death in remission (TRM). Table 4.8 provides the estimates for the three probabilities. The estimated standard error for the relapse cumulative incidence at 1 year is 0.069 so an approximate 95% confidence interval for the probability of relapsing before death is $0.238 \pm 1.96 \times 0.069 =$

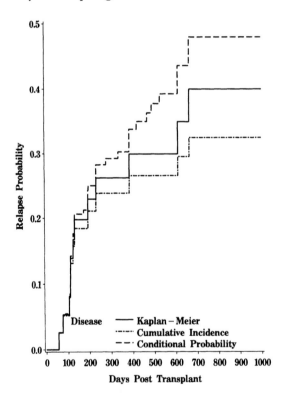

Figure 4.12 *Comparison estimated probability of relapse for ALL patients. Complement of Kaplan-Meier* (——), *cumulative incidence* (−·−·), *conditional probability* (- - -)

(0.103, 0.373). The estimated conditional probability of relapse at 1 year was 0.302 with a standard error of 0.087. A 95% confidence interval for the conditional probability of relapse is $0.302 \pm 1.96 \times 0.087 = (0.131, 0.473)$.

Figures 4.12 and 4.13 show the estimated probabilities for relapse and death in remission, respectively. Note that the conditional probability curve changes value at the occurrence of either of the two competing risks. The probabilities have the characteristic property of the conditional probability estimate being the largest and the cumulative incidence estimate the smallest.

It is important that summary curves for all the competing risks be presented since changes in the likelihood of one event cause changes in the probabilities for the other events. A nice summary curve is shown in Figure 4.14. Here we plot the relapse cumulative incidence and the sum of the relapse and death in remission cumulative incidences. The complement of the sum of the two cumulative incidences is the disease free survival probability found in section 4.2. At a given time the height of the first curve is the probability of relapsing, the distance between

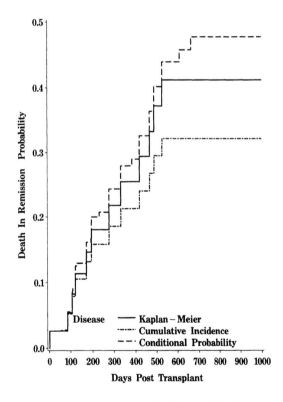

Figure 4.13 *Comparison estimated probability of death in remission for ALL patients. Complement of Kaplan-Meier (———), cumulative incidence (—·—·), conditional probability (- - -)*

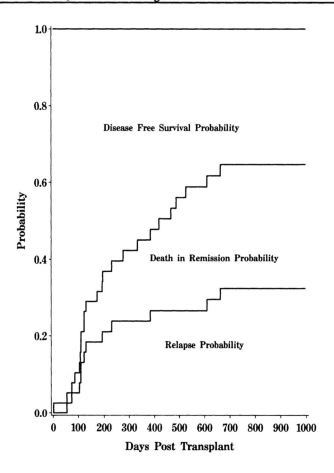

Figure 4.14 *Interaction between the relapse and death in remission*

the first and second curves the probability of death in remission, and the distance of the second curve from 1.0 the disease free survival function. For example, at 400 days the relapse probability is 0.2654, the death in remission probability is $0.4982 - 0.2654 = 0.2328$, and the disease-free survival function is $1 - 0.4982 = 0.5018$. This graph allows us dynamically to access the relationship between the competing risks.

Theoretical Notes

1. Suppose we have two competing risks X and Y and let $T = \min(X, Y)$ and $I = 1$ if $X < Y$, 0 if $X > Y$. The cause-specific hazard rate for X is

$$\lambda_X(t) = P[t \le X < t + \Delta t \mid \min(X, Y) > t]\Delta t.$$

The Kaplan-Meier estimator obtained by treating times with $I = 0$ as censored observations provides a consistent estimator of

$\exp\{-\int_0^t \lambda_X(u)\,du\}$. This quantity has no interpretation as a probability.

2. The cumulative incidence estimator was first proposed by Kalbfleisch and Prentice (1980). The estimator can be derived using techniques described in Andersen et al. (1993) as a special case of a more general theory for product-limit estimators for the transitions of a non-homogeneous Markov process.

3. Pepe and Mori (1993), Pepe et al. (1993), and Gooley et al. (1999) provide a nice discussion of these three estimates and present alternative derivations of the variance estimates.

Practical Note

1. A SAS macro to compute the cumulative incidence curves can be found on our web site.

4.8 Exercises

4.1 In section 1.11 we discussed a study of the effect of ploidy on the survival of patients with cancer of the tongue. Using the data on aneuploid tumors found in Table 1.6.

(a) Estimate the survival function at one (12 months) and five years (60 months) after transplant. Find the standard errors for your estimates.

(b) Estimate the cumulative hazard rate, $H(t)$, at 60 months. Find the standard error of $\hat{H}(t)$. Estimate $S(60)$ by $\exp\{-\hat{H}(t)\}$ and compare to your estimate in part a.

(c) Find a 95% linear confidence interval for $S(60)$.

(d) Find a 95% log-transformed confidence interval for $S(60)$.

(e) Find a 95% arcsine-square root confidence interval for $S(60)$.

(f) Using the log transformation find a 95% EP confidence band for the survival function over the range three years to six years (i.e., 36–72 months).

(g) Using the log transformation find a 95% Hall-Wellner confidence band for the survival function over the range three years to six years (i.e., 36–72 months).

(h) Estimate the mean survival time restricted to 400 months. Also provide a 95% confidence interval for the restricted mean survival time.

(i) Estimate the median time to death and find a 95% confidence interval for the median survival time based on a linear confidence interval.

4.2   Using the data reported in section 1.3, find the quantities specified below for the AML low risk and AML high risk groups. Note that most of these quantities are worked out in detail in Example 4.2 and its continuations for the ALL group.

(a) Estimate the survival functions and their standard errors for the AML low risk and AML high risk groups.

(b) Estimate the cumulative hazard rates and their standard errors for the AML low risk and AML high risk groups.

(c) Provide a crude estimate of the hazard rates for each group based on the estimates obtained in (b).

(d) Estimate the mean time to death and find 95% confidence intervals for the mean survival time for both the AML low risk and AML high risk groups. (Answers are given in section 4.5.)

(e) Work out estimates of the median time to death and find 95% confidence intervals for the median survival time for both the AML low risk and AML high risk groups using the linear, log-transformed, and arcsine formulas. (Answers are given in section 4.5.)

(f) Find 95% confidence intervals for the survival functions at 300 days post-transplant for both the AML low risk and AML high risk groups using the log- and arcsine-transformed formulas.

(g) Find 95% EP confidence bands for the survival functions over the range 100–400 days post-transplant for both the AML low risk and AML high risk groups using the linear, log-transformed, and arcsine-transformed formulas.

(h) Find 95% HW confidence bands for the survival functions over the range 100–400 days post-transplant for both the AML low risk and AML high risk groups using the linear, log-transformed, and arcsine-transformed formulas.

(i) Based on the results above and those discussed in Example 4.2 and its continuations, how do the survival experiences of the ALL, AML low risk, and AML high risk groups compare?

4.3   The following table contains data on the survival times of 25 patients with inoperable lung cancer entered on a study between November 1, 1979, and December 23, 1979. Complete follow-up was obtained on all patients so that the exact date of death was known. The study had one interim analysis conducted on March 31, 1980, by which time only 13 patients had died.

(a) Estimate the survival function based on the available sample information at the time of the interim analysis on 3/31/80. Provide the standard error of your estimate.

(b) Use the Brown, Hollandar, and Kowar technique (Practical Note 2 of section 4.1) to complete the right-hand tail of the product-limit estimate found in part a.

Patient	Date of Diagnosis	Date of Death	Days to death	Days to 3/31/80(Status)
1	1/11/79	5/30/79	139	139(Dead)
2	1/23/79	1/21/80	363	363(Dead)
3	2/15/79	3/27/80	559	410(Alive)
4	3/7/79	11/10/79	248	248(Dead)
5	3/12/79	4/8/79	27	27(Dead)
6	3/25/79	10/21/79	210	210(Dead)
7	4/4/79	8/16/79	134	134(Dead)
8	4/30/79	11/19/79	203	203(Dead)
9	5/16/79	5/9/81	724	320 (Alive)
10	5/26/79	7/15/79	50	50(Dead)
11	5/30/79	10/22/80	511	306(Alive)
12	6/3/79	6/25/79	22	22(Dead)
13	6/15/79	12/27/80	561	290(Alive)
14	6/29/79	1/29/81	580	276(Alive)
15	7/1/79	11/14/79	136	136(Dead)
16	8/13/79	6/16/80	308	231(Alive)
17	8/27/79	4/7/80	224	217(Alive)
18	9/15/79	1/9/81	482	198(Alive)
19	9/27/79	4/5/80	191	186(Alive)
20	10/11/79	3/3/80	144	144(Dead)
21	11/17/79	1/24/80	68	68(Dead)
22	11/21/79	10/4/81	683	131(Alive)
23	12/1/79	8/13/80	256	121(Alive)
24	12/14/79	2/27/81	441	108(Alive)
25	12/23/79	4/2/80	101	99(Alive)

(c) Compute the estimate of the survival function and an estimate of its standard error using the complete follow-up on each patient. Compare this estimate to that found in part a.

(d) Estimate the mean time to death restricted to 683 days based on the product-limit estimator found in part c.

(e) Estimate the mean time to death by finding the area under the survival curve found in part c. Find the standard error of your estimate.

(f) Compute the usual estimate of the time to death based on complete follow-up data by finding the arithmetic mean of the complete follow-up data. Find the standard error of this estimate in the usual way as the sample standard deviation divided by the square root of the sample size. Compare your answers to those obtained in part e.

4.4 In section 1.4 the times to first exit site infection (in months) of patients with renal insufficiency was reported. In the study 43 patients had a surgically placed catheter (Group 1) and 76 patients had a percutaneous placement of their catheter (Group 0).

(a) For each group plot the estimated survival function. Which technique seems better in delaying the time to infection?

(b) Estimate the cumulative hazard rate for each group of patients. Provide a crude estimate of the hazard rate at 5 months after placement of the catheter in each group.

(c) Find a 95% confidence interval for the mean time to first exit site infection restricted to 36 months for both groups.

4.5   Using the survival times of 59 black females given a kidney transplant at the OSU transplant center discussed in section 1.7—

(a) Estimate the distribution of the time to death, measured from transplant, for black female kidney transplant patients. Provide the standard error of the estimated survival function.

(b) Find a 95% confidence interval, based on the linear transformation, for the probability a black female will survive at least 12 months (365 days) after transplantation.

(c) Repeat b using the log-transformed confidence interval.

(d) Repeat c using the arcsine-transformed confidence interval. Compare the intervals found in parts c–e.

4.6   In section 1.6 a study is described to evaluate a protocol change in disinfectant practice in a large midwestern university medical center. Control of infection is the primary concern for the 155 patients entered into the burn unit with varying degrees of burns. The outcome variable is the time until infection from admission to the unit. Censoring variables are discharge from the hospital without an infection or death without an infection. Eighty-four patients were in the group which had chlorhexidine as the disinfectant and 70 patients received the routine disinfectant povidone-iodine.

(a) Estimate the survival (infection-free) functions and their standard errors for the chlorhexidine and povidone-iodine groups.

(b) Estimate the cumulative hazard rates and their standard errors for the chlorhexidine and povidone-iodine groups. Plot these estimates. Does it appear that the two cumulative hazard rates are proportional to each other?

(c) Provide estimates of the median time to infection and find 95% confidence intervals for the median time to infection for both the chlorhexidine and povidone-iodine groups using the linear, log-transformed, and arcsine formulas.

(d) Find 95% confidence intervals for the survival (infection-free) functions at 10 days postadmission for both the chlorhexidine and povidone-iodine groups using the log transformed and arcsine transformed formulas.

(e) Find 95% confidence bands for the infection-free functions over the range 8–20 days postinfection for both the chlorhexidine and povidone-iodine groups using the linear, log transformed, and arcsine transformed formulas.

(f) Find 95% HW confidence bands for the infection-free functions over the range 8–20 days postinfection for both the chlorhexidine and povidone-iodine.

(g) Based on the results above, how does the infection experience of the chlorhexidine and povidone-iodine groups compare?

4.7   Consider a hypothetical study of the mortality experience of diabetics. Thirty diabetic subjects are recruited at a clinic and followed until death or the end of the study. The subject's age at entry into the study and their age at the end of study or death are given in the table below. Of interest is estimating the survival curve for a 60- or for a 70-year-old diabetic.

(a) Since the diabetics needed to survive long enough from birth until the study began, the data is left truncated. Construct a table showing the number of subjects at risk, Y, as a function of age.

(b) Estimate the conditional survival function for the age of death of a diabetic patient who has survived to age 60.

(c) Estimate the conditional survival function for the age of death of a diabetic patient who has survived to age 65.

(d) Suppose an investigator incorrectly ignored the left truncation and simply treated the data as right censored. Repeat parts a–c.

Entry Age	Exit Age	Death Indicator	Entry Age	Exit Age	Death Indicator
58	60	1	67	70	1
58	63	1	67	77	1
59	69	0	67	69	1
60	62	1	68	72	1
60	65	1	69	79	0
61	72	0	69	72	1
61	69	0	69	70	1
62	73	0	70	76	0
62	66	1	70	71	1
62	65	1	70	78	0
63	68	1	71	79	0
63	74	0	72	76	1
64	71	1	72	73	1
66	68	1	73	80	0
66	69	1	73	74	1

4.8   Table 1.7 reports the results of a study on the survival times of patients admitted to a psychiatric hospital. In this data set patients were admitted to the hospital at a random age and followed until death or the end of the study. Let X be the patient's age at death. Note that the data we have on X is left truncated by the patient's age at entry into the hospital and right censored by the end of the study.

(a) Plot the number at risk, Y_i, as a function of age.

(b) Estimate the conditional survival function for a psychiatric patient who has survived to age 30 without entering a psychiatric hospital.

4.9   Hoel and Walburg (1972) report results of an experiment to study the effects of radiation on life lengths of mice. Mice were given a dose of 300 rads of radiation at 5–6 weeks of age and followed to death. At death each mouse was necropsied to determine if the cause of death was thymic lymphoma, reticulum cell sarcoma, or another cause. The ages of the mice at death are shown below:

Cause of Death	Age at Death (Days)
Thymic lymphoma	158, 192, 193, 194, 195, 202, 212, 215, 229, 230, 237, 240, 244, 247, 259, 300, 301, 337, 415, 444, 485, 496, 529, 537, 624, 707, 800
Reticulum cell sarcoma	430, 590, 606, 638, 655, 679, 691, 693, 696, 747, 752, 760, 778, 821, 986
Other causes	136, 246, 255, 376, 421, 565, 616, 617, 652, 655, 658, 660, 662, 675, 681, 734, 736, 737, 757, 769, 777, 801, 807, 825, 855, 857, 864, 868, 870, 873, 882, 895, 910, 934, 942, 1,015, 1,019

(a) For each of the three competing risks estimate the cumulative incidence function at 200, 300, . . . , 1,000 days by considering the two other risks as a single competing risk.

(b) Show that the sum of the three cumulative incidence functions in part a is equal to the 1 minus Kaplan-Meier estimate of the overall survival function for this set of data.

(c) Repeat part a using the complement of the marginal Kaplan-Meier estimates. What are the quantities estimating and how different from the results found in part a are these estimates?

(d) Compute the conditional probability function for thymic lymphoma at 500 and 800 days. What are the quantities estimating?

4.10   Using the data reported in section 1.3 for the AML low risk and AML high risk groups, find the following quantities for the two competing risks of relapse and death:

(a) The estimated cumulative incidence at one year.

(b) The standard errors of the two estimates in part a.

(c) The estimated conditional probabilities of relapse and of death in remission.

(d) The standard errors of the probabilities found in part c.

(e) Graphically express the development of relapse and death in remission for these two disease groups.

5
Estimation of Basic Quantities for Other Sampling Schemes

5.1 Introduction

In Chapter 4, we examined techniques for estimating the survival function for right-censored data in sections 4.2-4.5 and for left-truncated data in section 4.6. In this chapter, we discuss how to estimate the survival function for other sampling schemes, namely, left, double, and interval censoring, right-truncation, and grouped data. Each sampling scheme provides different information about the survival function and requires a different technique for estimation.

In section 5.2, we examine estimating for three censoring schemes. In the first scheme, left censoring, censored individuals provide information indicating only that the event has occurred prior to entry into the study. Double-censored samples include some individuals that are left-censored and some individuals that are right-censored. In both situations, some individuals with exact event times are observed. The last censoring scheme considered in this section is interval censoring, where individual event times are known to occur only within an interval.

In section 5.3, we present an estimator of the survival function for right-truncated data. Such samples arise when one samples individuals from event records and, retrospectively determines the time to event.

In section 5.4, we consider estimation techniques for grouped data. In elementary statistics books, the relative frequency histogram is often used to describe such data. In survival analysis, however, the complicating feature of censoring renders this simple technique ineffective because we will not know exactly how many events would have occurred in each interval had all subjects been observed long enough for them to have experienced the event. The life table methodology extends these elementary techniques to censored data.

Grouped survival data arises in two different situations. In the first, discussed in section 5.4, we follow a large group of individuals with a common starting time. The data consists of only the number who die or are lost within various time intervals. The basic survival quantities are estimated using a *cohort* (sometimes called a generation) life table.

In the second, a different sampling scheme is considered. Here a cross-sectional sample of the number of events and number at risk at different ages in various time intervals are recorded. In this instance, the cohort life table, which is based on longitudinal data, is not appropriate, and the basic survival quantities are estimated by the *current* life table. We refer the reader to Chiang (1984) for details of constructing this type of life table.

5.2   Estimation of the Survival Function for Left, Double, and Interval Censoring

In this section we shall present analogues of the Product-Limit estimator of the survival function for left-, double-, and interval-censored data. As discussed in section 3.3, left-censoring occurs when some individuals have experienced the event of interest prior to the start of the period of observation, while interval censoring occurs when all that is known is that the event of interest occurs between two known times. Double censoring occurs when both left censoring and right censoring are present. In addition some exact event times are observed. Each censoring scheme requires a distinct construction of the survival function.

For left censoring for some individuals, all we know is that they have experienced the event of interest prior to their observed study time, while for others their exact event time is known. This type of censoring is handled quite easily by reversing the time scale. That is, instead of measuring time from the origin we fix a large time τ and define new times by τ minus the original times. The data set based on these reverse times is now right-censored and the estimators in sections 4.2–4.4 can be applied directly. Note that the Product-Limit estimator in this case is estimating $P[\tau - X > t] = P[X < \tau - t]$. Examples of this procedure are found in Ware and Demets (1976).

Examples of pure left censoring are rare. More common are samples which include both left and right censoring. In this case a modified Product-Limit estimator has been suggested by Turnbull (1974). This estimator, which has no closed form, is based on an iterative procedure which extends the notion of a self-consistent estimator discussed in Theoretical Note 3 of section 4.2. To construct this estimator we assume that there is a grid of time points $0 = t_0 < t_1 < t_2 < \cdots < t_m$ at which subjects are observed. Let d_i be the number of deaths at time t_i (note here the t_i's are not event times, so d_i may be zero for some points). Let r_i be the number of individuals right-censored at time t_i (i.e., the number of individuals withdrawn from the study without experiencing the event at t_i), and let c_i be the number of left-censored observations at t_i (i.e., the number for which the only information is that they experienced the event prior to t_i). The only information the left-censored observations at t_i give us is that the event of interest has occurred at some $t_j \le t_i$. The self-consistent estimator estimates the probability that this event occurred at each possible t_j less than t_i based on an initial estimate of the survival function. Using this estimate, we compute an expected number of deaths at t_j, which is then used to update the estimate of the survival function and the procedure is repeated until the estimated survival function stabilizes. The algorithm is as follows:

Step 0: Produce an initial estimate of the survival function at each t_j, $S_o(t_j)$. Note any legitimate estimate will work. Turnbull's suggestion is to use the Product-Limit estimate obtained by ignoring the left-censored observations.

Step (K + 1)1: Using the current estimate of S, estimate $p_{ij} = P[t_{j-1} < X \le t_j \mid X \le t_i]$ by $\frac{S_K(t_{j-1}) - S_K(t_j)}{1 - S_K(t_i)}$, for $j \le i$.

Step (K + 1)2: Using the results of the previous step, estimate the number of events at time t_j by $\hat{d}_j = d_j + \sum_{i=j}^{m} c_i p_{ij}$.

Step (K + 1)3: Compute the usual Product-Limit estimator (4.2.1) based on the estimated right-censored data with $\hat{d}_j$ events and r_j right-censored observations at t_j, ignoring the left-censored data. If this estimate, $S_{K+1}(t)$, is close to $S_K(t)$ for all t_i, stop the procedure; if not, go to step 1.

EXAMPLE 5.1   To illustrate Turnbull's algorithm, consider the data in section 1.17 on the time at which California high school boys first smoked marijuana. Here left censoring occurs when boys respond that they have used

TABLE 5.1
Initial Estimate of the Survival Function Formed by Ignoring the Left-Censored Observations

i	Age t_i	Number Left-Censored c_i	Number of Events d_i	Number Right-Censored r_i	$Y_i = \sum_{j=i}^{m} d_j + r_j$	$S_o(t_i)$
0	0					1.000
1	10	0	4	0	179	0.978
2	11	0	12	0	175	0.911
3	12	0	19	2	163	0.804
4	13	1	24	15	142	0.669
5	14	2	20	24	103	0.539
6	15	3	13	18	59	0.420
7	16	2	3	14	28	0.375
8	17	3	1	6	11	0.341
9	18	1	0	0	4	0.341
10	>18	0	4	0	4	0.000
Total		12	100	79	0	

marijuana but can not recall the age of first use, while right-censored observations occur when boys have never used marijuana. Table 5.1 shows the data and the initial Product-Limit estimator, S_o, obtained by ignoring the left-censored observations.

In step 1, we estimate the p_{ij}'s. Note we only need estimates for those i with $c_i > 0$ for the computations in step 2. For the left-censored observation at t_4 we have

$$p_{41} = \frac{1.000 - 0.978}{1 - 0.669} = 0.067; \quad p_{42} = \frac{0.978 - 0.911}{1 - 0.669} = 0.202;$$

$$p_{43} = \frac{0.911 - 0.804}{1 - 0.669} = 0.320; \quad p_{44} = \frac{0.804 - 0.669}{1 - 0.669} = 0.410.$$

Similar computations yield the values for p_{ij} in Table 5.2.

Using these values, we have $\hat{d}_1 = 4 + 0.067 \times 1 + 0.048 \times 2 + 0.039 \times 3 + 0.036 \times 2 + 0.034 \times 3 + 0.034 \times 1 = 4.487$, $\hat{d}_2 = 13.461$, $\hat{d}_3 = 21.313$, $\hat{d}_4 = 26.963$, $\hat{d}_5 = 22.437$, $\hat{d}_6 = 14.714$, $\hat{d}_7 = 3.417$, $\hat{d}_8 = 1.206$, $\hat{d}_9 = 0$, and $\hat{d}_{10} = 4$. These values are then used in Table 5.3 to compute the updated estimate of the survival function, $S_1(t)$.

Then using these estimates of the survival function the p_{ij}'s are re-computed, the $\hat{d}$'s are re-estimated, and the second step estimator $S_2(t_i)$ is computed. This estimate is found to be within 0.001 of S_1 for all t_i, so

TABLE 5.2

Values of p_{ij} in Step 1

i/j	4	5	6	7	8	9
1	0.067	0.048	0.039	0.036	0.034	0.034
2	0.202	0.145	0.116	0.107	0.102	0.102
3	0.320	0.230	0.183	0.170	0.161	0.161
4	0.410	0.295	0.234	0.218	0.206	0.206
5		0.281	0.224	0.208	0.197	0.197
6			0.205	0.190	0.180	0.180
7				0.072	0.068	0.068
8					0.052	0.052
9						0.000

TABLE 5.3

First Step of the Self-Consistency Algorithm

t_i	$\hat{d}$	r_i	Y_i	$S_1(t_i)$
0				1.000
10	4.487	0	191.000	0.977
11	13.461	0	186.513	0.906
12	21.313	2	173.052	0.794
13	26.963	15	149.739	0.651
14	22.437	24	107.775	0.516
15	14.714	18	61.338	0.392
16	3.417	14	28.624	0.345
17	1.207	6	11.207	0.308
18	0.000	0	4.000	0.308
>18	4.000	0	4.000	0.000

the iterative process stops. The final estimate of the survival function, to three decimal places, is given by $S_1(t)$ in Table 5.3.

In some applications the data may be interval-censored. Here the only information we have for each individual is that their event time falls in an interval $(L_i, R_i]$, $i = 1, \ldots, n$, but the exact time is unknown. An estimate of the survival function can be found by a modification of above iterative procedure as proposed by Turnbull (1976). Let $0 = \tau_0 < \tau_1 < \cdots < \tau_m$ be a grid of time points which includes all the points L_i, R_i for $i = 1, \ldots, n$. For the ith observation, define a weight α_{ij} to be 1 if the interval $(\tau_{j-1}, \tau_j]$ is contained in the interval $(L_i, R_i]$, and 0 otherwise. Note that α_{ij} is an indicator of whether the event which occurs in the interval $(L_i, R_i]$ could have occurred at τ_j. An initial guess at $S(\tau_j)$ is made. The algorithm is as follows:

Step 1: Compute the probability of an event's occurring at time τ_j, $p_j = S(\tau_{j-1}) - S(\tau_j)$, $j = 1, \ldots, m$.

Step 2: Estimate the number of events which occurred at τ_j by

$$d_j = \sum_{i=1}^{n} \frac{\alpha_{ij}p_j}{\sum_k \alpha_{ik}p_k}.$$

Note the denominator is the total probability assigned to possible event times in the interval $(L_i, R_i]$.

Step 3: Compute the estimated number at risk at time τ_j by $Y_i = \sum_{k=j}^{m} d_k$.

Step 4: Compute the updated Product-Limit estimator using the pseudo data found in steps 2 and 3. If the updated estimate of S is close to the old version of S for all τ_i's, stop the iterative process, otherwise repeat steps 1–3 using the updated estimate of S.

EXAMPLE 5.2 To illustrate the estimation procedure for interval-censored data consider the data on time to cosmetic deterioration for early breast cancer patients presented in section 1.18.

Consider first the 46 individuals given radiation therapy only. The end points of the intervals for the individuals form the τ_i's as listed in Table 5.4. An initial estimate is found by distributing the mass of $1/46$ for the ith individual equally to each possible value of τ contained in $(L_i, R_i]$. For example, the individual whose event time is in the interval $(0, 7]$ contributes a value of $(1/46)(1/4)$ to the probability of the event's occurring at 4, 5, 6, and 7 months, respectively. Using this initial approximation in step 1, we can compute the p_j's. Here, for example, we have $p_1 = 1. - 0.979 = 0.021$, $p_2 = 0.979 - 0.955 = 0.024$, $p_3 = 0.0214$, etc. The estimated number of deaths as shown in Table 5.4 is then computed. As an example, at $\tau = 4$ we have $d_1 = 0.021/(0.021 + 0.024 + 0.021 + 0.029 + 0.031) + 0.021/(0.021 + 0.024 + 0.021 + 0.029) + 0.021/(0.021 + 0.024) = 0.842$. These estimates are then used to compute the estimated number at risk at each τ_i.

Using the estimated number of deaths and number at risk we compute the updated estimate of the survival function, as shown in Table 5.4. This revised estimate is then used to re-estimate the number of deaths, and the process continues until the maximum change in the estimate is less than 10^{-7}. This requires, in this case, 305 iterations of the process. The final estimate is shown in the second half of Table 5.4.

Figure 5.1 shows the estimated survival functions for the radiotherapy only and combination radiotherapy and chemotherapy groups. The

TABLE 5.4

Calculations for Estimating the Survival Function Based on Interval-Censored Data

τ	Initial $S(t)$	Estimated Number of Deaths d	Estimated Number at Risk Y	Updated $S(t)$	Change
0	1.000	0.000	46.000	1.000	0.000
4	0.979	0.842	46.000	0.982	−0.002
5	0.955	1.151	45.158	0.957	−0.002
6	0.934	0.852	44.007	0.938	−0.005
7	0.905	1.475	43.156	0.906	−0.001
8	0.874	1.742	41.680	0.868	0.006
10	0.848	1.286	39.938	0.840	0.008
11	0.829	0.709	38.653	0.825	0.004
12	0.807	1.171	37.944	0.799	0.008
14	0.789	0.854	36.773	0.781	0.008
15	0.775	0.531	35.919	0.769	0.006
16	0.767	0.162	35.388	0.766	0.001
17	0.762	0.063	35.226	0.764	−0.002
18	0.748	0.528	35.163	0.753	−0.005
19	0.732	0.589	34.635	0.740	−0.009
22	0.713	0.775	34.045	0.723	−0.011
24	0.692	0.860	33.270	0.705	−0.012
25	0.669	1.050	32.410	0.682	−0.012
26	0.652	0.505	31.360	0.671	−0.019
27	0.637	0.346	30.856	0.663	−0.026
32	0.615	0.817	30.510	0.646	−0.031
33	0.590	0.928	29.693	0.625	−0.035
34	0.564	1.056	28.765	0.602	−0.039
35	0.542	0.606	27.709	0.589	−0.047
36	0.523	0.437	27.103	0.580	−0.057
37	0.488	1.142	26.666	0.555	−0.066
38	0.439	1.997	25.524	0.512	−0.073
40	0.385	2.295	23.527	0.462	−0.077
44	0.328	2.358	21.233	0.410	−0.082
45	0.284	1.329	18.874	0.381	−0.097
46	0.229	1.850	17.545	0.341	−0.112
48	0.000	15.695	15.695	0.000	0.000

Interval	Survival Probability
0–4	1.000
5–6	0.954
7	0.920
8–11	0.832
12–24	0.761
25–33	0.668
34–38	0.586
40–48	0.467
≥48	0.000

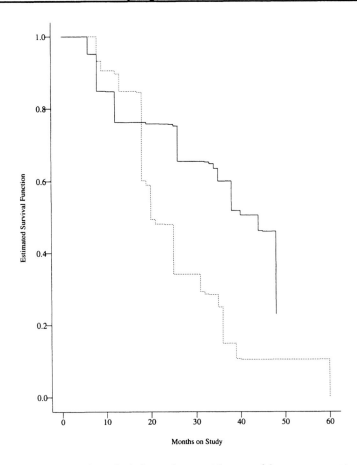

Figure 5.1 *Estimated probability of no evidence of breast retraction based on interval-censored data. Radiation only group (————). Radiation plus chemotherapy group (------)*

figure seems to indicate a longer time to cosmetic deterioration for patients given only radiotherapy.

Practical Notes

1. An example of the left-censored Product-Limit estimator is found in Ware and DeMets (1976). A key assumption needed for these calculations is that the death and censoring times are independent.
2. For the case of combined left and right censoring, Turnbull (1974) shows that an estimator of the variance-covariance matrix of $\hat{S}(t)$ is given by the matrix $\mathbf{V} = (V_{ij})$, where $V_{ij} = \text{Cov}[\hat{S}(t_i), \hat{S}(t_j)]$, constructed as follows. Let

$$A_{ii} = \frac{d_i}{[\hat{S}(t_{i-1}) - \hat{S}(t_i)]^2} + \frac{d_{i+1}}{[\hat{S}(t_i) - \hat{S}(t_{i+1})]^2} + \frac{r_i}{\hat{S}(t_i)^2} + \frac{c_i}{(1 - \hat{S}(t_i))^2}, \; i = 1, 2, \ldots, m - 1;$$

$$A_{mm} = \frac{d_m}{[\hat{S}(t_{m-1}) - \hat{S}(t_m)]^2} + \frac{r_m}{\hat{S}(t_m)^2} + \frac{c_m}{[1 - \hat{S}(t_m)]^2};$$

$$A_{i+1,i} = A_{i,i+1} = -\frac{d_{i+1}}{[\hat{S}(t_i) - \hat{S}(t_{i+1})]^2}, \; i = 1, 2, \ldots, m - 1; \text{and}$$

$$A_{ij} = 0 \text{ for } |i - j| \geq 2.$$

Define the matrix $\mathbf{J}$ to be the symmetric matrix given by

$$\mathbf{J} = \begin{pmatrix} b_1 & q_1 & 0 & 0 & \cdots & 0 & 0 \\ q_1 & b_2 & q_2 & 0 & \cdots & 0 & 0 \\ 0 & q_2 & b_3 & q_3 & \cdots & 0 & 0 \\ 0 & 0 & q_3 & b_4 & \cdots & 0 & 0 \\ \cdot & \cdot & \cdot & \cdot & & 0 & 0 \\ \cdot & \cdot & \cdot & \cdot & & 0 & 0 \\ \cdot & \cdot & \cdot & \cdot & & 0 & 0 \\ 0 & 0 & 0 & 0 & \cdots & b_{m-1} & q_{m-1} \\ 0 & 0 & 0 & 0 & \cdots & q_{m-1} & b_m \end{pmatrix}$$

where $b_i = A_{ii}$ and $q_i = A_{i,i+1}$. The inverse of this matrix is the estimated covariance matrix $\mathbf{V}$.

Using the data in Example 5.1, we find that

$$\mathbf{J} = \begin{pmatrix} 9941.9 & -2380.5 & 0 & 0 & 0 & 0 & 0 & 0 & 0 & 0 \\ -2380.5 & 3895.1 & -1514.7 & 0 & 0 & 0 & 0 & 0 & 0 & 0 \\ 0 & -1514.7 & 2691.5 & -1173.7 & 0 & 0 & 0 & 0 & 0 & 0 \\ 0 & 0 & -1173.7 & 2314.6 & -1097.4 & 0 & 0 & 0 & 0 & 0 \\ 0 & 0 & 0 & -1097.4 & 2041.5 & -845.5 & 0 & 0 & 0 & 0 \\ 0 & 0 & 0 & 0 & -845.5 & 2328.8 & -1358.1 & 0 & 0 & 0 \\ 0 & 0 & 0 & 0 & 0 & -1358.1 & 2210.8 & -730.5 & 0 & 0 \\ 0 & 0 & 0 & 0 & 0 & 0 & -730.5 & 800 & 0 & 0 \\ 0 & 0 & 0 & 0 & 0 & 0 & 0 & 0 & 44.3 & -42.2 \\ 0 & 0 & 0 & 0 & 0 & 0 & 0 & 0 & -42.2 & 42.2 \end{pmatrix}$$

and the matrix $\mathbf{V}$ is

$$\begin{pmatrix} 0.0001 & 0.0001 & 0.0001 & 0.0001 & 0.0001 & 0.0000 & 0.0000 & 0.0000 & 0.0000 & 0.0000 \\ 0.0001 & 0.0005 & 0.0004 & 0.0003 & 0.0003 & 0.0002 & 0.0002 & 0.0002 & 0.0000 & 0.0000 \\ 0.0001 & 0.0004 & 0.0009 & 0.0008 & 0.0006 & 0.0004 & 0.0004 & 0.0004 & 0.0000 & 0.0000 \\ 0.0001 & 0.0003 & 0.0008 & 0.0013 & 0.0010 & 0.0007 & 0.0007 & 0.0006 & 0.0000 & 0.0000 \\ 0.0001 & 0.0003 & 0.0006 & 0.0010 & 0.0015 & 0.0011 & 0.0010 & 0.0009 & 0.0000 & 0.0000 \\ 0.0000 & 0.0002 & 0.0004 & 0.0007 & 0.0011 & 0.0017 & 0.0015 & 0.0014 & 0.0000 & 0.0000 \\ 0.0000 & 0.0002 & 0.0004 & 0.0007 & 0.0010 & 0.0015 & 0.0020 & 0.0018 & 0.0000 & 0.0000 \\ 0.0000 & 0.0002 & 0.0004 & 0.0006 & 0.0009 & 0.0014 & 0.0018 & 0.0029 & 0.0000 & 0.0000 \\ 0.0000 & 0.0000 & 0.0000 & 0.0000 & 0.0000 & 0.0000 & 0.0000 & 0.0000 & 0.3263 & 0.3187 \\ 0.0000 & 0.0000 & 0.0000 & 0.0000 & 0.0000 & 0.0000 & 0.0000 & 0.0000 & 0.3187 & 0.3345 \end{pmatrix}$$

Thus, the estimated survival function and its standard error are obtained from the square root of the diagonal elements of the matrix **V**. In this example,

Age	$\hat{S}(t)$	Standard Error
0.000	1.000	0.000
10.000	0.977	0.011
11.000	0.906	0.022
12.000	0.794	0.031
13.000	0.651	0.036
14.000	0.516	0.039
15.000	0.392	0.041
16.000	0.345	0.044
17.000	0.308	0.054
18.000	0.308	0.571
> 18	0.000	0.578

3. Standard errors for the estimator of the survival function based on interval-censored data are found in Turnbull (1976) or Finkelstein and Wolfe (1985). Finkelstein (1986) and Finkelstein and Wolfe (1985) provide algorithms for adjusting these estimates for possible covariate effects.

Theoretical Notes

1. For left-censored data, Gomez et al. (1992) discuss the derivation of the left-censored Kaplan–Meier estimator and a "Nelson–Aalen" estimator of the cumulative backward hazard function defined by $G(t) = \int_t^\infty \frac{f(x)}{F(x)}\,dx$. These derivations are similar to those discussed in the notes after section 4.2. Further derivations are found in Andersen et al. (1993), using a counting process approach.
2. The estimator of the survival function, based on Turnbull's algorithms for combined left and right censoring or for interval censoring, are generalized maximum likelihood estimators. They can be derived by a self-consistency argument or by using a modified EM algorithm. For both types of censoring, standard counting process techniques have yet to be employed for deriving results.

5.3   Estimation of the Survival Function for Right-Truncated Data

For right-truncated data, only individuals for which the event has occurred by a given date are included in the study. Right truncation arises
commonly in the study of infectious diseases. Let T_i denote the chrono-
logical time at which the ith individual is infected and X_i the time
between infection and the onset of disease. Sampling consists of observing (T_i, X_i) for patients over the period (0 to τ). Note that only
patients who have the disease prior to τ are included in the study. Estimation for this type of data proceeds by reversing the time axis. Let
$R_i = \tau - X_i$. The R_i's are now left truncated in that only individuals with
values of $T_i \leq R_i$ are included in the sample. Using the method discussed in section 4.6 for left-truncated data, the Product-Limit estimator
of $Pr[R > t \mid R \geq 0]$ can be constructed. In the original time scale, this
is an estimator of $Pr[X < \tau - t \mid X \leq \tau)]$. Example 5.3 shows that this
procedure is useful in estimating the induction time for AIDS.

EXAMPLE 5.3   To illustrate the analysis of right-truncated data, consider the data on the
induction time for 37 children with transfusion-related AIDS, described
in section 1.19. The data for each child consists of the time of infection T_i
(in quarter of years from April 1, 1978) and the waiting time to induction
X_i. The data was based on an eight year observational window, so $\tau = 8$
years.

Table 5.5 shows the calculations needed to construct the estimate of
the waiting time to infection distribution. Here $R_i = 8 - X_i$. The column
headed d_i is the number of individuals with the given value of R_i or, in
the original time scale, the number with an induction time of X_i. The
number at risk column, Y_i, is the number of individuals with a value
of R between X_i and R_i or, in the original time scale, the number of
individuals with induction times no greater than X_i and infection times
no greater than $8 - X_i$. For example, when $X_i = 1.0$ ($R_i = 7.0$) in
the original time scale, there are 19 individuals with induction times
greater than 1 and one individual with an infection time greater than
7, so $Y_i = 37 - 19 - 1 = 17$. The final column of Table 5.5 is the
Product-Limit estimator for R_i based on d_i and Y_i. This is an estimate
of the probability that the waiting time to AIDS is less than x, given X
is less than 8 years, $G(t) = Pr[X < x \mid X \leq 8]$. Figure 5.2 shows the
estimated distribution of the induction time for AIDS for the 37 children
and 258 adults.

TABLE 5.5

Estimation of the Distribution of the Induction Time to AIDS Based on Right-Truncated Data

T_i	X_i	R_i	d_i	Y_i	$\hat{Pr}[X < x_i \mid X \leq 8]$
5.25	0.25	7.75			
7.25	0.25	7.75	2	2	0.0000
5.00	0.50	7.50			
5.50	0.50	7.50			
6.00	0.50	7.50			
6.25	0.50	7.50			
6.75	0.50	7.50	5	7	0.0243
3.50	0.75	7.25			
3.75	0.75	7.25			
5.00	0.75	7.25			
6.50	0.75	7.25			
6.75	0.75	7.25			
7.00	0.75	7.25	6	13	0.0850
2.75	1.00	7.00			
3.75	1.00	7.00			
4.00	1.00	7.00			
4.75	1.00	7.00			
5.50	1.00	7.00	5	17	0.1579
6.00	1.25	6.75			
6.25	1.25	6.75	2	18	0.2237
5.00	1.50	6.50			
5.25	1.50	6.50			
5.50	1.50	6.50	3	19	0.2516
3.00	1.75	6.25			
4.25	1.75	6.25			
5.75	1.75	6.25	3	21	0.2988
1.50	2.25	5.75			
4.75	2.25	5.75	2	19	0.3486
5.00	2.50	5.50			
5.25	2.50	5.50	2	20	0.3896
3.75	2.75	5.25	1	18	0.4329
2.25	3.00	5.00			
3.75	3.00	5.00	2	17	0.4584
4.50	3.25	4.75	1	14	0.5195
3.75	3.50	4.50	1	13	0.5594
3.75	4.25	3.75	1	11	0.6061
1.00	5.50	2.50	1	3	0.6667

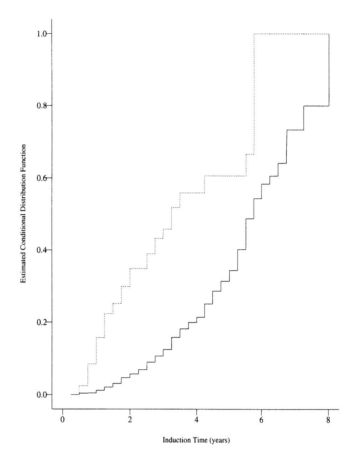

Figure 5.2 *Estimated conditional distribution of the induction time for AIDS for the 258 adults (———) and 37 children (------)*

Practical Note

1. For right-truncated data, standard errors of the survival estimator function follow directly by using Greenwood's formula. Lagakos et al. (1988) discuss techniques for comparing two samples based on right-truncated data. Gross and Huber–Carol (1992) discuss regression models for right-truncated data.

Theoretical Note

1. For right-truncated data, as for left censoring, the reversal of time allows direct estimation of the cumulative backward hazard function. Keiding and Gill (1990) discuss the large-sample-size properties of the estimated survival function for this type of data, using a counting process approach.

5.4 Estimation of Survival in the Cohort Life Table

A "cohort" is a group of individuals who have some common origin from which the event time will be calculated. They are followed over time and their event time or censoring time is recorded to fall in one of $k + 1$ adjacent, nonoverlapping intervals, $(a_{j-1}, a_j]$, $j = 1, \ldots, k + 1$. A traditional cohort life table presents the actual mortality experience of the cohort from the birth of each individual to the death of the last surviving member of the cohort. Censoring may occur because some individuals may migrate out of the study area, drop out of observation, or die unrecorded.

The cohort life table has applications in assessing survival in animal or human populations. The event need not be death. Other human studies may have, as an end point, the first evidence of a particular disease or symptom, divorce, conception, cessation of smoking, or weaning of breast-fed newborns, to name a few.

The basic construction of the cohort life table is described below:

1. The first column gives the adjacent and nonoverlapping fixed intervals, $I_j = (a_{j-1}, a_j]$, $j = 1, \ldots, k + 1$, with $a_0 = 0$ and $a_{k+1} = \infty$. Event and censoring times will fall into one and only one of these intervals. The lower limit is in the interval and the upper limit is the start of the next interval.

2. The second column gives the number of subjects Y'_j, entering the jth interval who have not experienced the event.

3. The third column gives the number of individuals W_j lost to follow-up or withdrawn alive, for whatever reason, in the jth interval. As for the product limit estimator, the censoring times must be independent of the event times.

4. The fourth column gives an estimate of the number of individuals Y_j at risk of experiencing the event in the jth interval, assuming that censoring times are uniformly distributed over the interval $Y_j = Y'_j - W_j/2$.

5. The fifth column reports the number of individuals d_j who experienced the event in the jth interval.

6. The sixth column gives the estimated survival function at the start of the jth interval $\hat{S}(a_{j-1})$. For the first interval, $\hat{S}(a_0) = 1$. Analogous to the product-limit estimator for successive intervals (see 4.2.1),

$$\hat{S}(a_j) = \hat{S}(a_{j-1})[1 - d_j/Y_j]. \qquad (5.4.1)$$

$$= \prod_{i=1}^{j}(1 - d_i/Y_i)$$

7. The seventh column gives the estimated probability density function $\hat{f}(a_{mj})$ at the midpoint of the jth interval, $a_{mj} = (a_j + a_{j-1})/2$. This quantity is defined as the probability of having the event in the jth interval per unit time, i.e.,

$$\hat{f}(a_{mj}) = [\hat{S}(a_{j-1}) - \hat{S}(a_j)]/(a_j - a_{j-1}) \qquad (5.4.2)$$

8. The eighth column gives the estimated hazard rate, $\hat{h}(a_{mj})$ at the midpoint of the jth interval, a_{mj}. Based on (2.3.2), this quantity is defined in the usual way as

$$\hat{h}(a_{mj}) = \hat{f}(a_{mj})/\hat{S}(a_{mj})$$

$$= \hat{f}(a_{mj})/\{\hat{S}(a_j) + [\hat{S}(a_{j-1}) - \hat{S}(a_j)]/2\}$$

$$= \frac{2\hat{f}(a_{mj})}{[\hat{S}(a_j) + \hat{S}(a_{j-1})]} \qquad (5.4.3)$$

Note that $\hat{S}(a_{mj})$ is based on a linear approximation between the estimate of S at the endpoints of the interval.

It may also be calculated as the number of events per person-units, i.e.,

$$\hat{h}(a_{mj}) = d_j/[(a_j - a_{j-1})(Y_j - d_j/2)]. \qquad (5.4.4)$$

Because the last interval is theoretically infinite, no estimate of the hazard or probability density function (and, of course, their standard errors) may be obtained for this interval.

Other useful quantities in subsequent calculations are the estimated conditional probability of experiencing the event in the jth interval, $\hat{q}_j = d_j/Y_j$, and the conditional probability of surviving through the jth interval, $\hat{p}_j = 1 - \hat{q}_j = 1 - d_j/Y_j$. Specifically, we could write (5.4.1) as

$$\hat{S}(a_j) = \hat{S}(a_{j-1})\hat{p}_j.$$

Note, also, that (5.4.2) and (5.4.3) could be written as

$$\hat{f}(a_{mj}) = \hat{S}(a_{j-1})\hat{q}_j/(a_j - a_{j-1}) \text{ and}$$

$$\hat{h}(a_{mj}) = 2\hat{q}_j/[(a_j - a_{j-1})(1 + \hat{p}_j)],$$

respectively.

9. The ninth column gives the estimated standard deviation of survival at the beginning of the jth interval (see Greenwood, 1926) which is

approximately equal to

$$\hat{S}(a_{j-1})\sqrt{\sum_{i=1}^{j-1} \frac{\hat{q}_i}{Y_i\hat{p}_i}} = \hat{S}(a_{j-1})\sqrt{\sum_{i=1}^{j-1} \frac{d_i}{Y_i(Y_i - d_i)}}, \qquad (5.4.5)$$

for $j = 2, \ldots, k+1$, and, of course, the estimated standard deviation of the constant $\hat{S}(a_o) = 1$ is 0. Note that this estimated standard error is identical to the standard error obtained for the product limit estimator in (4.2.2).

10. The tenth column shows the estimated standard deviation of the probability density function at the midpoint of the jth interval which is approximately equal to

$$\left[\frac{\hat{S}(a_{j-1})\hat{q}_j}{(a_j - a_{j-1})} \sqrt{\sum_{i=1}^{j-1} [\hat{q}_i/(Y_i\hat{p}_i)] + [\hat{p}_j/(Y_j\hat{q}_j)]} \right] \qquad (5.4.6)$$

11. The last column gives the estimated standard deviation of the hazard function at the midpoint of the jth interval which is approximately equal to

$$\left\{ \frac{1 - [\hat{h}(a_{mj})(a_j - a_{j-1})/2]^2}{Y_j q_j \ d_j} \right\}^{1/2} \cdot \hat{h}(a_{mj}) \qquad (5.4.7)$$

As noted in Chapter 2, the mean would be computed as in formula (2.4.2) with $S(x)$ replaced by $\hat{S}(x)$. There is some ambiguity regarding the mean lifetime because $\hat{S}(x)$ may be defined in several ways, as explained in Chapter 4, when the largest observation is a censored observation. For this reason, the median lifetime is often used. The median survival time may be determined by using relationship (2.4.4). For life tables, one first determines the interval where $\hat{S}(a_j) \leq 0.5$ and $\hat{S}(a_{j-1}) \geq 0.5$. Then, the median survival time can be estimated by linear interpolation as follows:

$$\hat{x}_{0.5} = a_{j-1} + [\hat{S}(a_{j-1}) - 0.5](a_j - a_{j-1})/[\hat{S}(a_{j-1}) - \hat{S}(a_j)] \quad (5.4.8)$$

$$= a_{j-1} + [\hat{S}(a_{j-1}) - 0.5]/\hat{f}(a_{mj})$$

Because we are often interested in the amount of life remaining after a particular time, the mean residual lifetime and the median residual lifetime are descriptive statistics that will estimate this quantity. For reasons stated above, the median residual lifetime at time x is often the preferable quantity. If the mean residual lifetime can be estimated without ambiguity, then, formula (2.4.1) with $S(x)$ replaced by $\hat{S}(x)$ is used. If the proportion of individuals surviving at time a_{i-1} is $S(a_{i-1})$, then the median residual lifetime is the amount of time that needs to be added to a_{i-1} so that $S(a_{i-1})/2 = S(a_{i-1} + \text{mdrl}(a_{i-1}))$, i.e., the $\text{mdrl}(a_{i-1})$ is the increment of time at which half of those alive at time a_{i-1} are expected to survive beyond. Suppose the jth interval contains the survival probability $S(a_{i-1} + \text{mdrl}(a_{i-1}))$, then an estimate of $\text{mdrl}(a_{i-1})$, determined in a similar fashion as (5.4.8) is given by

$$\widehat{\text{mdrl}}(a_{i-1}) = \tag{5.4.9}$$
$$(a_{j-1} - a_{i-1}) + [\hat{S}(a_{j-1}) - \hat{S}(a_{i-1})/2](a_j - a_{j-1})/[\hat{S}(a_{j-1}) - \hat{S}(a_j)]$$

Hence the median residual lifetime at time 0 will, in fact, be the median lifetime of the distribution.

The variance of this estimate is approximately

$$\widehat{\text{Var}}\,[\widehat{\text{mdrl}}(a_{i-1})] = \frac{[\hat{S}(a_{i-1})]^2}{4Y_i[\hat{f}(a_{mj})]^2} \tag{5.4.10}$$

Some major statistical packages will provide the median residual lifetime and its standard error at the beginning of each interval.

EXAMPLE 5.4   Consider The National Labor Survey of Youth (NLSY) data set discussed in section 1.14. Beginning in 1983, females in the survey were asked about any pregnancies that have occurred since they were last interviewed (pregnancies before 1983 were also documented). Questions regarding breast feeding are included in the questionnaire.

This data set consists of the information from 927 first-born children to mothers who chose to breast feed their child and who have complete information for all the variables of interest. The universe was restricted to children born after 1978 and whose gestation was between 20 and 45 weeks. The year of birth restriction was included in an attempt to eliminate recall problems.

The response variable in the data set is the duration of breast feeding in weeks, followed by an indicator if the breast feeding is completed (i.e., the infant is weaned).

The quantities described above are shown in Table 5.6 for this data set. Because none of the mothers claimed to wean their child before one week, the first interval will be from birth to two weeks. As always, when data are grouped, the selection of the intervals is a major decision. Generally, guidelines used in selecting intervals for frequency histograms apply, namely, the number of intervals should be reasonable, there should be enough observations within each interval to adequately represent that interval, and the intervals should be chosen to reflect the nature of the data. For example, in this data set, it is of interest to examine early weaners in smaller intervals and later weaners in broader intervals. This principle is also true in most population mortality studies where one wishes to study infant mortality in smaller intervals and later mortality may be studied in broader intervals.

TABLE 5.6

Life Table for Weaning Example

Week weaned [lower, upper]	Number of infants not weaned entering interval	Number lost to follow-up or withdrawn without being weaned	Number exposed to weaning	Number weaned	Est. Cum. proportion not weaned at beginning of interval	Est. p.d.f. at middle of interval	Est. hazard at middle of interval	Est. stand. dev. of survival at beginning of interval	Est. stand. dev of. p.d.f. at middle of interval	Est. stand. dev of. hazard at middle of interval
0– 2	927	2	926	77	1.0000	0.0416	0.0434	0	0.0045	0.0049
2– 3	848	3	846.5	71	0.9168	0.0769	0.0875	0.0091	0.0088	0.0104
3– 5	774	6	771	119	0.8399	0.0648	0.0836	0.0121	0.0055	0.0076
5– 7	649	9	644.5	75	0.7103	0.0413	0.0618	0.0149	0.0046	0.0071
7–11	565	7	561.5	109	0.6276	0.0305	0.0537	0.0160	0.0027	0.0051
11–17	449	5	446.5	148	0.5058	0.0279	0.0662	0.0166	0.0021	0.0053
17–25	296	3	294.5	107	0.3381	0.0154	0.0555	0.0158	0.0014	0.0052
25–37	186	0	186	74	0.2153	0.0071	0.0414	0.0138	0.0008	0.0047
37–53	112	0	112	85	0.1296	0.0061	0.0764	0.0114	0.0006	0.0066
53–	27	0	27	27	0.0313			0.0059		

An interesting feature of these data is that the hazard rate for weaning is high initially (many mothers stop breastfeeding between 1 and 5 weeks), levels off between 5 and 37 weeks, and begins to rise after 37 weeks as can be seen in Figure 5.3.

The median weaning time for all mothers starting to breast-feed is determined from (5.4.8) to be 11.21 weeks (with a standard error of 0.5678 weeks) and the median residual weaning time at 25 weeks is 15.40 weeks (with a standard error of 1.294 weeks).

Practical Notes

1. Summarizing the assumptions made in the life table methodology, we have seen that i) censored event times (including loss or withdrawal) are assumed to be independent of the time those individuals would have realized the event had they been observed until the event occurred, ii) the censoring times and death times are assumed to be uniformly distributed within each interval, (hence $Y_j' - W_j/2$ is taken to be the number exposed (or at risk) in the jth interval (see the number of people at risk in column 4 and the calculation of the number of person-units in the denominator of eq. (5.4.4), and iii) the hazard rate is assumed constant within intervals.

2. Individuals lost to follow-up are lost to observation if they move, fail to return for treatment, or, for some other reason, their survival status becomes unknown in the jth interval. On the other hand, individuals withdrawn alive are those known to be alive at the closing date of the

study. Such observations typically arise in cohort studies or clinical trials. One assumption, as stated in the preceding note, is that the survival experience after the date of last contact of those lost to follow-up and withdrawn alive is similar to that of the individuals who remain under observation. Cutler and Ederer (1958) point out that the survival experience of lost individuals may be better than, the same as, or worse than individuals continuing under observation. Thus, every attempt should be made to trace such individuals and to minimize the number of individuals lost.

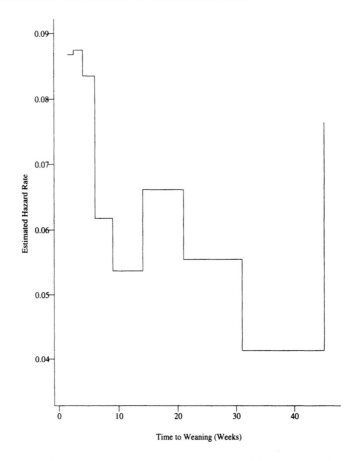

Figure 5.3   *Life table estimate of the hazard rate of the time to infant weaning*

3. SAS and SPSS have routines which reproduce the cohort life table.

Theoretical Notes

1. An alternative estimator of the hazard function is given by Sacher (1956) assuming that the hazard rate is constant within each interval

but is allowed to vary between intervals. This estimator is given by $\hat{h}(a_{mj}) = (-\ln \hat{p}_j)/(a_j - a_{j-1})$, which, Gehan and Siddiqui (1973) show, is slightly more biased than (5.4.3).

2. If the lengths of the grouping intervals approach 0, then, the life table estimates of the survival function are equivalent to the Kaplan–Meier estimate (Thompson, 1977). This limiting process provides a framework to link life table estimates with those using exact lifetimes, even in the presence of covariates.

5.5   Exercises

5.1   A study of 190 first-year medical students asked the question, How old were you when you first smoked a cigarette? Responses were either the exact ages at which they started smoking, that they never smoked, or that they currently smoke but cannot remember when they started. The data is summarized below. Using this sample, estimate the survival function to within 0.001.

Age (t)	Number Who Started Smoking at Age t	Number of Age t Who Smoke Now but Do Not Know the Age They Started	Number of Age t Who Do Not Smoke
14	2	0	0
15	3	0	0
16	10	0	0
17	13	0	0
18	5	0	0
19	3	0	1
20	2	4	13
21	1	6	44
22	2	8	39
23	1	2	19
24	0	0	3
25	0	0	4
26	1	0	4
Total	43	20	127

5.2   A study involving 100 veterinarians was performed to estimate the time until their first needlestick injury. They completed a survey which asked, How many years after graduation from veterinarian school did you experience your first needlestick injury? Many of them could remember or determine from their records the month and year their first injury occurred, but others could only say that it happened before a certain time. Others had had no needlestick injury yet at the time of the survey. The data below reflects these times after graduation.

Time (t) After Graduation (in Months)	Number Who Had Needlestick Injury at Time t	Number Who Had Needlestick Injury Prior to Time t	Number Who Never Had Needlestick Injury at Time t
2	3	0	0
4	2	0	0
8	1	0	0
10	2	1	0
12	4	2	0
15	6	2	1
20	3	4	1
24	3	3	2
28	2	3	3
34	1	4	5
41	0	2	3
62	0	3	4
69	0	2	6
75	0	1	6
79	0	2	3
86	0	3	7
Total	27	32	41

Estimate the survival (injury-free) function to an accuracy of three decimal places.

5.3 Eighteen elderly individuals who had entered a nursing home in the past five years were asked when they experienced their first fall (post-admittance). Some of the individuals only indicated that it occurred within a certain time period (in months), whereas others said they had never had a fall. The data (in months post-admittance) is as follows:

Falls occurred in (6–12], (48–60], (24–36], (12–24], (18–24], (9–12], (36–42], (12–36]

Times since admittance for individuals who never had a fall: 23, 41, 13, 25, 59, 39, 22, 18, 49, 38.

Estimate the survival function of the time from admittance to first fall to within three decimal places.

5.4 Twenty women who had a lumpectomy as a primary treatment for breast cancer were followed periodically for the detection of a metastasis. When a metastasis was detected it was only known that the time of the clinical appearance of the metastasis was between the times of the last two visits to the physician, so the data is interval-censored. Suppose the data is as follows:

Times in months between which a metastasis could be detected: (12, 18], (20, 24], (10, 13], (14, 15], (25, 33], (33, 44], (18, 22], (19, 25], (13, 22], (11, 15].

Times last seen for patients disease free at the end of study: 25, 27, 33, 36, 30, 29, 35, 44, 44, 44.

Estimate the survival time for the distribution of the time from surgery to first clinical evidence of a metastasis.

5.5   A study was performed to estimate the distribution of incubation times of individuals known to have a sexually transmitted disease (STD). Twenty-five patients with a confirmed diagnosis of STD at a clinic were identified on June 1, 1996. All subjects had been sexually active with a partner who also had a confirmed diagnosis of a STD at some point after January 1, 1993 (hence $\tau = 42$ months). For each subject the date of the first encounter was recorded as well as the time in months from that first encounter to the clinical confirmation of the STD diagnosis. Based on this right-truncated sample, compute an estimate of the probability that

Date of First Encounter	Months From 1/93 to Encounter	Time (in months) until STD Diagnosed in Clinic
2/93	2	30
4/93	4	27
7/93	7	25
2/94	14	19
8/94	20	18
6/94	18	17
8/93	8	16
1/94	13	16
5/94	17	15
2/95	26	15
8/94	20	15
3/94	15	13
11/94	23	13
5/93	5	12
4/94	16	11
3/94	15	9
11/93	11	8
6/93	6	8
9/95	33	8
4/93	4	7
8/93	8	6
11/95	35	6
10/93	10	6
12/95	36	4
1/95	25	4

the infection period is less than x months conditional on the infection period's being less than 42 months.

Estimate the distribution of infection-free time (survival).

5.6 Using the data on 258 adults with AIDS reported in section 1.19, estimate the probability that the waiting time to AIDS is less than x, given the waiting time is less than eight years.

5.7 The following data is based on a cohort of 1,571 men in the Framingham Heart Study who were disease free at age 40 and followed for a period of 40 years. (See Klein, Keiding, and Kreiner (1995) for a detailed description of the cohort.) Of interest is the distribution of the time to development or coronary heart disease (CHD). The following life table data is available to estimate this distribution.

Age Interval	Number of CHD Events	Number Lost to Follow-Up
45–50	17	29
50–55	36	60
55–60	62	83
60–65	76	441
65–70	50	439
70–75	9	262
75–80	0	7

Construct a cohort life table for this data.

5.8 Individuals seen at a large city sexually transmitted disease (STD) clinic are considered at high risk for acquiring HIV. The following data is recorded on 100 high-risk individuals who are infected with some STD, but did not have HIV, when they were seen at the clinic in 1980. Their records were checked at subsequent visits to determine the time that HIV was first detected.

Year Intervals	Number of HIV-Positive	Number Lost to Follow-Up
0–2	2	3
2–4	1	2
4–6	4	8
6–8	3	10
8–10	2	18
10–12	2	21
12–14	3	21

Construct a cohort life table for the incidence of HIV.

5.9 An investigator, performing an animal study on mammary carcinogenesis risk, wants to describe the distribution of times (in days) until the onset of the first palpable tumor for rats fed a control diet. Mammary tumors were induced by a single oral dose (5 mg dissolved in 1.0 ml. com oil) of 7,12-dimethylbenz(a)anthracene (DMBA) administered by intragastric intubation when the animals were seven weeks old. Starting six weeks after DMBA administration, each rat was examined once daily and the time (in days) until the onset of the first palpable tumor was recorded. Three rats had a palpable tumor when the first examination was made at day 62. The remaining times when the first palpable tumor was detected are below.

Times (in days) when the first palpable tumor was detected: 46, 49, 54, 61, 62, 64, 68, 120, 150, 160.

Estimate the survival time for the distribution of the time from DBMA administration until the first palpable evidence of a tumor occurred.

5.10 Wagner and Altmann (1973) report data from a study conducted in the Amboseli Reserve in Kenya on the time of the day at which members of a baboon troop descend from the trees in which they sleep. The time is defined as the time at which half of the troop has descended and begun that day's foraging. On some days the observers arrived at the site early enough to observe at what time this event occurred, whereas on other days they arrived after this median descent time, so that day's observation was left censored at their arrival time. That data is in the following tables. By reversing the time scale to be the number of minutes from midnight (2400 hours), estimate the distribution of the time to descent for a randomly selected troop of baboons.

Observed Time of Day When Half of the Troop Descended from the Trees

Day	Date	Descent Time	Day	Descent Date	Time	Descent Day	Date	Time
1	25/11/63	0656	20	12/7/64	0827	39	10/6/64	0859
2	29/10/63	0659	21	30/6/64	0828	40	11/3/64	0900
3	5/11/63	0720	22	5/5/64	0831	41	23/7/64	0904
4	12/2/64	0721	23	12/5/64	0832	42	27/2/64	0905
5	29/3/64	0743	24	25/4/64	0832	43	31/3/64	0905
6	14/2/64	0747	25	26/3/64	0833	44	10/4/64	0907
7	18/2/64	0750	26	18/3/64	0836	45	22/4/64	0908
8	1 /4/64	0751	27	15/3/64	0840	46	7/3/64	0910
9	8/2/64	0754	28	6/3/64	0842	47	29/2/64	0910
10	26/5/64	0758	29	11/5/64	0844	48	13/5/64	0915
11	19/2/64	0805	30	5/6/64	0844	49	20/4/64	0920
12	7/6/64	0808	31	17/7/64	0845	50	27/4/64	0930
13	22/6/64	0810	32	12/6/64	0846	51	28/4/64	0930
14	24/5/64	0811	33	28/2/64	0848	52	23/4/64	0932
I5	21/2/64	0815	34	14/5/64	0850	53	4/3/64	0935
16	13/2/64	0815	35	7/7/64	0855	54	6/5/64	0935
17	11/6/64	0820	36	6/7/64	0858	55	26/6/64	0945
18	21/6/64	0820	37	2/7/64	0858	56	25/3/64	0948
19	13/3/64	0825	38	17/3/64	0859	57	8/7/64	0952
						58	21/4/64	1027

Observer Arrival Time on Days Where the Descent Time Was Not Observed

Day	Date	Arrival Time	Day	Date	Arrival Time	Day	Date	Arrival Time
1	1/12/63	0705	32	13/10/63	0840	63	2/5/64	1012
2	6/11/63	0710	33	4/7/64	0845	64	1/3/64	1018
3	24/10/63	0715	34	3/5/64	0850	65	17/10/63	1020
4	26/11/63	0720	35	25/5/64	0851	66	23/10/63	1020
5	18/10/63	0720	36	24/11/63	0853	67	25/7/64	1020
6	7/5/64	0730	37	15/7/64	0855	68	13/7/64	1031
7	7/11/63	0740	38	16/2/64	0856	69	8/6/64	1050
8	23/11/63	0750	39	10/3/64	0857	70	9/3/64	1050
9	28/11/63	0750	40	28/7/64	0858	71	26/4/64	1100
10	27/11/63	0753	41	18/6/64	0858	72	14/10/63	1205
11	28/5/64	0755	42	20/2/64	0858	73	18/11/63	1245
12	5/7/64	0757	43	2/8/64	0859	74	2/3/64	1250
13	28/3/64	0800	44	27/5/64	0900	75	8/5/64	1405
14	23/3/64	0805	45	28/10/64	0905	76	1/7/64	1407
15	26/10/63	0805	46	15/5/64	0907	77	12/10/63	1500
16	11/7/64	0805	47	10/5/64	0908	78	31/7/64	1531
17	27/7/64	0807	48	27/6/64	0915	79	6/10/63	1535
18	9/6/64	0810	49	11/10/63	0915	80	19/6/64	1556
19	24/6/64	0812	50	17/2/64	0920	81	29/6/64	1603
20	16/ 10/63	0812	51	22/10/63	0920	82	9/5/64	1605
21	25/2/64	0813	52	10/7/64	0925	83	9/10/63	1625
22	6/6/64	0814	53	14/7/64	0926	84	8/3/64	1625
23	22/11/63	0815	54	11/4/64	0931	85	11/2/64	1653
24	10/10/63	0815	55	23/5/64	0933	86	30/5/64	1705
25	2/11/63	0815	56	30/7/64	0943	87	5/3/64	1708
26	23/6/64	0817	57	18/7/64	0945	88	26/2/64	1722
27	24/4/64	0823	58	29/7/64	0946	89	4/5/64	1728
28	3/7/64	0830	59	16/7/64	0950	90	12/3/64	1730
29	29/4/64	0831	60	22/7/64	0955	91	25/10/63	1730
30	4/8/63	0838	61	15/10/63	0955	92	29/11/63	1750
31	7/10/63	0840	62	19/10/63	1005	93	22/2/64	1801
						94	22/3/64	1829

6
Topics in Univariate
Estimation

6.1 Introduction

In Chapter 4, we presented two techniques for providing summary curves which tell us about the survival experience of a cohort of individuals. These two estimators were the Kaplan–Meier estimator, which provides an estimate of the survival function, and the Nelson–Aalan estimator, which provides an estimate of the cumulative hazard rate. These statistics are readily available in many statistical packages.

Although these two statistics provide an investigator with important information about the eventual death time of an individual, they provide only limited information about the mechanism of the process under study, as summarized by the hazard rate. The slope of the Nelson–Aalan estimator provides a crude estimate of the hazard rate, but this estimate is often hard to interpret. In section 6.2, we discuss how these crude estimates of the hazard rate can be smoothed to provide a better estimator of the hazard rate by using a kernel-smoothing technique.

In some applications of survival analysis, an investigator has available very precise information about the mortality rates in a historical control or standard population. It is of interest to compare the hazard rates in the sample group to the known hazard rates in the reference population to determine how the mortality experience of the experimental subjects differs. The "excess" mortality in the experimental group can have either a multiplicative or additive effect on the reference hazard rate. In section 6.3, estimation techniques for both the additive and multiplicative models for excess mortality are developed.

In section 6.4, the problem of estimation of the survival function for right censored data is considered from a Bayesian perspective. In this framework, an investigator has some prior information on the survival function from results of similar studies, from a group of experts, or from some reference population. The prior information is combined with sample data to provide a posterior distribution of the survival function on which the estimation is based. The combination of prior and sample information can be done analytically by Bayes theorem or by a Monte Carlo method via the Gibbs sampler. Both methods are illustrated.

6.2   Estimating the Hazard Function

The Nelson–Aalen estimator $\tilde{H}(t)$, discussed in sections 4.2 or 4.6, provides an efficient means of estimating the cumulative hazard function $H(t)$. In most applications, the parameter of interest is not $H(t)$, but rather its derivative $h(t)$, the hazard rate. As noted earlier, the slope of the Nelson–Aalen estimator provides a crude estimate of the hazard rate $h(t)$. Several techniques have been proposed in the literature to estimate $h(t)$. In this section, we shall concentrate on the use of kernel smoothing to estimate $h(t)$.

Kernel-smoothed estimators of $h(t)$ are based on the Nelson–Aalen estimator $\tilde{H}(t)$ and its variance $\hat{V}[\tilde{H}(t)]$. The estimator $\tilde{H}(t)$ can be based on right-censored data (see section 4.2) or on left-truncated data (see section 4.6). Recall that, in either case, $\tilde{H}(t)$ is a step function with jumps at the event times, $0 = t_0 < t_1 < t_2 < \cdots < t_D$. Let $\Delta \tilde{H}(t_i) = \tilde{H}(t_i) - \tilde{H}(t_{i-1})$ and $\Delta \hat{V}[\tilde{H}(t_i)] = \hat{V}[\tilde{H}(t_i)] - \hat{V}[\tilde{H}(t_{i-1})]$ denote the magnitude of the jumps in $\tilde{H}(t_i)$ and $\hat{V}[\tilde{H}(t_i)]$ at time t_i. Note that $\Delta \tilde{H}(t_i)$ provides a crude estimator of $h(t)$ at the death times. The kernel-smoothed estimator of $h(t)$ is a weighted average of these crude estimates over event times close to t. Closeness is determined by a bandwidth b, so that event times in the range $t - b$ to $t + b$ are included in the weighted average which estimates $h(t)$. The bandwidth is chosen either to minimize some measure of the mean-squared error or to give a desired degree of smoothness, as illustrated in Example 6.2. The weights are controlled by the choice of a kernel function, $K()$, defined on the interval $[-1, +1]$, which determines how much weight is given to points at a distance from t. Common choices for the kernel are the uniform kernel with

$$K(x) = 1/2 \quad \text{for } -1 \leq x \leq 1, \qquad (6.2.1)$$

the Epanechnikov kernel with

$$K(x) = 0.75(1 - x^2) \quad \text{for } -1 \leq x \leq 1, \qquad (6.2.2)$$

and the biweight kernel with

$$K(x) = \frac{15}{16}(1 - x^2)^2 \quad \text{for } -1 \leq x \leq 1. \qquad (6.2.3)$$

The uniform kernel gives equal weight to all deaths in the interval $t - b$ to $t + b$, whereas the other two kernels give progressively heavier weight to points close to t.

The kernel-smoothed hazard rate estimator is defined for all time points $t > 0$. For time points t for which $b \leq t \leq t_D - b$, the kernel-smoothed estimator of $h(t)$ based on the kernel $K()$ is given by

$$\hat{h}(t) = b^{-1} \sum_{i=1}^{D} K\left(\frac{t - t_i}{b}\right) \Delta \tilde{H}(t_i). \qquad (6.2.4)$$

The variance of $\hat{h}(t)$ is estimated by the quantity

$$\sigma^2[\hat{h}(t)] = b^{-2} \sum_{i=1}^{D} K\left(\frac{t - t_i}{b}\right)^2 \Delta \hat{V}[\tilde{H}(t_i)]. \qquad (6.2.5)$$

When t is smaller than b, the symmetric kernels described in (6.2.1)–(6.2.3) are not appropriate because no event times less than 0 are observable. In this region, the use of an asymmetric kernel is suggested. Let $q = t/b$. We define a modified kernel which accounts for the restricted range of the data. Following Gasser and Müller (1979) these modified kernels, for the uniform kernel (6.2.1), are expressed by

$$K_q(x) = \frac{4(1 + q^3)}{(1 + q)^4} + \frac{6(1 - q)}{(1 + q)^3}x, \quad \text{for } -1 \leq x \leq q, \qquad (6.2.6)$$

for the Epanechnikov kernel (6.2.2),

$$K_q(x) = K(x)(\alpha_E + \beta_E x), \quad \text{for } -1 \leq x \leq q, \qquad (6.2.7)$$

where

$$\alpha_E = \frac{64(2 - 4q + 6q^2 - 3q^3)}{(1 + q)^4(19 - 18q + 3q^2)}$$

and

$$\beta_E = \frac{240(1 - q)^2}{(1 + q)^4(19 - 18q + 3q^2)},$$

and for the biweight kernel (6.2.3),

$$K_q(x) = K(x)(\alpha_{BW} + \beta_{BW}x), \quad \text{for } -1 \le x \le q, \qquad (6.2.8)$$

where

$$\alpha_{BW} = \frac{64(8 - 24q + 48q^2 - 45q^3 + 15q^4)}{(1 + q)^5(81 - 168q + 126q^2 - 40q^3 + 5q^4)}$$

and

$$\beta_{BW} = \frac{1120(1 - q)^3}{(1 + q)^5(81 - 168q + 126q^2 - 40q^3 + 5q^4)}.$$

For time points in the right-hand tail ($t_D - b < t < t_D$) let $q = (t_D - t)/b$. The asymmetric kernel $K_q(x)$ in (6.2.6)–(6.2.8) is used with x replaced by $-x$. The estimated, smoothed, hazard rate and its variance are given by (6.2.4) and (6.2.5), respectively, using the kernel K_q.

Confidence intervals or confidence bands for the hazard rate, based on the smoothed hazard rate estimate, can be constructed similarly to those for the cumulative hazard rate discussed in Chapter 4. For example a $(1 - \alpha) \times 100\%$ pointwise confidence interval for the hazard rate, based on a log transformation, is expressed as

$$\hat{h}(t) \exp\left[\pm \frac{Z_{1-\alpha/2}\sigma(\hat{h}(t))}{\hat{h}(t)}\right].$$

Some care in interpreting this interval must be taken because the estimator $\hat{h}(t)$ may be quite biased (See Practical Note 1).

EXAMPLE 6.1

We shall find the smoothed hazard rate estimates, in the three disease categories, for the disease-free survival times of bone marrow transplant patients discussed in section 1.3. To illustrate the calculations, consider the group of ALL patients. In Example 4.2 the Nelson–Aalen estimator of the cumulative hazard rate of the disease-free survival time was found (see Table 4.3). For illustrative purposes, we shall use the Epanechnikov kernel with a bandwidth of 100 days. An estimate of $h(t)$ over the first two years (730 days) after transplant is desired.

Table 6.1 shows some of the calculations needed to construct the estimate. First, consider the estimate at $t = 150$ days. Here, t is in the interval b to $t_D - b$ (662–100), so that the symmetric kernel (6.2.2) is used. The estimate of the hazard rate is given by $\hat{h}(150) = [0.0270 \times 0.0731 + 0.0278 \times 0.3168 + 0.0286 \times 0.4428 + 0.0294 \times 0.5913 + 0.0303 \times 0.6113 + 0.0313 \times 0.6239 + 0.0322 \times 0.6300 + 0.0667 \times 0.6912 + 0.0357 \times 0.7169 + 0.0370 \times 0.7137 + 0.0385 \times 0.6177 + 0.0400 \times 0.6048 + 0.0435 \times 0.2700]/100 = 0.00257$. Similar calculations, using (6.2.6), yield an estimated standard error of $\sigma(\hat{h}(150)) = 0.00073$.

TABLE 6.1

Weights Used in Smoothing the Nelson–Aalen Estimator for the ALL Group

t_i	$\Delta \tilde{H}(t_i)]$	$\Delta \hat{V}[\tilde{H}(t_i)]$	$\dfrac{150 - t_i}{100}$	$K\left(\dfrac{150 - t_i}{100}\right)$	$\dfrac{50 - t_i}{100}$	$K\left(\dfrac{50 - t_i}{100}\right)$	$\dfrac{600 - t_i}{100}$	$K\left(\dfrac{600 - t_i}{100}\right)$
1	0.0263	0.00069	1.49	0.0000	0.49	1.0618	5.99	0.0000
55	0.0270	0.00073	0.95	0.0731	−0.05	0.9485	5.45	0.0000
74	0.0278	0.00077	0.76	0.3168	−0.24	0.7482	5.26	0.0000
86	0.0286	0.00082	0.64	0.4428	−0.36	0.6047	5.14	0.0000
104	0.0294	0.00087	0.46	0.5913	−0.54	0.3867	4.96	0.0000
107	0.0303	0.00091	0.43	0.6113	−0.57	0.3518	4.93	0.0000
109	0.0313	0.00099	0.41	0.6239	−0.59	0.3290	4.91	0.0000
110	0.0322	0.00103	0.40	0.6300	−0.60	0.3177	4.90	0.0000
122	0.0667	0.00222	0.28	0.6912	−0.72	0.1913	4.78	0.0000
129	0.0357	0.00128	0.21	0.7169	−0.79	0.1275	4.71	0.0000
172	0.0370	0.00138	−0.22	0.7137	−1.22	0.0000	4.28	0.0000
192	0.0385	0.00147	−0.42	0.6177	−1.42	0.0000	4.08	0.0000
194	0.0400	0.00161	−0.44	0.6048	−1.44	0.0000	4.06	0.0000
230	0.0435	0.00188	−0.80	0.2700	−1.80	0.0000	3.70	0.0000
276	0.0454	0.00207	−1.26	0.0000	−2.26	0.0000	3.24	0.0000
332	0.0476	0.00228	−1.82	0.0000	−2.82	0.0000	2.68	0.0000
383	0.0500	0.00247	−2.33	0.0000	−3.33	0.0000	2.17	0.0000
418	0.0527	0.00277	−2.68	0.0000	−3.68	0.0000	1.82	0.0000
468	0.0555	0.00310	−3.18	0.0000	−4.18	0.0000	1.32	0.0000
487	0.0589	0.00345	−3.37	0.0000	−4.37	0.0000	1.13	0.0000
526	0.0625	0.00391	−3.76	0.0000	−4.76	0.0000	0.74	0.2492
609	0.0714	0.00511	−4.59	0.0000	−5.59	0.0000	−0.09	0.8918
662	0.0769	0.00592	−5.12	0.0000	−6.12	0.0000	−0.62	0.6904

At $t = 50$ days, the asymmetric kernel (6.2.7) is used with $q = 50/100 = 0.5$. We have $\alpha_E = 64(2 - 4 \times 0.5 + 6 \times 0.5^2 - 3 \times 0.5^3)/[(1 + 0.5)^4(19 - 18 \times 0.5 + 3 \times 0.5^2)] = 1.323$ and $\beta_E = 240(1 - 0.5)^2/[(1 + .5)^4(19 - 18 \times 0.5 + 3 \times 0.5^2)] = 1.102$. Thus $K_{0.5}(-0.05) = 0.75[1.323 + 1.102(-0.05)] \times (1 - 0.05^2) = 0.9485$. Applying formulas (6.2.4) and (6.2.5) yields $\hat{h}(50) = 0.0015$ and $\sigma[\hat{h}(50)] = 0.00052$. Note that the tail adjustment using this kernel gives a higher weight to estimates of $\Delta \tilde{H}$ smaller than 50 to compensate for the fact that we can not observe any estimates in the range −50 to 0.

At $t = 600$ days we make the upper tail correction. Here $q = (662 - 600)/100 = 0.62$, which yields $\alpha_E = 1.148$ and $\beta_E = 0.560$. Only deaths in the range 500–662 have a nonzero value of the kernel. For $t_i = 609$ days ($x = -0.09$) the weight is $K(-0.09) = 0.75[1.148 + 0.560(0.09)](1 - 0.09^2) = 0.8918$. Note that, because we are estimating h in the right-hand tail, we have replaced −0.09 by 0.09. Applying (6.2.4) and (6.2.5) yields $\hat{h}(600) = 0.0013$ and $\sigma[\hat{h}(600)] = 0.00084$.

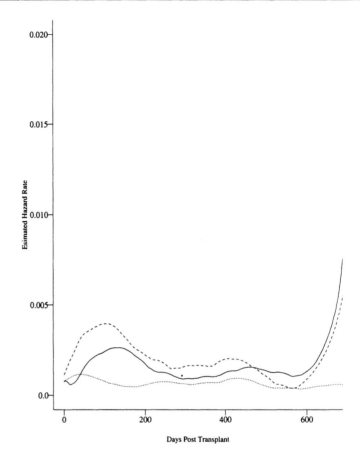

Figure 6.1   *Smoothed estimates of the hazard rates for bone marrow transplant patients based on the Epanechnikov kernel with a bandwidth of 100 days. ALL (———); AML-Low risk (--------); AML-High risk (— — —).*

Figure 6.1 shows the estimated hazard rates for the three disease groups, indicating that the risk of relapse or death increases in the first 150 days after transplant after which the hazard rate decreases. The initial peak is higher for AML high-risk patients. The estimated hazard rates again confirm the impression that AML low-risk patients have the lowest rate of relapse or death.

EXAMPLE 6.2   We shall illustrate the effects of changing the bandwidth and the choice of kernel on the kidney transplant data in section 1.7. Here, we shall ignore the age and race of the patient at the time of transplant. The estimate of the hazard rate constructed serves as the unadjusted mortality rate for these transplant patients.

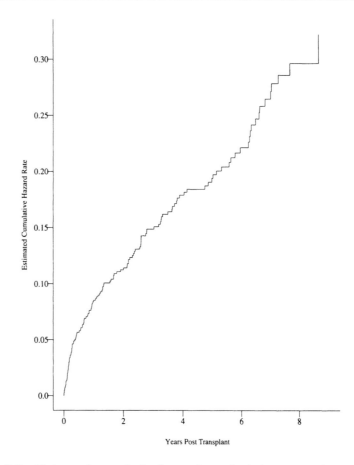

Figure 6.2   *Estimated cumulative hazard rate for kidney transplant patients*

Figure 6.2 shows the Nelson–Aalen estimate of the cumulative hazard rate on which the smoothed hazard rate estimator is based. Figure 6.3 shows the estimated hazard rate based on a bandwidth of 1 year for the uniform, Epanechnikov, and biweight kernels. Note that the kernels provide different degrees of smoothness. The biweight kernel is the smoothest, whereas the uniform kernel is rather jagged, typical of the performance of these kernels.

Figure 6.4 shows the effects of changing the bandwidth on the estimate of $b(t)$. In this figure, based on the Epanechnikov kernel, we see that increasing the bandwidth provides smoother estimates of the hazard rate. This increase in smoothness is at the expense of an increase in the bias of the estimate (see Practical Note 1).

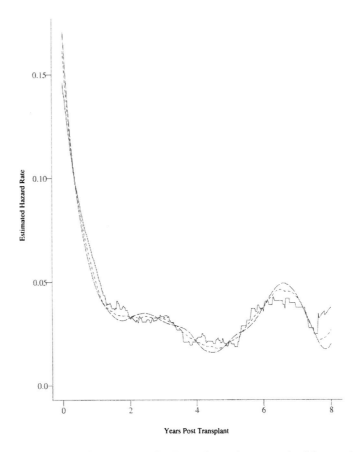

Figure 6.3 *Effects of changing the kernel on the smoothed hazard rate estimates for kidney transplant patients using a bandwidth of 1 year. Uniform kernel (———); Epanechnikov kernel (------) Biweight kernel (— — —)*

One problem in using kernel smoothing to obtain an estimate of the hazard rate is the selection of the proper bandwidth. One way to pick a good bandwidth is to use a cross-validation technique for determining the bandwidth that minimizes some measure of how well the estimator performs. One such measure is the mean integrated squared error (MISE) of $\hat{h}$ over the range τ_L to τ_U defined by

$$
MISE(b) = E \int_{\tau_L}^{\tau_U} [\hat{h}(u) - h(u)]^2 \, du
$$

$$
= E \int_{\tau_L}^{\tau_U} \hat{h}^2(u) \, du - 2E \int_{\tau_L}^{\tau_U} \hat{h}(u)h(u) \, du + E \int_{\tau_L}^{\tau_U} h^2(u) \, du.
$$

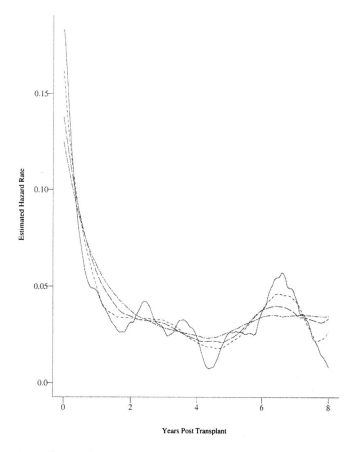

Figure 6.4 *Effects of changing the bandwidth on the smoothed hazard rate estimates for kidney transplant patients using the Epanechnikov kernel. bandwidth = 0.5 years (————) bandwidth = 1.0 years (-------) bandwidth = 1.5 years (— — —) bandwidth = 2.0 years (− · − · −)*

This function depends both on the kernel used to estimate b and on the bandwidth b. Note that, although the last term depends on the unknown hazard rate, it is independent of the choice of the kernel and the bandwidth and can be ignored when finding the best value of b. The first term can be estimated by $\int_{\tau_L}^{\tau_U} \hat{b}^2(u)\,du$. If we evaluate $\hat{b}$ at a grid of points $\tau_L = u_1 < \cdots < u_M = \tau_U$, then, an approximation to this integral by the trapezoid rule is $\sum_{i=1}^{M-1}\left(\frac{u_{i+1}-u_i}{2}\right)[\hat{b}^2(u_i) + \hat{b}^2(u_{i+1})]$. The second term can be estimated by a cross-validation estimate suggested by Ramlau–Hansen (1983a and b). This estimate is $b^{-1}\sum_{i\neq j} K(\frac{t_i - t_j}{b})\Delta\tilde{H}(t_i)\Delta\tilde{H}(t_j)$, where the sum is over the event times between τ_L and τ_U. Thus, to find the best value of b which minimizes the MISE for a fixed kernel, we find b which minimizes the function

$$g(b) = \sum_{i=1}^{M-1} \left(\frac{u_{i+1} - u_i}{2}\right) [\hat{h}^2(u_i) + \hat{h}^2(u_{i+1})]$$

$$- 2b^{-1} \sum_{i \neq j} K \left(\frac{t_i - t_j}{b}\right) \Delta \tilde{H}(t_i) \Delta \tilde{H}(t_j).$$

EXAMPLE 6.2   *(continued)*   To find the best bandwidth for the kidney transplant patients, in Figure 6.5 we show a plot of b versus $g(b)$ for the three kernels with $\tau_L = 0$, $\tau_U = 6$ years. This figure is based on a grid of 100 equally spaced values for b over the range 0.01–1.00. The optimal values of b are 0.17 for the uniform kernel, 0.20 for the Epanechnikov kernel and

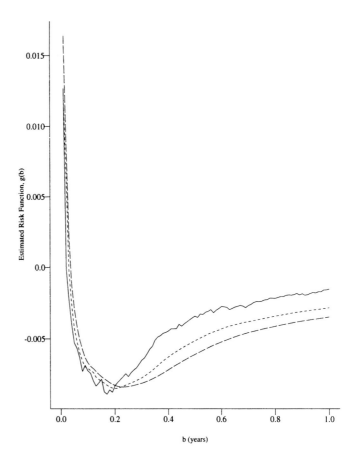

Figure 6.5   *Estimated risk function, g(b), for use in determination of the best bandwidth for the kidney transplant data. Uniform kernel (———); Epanechnikov kernel (------) Biweight kernel (———).*

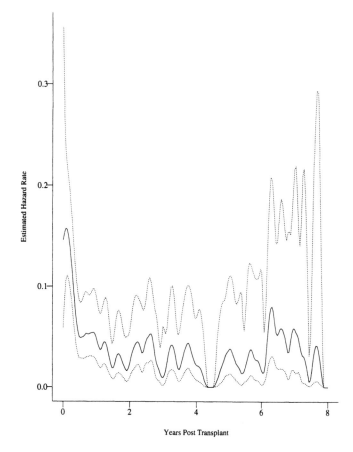

Figure 6.6   *Smoothed estimate of the hazard rate (———) and 95% confidence interval (------) for the time to death following a kidney transplant based on the biweight kernel and the best bandwidth.*

0.23 for the biweight kernel. Figure 6.6 shows the estimated hazard rate and a 95% pointwise confidence interval based on the biweight kernel with this optimal bandwidth.

EXAMPLE 6.1   *(continued)* The cross validation technique yields optimal bandwidths, based on the Epanechnikov kernel, of 161 days for the ALL group, 50 days for the AML low-risk group, and 112 days for the AML high-risk group for the estimates of the hazard rate over the range 0–730 days. Figure 6.7 shows the estimated hazard rates using these values of *b*.

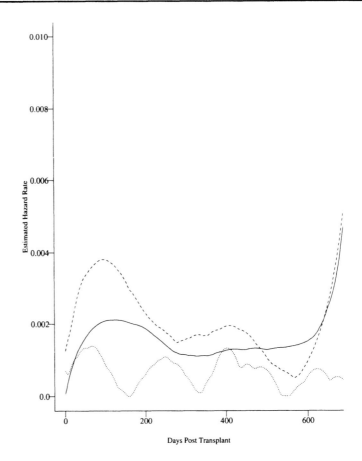

Figure 6.7   *Smoothed estimates of the hazard rates for bone marrow transplant patients based on the Epanechnikov kernel using optimal bandwidths. AML-Low risk (------) AML-High risk (— — —) ALL (————)*

Practical Notes

1. One must be very careful in interpreting the kernel-smoothed estimates constructed by these techniques. What these statistics are estimating is not the hazard rate $h(t)$, but rather a smoothed version of the hazard rate $h^*(t)$. This quantity is defined by $h^*(t) = b^{-1} \int K(\frac{t-u}{b})h(u)du$. It depends on both the bandwidth b and the kernel used in estimation. The confidence interval formula is, in fact, a confidence interval for h^*.
2. All that is required to apply the techniques in this section is an estimator of the cumulative hazard rate and its variance. Hence, these techniques apply equally well to right-censored or left-truncated data.

3. The smoothed estimator of the hazard rate was first introduced in Ramlau–Hansen (1983a and b). A detailed discussion of the large-sample properties of this estimator can be found in Andersen et al. (1993). A good general survey of smoothing techniques is found in Izenman (1991).

Theoretical Notes

1. The mean integrated squared error (MISE) measures $E\{\int_{\tau_L}^{\tau_U}[\hat{h}(u) - h(u)]^2 du\}$. This quantity is asymptotically approximately equal to the sum of a "bias" term, $\int\{h^*(u) - h(u)\}^2 du$ and a "variance" term $\int E\{[\hat{h}(u) - h^*(u)]^2\}du$. A small bandwidth produces a small bias term, but a large variance term, whereas the reverse holds for a large bandwidth. The optimal bandwidth is a trade-off between the two terms.
2. The bias of the smoothed hazard rate estimator, for large n, is approximately, $0.5b^2h''(t)k^*$, where h'' is the second derivative of h and $k^* = \int_{-1}^1 s^2 K(s)ds$.

6.3   Estimation of Excess Mortality

In some applications of survival analysis techniques, it is of interest to compare the mortality experience of a group of individuals to a known standard survival curve. The reference survival curve, which may be different for each individual in the sample, could be drawn from published mortality tables or other population-based mortality studies. Two simple models have been proposed to provide an inference on how the study population's mortality differs from that in the reference population.

Suppose we have data on n individuals. Let $\theta_j(t)$ be the reference hazard rate for the jth individual in the study. This known reference hazard rate typically depends on the characteristics of the jth patient, such as race, sex, age, etc. The first model for excess mortality, commonly known as the relative mortality model, assumes that the hazard rate at time t for the jth patient under study conditions, $h_j(t)$, is a multiple, $\beta(t)$, of the reference hazard rate for this individual, that is,

$$h_j(t) = \beta(t)\theta_j(t), \quad j = 1, 2, \ldots, n. \tag{6.3.1}$$

Here, if $\beta(t)$ is greater than 1, then, individuals in the study group are experiencing the event of interest at a faster rate than comparable individuals in the reference population. Let $B(t) = \int_0^t \beta(u)\,du$ be the cumulative relative excess mortality.

The data available for estimating $B(t)$, for each individual, consists of study times and death indicators. For the jth individual, let $Y_j(t)$ be 1 if the individual is at risk at time t and 0, otherwise. Note that this definition of $Y_j(t)$ allows for left-truncated and right-censored data. Define the function $Q(t) = \sum_{j=1}^n \theta_j(t)Y_j(t)$. To allow for ties in the data, let $t_1 < t_2 < \cdots < t_D$ be the times at which the events occur and d_i the number of events observed at time t_i. The estimator of $B(t)$ is given by

$$\hat{B}(t) = \sum_{t_i \le t} \frac{d_i}{Q(t_i)}. \tag{6.3.2}$$

An estimator of the variance of $\hat{B}(t)$ is given by

$$\hat{V}[\hat{B}(t)] = \sum_{t_i \le t} \frac{d_i}{Q(t_i)^2}. \tag{6.3.3}$$

The statistic $\hat{B}(t)$ has a large-sample normal distribution so that confidence intervals or confidence bands for the cumulative relative mortality can be constructed by replacing the Nelson–Aalen estimator and its variance by $\hat{B}(t)$ and its variance in the appropriate formulas in sections 4.4 and 4.5. A crude estimator of the relative risk function $\beta(t)$ is given by the slope of the estimated cumulative relative mortality estimator. An improved estimator of $\hat{B}(t)$ can be found by a kernel smoothing of $\hat{B}(t)$ similar to that developed for the estimated cumulative hazard rate discussed in the previous section.

EXAMPLE 6.3   To illustrate the estimation of the relative mortality function consider the data on the 26 psychiatric patients in Iowa described in section 1.15. We shall use the 1959–1961 Iowa State life tables (US Dept. of Health and Human Services (1959)) as the reference population. This life table in Table 6.2 is based on the 1960 census and the average number of deaths in the period 1959–1961 and provides the population survival functions $S(\ )$ for males and females. For the population hazard rates, we assume that the hazard rates are constant over each one year interval reported in the table, so that the hazard rate at age a is $\lambda(a) = \ln S(a) - \ln S(a + 1)$. Table 6.2 shows values of the estimated hazard rates for males and female for $a = 18, 19, \ldots, 77$.

The time scale used in this example is the time on study for each patient. A patient who enters the study at age a has $\theta_i(t)$ found by using the hazard rate in the $(a + t)$th row of Table 6.2. For example, the female who entered the study at age 36 has $\theta(1) = \lambda_F(36+1) = 0.00130$, $\theta(2) = \lambda_F(38) = 0.00140$, etc. Table 6.3 shows the estimate of $B(t)$

TABLE 6.2

1960 Iowa Standard Mortality

	Males				
Age	Survival Function	Hazard Rate	Age	Survival Function	Hazard Rate
18–19	0.96394	0.00154	48–49	0.89596	0.00694
19–20	0.96246	0.00164	49–50	0.88976	0.00751
20–21	0.96088	0.00176	50–51	0.88310	0.00810
21–22	0.95919	0.00188	51–52	0.87598	0.00877
22–23	0.95739	0.00190	52–53	0.86833	0.00956
23–24	0.95557	0.00185	53–54	0.86007	0.01052
24–25	0.95380	0.00173	54–55	0.85107	0.01159
25–26	0.95215	0.00158	55–56	0.84126	0.01278
26–27	0.95065	0.00145	56–57	0.83058	0.01402
27–28	0.94927	0.00137	57–58	0.81902	0.01536
28–29	0.94797	0.00134	58–59	0.80654	0.01683
29–30	0.94670	0.00136	59–60	0.79308	0.01844
30–31	0.94541	0.00141	60–61	0.77859	0.02013
31–32	0.94408	0.00146	61–62	0.76307	0.02195
32–33	0.94270	0.00153	62–63	0.74650	0.02386
33–34	0.94126	0.00159	63–64	0.72890	0.02586
34–35	0.93976	0.00170	64–65	0.71029	0.02795
35–36	0.93816	0.00181	65–66	0.69071	0.03020
36–37	0.93646	0.00198	66–67	0.67016	0.03262
37–38	0.93461	0.00215	67–68	0.64865	0.03521
38–39	0.93260	0.00235	68–69	0.62621	0.03800
39–40	0.93041	0.00258	69–70	0.60286	0.04102
40–41	0.92801	0.00284	70–71	0.57863	0.04424
41–42	0.92538	0.00312	71–72	0.55359	0.04773
42–43	0.92250	0.00350	72–73	0.52779	0.05175
43–44	0.91928	0.00397	73–74	0.50117	0.05646
44–45	0.91564	0.00450	74–75	0.47366	0.06188
45–46	0.91153	0.00511	75–76	0.44524	0.06795
46–47	0.90688	0.00575	76–77	0.41599	0.07454
47–48	0.90168	0.00636	77–78	0.38611	0.08181

and its standard error. Figure 6.8 shows the estimated value of $B(t)$ and a 95% pointwise confidence interval for $B(t)$ based on the log-transformed confidence interval formula for the cumulative hazard rate. (See Practical Note 1 in section 4.3. Here we use Eq. 4.3.5 and replace $\tilde{H}(t_0)$ by $\hat{B}(t)$ and $\sigma_H(t_0)$ by the standard error of $\hat{B}(t)$.)

The slope of $\hat{B}(t)$ in Figure 6.8 provides a crude estimate of $\beta(t)$. Here, we see that, in the first two years of observation, psychiatric patients were 20–30 times more likely to die than comparable individuals

TABLE 6.2
1960 Iowa Standard Mortality

	Females				
Age	Survival Function	Hazard Rate	Age	Survival Function	Hazard Rate
18–19	0.97372	0.00057	48–49	0.93827	0.00352
19–20	0.97317	0.00056	49–50	0.93497	0.00381
20–21	0.97263	0.00055	50–51	0.93141	0.00414
21–22	0.97210	0.00054	51–52	0.92756	0.00448
22–23	0.97158	0.00054	52–53	0.92341	0.00481
23–24	0.97106	0.00056	53–54	0.91898	0.00509
24–25	0.97052	0.00059	54–55	0.91431	0.00536
25–26	0.96995	0.00062	55–56	0.90942	0.00565
26–27	0.96935	0.00065	56–57	0.90430	0.00600
27–28	0.96872	0.00069	57–58	0.89889	0.00653
28–29	0.96805	0.00072	58–59	0.89304	0.00724
29–30	0.96735	0.00075	59–60	0.88660	0.00812
30–31	0.96662	0.00079	60–61	0.87943	0.00912
31–32	0.96586	0.00084	61–62	0.87145	0.01020
32–33	0.96505	0.00088	62–63	0.86261	0.01132
33–34	0.96420	0.00095	63–64	0.85290	0.01251
34–35	0.96328	0.00103	64–65	0.84230	0.01376
35–36	0.96229	0.00110	65–66	0.83079	0.01515
36–37	0.96123	0.00121	66–67	0.81830	0.01671
37–38	0.96007	0.00130	67–68	0.80474	0.01846
38–39	0.95882	0.00140	68–69	0.79002	0.02040
39–40	0.95748	0.00152	69–70	0.77407	0.02259
40–41	0.95603	0.00162	70–71	0.75678	0.02494
41–42	0.95448	0.00176	71–72	0.73814	0.02754
42–43	0.95280	0.00193	72–73	0.71809	0.03067
43–44	0.95096	0.00216	73–74	0.69640	0.03446
44–45	0.94891	0.00240	74–75	0.67281	0.03890
45–46	0.94664	0.00268	75–76	0.64714	0.04376
46–47	0.94411	0.00296	76–77	0.61943	0.04902
47–48	0.94132	0.00325	77–78	0.58980	0.05499

in the standard population. In years 3–40, the patients were between 2–5 times more likely to die.

A second model, which can be used for comparing the study population to a reference population is the excess or additive mortality model. Here, we assume that the hazard rate at time t for the jth individual under study is a sum of the population mortality rate $\theta_j(t)$ and an excess mortality function $\alpha(t)$. The function $\alpha(t)$, which is assumed to be

TABLE 6.3

Computation of Cumulative Relative Mortality for 26 Psychiatric Patients

t_i	d_i	$Q(t_i)$	$\hat{B}(t)$	$\hat{V}[\hat{B}(t)]$	$\sqrt{\hat{V}[\hat{B}(t)]}$
1	2	0.05932	33.72	568.44	23.84
2	1	0.04964	53.86	974.20	31.21
11	1	0.08524	65.59	1111.84	33.34
14	1	0.10278	75.32	1206.51	34.73
22	2	0.19232	85.72	1260.58	35.50
24	1	0.19571	90.83	1286.69	35.87
25	1	0.18990	96.10	1314.42	36.25
26	1	0.18447	101.52	1343.81	36.66
28	1	0.19428	106.67	1370.30	37.02
32	1	0.18562	112.05	1399.32	37.41
35	1	0.16755	118.02	1434.94	37.88
40	1	0.04902	138.42	1851.16	43.03

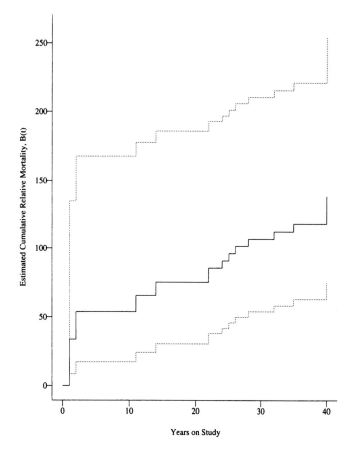

Figure 6.8 *Estimated cumulative relative mortality (solid line) and 95% point-wise confidence interval (dashed line) for Iowa psychiatric patients*

the same for all individuals in the study group, can be positive when study patients are dying faster than those in the reference population or be negative when the study group has a better survival rate than the reference population. The model is

$$b_j(t) = \alpha(t) + \theta_j(t), \quad j = 1, \dots, n. \tag{6.3.4}$$

As in the case of the multiplicative model, direct estimation of $\alpha()$ is difficult. Instead, we estimated the cumulative excess mortality function $A(t) = \int_0^t \alpha(u)\,du$. The estimator of $A(t)$ is constructed from the difference of the observed hazard rate, estimated by the ordinary Nelson–Aalen estimator (see section 4.2) $\tilde{H}(t)$ and an "expected" cumulative hazard rate $\Theta(t)$ based on the reference hazard rates. The expected cumulative hazard rate is a weighted average of the reference cumulative hazard rates at each time, where the weights are based on the fraction of individuals at risk at time t, that is,

$$\Theta(t) = \sum_{j=1}^{n} \int_0^t \theta_j(u) \frac{Y_j(u)}{Y(u)}\,du, \tag{6.3.5}$$

where $Y(t) = \sum_{j=1}^{n} Y_j(t)$ is the number at risk at time t.

The estimated excess mortality is given by

$$\hat{A}(t) = \sum_{t_i \le t} \frac{d_i}{Y(t)} - \Theta(t). \tag{6.3.6}$$

The estimated variance of the cumulative excess mortality function is given by the variance of the Nelson–Aalen estimator, namely,

$$\hat{V}[\hat{A}(t)] = \sum_{t_i \le t} \frac{d_i}{Y(t)^2}. \tag{6.3.7}$$

As for the relative mortality model, confidence intervals and confidence bands for $A(t)$ can be computed using the techniques in sections 4.3 and 4.4, and smoothed estimates of $\alpha(t)$ can be constructed using the methods of the previous section.

The $\hat{A}(t)$ may be either negative or positive. It will be decreasing and negative for times smaller than the smallest death time. With this caution in mind, one may use these estimates to construct "corrected" survival curves. The Kaplan–Meier estimator, $\hat{S}(t)$, provides an estimate of the observed or uncorrected survival curve. The survival curve, $S^*(t) = \exp[-\Theta(t)]$, provides an estimate of the expected survival curve if the reference mortality model is the same as the study population. The ratio of these two survival functions, $S^C(t) = \hat{S}(t)/S^*(t)$, is taken as a "corrected" survival function estimate for the study population. Care must be taken in using this curve because the ratio of the two curves may be greater than one (especially for small t) and the corrected survival curve need not be nonincreasing.

EXAMPLE 6.3

(continued) To estimate the expected hazard rate using the standard Iowa mortality data, we first compute $\Theta(t)$. Here, we assume, again, that the hazard rates are constant over each age interval of unit length which simplifies computations. At one year after entry into the study, $\Theta(1) = \sum_{j=1}^{n} \lambda_S(a_j)/26$, where a_j is the age of the jth individual at entry into the study and $\lambda_S()$ is the value of the hazard rate from Table 6.2 for the patient's sex. For an integer age $t > 1$, $\Theta(t) = \Theta(t-1) + \sum_t \lambda_S(a_j + t - 1)/Y(t)$, where the sum is over all patients under observation in the interval $[t-1, t)$. For noninteger times, $\Theta(t)$ is found by linear interpolation.

Table 6.4 shows the results of the computations. Figure 6.9 shows the observed cumulative hazard rate $[\tilde{H}(t)]$, the expected cumulative hazard rate $[\Theta(t)]$ and the cumulative excess mortality $[\hat{A}(t)]$. Notice that the expected cumulative hazard function is a smooth function of the number of years on study, whereas the Nelson–Aalen estimator is a step function with jumps at the observed death times. The excess mortality function has jumps at the death times and is decreasing between the death times. From this figure, we see that a crude estimate of $\alpha(t)$, given by the slope of $\hat{A}(t)$, is a function which is about 0.05 for $t < 2$, about 0 for $2 < t < 21$, and, then, about 0.05 for $t > 21$. After 30 years on study, the cumulative excess mortality is about 0.35, so we estimate that, in a group of 100 patients, we would see 35 more deaths after 30 years than we would expect to see in a standard population. A crude 95% confidence interval for the excess number of deaths after 30 years is $0.3592 \pm 1.96(0.1625)$ or $(0.0407, 0.6777)$. These estimates are a bit imprecise due to the relatively small sample size of this study.

Figure 6.10 depicts the adjusted survival curves for this study. Again, the expected survival function is a smooth curve, and the observed survival curve is a step function. It is of interest here to note that the "corrected" survival curve is not monotone decreasing and, as such, is not strictly a survival curve. A better graphical representation is to plot this function by connecting the points $\hat{S}(t_i)/S^*(t_i)$ only at the death times.

Practical Notes

1. The estimator of relative mortality is a time-varying extension of the standard mortality ratio (SMR) estimator (Breslow, 1975) which assumes a constant relative mortality over time. For this estimator, one computes $E(t) = \int_0^t Q(u)du$, which is thought of as the expected number of deaths before time t. If $\beta(t) = \beta_0$, a constant, then, the

TABLE 6.4

Computation for the Excess Mortality Model

t_i	d_i	$Y(t_i)$	$\tilde{H}(t_i)$	$\Theta(t_i)$	$\hat{A}(t)$	$SE[\hat{A}(t)]$	$\hat{S}(t)$	$S^*(t_i)$	$\dfrac{\hat{S}(t_i)}{S^*(t_i)}$
1	2	26	0.0769	0.0021	0.0748	0.0544	0.9231	0.9979	0.9250
2	1	24	0.1186	0.0041	0.1145	0.0685	0.8846	0.9959	0.8882
3	0	23	0.1186	0.0059	0.1127	0.0685	0.8846	0.9941	0.8899
4	0	23	0.1186	0.0079	0.1107	0.0685	0.8846	0.9921	0.8917
5	0	23	0.1186	0.0101	0.1085	0.0685	0.8846	0.9900	0.8936
6	0	23	0.1186	0.0124	0.1062	0.0685	0.8846	0.9877	0.8956
7	0	23	0.1186	0.0148	0.1038	0.0685	0.8846	0.9853	0.8978
8	0	23	0.1186	0.0175	0.1011	0.0685	0.8846	0.9827	0.9002
9	0	23	0.1186	0.0203	0.0983	0.0685	0.8846	0.9799	0.9028
10	0	23	0.1186	0.0234	0.0952	0.0685	0.8846	0.9769	0.9056
11	1	23	0.1621	0.0268	0.1353	0.0811	0.8462	0.9736	0.8691
12	0	22	0.1621	0.0303	0.1318	0.0811	0.8462	0.9702	0.8722
13	0	22	0.1621	0.0341	0.1279	0.0811	0.8462	0.9664	0.8755
14	1	22	0.2075	0.0384	0.1691	0.0930	0.8077	0.9623	0.8393
15	0	21	0.2075	0.0428	0.1647	0.0930	0.8077	0.9581	0.8430
16	0	21	0.2075	0.0476	0.1599	0.0930	0.8077	0.9535	0.8471
17	0	21	0.2075	0.0530	0.1546	0.0930	0.8077	0.9484	0.8516
18	0	21	0.2075	0.0588	0.1487	0.0930	0.8077	0.9429	0.8566
19	0	21	0.2075	0.0652	0.1423	0.0930	0.8077	0.9369	0.8621
20	0	21	0.2075	0.0722	0.1353	0.0930	0.8077	0.9303	0.8682
21	0	21	0.2075	0.0799	0.1276	0.0930	0.8077	0.9232	0.8749
22	2	21	0.3028	0.0882	0.2145	0.1148	0.7308	0.9155	0.7982
23	0	19	0.3028	0.0968	0.2060	0.1148	0.7308	0.9078	0.8050
24	1	19	0.3554	0.1062	0.2492	0.1263	0.6923	0.8993	0.7698
25	1	18	0.4109	0.1158	0.2952	0.1380	0.6538	0.8907	0.7341
26	1	17	0.4698	0.1257	0.3441	0.1500	0.6154	0.8819	0.6978
27	0	16	0.4698	0.1358	0.3340	0.1500	0.6154	0.8730	0.7049
28	1	16	0.5323	0.1469	0.3854	0.1625	0.5769	0.8634	0.6682
29	0	15	0.5323	0.1594	0.3729	0.1625	0.5769	0.8527	0.6766
30	0	15	0.5323	0.1731	0.3592	0.1625	0.5769	0.8411	0.6860
31	0	13	0.5323	0.1874	0.3449	0.1625	0.5769	0.8291	0.6958
32	1	11	0.6232	0.2028	0.4204	0.1862	0.5245	0.8164	0.6424
33	0	10	0.6232	0.2207	0.4025	0.1862	0.5245	0.8019	0.6540
34	0	8	0.6232	0.2412	0.3820	0.1862	0.5245	0.7857	0.6676
35	1	7	0.7660	0.2631	0.5029	0.2347	0.4496	0.7687	0.5849
36	0	4	0.7660	0.2848	0.4812	0.2347	0.4496	0.7522	0.5977
37	0	3	0.7660	0.3133	0.4527	0.2347	0.4496	0.7310	0.6150
38	0	2	0.7660	0.3510	0.4150	0.2347	0.4496	0.7040	0.6387
39	0	2	0.7660	0.3926	0.3734	0.2347	0.4496	0.6753	0.6658
40	1	1	1.7660	0.4363	1.3297	1.0272	0.0000	0.6464	0.0000

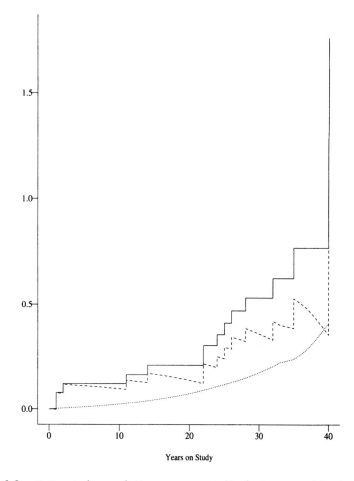

Figure 6.9 *Estimated cumulative excess mortality for Iowa psychiatric patients. Nelson–Aalen estimator (———) Expected cumulative hazard (------) Cumulative excess mortality (— — —)*

maximum likelihood estimator of β_0 is the total number of deaths divided by $E(t_{MAX})$, where t_{MAX} is the largest on study time. The SMR is 100 times this value. If the constant mortality model holds, then, a plot of $\hat{B}(t)$ versus t should be a straight line through the origin. Andersen and Væth (1989) present a test for constant mortality and a version of the total time on test plot which can be used as a graphical check of the assumption of constant relative mortality.

2. An estimator of constant excess mortality $\alpha(t) = \alpha_0$ was proposed by Buckley (1984). For this estimator, let $T(t) = \int_0^t Y(u)du$ be the total time on test at time t, that is, the number of person-years of

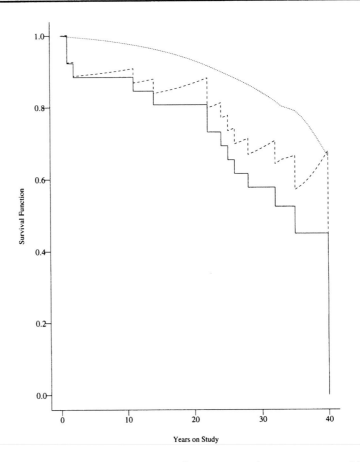

Figure 6.10   *Adjusted survival curves for Iowa psychiatric patients. Observed survival (———) Expected survival (------) Corrected survival (— — —)*

observation prior to time t. At the largest study time, t_{MAX}, $T(t_{MAX})$ is the total years of exposure of all study individuals. The statistic $\frac{D - E(t_{MAX})}{T(t_{MAX})}$ estimates α_0. Here D is the total number of deaths. This estimate is the difference between the occurrence/exposure rate and the expected number of deaths per time on test. Buckley also presents a maximum likelihood estimator that must be found numerically. Again, the constant excess mortality model is reasonable if the plot of $\hat{A}(t)$ versus t is linear. Andersen and Væth (1989) present a formal test.

3. A more general model for excess mortality is a mixed model. Here, $h_j(t) = \beta(t)\theta_j(t) + \alpha(t)$. This model can be fit into an additive regression formulation discussed in Chapter 10.

Theoretical Note

1. Detailed derivation of the estimators for excess and relative mortality are found in Andersen and Væth (1989). These statistics can be derived using a counting process technique as discussed in Andersen et al. (1993).

6.4   Bayesian Nonparametric Methods

An alternative to the classical nonparametric approach to estimating the survival function discussed in Chapters 4 and 5 is to use Bayesian nonparametric methods. In applying these methods, an investigator's a priori belief in the shape of the survival function is combined with the data to provide an estimated survival function. The prior information, which may be based on previous experience with the process under observation or based on expert opinion, is reflected in a prior distribution for the survival function. The sample information is contained in the likelihood function. These two distinct pieces of information are combined by Bayes' theorem to obtain an a posteriori distribution of the survival function which is the distribution of the survival function, given the data.

In the Bayesian approach, the parameters of the model are treated as random variables selected from the prior distribution. This prior distribution, which is a multivariate distribution on the parameters, is selected to reflect the investigator's prior belief in the values of the parameters. Typically, the prior means reflect the investigators best guess, before seeing any data, of the value of the parameters, and the prior variance is a measure of the investigator's uncertainty in his prior means. Often one can think of the prior variance as being inversely proportional to the amount of sample information to be represented by the prior.

In our problem, the parameter of interest is the survival function or, equivalently, the cumulative hazard function. This is to be treated as a random quantity sampled from some stochastic process. Nature picks a sample path from this stochastic process, and this is our survival function. We, then, have data sampled from a population with this survival function which we shall combine with our prior to obtain the distribution of the survival function, given the data.

To obtain an estimate of the survival function, we need to specify a loss function on which to base the decision rule. Analogous to the simple parametric case, we shall use the squared-error loss function

$$L(S, \hat{S}) = \int_0^\infty [\hat{S}(t) - S(t)]^2 \, dw(t),$$

where $w(t)$ is a weight function. This loss function is the weighted integrated difference between the true value of the survival function and our estimated value. For this loss function, the value of $\hat{S}$, which minimizes the posterior expected value of $L(S, \hat{S})$, is the posterior mean and the Bayes risk $E[L(S, \hat{S}) \mid \text{DATA}]$ is the posterior variance.

Two classes of prior distributions have been suggested for this problem. Both lead to closed form estimates of the survival function using the squared-error loss function. These priors are chosen because they are conjugate priors for either the survival function or the cumulative hazard function. For a conjugate prior, the prior and posterior distributions are in the same family.

The first prior is for the survival function. For this prior, we assume that the survival function is sampled from a Dirichlet process with a parameter function α. A Dirichlet process, defined on the positive real line, has the property that, for any set of intervals $A_1, \ldots, A_k$, which partition the positive real line, the joint distribution of the prior probabilities $Pr[X \in A_1] = W_1, \ldots, Pr[X \in A_k] = W_k$ has a k dimensional Dirichlet distribution with parameters $[\alpha(A_1), \ldots, \alpha(A_k)]$. This property must hold for any such set of intervals and any k. A k vector $(W_1, \ldots, W_k)$ has a k-dimensional Dirichlet distribution with parameters $(\alpha_1, \ldots, \alpha_k)$ if $W_i = Z_i / \sum_{i=1}^k Z_i$ where the Z_i's are independent gamma random variables with shape parameter α_i. The joint density function of $(W_1, \ldots, W_{k-1})$ is given by

$$f(w_1, \ldots, w_{k-1}) = \frac{\Gamma[\alpha_1 + \cdots + \alpha_k]}{\Gamma[\alpha_1] \cdots \Gamma[\alpha_k]} \left[\prod_{i=1}^{k-1} w_i^{\alpha_i - 1} \right] \left[1 - \sum_{i=1}^{k-1} w_i \right]^{\alpha_k - 1}.$$

The mean of W_i is α_i / α and the variance is $(\alpha - \alpha_i)\alpha_i / (\alpha^2 + \alpha^3)$ where $\alpha = \sum_{i=1}^k \alpha_i$. When $k = 2$ the Dirichlet distribution reduces to the beta distribution with parameters (α_1, α_2).

To assign a prior distribution to the survival function, we assume that $S(t)$ follows a Dirichlet distribution with parameter function α. Typically, we take the parameter function to be of the form $\alpha([t, \infty)) = cS_0(t)$ where $S_0(t)$ is our prior guess at the survival function and c is a measure of how much weight to put on our prior guess. With this prior distribution for $S(t)$, the prior mean is expressed by

$$E[S(t)] = \frac{\alpha(t, \infty)}{\alpha(0, \infty)} = \frac{cS_0(t)}{cS_0(0)} = S_0(t),$$

and the prior variance is given by

$$V[S(t)] = \frac{[\alpha(0, \infty) - \alpha(t, \infty)]\alpha(t, \infty)}{[\alpha(0, \infty)^2 + \alpha(0, \infty)^3]} = \frac{S_0(t)[1 - S_0(t)]}{c + 1}.$$

Note that the prior variance is the equivalent to the sample variance one would have if we had an uncensored sample of size $c + 1$ from a population with a survival function $S_0(t)$. To illustrate what the sample paths of the prior distribution for S look like, we have simulated 10 sample paths for a Dirichlet prior with $S_0(t) = \exp(-0.1t)$ and $c = 5$. These are plotted as dashed lines in Figure 6.11 along with their mean $S_0(t)$, which is plotted as a solid line. Here we see that each sample path is a nonincreasing function with a value of 1 at 0. Note that, although the curves are continuous functions, they are not too smooth in this example. As the value of c increases, the curves will become smoother.

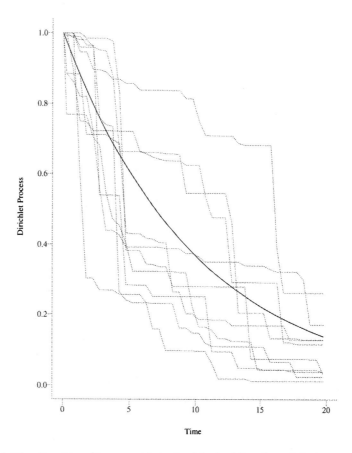

Figure 6.11 *Sample of ten sample paths (dashed lines) and their mean (solid line) for samples from a Dirichlet prior with $S_0(t) = \exp(-0.1t)$ and $c = 5$.*

The data we have available to combine with our prior consists of the on study times T_j and the event indicator, δ_j. To simplify calculations let $0 = t_o < t_1 < \cdots < t_M < t_{M+1} = \infty$, denote the M distinct times (censored or uncensored). At time t_i, let Y_i be the number of individuals at risk, d_i the number of deaths and λ_i the number of censored observations. Let Δ_i be 1 if $d_i > 0$ and 0 if $d_i = 0$.

Combining this data with the prior, we find that the posterior distribution of S is also Dirichlet. The parameter of the posterior distribution, α^*, is the original α parameter plus a point mass of one at points where deaths occur. That is, for any interval (a, b),

$$\alpha^*((a, b)) = \alpha((a, b)) + \sum_{j=1}^{n} I[\delta_j > 0, \ a < T_j < b],$$

where $I[\ ]$ is the indicator function.

The Bayes estimator of the survival function is

$$\tilde{S}_D(t) = \frac{\alpha(t, \infty) + Y_{i+1}}{\alpha(0, \infty) + n} \prod_{k=1}^{i} \frac{\alpha(t_k, \infty) + Y_{k+1} + \lambda_k}{\alpha(t_k, \infty) + Y_{k+1}} \qquad (6.4.1)$$

$$\text{for } t_i \leq t < t_{i+1}, \ i = 0, \ldots, M$$

The Bayes estimator is a continuous function between the distinct death times and has jumps at these death times. For large n this reduces to the Kaplan–Meier estimator, so that the prior information plays no role in the estimate. For small samples, the prior will dominate, and the estimator will be close to the prior guess at S.

A second approach to modeling prior information for survival data is to provide a prior distribution for the cumulative hazard function $H(t) = -\ln[S(t)]$. Here, we shall use a beta process prior. This prior depends on two parameters, $H_0(t)$ and $c(t)$. $H_0(t)$ is a prior guess at the value of the cumulative hazard rate $H(t)$, and $c(t)$ is a measure of how much weight to put on the prior guess at the function $H(t)$ at time t. For this prior, if we let $A_i = [a_{i-1}, a_i)$, $i = 1, \ldots, k$ be a series of nonoverlapping intervals with $0 = a_0 < a_1 < a_2 < \cdots < a_k$, then, a priori, $W_1 = H(a_1) - H(a_0), \ldots W_k = H(a_k) - H(a_{k-1})$ are independent beta random variables with parameters $p_i = c([a_i + a_{i-1}]/2)[H_0(a_i) - H_0(a_{i-1})]$ and $q_i = c([a_i + a_{i-1}]/2)\{1 - [H_0(a_i) - H_0(a_{i-1})]\}$. The prior mean of $H(a_i) - H(a_{i-1})$ is $H_0(a_i) - H_0(a_{i-1})$ and the prior variance is

$$V(W_i) = \frac{\{H_0(a_i) - H_0(a_{i-1})\}\{1 - [H_0(a_i) - H_0(a_{i-1})]\}}{c([a_i + a_{i-1}]/2) + 1}.$$

Here, $c(t)$ can be thought of as the weight to be given to our prior guess at $H_0(a_i) - H_0(a_{i-1})$ at the time $[a_i + a_{i-1}]/2$. The beta process prior is obtained by letting the number of subintervals increase to infinity, so that the interval lengths go to zero. Roughly speaking, $H(t)$ has a beta process if $dH(s)$ has a beta distribution with parameters $c(s)h_0(s)$ and

$c(s)[1 - h_0(s)]$, and $dH(s)$ is independent of $dH(u)$ for $u \neq s$. (Here, $dH(s) = [H(s + ds) - H(s)]ds$ for a very small increment of time, and $h_0(t) = dH_0(t)/dt$.)

To illustrate what the sample paths of the prior distribution, for S based on a beta process prior, look like, we have simulated 10 sample paths for a beta process prior with $H_0(t) = 0.1t$ and $c(t) = 5$. These are plotted as dashed lines in Figure 6.12 along with the prior guess at the survival function, $\exp(-0.1t)$. Here we see that each sample path is a nondecreasing function with a value of 1 at 0. As for the Dirichlet process prior, the sample paths are continuous and nondecreasing. As compared to the Dirichlet, the sample paths for the beta process prior are less variable, especially, in the middle section of the curve.

When the data is right-censored with $D(t)$ deaths observed at or prior to time t and $Y(t)$ individuals at risk at time t and a beta process prior is used, then, the posterior distribution of $H(t)$ is a beta process with

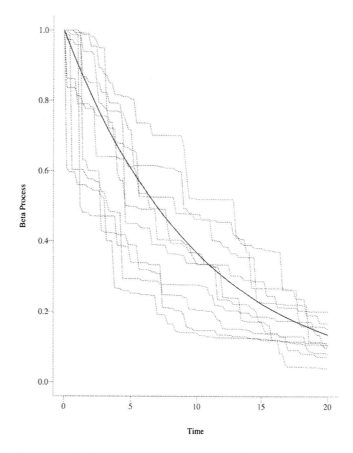

Figure 6.12   *Sample of ten sample paths (dashed lines) and their mean (solid line) for samples from a Beta process prior with $H_0(t) = 0.1t$ and $c(t) = 5$.*

parameters $[c(t)A_0(t) + Y(t)D(t)]/(c(t) + Y(t))$ and $c(t) + Y(t)$. Under squared-error loss the Bayes estimator of the survival function is given by

$$\tilde{S}_B(t) = \exp\left\{-\sum_{k=1}^{i}\int_{t_{k-1}}^{t_k}\frac{c(u)h_0(u)}{c(u)+Y_k} - \int_{t_i}^{t}\frac{c(u)h_0(u)}{c(u)+Y_{i+1}}du\right\} \quad (6.4.2)$$

$$\times \prod_{k:t_k\le t}\left[1 - \frac{c(t_k)h_0(t_k)+d_k}{c(t_k)+Y_k}\right]^{\Delta_k}, \quad \text{for } t_i \le t < t_{i+1}.$$

When $c(t)$ is a constant c, this reduces to

$$\tilde{S}_B(t) = \exp\left\{-\sum_{k=1}^{i}\frac{c[H_0(t_k)-H_0(t_{k-1})]}{c+Y_k} - \frac{c[H_0(t)-H_0(t_i)]}{c+Y_{i+1}}\right\}$$

$$\times \prod_{k:t_k\le t}\left[1 - \frac{ch_0(t_k)+d_k}{c+Y_k}\right]^{\Delta_k}, \quad \text{if } t_i \le t < t_{i+1}.$$

The estimator based on the Dirichlet prior has jumps at the death times and is continuous between deaths. Note that, as $c(t) \to 0$ this estimator reduces to the Kaplan–Meier estimator.

EXAMPLE 6.4   We shall illustrate these Bayesian estimators, using the data on remission duration for patients given the drug 6-MP, which was presented in section 1.2. For the Dirichlet prior, we shall use a prior guess at $S_0(t)$ of $\alpha(t,\infty)/\alpha(0,\infty) = e^{-0.1t}$. This prior estimate was chosen so that the a priori mean of the 6-MP group is the same as the control group. Our degree of belief in this prior estimate is that it is worth about $C = 5$ observations, so that $\alpha(0,\infty) = 5$ and $\alpha(t,\infty) = 5e^{-0.1t}$. For the beta process prior, we shall assume the same prior estimate of the survival function and degree of belief in the validity of this guess, so $H_0(t) = 0.1t$ and $c(t) = 5$. Figures 6.11 and 6.12 show samples of sample paths from these two priors.

From the data, we have the following information:

T_i	6	7	9	10	11	13	16	17	19	20	22	23	25	32	34	35
Y_i	21	17	16	15	13	12	11	10	9	8	7	6	5	4	2	1
d_i	3	1	0	1	0	1	1	0	0	0	1	1	0	0	0	0
c_i	1	0	1	1	1	0	0	1	1	1	0	0	1	2	1	1
Δ_i	1	1	0	1	0	1	1	0	0	0	1	1	0	0	0	0.

To illustrate the calculations, first consider a t in the interval $[0,6)$. For the Dirichlet prior,

$$\hat{S}_D(t) = \left[\frac{5e^{-0.1t} + 21}{5 + 21}\right] = \left[\frac{5e^{-0.1t} + 21}{26}\right],$$

whereas, for the beta process prior,

$$\hat{S}_B(t) = \exp\left[-\frac{5[0.1(t) - 0.1(0)]}{5 + 21}\right] = \exp\left(-\frac{0.5t}{26}\right).$$

For a t in the interval $[6,7)$,

$$\hat{S}_D(t) = \left[\frac{5e^{-0.1t} + 17}{5 + 21}\right]\left\{\frac{5e^{-0.6} + 18}{5e^{-0.6} + 17}\right\},$$

whereas, for the beta process prior,

$$\hat{S}_B(t) = \exp\left\{-\frac{5[0.1(6) - 0.1(0)]}{5 + 21} - \frac{5[0.1(t) - 0.1(6)]}{5 + 17}\right\}\left[1 - \frac{5(0.1) + 3}{5 + 21}\right].$$

Figure 6.13 shows the two Bayes estimates, the Kaplan–Meier estimator, and the prior estimate of the survival function. Here, we note that the beta process prior estimate is closer to the prior mean, which is to be expected, because the beta process has sample paths which tend to lie closer to the hypothesized prior guess at the survival function.

The third approach to Bayesian estimation of the survival function is by Monte Carlo Bayesian methods or the Gibbs sampler. This approach is more flexible than the other two approaches. For right-censored data, for which we will describe the procedure, closed form estimates of the survival function are available. For other censoring or truncation schemes, such simple estimates are not available, and the Gibbs sample provides a way of simulating the desired posterior distribution of the survival function. This approach can also be extended to more complicated problems, such as the regression problems discussed in Chapter 8.

To illustrate how this method works, we shall focus on the right-censored data problem. We let $0 < t_1 < \cdots < t_M$ be M time points. Let d_j be the number of deaths in the interval $(t_{j-1}, t_j]$ and λ_j the number of right-censored observations at t_j. Let $P_j = S(t_j)$ be the survival function at time t_j, so the likelihood function is proportional to $\prod_{j=1}^{M}(P_{j-1} - P_j)^{d_j}P_j^{\lambda_j}$. Let $\theta_j = P_{j-1} - P_j$, for $j = 1, \ldots, M$ and $\theta_{M+1} = P_M$. For a prior distribution, we assume that the joint distribution

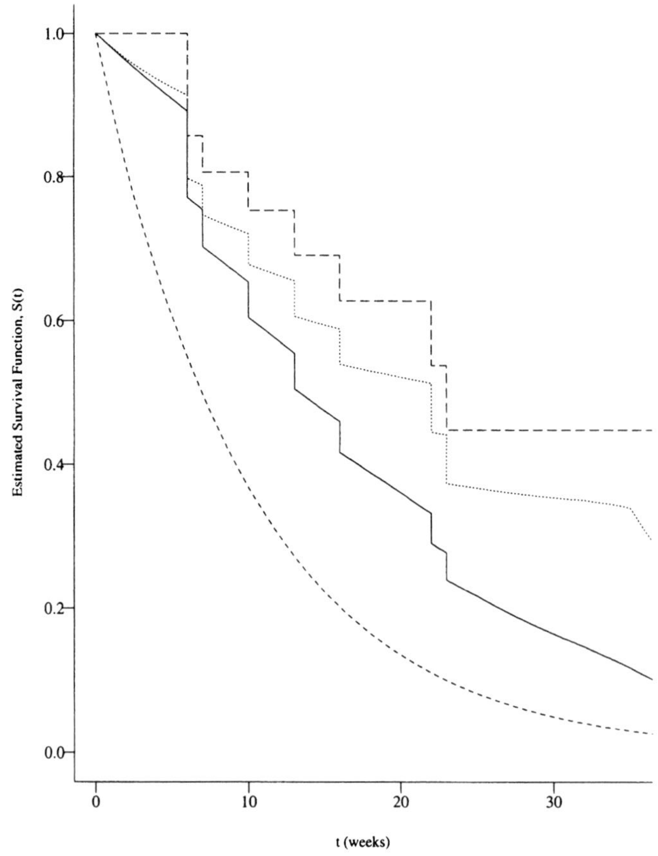

Figure 6.13 *Bayes estimates of the survival function for the 6-MP group. Beta process prior (———) Dirichlet process prior (------) Prior (– – – – –) Kaplan–Meier estimate (— — —)*

of the θ's is the Dirichlet distribution with density function

$$\pi(\theta, \ldots, \theta_m) = \text{Constant} \prod_{j=0}^{M+1} (\theta_j)^{\alpha_{j-1}}, \qquad (6.4.3)$$

where $\alpha_j = C[S_0(t_{j-1}) - S_0(t_j)]$ for $j = 1, \ldots, M + 1$ with $S_0(t_{M+1}) = 0$ and the constant in (6.4.3) is

$$\frac{\Gamma(C)}{\prod_{j=1}^{M+1} \Gamma(\alpha_j)}.$$

The Gibbs sampling approach to Bayesian estimation approximates the posterior distribution via a Monte Carlo simulation. Here, we treat the censored observations as unknown parameters, and we simulate death times for each censored observation. Using these values with the death information, one simulates the parameters θ_j. These new θ's are used to generate new death times for the censored observations, and so forth. Gelfand and Smith (1990) have shown that this procedure converges to a realization of θ drawn from the posterior distribution θ, given the data. This process is repeated a large number of times to obtain a sample from the posterior distribution of θ, given the data which is analyzed to provide the Bayes estimator.

For our censored data problem, a single Gibbs sample is generated as follows. If $\lambda_j > 0$, let $Z_{j+1,j}, \ldots, Z_{M+1,j}$ denote the number of observations out of the λ_j that may have been deaths in the intervals $(t_j, t_{j+1}], \ldots, (t_{M-1}, t_M], (t_M, \infty)$, respectively. Note that $\lambda_j = \sum_{k=j+1}^{M+1} Z_{k,j}$. Suppose that, at the ith iteration, we have a realization of $\boldsymbol{\theta}^i = (\theta_1^i, \theta_2^i, \ldots, \theta_{M+1}^i)$ which sums to 1. We sample $Z_{j+1,j}, \ldots, Z_{M+1,j}$ from a multinomial with sample size λ_j and probabilities

$$\rho_k = \frac{\theta_k^i}{\sum_{b=j+1}^{M+1} \theta_b^i}.$$

Having sampled the Z's, new θ's are generated from the Dirichlet by first computing

$$R_b^{i+1} = \alpha_b + d_b + \sum_{j=1}^{M} Z_{b,j}$$

and, then, sampling $\theta^{i+1} = (\theta_1^{i+1}, \theta_2^{i+1}, \ldots, \theta_{M+1}^{i+1})$ for a Dirichlet distribution with parameters $(R_1^{i+1}, R_2^{i+1}, \ldots, R_{M+1}^{i+1})$.

The procedure above yields a single realization of θ and R after i steps. Typically i is relatively small, of the order 10 or 20. This process is repeated S times where S is typically of the order 1000–10,000. The posterior estimate of θ_b is, then, given by

$$\tilde{\theta}_b = S^{-1} \sum_{s=1}^{S} \frac{R_{bs}^i}{\sum_{k=1}^{M+1} R_{ks}^i}. \tag{6.4.4}$$

EXAMPLE 6.4   *(continued)*   We shall apply the Gibbs sampling approach to the data in Example 6.4. As in that example, we assume, a priori, that $S_0(t) = e^{-0.1t}$ and that our prior belief in the accuracy of our prior guess is $C = 5$ observations. Intervals are formed by taking t_j to be the death and censoring times. For a death time T, we include an "interval" $(T^-, T]$ with a θ_b representing the point mass at time T. (That is, θ_b is the jump in the estimated survival function at an observed death.) The following Table 6.5 shows the 24 intervals needed for this problem and the values of α_j from our prior.

To generate the first Gibbs observation, we generated θ_b^0, $b = 1, \ldots, 24$ from the prior distribution (6.4.3) which is Dirichlet

TABLE 6.5

Estimates Based on Gibbs Sampling

j	$(t_j - 1, t_j]$	d_j	λ_j	α_j	θ^0	Revised Death Count Iteration 1	Posterior Probability (SE)
1	$(0, 6^-]$	0	0	2.256	0.3378	0	0.0867 (0)
2	$(6^-, 6]$	3	1	0.000	0	3	0.1154 (0)
3	$(6, 7^-]$	0	0	0.261	0.0867	0	0.0105 (0.0001)
4	$(7^-, 7]$	1	0	0.000	0	1	0.0408 (0.0003)
5	$(7, 9]$	0	1	0.450	0.0228	0	0.0182 (0.0002)
6	$(9, 10^-]$	0	0	0.193	0.0001	0	0.0083 (0.0002)
7	$(10^-, 10]$	1	1	0.000	0	1	0.0430 (0.0004)
8	$(10, 11]$	0	1	0.175	0.0428	0	0.0077 (0.0002)
9	$(11, 13^-]$	0	0	0.302	0.0001	0	0.0148 (0.0004)
10	$(13^-, 13]$	1	0	0.000	0	1	0.0500 (0.0007)
11	$(13, 16^-]$	0	0	0.353	0.2673	2	0.0169 (0.0004)
12	$(16^-, 16]$	1	0	0.000	0	1	0.0492 (0.0007)
13	$(16, 17]$	0	1	0.096	0.0000	0	0.0050 (0.0002)
14	$(17, 19]$	0	1	0.166	0.0028	0	0.0091 (0.0004)
15	$(19, 20]$	0	1	0.071	0.0721	1	0.0042 (0.0003)
16	$(20, 22^-]$	0	0	0.123	0.0058	0	0.0080 (0.0004)
17	$(22^-, 22]$	1	0	0.000	0	1	0.0678 (0.0012)
18	$(22, 23^-]$	0	0	0.053	0.0045	0	0.0038 (0.0003)
19	$(23^-, 23]$	1	0	0.000	0	1	0.0662 (0.0381)
20	$(23, 25]$	0	1	0.091	0.0003	0	0.0066 (0.0005)
21	$(25, 32]$	0	2	0.207	0.1570	5	0.0183 (0.0008)
22	$(32, 34]$	0	1	0.037	0.0000	0	0.0072 (0.0008)
23	$(34, 35]$	0	1	0.016	0.0000	0	0.0117 (0.0014)
24	$(35, \infty)$	0	0	0.151	0.0000	4	0.3306 (0.0024)

$(\alpha_1, \ldots, \alpha_{24})$. To generate observations from the Dirichlet distribution, one generates $W_1, \ldots, W_{24}$ as independent gamma random variables with parameters α_b and $\beta = 1$ (i.e., $f(w_b) = w_b^{\alpha_b - 1} \exp\{-w_b\}/\Gamma(\alpha_b)$, and, then, $\theta_b = W_b/\Sigma W_j$. The first realization of the θ's is included in the table. Using these values, we, then, generate the Z's. For example, we generate $Z_{3,2}, Z_{4,2}, \ldots, Z_{12,2}$ from the appropriate multinomial distribution. In our example, this corresponds to picking an interval $(t_{b-1}, t_b]$ with $b > j$ in which each censored observation at t_j is to be placed and counted as a death. The table includes entries which give the revised death count at t_b of $d_b + \sum_{j=1}^{M} Z_{b,j}$, for the first iteration. These revised death counts are used to update the values of α_b by $Y_b = \alpha_b + d_b + \sum_{j=1}^{M} Z_{b,j}$, also given in Table 6.5. The procedure continues through a total of 10 cycles to produce the Gibbs iterate Y_{b1}^{10}.

This is repeated 1000 times. The final column of the table provides the posterior means of the θ's from (6.4.4) and, for reference, the sample standard errors of the standardized R_{hs}^i which provide some information on the rate of convergence of the algorithm. Notice that the posterior mean estimates in this table are precisely what we would obtain from the Dirichlet process prior, discussed earlier.

Practical Notes

1. The Bayesian estimator of the survival function obtained from a right-censored sample from the Dirichlet process prior model can be extended to other censoring schemes. Johnson and Christensen (1986) developed the estimation procedure for grouped data as found in a life table. Cornfield and Detre (1977) also consider a Bayes estimator of the survival function for life table data which is based on a Dirichlet-like prior.

2. Using the Gibbs sampling approach, additional censoring schemes can be handled quite easily. For example, Kuo and Smith (1992) show how to handle combined right- and left-censored data. This flexibility of the Monte Carlo Bayesian approach is one of the major strengths of the technique.

3. The Gibbs sampling approach presented here generates a Gibbs sample based on a large number of short runs of the algorithm. An alternative is to run a single realization of the algorithm until the successive iterations have the desired posterior distribution and, then, take, as the Gibbs sample, successive θ's generated by the algorithm. The approach suggested here, although requiring a bit more computation time, has the advantage of producing independent replications of the posterior distribution. (See Gelfand and Smith (1990) for a discussion of the merits of the two approaches.)

4. The posterior estimator of the survival function from the Gibbs sample, (6.4.4), is based on the fact that the posterior distribution of θ_h is a mixture of a beta random variable with parameters Y_h and $\sum_{k \neq h} Y_k$. An alternative technique to estimate the posterior distribution of θ_h is to use the empirical distribution function of the simulated values of θ, θ_{hs}^i, $s = 1, \ldots, S$. This would give a posterior estimator of θ_h of the sample mean of S replicates, θ_{hs}^i. To achieve the same precision as found by (6.4.4) for this approach, a larger value of S is required. By this approach, however, one can routinely provide an estimate of any functional of the posterior distribution of θ, by the appropriate functional of the empirical distribution of the simulated θ's.

5. Hjort (1992) discusses how the beta process prior can be used in more complicated censoring schemes and in making adjustments to

the survival function to account for covariates. He provides a Bayes approach to the proportional hazard regression problem discussed in Chapter 8.

Theoretical Notes

1. The Dirichlet process prior estimator of the survival function was first proposed by Ferguson (1973) for uncensored data. Susarla and Van Ryzin (1976) and Ferguson and Phadia (1979) extend the estimation process to right censored data.
2. The beta process prior was introduced in this context by Hjort (1990).
3. Both the Dirichlet and beta process prior estimates converge to the Product-Limit estimator for large samples for any nontrivial prior distribution. By an appropriate choice of the prior distribution, the Product-Limit estimator is a Bayes estimator for any n for both of these priors.
4. If one chooses $c(t) = kS_0(t)$, where $S_0(t) = \exp[-H_0(t)]$ for the weight parameter of the beta process, then, the beta process prior on H is the same as a Dirichlet process prior with parameters $S_0(t)$ and k. Thus, the beta process prior is a more general class of priors than the class of Dirichlet priors.
5. Kuo and Smith (1992) have introduced the use of Monte Carlo Bayesian methods to survival analysis.

6.5   Exercises

6.1   (a) Using the data on the time to relapse of 6-MP patients found in section 1.2, estimate the hazard rate at 12 months using the uniform kernel with a bandwidth of 6 months. Provide the standard error of your estimate.

(b) Compare the estimates obtained in part a to the estimate of $h(12)$ obtained using the Epanechnikov kernel.

(c) Repeat part b using the biweight kernel.

(d) Estimate the hazard rate at 5 months using all three kernels.

6.2   Using the data on the leukemia-free survival times of allogeneic bone marrow transplants in Table 1.4 of Chapter 1 (See Exercise 7 of Chapter 4), estimate the hazard rate at 1, 3, 5, 7, 9, 11, and 13 months using a uniform kernel with a bandwidth of 5 months. Plot your estimates and interpret the shape of the estimated hazard rate.

6.3   (a) Using the data on the infection times of kidney dialysis patients in section 1.4, estimate the hazard rate using a biweight kernel with a bandwidth of 5 months at 3 months for each of the two groups.

(b) Using the same bandwidth and kernel estimate the hazard rate at 10 months in both groups.

6.4 In section 1.7 a study of the death times (in years) and the age (in years) at transplant of 59 black female kidney transplant patients is reported. From this data, compute the patients' age in years at death or at the end of the study. The survival experience of this sample of patients is to be compared to the standard mortality rates of black females found in the 1990 U.S. census using the all-cause mortality for the U.S. population in 1990 found in Table 2.1 of Chapter 2.

(a) Estimate the cumulative relative mortality, $B(t)$, for this group of patients.

(b) Find the standard error of your estimate in part a.

(c) Estimate the excess mortality, $A(t)$, for this group of patients.

(d) Find the standard error of your estimate in part c.

(e) Plot the Kaplan–Meier estimate of the survival function, the expected survival curve, and the corrected survival curve for this group of patients.

6.5 An alternative to autologous bone marrow transplantation for leukemia is chemotherapy. Suppose that it is known that for chemotherapy patients the time from diagnosis to relapse or death has an exponential distribution with survival function hazard rate $\lambda = 0.045$. Assume that this rate is the same for all patients. To compare the survival experience of this reference population to autologous bone marrow transplant patients use the data on autologous transplants in Table 1.4 of Chapter 1 (see Problem 7 of Chapter 4).

(a) Estimate the cumulative relative mortality, $B(t)$, for this group of patients.

(b) Find the standard error of your estimate in part a.

(c) Estimate the excess mortality, $A(t)$, for this group of patients.

(d) Find the standard error of your estimate in part c.

6.6 Table 1.3 of section 1.5 provides data on the time to death (in months) of nine immunoperoxidase-positive breast-cancer patients.

(a) Using a Dirichlet prior for $S(t)$ with $\alpha(t, \infty) = 6\exp(-0.1t^{0.5})$, find the Bayes estimate of the survival function under squared-error loss.

(b) Using a beta prior for $H(t)$ with $q = 6$ and $H_0(t) = 0.1t^{0.5}$ find the Bayes estimate of the survival function under squared-error loss.

(c) Compare the estimates found in parts a and b to the usual Kaplan–Meier estimate of the survival function.

6.7 Table 1.6 of section 1.11 gives data on the times in weeks from diagnosis to death of 28 patients with diploid cancers of the tongue.

(a) Using a Dirichlet prior for $S(t)$ with $\alpha(t, \infty) = 4/(1 + 0.15t^{0.5})$, find the Bayes estimate of the survival function under squared-error loss.

(b) Using a beta prior for $H(t)$ with $q = 4$ and $H_0(t) = \ln(1 + 0.15t^{0.5})$, find the Bayes estimate of the survival function under squared-error loss.

(c) Compare the estimates found in parts a and b to the usual Kaplan–Meier estimate of the survival function.

7

Hypothesis Testing

7.1 Introduction

As we have seen in Chapters 4–6, the Nelson–Aalen estimator of the cumulative hazard rate is a basic quantity in describing the survival experience of a population. In Chapter 4, we used this estimator along with the closely related Product-Limit estimator to make crude comparisons between the disease-free survival curves of bone marrow transplant patients with different types of leukemia, and in section 6.3, we used this statistic as the basis for estimating excess mortality of Iowa psychiatric patients.

In this chapter, we shall focus on hypothesis tests that are based on comparing the Nelson–Aalen estimator, obtained directly from the data, to an expected estimator of the cumulative hazard rate, based on the assumed model under the null hypothesis. Rather than a direct comparison of these two rates, we shall examine tests that look at weighted differences between the observed and expected hazard rates. The weights will allow us to put more emphasis on certain parts of the curves. Different weights will allow us to present tests that are most sensitive to early or late departures from the hypothesized relationship between samples as specified by the null hypothesis.

In section 7.2, we shall consider the single sample problem. Here, we wish to test if the sample comes from a population with a prespecified hazard rate $h_0(t)$. In section 7.3, we will look at tests of the null hypothesis of no difference in survival between K treatments against a global alternative that at least one treatment has a different survival rate. Here, for example, we will discuss censored data versions of the Wilcoxon or Kruskal–Wallis test and log-rank or Savage test. In section 7.4, we look at K sample tests that have power to detect ordered alternatives. A censored data version of the Jonckheere–Terpstra test will be presented. In section 7.5, we will see how these tests can be modified to handle stratification on covariates which may confound the analysis. We shall see how this approach can be used to handle matched data, and we will have a censored-data version of the sign test. In section 7.6, we will look at tests based on the maximum of the sequential evaluation of these tests at each death time. These tests have the ability to detect alternatives where the hazard rates cross and are extensions of the usual Kolmogorov–Smirnov test. Finally, in section 7.7, we present three other tests which have been proposed to detect crossing hazard rates, a censored-data version of the Cramer–von Mises test, a test based on weighted differences in the Kaplan–Meier estimators, and a censored-data version of the median test.

The methods of this chapter can be applied equally well to right-censored data or to samples that are right-censored and left-truncated. As we shall see, the key statistics needed to compute the tests are the number of deaths at each time point and the number of individuals at risk at these death times. Both quantities are readily observed with left-truncated and right-censored data.

7.2   One-Sample Tests

Suppose that we have a censored sample of size n from some population. We wish to test the hypothesis that the population hazard rate is $h_0(t)$ for all $t \leq \tau$ against the alternative that the hazard rate is not $h_0(t)$ for some $t \leq \tau$. Here $h_0(t)$ is a completely specified function over the range 0 to τ. Typically, we shall take τ to be the largest of the observed study times.

An estimate of the cumulative hazard function $H(t)$ is the Nelson–Aalen estimator, (4.2.3), given by $\sum_{t_i \leq t} \frac{d_i}{Y(t_i)}$, where d_i is the number of events at the observed event times, $t_1, \ldots, t_D$ and $Y(t_i)$ is the number of individuals under study just prior to the observed event time t_i. The quantity $\frac{d_i}{Y(t_i)}$ gives a crude estimate of the hazard rate at an event time t_i. When the null hypothesis is true, the expected hazard rate at t_i is

$h_0(t_i)$. We shall compare the sum of weighted differences between the observed and expected hazard rates to test the null hypothesis.

Let $W(t)$ be a weight function with the property that $W(t)$ is zero whenever $Y(t)$ is zero. The test statistic is

$$Z(\tau) = O(\tau) - E(\tau) = \sum_{i=1}^{D} W(t_i)\frac{d_i}{Y(t_i)} - \int_0^{\tau} W(s)h_0(s)\,ds. \quad (7.2.1)$$

When the null hypothesis is true, the sample variance of this statistic is given by

$$V[Z(\tau)] = \int_0^{\tau} W^2(s)\frac{h_0(s)}{Y(s)}\,ds. \quad (7.2.2)$$

For large samples, the statistic $Z(\tau)^2/V[Z(\tau)]$ has a central chi-squared distribution when the null hypothesis is true.

The statistic $Z(\tau)/V[Z(\tau)]^{1/2}$ is used to test the one sided alternative hypothesis that $h(t) > h_0(t)$. When the null hypothesis is true and the sample size is large, this statistic has a standard normal distribution. The null hypothesis is rejected for large values of the statistic.

The most popular choice of a weight function is the weight $W(t) = Y(t)$ which yields the one-sample log-rank test. To allow for possible left truncation, let T_j be the time on study and L_j be the delayed entry time for the jth patient. When τ is equal to the largest time on study,

$$O(\tau) = \text{observed number of events at or prior to time } \tau, \quad (7.2.3)$$

and

$$E(\tau) = V[Z(\tau)] = \sum_{j=1}^{n}[H_0(T_j) - H_0(L_j)] \quad (7.2.4)$$

where $H_0(t)$ is the cumulative hazard under the null hypothesis.

Other weight functions proposed in the literature include the Harrington and Fleming (1982) family $W_{HF}(t) = Y(t)S_0(t)^p[1 - S_0(t)]^q$, $p \geq 0$, $q \geq 0$, where $S_0(t) = \exp[-H_0(t)]$ is the hypothesized survival function. By choice of p and q, one can put more weight on early departures from the null hypothesis (p much larger than q), late departures from the null hypothesis (p much smaller than q), or on departures in the mid-range ($p = q > 0$). The log-rank weight is a special case of this model with $p = q = 0$.

EXAMPLE 7.1   In section 6.3, we examined models for excess and relative mortality in a sample of 26 Iowa psychiatric patients described in section 1.15. We shall now use the one-sample log-rank statistic to test the hypothesis that the hazard rate of this group of patients is the same as the hazard rate in the general Iowa population, given by the standard 1960 Iowa mortality table. To perform this test, we will use the sex specific survival rates. Time T_j is taken as the jth individual's age at death or the end of the study, and the left-truncation time L_j, is this individual's age at entry into the study. We obtain $H(t)$ as $-\ln[S(t)]$ from the appropriate column of Table 6.2. Table 7.1 shows the calculations to compute $O(71)$ and $E(71)$.

The test statistic is $\chi^2 = (15 - 4.4740)^2/4.4740 = 24.7645$ which has a chi-squared distribution with one degree of freedom. Here the p-value of this test is close to zero, and we can conclude that the mortality rates of the psychiatric patients differ from those of the general public.

TABLE 7.1

Computation of One-Sample, Log-Rank Test

Subject j	Sex	Status d_i	Age at Entry L_i	Age at Exit T_j	$H_0(L_j)$	$H_0(T_j)$	$H_0(T_j) - H_0(L_j)$
1	f	1	51	52	0.0752	0.0797	0.0045
2	f	1	58	59	0.1131	0.1204	0.0073
3	f	1	55	57	0.0949	0.1066	0.0117
4	f	1	28	50	0.0325	0.0711	0.0386
5	m	0	21	51	0.0417	0.1324	0.0907
6	m	1	19	47	0.0383	0.1035	0.0652
7	f	1	25	57	0.0305	0.1066	0.0761
8	f	1	48	59	0.0637	0.1204	0.0567
9	f	1	47	61	0.0606	0.1376	0.0770
10	f	1	25	61	0.0305	0.1376	0.1071
11	f	0	31	62	0.0347	0.1478	0.1131
12	m	0	24	57	0.0473	0.1996	0.1523
13	m	0	25	58	0.0490	0.2150	0.1660
14	f	0	30	67	0.0339	0.2172	0.1833
15	f	0	33	68	0.0365	0.2357	0.1992
16	m	1	36	61	0.0656	0.2704	0.2048
17	m	0	30	61	0.0561	0.2704	0.2143
18	m	1	41	63	0.0776	0.3162	0.2386
19	f	1	43	69	0.0503	0.2561	0.2058
20	f	1	45	69	0.0548	0.2561	0.2013
21	f	0	35	65	0.0384	0.1854	0.1470
22	m	0	29	63	0.0548	0.3162	0.2614
23	m	0	35	65	0.0638	0.3700	0.3062
24	m	1	32	67	0.0590	0.4329	0.3739
25	f	1	36	76	0.0395	0.4790	0.4395
26	m	0	32	71	0.0590	0.5913	0.5323
Total		15					4.4740

Practical Notes

1. An alternate estimator of the variance of $Z(\tau)$ is given by $V[Z(\tau)] = \sum_{i=1}^{D} W(t_i)^2 \frac{d_i}{Y(t_i)^2}$ which uses the empirical estimator of $h_0(t)$ rather than the hypothesized value. When the alternative hypothesis $h(t) > h_0(t)$ is true, for some $t \leq \tau$, this variance estimator is expected to be larger than (7.2.2), and the test is less powerful using this value. On the other hand, if $h(t) < h_0(t)$, then, this variance estimator will tend to be smaller, and the test will be more powerful.

2. The statistic $O(\tau)/E(\tau)$ based on the log-rank weights is called the standardized mortality ratio (SMR).

3. A weight function suggested by Gatsonis et al. (1985) is $W(t) = (1 + \{\log[1 - S_0(t)]\}/S_0(t)) Y(t)$.

Theoretical Notes

1. In this class of tests, the one-sample, log-rank test is the locally most powerful test against a shift alternative of the extreme value distribution. The weight function $W_{HF}(t) = Y(t)S_0(t)$ is the locally most powerful test for the logistic distribution. Because the one-sample Wilcoxon test also has this property, this choice of weights leads to a censored-data, one-sample, Wilcoxon test. See Andersen et al. (1993) for details.

2. These one-sample tests arise quite naturally from the theory of counting processes. Under the null hypothesis, using the notation in section 3.6, $\int_0^\tau [J(u)/Y(u)]dN(u) - \int_0^\tau J(u)h_0(u)\,du$ is a martingale. The statistic $Z(\tau)$ is a stochastic integral of the weight function $W(t)$ with respect to this martingale, and $\text{Var}[Z(\tau)]$ is the predictable variation process of this stochastic integral. The asymptotic chi-squared distribution follows by the martingale central limit theorem.

3. The one-sample, log-rank test was first proposed by Breslow (1975) and generalized to left truncation by Hyde (1977) and Woolson (1981).

7.3   Tests for Two or More Samples

In section 7.2, we looked at one-sample tests that made a weighted comparison between the estimated hazard rate and the hypothesized hazard rates. We now shall extend these methods to the problem of comparing hazard rates of K ($K \geq 2$) populations, that is, we shall test the following set of hypotheses:

$$H_0 : h_1(t) = h_2(t) = \cdots = h_K(t), \text{ for all } t \leq \tau, \text{ versus} \quad (7.3.1)$$

$$H_A : \text{at least one of the } h_j(t)\text{'s is different for some } t \leq \tau.$$

Here τ is the largest time at which all of the groups have at least one subject at risk.

As in section 7.2, our inference is to the hazard rates for all time points less than τ, which is, typically, the smallest of the largest time on study in each of the k groups. The alternative hypothesis is a global one in that we wish to reject the null hypothesis if, at least, one of the populations differs from the others at some time. In the next section, we will present tests that are more powerful in the case of ordered alternatives.

The data available to test the hypothesis (7.3.1) consists of independent right-censored and, possibly, left-truncated samples for each of the K populations. Let $t_1 < t_2 < \cdots < t_D$ be the distinct death times in the pooled sample. At time t_i we observe d_{ij} events in the jth sample out of Y_{ij} individuals at risk, $j = 1, \ldots, K$, $i = 1, \ldots, D$. Let $d_i = \sum_{j=1}^{K} d_{ij}$ and $Y_i = \sum_{j=1}^{K} Y_{ij}$ be the number of deaths and the number at risk in the combined sample at time t_i, $i = 1, \ldots, D$.

The test of H_0 is based on weighted comparisons of the estimated hazard rate of the jth population under the null and alternative hypotheses, based on the Nelson–Aalen estimator (4.2.3). If the null hypothesis is true, then, an estimator of the expected hazard rate in the jth population under H_0 is the pooled sample estimator of the hazard rate d_i / Y_i. Using only data from the jth sample, the estimator of the hazard rate is d_{ij} / Y_{ij}. To make comparisons, let $W_j(t)$ be a positive weight function with the property that $W_j(t_i)$ is zero whenever Y_{ij} is zero. The test of H_0 is based on the statistics

$$Z_j(\tau) = \sum_{i=1}^{D} W_j(t_i) \left\{ \frac{d_{ij}}{Y_{ij}} - \frac{d_i}{Y_i} \right\}, \quad j = 1, \ldots, K. \quad (7.3.2)$$

If all the $Z_j(\tau)$'s are close to zero, then, there is little evidence to believe that the null hypothesis is false, whereas, if one of the $Z_j(\tau)$'s is far from zero, then, there is evidence that this population has a hazard rate differing from that expected under the null hypothesis.

Although the general theory allows for different weight functions for each of the comparisons in (7.3.2), in practice, all of the commonly used tests have a weight function $W_j(t_i) = Y_{ij} W(t_i)$. Here, $W(t_i)$ is a common weight shared by each group, and Y_{ij} is the number at risk in the jth group at time t_i. With this choice of weight functions

$$Z_j(\tau) = \sum_{i=1}^{D} W(t_i) \left[d_{ij} - Y_{ij} \left(\frac{d_i}{Y_i} \right) \right], \quad j = 1, \ldots, K. \quad (7.3.3)$$

Note that with this class of weights the test statistic is the sum of the weighted difference between the observed number of deaths and the expected number of deaths under H_0 in the jth sample. The expected number of deaths in sample j at t_i is the proportion of individuals at risk Y_{ij}/Y_i that are in sample j at time t_i, multiplied by the number of deaths at time t_i.

The variance of $Z_j(\tau)$ in (7.3.3) is given by

$$\hat{\sigma}_{jj} = \sum_{i=1}^{D} W(t_i)^2 \frac{Y_{ij}}{Y_i}\left(1 - \frac{Y_{ij}}{Y_i}\right)\left(\frac{Y_i - d_i}{Y_i - 1}\right) d_i, \quad j = 1, \ldots, K \quad (7.3.4)$$

and the covariance of $Z_j(\tau), Z_g(\tau)$ is expressed by

$$\hat{\sigma}_{jg} = -\sum_{i=1}^{D} W(t_i)^2 \frac{Y_{ij}}{Y_i}\frac{Y_{ig}}{Y_i}\left(\frac{Y_i - d_i}{Y_i - 1}\right) d_i, \quad g \neq j. \quad (7.3.5)$$

The term $(Y_i - d_i)/(Y_i - 1)$, which equals one if no two individuals have a common event time, is a correction for ties. The terms $\frac{Y_{ij}}{Y_i}(1 - \frac{Y_{ij}}{Y_i})d_i$ and $-\frac{Y_{ij}}{Y_i}\frac{Y_{ig}}{Y_i}d_i$ arise from the variance and covariance of a multinomial random variable with parameters $d_i, p_j = Y_{ij}/Y_i, j = 1, \ldots, K$.

The components vector $(Z_1(\tau), \ldots, Z_K(\tau))$ are linearly dependent because $\sum_{j=1}^{K} Z_j(\tau)$ is zero. The test statistic is constructed by selecting any $K - 1$ of the Z_j's. The estimated variance-covariance matrix of these statistics is given by the $(K - 1) \times (K - 1)$ matrix Σ, formed by the appropriate $\hat{\sigma}_{jg}$'s. The test statistic is given by the quadratic form

$$\chi^2 = (Z_1(\tau), \ldots, Z_{K-1}(\tau))\Sigma^{-1}(Z_1(\tau), \ldots, Z_{K-1}(\tau))'. \quad (7.3.6)$$

When the null hypothesis is true, this statistic has a chi-squared distribution, for large samples with $K - 1$ degrees of freedom. An α level test of H_0 rejects when χ^2 is larger than the αth upper percentage point of a chi-squared, random variable with $K - 1$ degrees of freedom.

When $K = 2$ the test statistic can be written as

$$Z = \frac{\sum_{i=1}^{D} W(t_i)[d_{i1} - Y_{i1}(\frac{d_i}{Y_i})]}{\sqrt{\sum_{i=1}^{D} W(t_i)^2 \frac{Y_{i1}}{Y_i}(1 - \frac{Y_{i1}}{Y_i})(\frac{Y_i - d_i}{Y_i - 1})d_i}}, \quad (7.3.7)$$

which has a standard normal distribution for large samples when H_0 is true. Using this statistic, an α level test of the alternative hypothesis $H_A : h_1(t) > h_2(t)$, for some $t \leq \tau$, is rejected when $Z \geq Z_\alpha$, the αth upper percentage point of a standard normal distribution. The test of $H_A : h_1(t) \neq h_2(t)$, for some t, rejects when $|Z| > Z_{\alpha/2}$.

A variety of weight functions have been proposed in the literature. A common weight function, leading to a test available in most statistical packages, is $W(t) = 1$ for all t. This choice of weight function leads to the so-called log-rank test and has optimum power to detect alternatives where the hazard rates in the K populations are proportional to each other. A second choice of weights is $W(t_i) = Y_i$. This weight function yields Gehan's (1965) generalization of the two-sample Mann–Whitney–Wilcoxon test and Breslow's (1970) generalization of the Kruskal–Wallis test. Tarone and Ware (1977) suggest a class of tests where the weight function is $W(t_i) = f(Y_i)$, and f is a fixed function. They suggest a choice of $f(y) = y^{1/2}$. This class of weights gives more weight to differences between the observed and expected number of deaths in sample j at time points where there is the most data.

An alternate censored-data version of the Mann–Whitney–Wilcoxon test was proposed by Peto and Peto (1972) and Kalbfleisch and Prentice (1980). Here, we define an estimate of the common survival function by

$$\tilde{S}(t) = \prod_{t_i \le t} \left(1 - \frac{d_i}{Y_i + 1} \right), \qquad (7.3.8)$$

which is close to the pooled Product-Limit estimator. They suggest using $W(t_i) = \tilde{S}(t_i)$. Andersen et al. (1982) suggest that this weight should be modified slightly as $W(t_i) = \tilde{S}(t_i)Y_i/(Y_i + 1)$ (see Theoretical Note 2). Either of the weights depends on the combined survival experience in the pooled sample whereas the weight $W(t_i) = Y_i$ depends heavily on the event times and censoring distributions. Due to this fact, the Gehan-Breslow weights can have misleading results when the censoring patterns are different in the individual samples (see Prentice and Marek (1979) for a case study).

Fleming and Harrington (1981) propose a very general class of tests that includes, as special cases, the log-rank test and a version of the Mann–Whitney–Wilcoxon test, very close to that suggested by Peto and Peto (1972). Here, we let $\hat{S}(t)$ be the Product-Limit estimator (3.2.1) based on the combined sample. Their weight function is given by

$$W_{p,q}(t_i) = \hat{S}(t_{i-1})^p[1 - \hat{S}(t_{i-1})]^q, \quad p \ge 0, \ q \ge 0. \qquad (7.3.9)$$

Here, the survival function at the previous death time is used as a weight to ensure that these weights are known just prior to the time at which the comparison is to be made. Note that $S(t_0) = 1$ and we define $0^0 = 1$ for these weights. When $p = q = 0$ for this class, we have the log-rank test. When $p = 1$, $q = 0$, we have a version of the Mann–Whitney–Wilcoxon test. When $q = 0$ and $p > 0$, these weights give the most weight to early departures between the hazard rates in the K populations, whereas, when $p = 0$ and $q > 0$, these tests give most weight to departures which occur late in time. By an appropriate choice of p and q, one can construct tests which have the most power against alternatives which have the K hazard rates differing over any desired region. This is illustrated in the following example.

EXAMPLE 7.2   In section 1.4, data on a clinical trial of the effectiveness of two methods for placing catheters in kidney dialysis patients was presented. We are interested in testing if there is a difference in the time to cutaneous exit-site infection between patients whose catheter was placed surgically (group 1) as compared to patients who had their catheters placed percutaneously (group 2).

Figure 7.1 shows the survival curves for the two samples. Table 7.2 shows the calculations needed to construct the log-rank test. Here, $Z_{obs} = 3.964/\sqrt{6.211} = 1.59$ which has a p-value of $2Pr[Z > 1.59] = 0.1117$, so the log-rank test suggests no difference between the two procedures in the distribution of the time to exit-site infection.

To further investigate these two treatments, we shall apply some of the other weight functions discussed earlier. Table 7.3 summarizes

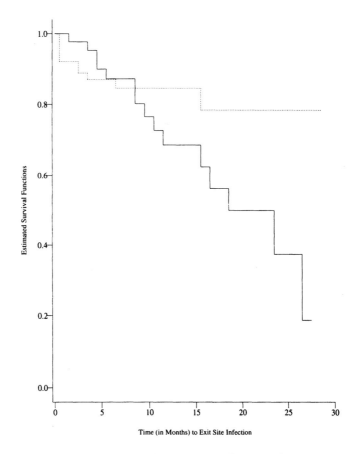

Figure 7.1   *Estimated (Infection-free) survival function for kidney dialysis patients with percutaneous (------) and surgical (———) placements of catheters.*

TABLE 7.2

Construction of Two-Sample, Log-Rank Test

t_i	Y_{i1}	d_{i1}	Y_{i2}	d_{i2}	Y_i	d_i	$Y_{i1}\left(\dfrac{d_i}{Y_i}\right)$	$d_{i1} - Y_{i1}\left(\dfrac{d_i}{Y_i}\right)$	$\dfrac{Y_{i1}}{Y_i}\left(1 - \dfrac{Y_{i1}}{Y_i}\right)\left(\dfrac{Y_i - d_i}{Y_i - 1}\right)d_i$
0.5	43	0	76	6	119	6	2.168	−2.168	1.326
1.5	43	1	60	0	103	1	0.417	0.583	0.243
2.5	42	0	56	2	98	2	0.857	−0.857	0.485
3.5	40	1	49	1	89	2	0.899	0.101	0.489
4.5	36	2	43	0	79	2	0.911	1.089	0.490
5.5	33	1	40	0	73	1	0.452	0.548	0.248
6.5	31	0	35	1	66	1	0.470	−0.470	0.249
8.5	25	2	30	0	55	2	0.909	1.091	0.487
9.5	22	1	27	0	49	1	0.449	0.551	0.247
10.5	20	1	25	0	45	1	0.444	0.556	0.247
11.5	18	1	22	0	40	1	0.450	0.550	0.248
15.5	11	1	14	1	25	2	0.880	0.120	0.472
16.5	10	1	13	0	23	1	0.435	0.565	0.246
18.5	9	1	11	0	20	1	0.450	0.550	0.248
23.5	4	1	5	0	9	1	0.444	0.556	0.247
26.5	2	1	3	0	5	1	0.400	0.600	0.240
SUM		15		11		26	11.036	3.964	6.211

TABLE 7.3

Comparison of Two-Sample Tests

Test	$W(t_i)$	$Z_1(\tau)$	σ_{11}^2	χ^2	p-value
Log-Rank	1.0	3.96	6.21	2.53	0.112
Gehan	Y_i	−9	38862	0.002	0.964
Tarone–Ware	$Y_i^{1/2}$	13.20	432.83	0.40	0.526
Peto–Peto	$\tilde{S}(t_i)$	2.47	4.36	1.40	0.237
Modified Peto–Peto	$\tilde{S}(t_i)Y_i/(Y_i + 1)$	2.31	4.20	1.28	0.259
Fleming–Harrington $p = 0, q = 1$	$[1 - \hat{S}(t_{i-1})]$	1.41	0.21	9.67	0.002
Fleming–Harrington $p = 1, q = 0$	$\hat{S}(t_{i-1})$	2.55	4.69	1.39	0.239
Fleming–Harrington $p = 1, q = 1$	$\hat{S}(t_{i-1})[1 - \hat{S}(t_{i-1})]$	1.02	0.11	9.83	0.002
Fleming–Harrington $p = 0.5, q = 0.5$	$\hat{S}(t_{i-1})^{0.5}[1 - \hat{S}(t_{i-1})]^{0.5}$	2.47	0.66	9.28	0.002
Fleming–Harrington $p = 0.5, q = 2$	$\hat{S}(t_{i-1})^{0.5}[1 - \hat{S}(t_{i-1})]^2$	0.32	0.01	8.18	0.004

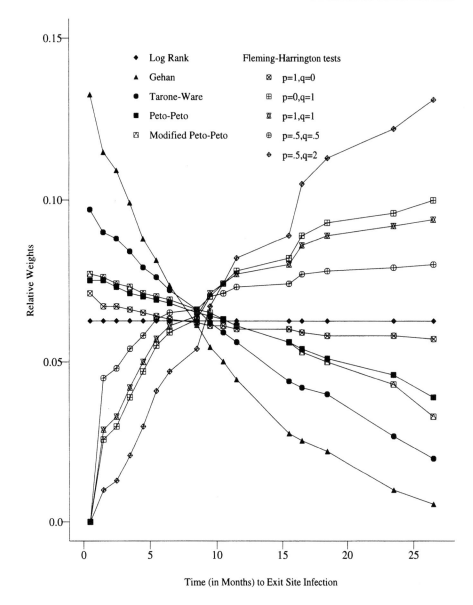

Figure 7.2 *Relative weights for comparison of observed and expected numbers of deaths for kidney dialysis patients.*

the results of these tests. Figure 7.2 shows the relative weights these tests give to the comparisons at each time point. $W(t_i)/\sum_{i=1}^{D} W(t_i)$ is plotted here. Note that Gehan's weight function gives very heavy weight to early comparisons at $t_i = 0.5$ and leads to a negative test statistic. The Fleming and Harrington tests, with $q > 0$, put more weight on late comparisons and lead to significant tests because the two survival curves diverge for larger values of t.

EXAMPLE 7.3

In section 1.16, data on 462 individuals who lived at the Channing House retirement center was reported. These data are left-truncated by the individual's entry time into the retirement center. In Example 4.3, survival curves were constructed for both males and females. We shall now apply the methods of this section to test the hypothesis that females tend to live longer than males. We test the hypothesis H_0 : $h_F(t) = h_M(t)$, 777 months $\leq t \leq$ 1152 months against the one sided hypothesis $H_A : h_F(t) \leq h_M(t)$ for all $t \in [777, 1152]$ and $h_F(t) < h_M(t)$ for some $t \in [777, 1152]$.

To perform this test, we need to compute Y_{iF} and Y_{iM} as the number of females and males, respectively, who were in the center at age t_i. The values of these quantities are depicted in Figure 4.10. The test will be based on the weighted difference between the observed and expected number of male deaths. Using the log-rank weights, we find $Z_M(1152) = 9.682$, $\hat{V}(Z_M(1152)) = 28.19$, so $Z_{\text{obs}} = 1.82$ and the one-sided p-value is 0.0341, which provides evidence that males are dying at a faster rate than females.

EXAMPLE 7.4

In Chapter 4, we investigated the relationship between the disease-free survival functions of 137 patients given a bone marrow transplant (see section 1.3 for details). Three groups were considered: Group 1 consisting of 38 ALL patients; Group 2 consisting of 54 AML low-risk patients and Group 3 consisting of 45 AML high-risk patients. The survival curves for these three groups are shown in Figure 4.2 in section 4.2.

We shall test the hypothesis that the disease-free survival functions of these three populations are the same over the range of observation, $t \leq$ 2204 days, versus the alternative that at least one of the populations has a different survival rate. Using the log-rank weights, we find $Z_1(2204) = 2.148$; $Z_2(2204) = -14.966$ and $Z_3(2204) = 12.818$, and the covariance matrix is

$$(\hat{\sigma}_{jg}, j, g = 1, \ldots, 3) = \begin{pmatrix} 15.9552 & -10.3451 & -5.6101 \\ -10.3451 & 20.3398 & -9.9947 \\ -5.6101 & -9.9947 & 15.6048 \end{pmatrix}.$$

Notice that the $Z_j(2204)$'s sum to zero and that the matrix $(\hat{\sigma}_{jg})$ is singular. The test is constructed by selecting any two of the $Z_j(2204)$'s, and constructing a quadratic form, using the appropriate rows and columns of the covariance matrix. The resulting statistic will be the same regardless of which $Z_j(2204)$'s are selected. The test statistic in this case is

$$\chi^2 = (2.148, -14.966) \begin{pmatrix} 15.9552 & -10.3451 \\ -10.3451 & 20.3398 \end{pmatrix}^{-1} \begin{pmatrix} 2.148 \\ -14.966 \end{pmatrix} = 13.8037.$$

When the null hypothesis is true, this statistic has an approximate chi-square distribution with two degrees of freedom which yields a p-value of 0.0010.

We can apply any of the weight functions discussed above to this problem. For example, Gehan's weight $W(t_i) = Y_i$ yields $\chi^2 = 16.2407$ ($p = 0.0003$); Tarone-Ware's weight $W(t_i) = Y_i^{1/2}$ yields $\chi^2 = 15.6529$ ($p = 0.0040$), Fleming and Harrington's weight, with $p = 1$, $q = 0$, yields $\chi^2 = 15.6725$ ($p = 0.0040$), with $p = 0$, $q = 1$, yields $\chi^2 = 6.1097$ ($p = 0.0471$), and with $p = q = 1$, yields $\chi^2 = 9.9331$ ($p = 0.0070$). All of these tests agree with the conclusion that the disease-free survival curves are not the same in these three disease categories.

An important consideration in applying the tests discussed in this section is the choice of the weight function to be used. In most applications of these methods, our strategy is to compute the statistics using the log-rank weights $W(t_i) = 1$ and the Gehan weight with $W(t_i) = Y_i$. Tests using these weights are available in most statistical packages which makes their application routine in most problems.

In some applications, one of the other weight functions may be more appropriate, based on the investigator's desire to emphasize either late or early departures between the hazard rates. For example, in comparing relapse-free survival between different regimes in bone marrow transplants for leukemia, a weight function giving higher weights to late differences between the hazard rates is often used. Such a function downweights early differences in the survival rates, which are often due to the toxicity of the preparative regimes, and emphasizes differences occurring late, which are due to differences in curing the leukemia. This is illustrated in the following example.

EXAMPLE 7.5   In section 1.9, data from a study of the efficacy of autologous (auto) versus allogeneic (allo) transplants for acute myelogenous leukemia was described. Of interest is a comparison of disease-free survival for these two types of transplants. Here, the event of interest is death or relapse, which ever comes first. In comparing these two types of transplants, it is well known that patients given an allogeneic transplant tend to have more complications early in their recovery process. The most critical of these complications is acute graft-versus-host disease which occurs within the first 100 days after the transplant and is often lethal. Because patients given an autologous transplant are not at risk of developing acute graft-versus-host disease, they tend to have a higher survival rate during this period. Of primary interest to most investigators in this area is comparing the treatment failure rate (death or relapse) among long-term survivors. To test this hypothesis, we shall use a test with the Fleming and Harrington weight function $W(t_i) = 1 - S(t_{i-1})$ (Eq. 7.3.9

with $p = 0$, $q = 1$). This function downweights events (primarily due to acute graft-versus-host disease) which occur early.

For these weights, we find that $Z_1(t) = -2.093$ and $\hat{\sigma}_{11}(\tau) = 1.02$, so that the chi-square statistic has a value of 4.20 which gives a p-value of 0.0404 . This suggest that there is a difference in the treatment failure rates for the two types of transplants.

By comparison, the log-rank test and Gehan's test have p-values of 0.5368 and 0.7556, respectively. These statistics have large p-values because the hazard rates of the two types of transplants cross at about 12 months, so that the late advantage of allogeneic transplants is negated by the high, early mortality of this type of transplant.

Practical Notes

1. The SAS procedure LIFETEST can be used to perform the log-rank test and Gehan's test for right-censored data. This procedure has two ways to perform the test. The first is to use the STRATA statement. This statement produces $Z_j(\tau)$, the matrix $(\hat{\sigma}_{jg})$ and the chi-square statistics. The second possibility for producing a test is to use the TEST statement. This statistic is equivalent to those obtained using the STRATA command when there is only one death at each time point. When there is more than one death at some time, it computes a statistic obtained as the average of the statistics one can obtain by breaking these ties in all possible ways. This leads to different statistics than those we present here. We recommend the use of the STRATA command for tests using SAS.

2. The S-Plus function surv.diff produces the Fleming and Harrington class of tests with $q = 0$. By choosing $p = 0$, the log-rank test can be obtained.

3. All of the tests described in this section are based on large-sample approximations to the distribution of the chi-square statistics. They also assume that the censoring distributions are independent of the failure distributions. Care should be used in interpreting these results when the sample sizes are small or when there are few events. (Cf. Kellerer and Chmelevsky 1983, Latta, 1981, or Peace and Flora, 1978, for the results of Monte Carlo studies of the small-sample power of these tests.)

4. Based on a Monte Carlo study, Kellerer and Chmelevsky (1983) conclude that, when the sample sizes are small for two-sample tests, the one-sided test must be used with caution. When the sample sizes are very different in the two groups and when the alternative hypothesis is that the hazard rate of the smaller sample is larger than the rate in the larger sample, these tests tend to falsely reject the null hypothesis too often. The tests are extremely conservative when the larger sample is associated with the larger hazard rate un-

der the alternative hypothesis. They and Prentice and Marek (1979) strongly recommend that only two-sided tests be used in making comparisons.

5. For the two-sample tests, the log-rank weights have optimal local power to detect differences in the hazard rates, when the hazard rates are proportional. This corresponds to the survival functions satisfying a Lehmann alternative $S_j(t) = S(t)^{\theta_j}$. These are also the optimal weights for the K sample test with proportional hazards when, for large samples, the numbers at risk in each sample at each time point are proportional in size. For the two-sample case, Fleming and Harrington's class of tests with $q = 0$ has optimal local power to detect the alternatives $h_2(t) = h_1(t)e^{\theta}[S_1(t)^p + [1 - S_1(t)]^q e^{\theta}]^{-1}$. See Fleming and Harrington (1981) or Andersen et al. (1993) for a more detailed discussion.

Theoretical Notes

1. The tests discussed in this section arise naturally using counting process theory. In section 3.6, we saw that the Nelson–Aalen estimator of the cumulative hazard rate in the jth sample was a stochastic integral of a predictable process with respect to a basic martingale and, as such, is itself a martingale. By a similar argument, when the null hypothesis is true, the Nelson–Aalen estimator of the common cumulative hazard rate is also a martingale. Furthermore, the difference of two martingales can be shown to also be a martingale. If $W_j(t)$ is a predictable weight function, then, $Z_j(\tau)$ is the integral of a predictable process with respect to a martingale and is, again, a martingale when the null hypothesis is true. The estimated variance of $Z_j(\tau)$ in (7.3.4) comes from the predictable variation process of $Z_j(\tau)$ using a version of the predictable variation process for the basic martingale which allows for discrete event times. More detailed descriptions of this derivation are found in Aalen (1975) and Gill (1980) for the two-sample case and in Andersen et al. (1982) for the K sample case.

2. The modification of Andersen et al. (1982) to the Peto and Peto weight, $W(t_i) = \tilde{S}(t_i)Y_i/(Y_i + 1)$ makes the weight function predictable in the sense discussed in section 3.6.

3. The statistics presented in this section are generalizations to censored data of linear rank statistics. For uncensored data, a linear rank statistic is constructed as $Z_j = \sum_{i=1}^{n_j} a_n(R_{ij})$, $j = 1, \ldots, K$.

    Here R_{ij} is the rank of the ith observation in the jth sample among the pooled observations. The scores $a_n(i)$ are obtained from a score function Ψ defined on the unit interval by $a_n(i) = E[\Psi(T_{(i)})]$, where $T_{(i)}$ is the ith order statistic from a uniform $[0, 1]$ sample of size n or by $a_n(i) = \Psi[i/(n + 1)]$. For a censored sample, these scores are generalized as follows: An uncensored observation is given a score of

$\Psi[1 - \tilde{S}(t)]$, with $\tilde{S}(t)$ given by (7.3.8); a censored observation is given a score of $\frac{1}{1-\Psi[1-\tilde{S}(t)]} \int_{\Psi[1-\tilde{S}(t)]}^{1} \Psi(u) du$. The score function $\Psi(u) = 2u - 1$, for example will yield the Peto and Peto version of Gehan's test. (See Kalbfleisch and Prentice, 1980 for additional development of this concept.)

4. Gill (1980) discusses the Pitman efficiency of these tests.

5. Gehan's two-sample test can be constructed as a generalization of the Mann–Whitney test as follows. Let (T_{ij}, δ_{ij}) denote the time on study and the event indicator for the ith observation in the jth sample. Define the scoring function $\phi[(T_{i1}, \delta_{i1}), (T_{h2}, \delta_{h2})]$ as follows:

$$\phi[(T_{i1}, \delta_{i1}), (T_{h2}, \delta_{h2})] = \begin{pmatrix} +1, \text{ if } T_{i1} \leq T_{h2}, \ \delta_{i1} = 1, \ \delta_{h2} = 0, \\ \qquad \text{or } T_{i1} < T_{h2}, \ \delta_{i1} = 1, \ \delta_{h2} = 1, \\ -1, \text{ if } T_{i1} \geq T_{h2}, \ \delta_{h2} = 1, \ \delta_{i1} = 0, \\ \qquad \text{or } T_{i1} > T_{h2}, \ \delta_{i1} = 1, \ \delta_{h2} = 1, \\ 0, \text{ otherwise} \end{pmatrix}$$

Then, $Z_1(\tau) = \sum_{i=1}^{n_1} \sum_{h=1}^{n_2} \phi[(T_{i1}, \delta_{i1}), (T_{h2}, \delta_{h2})]$ is the number of observations from the first sample that are definitely smaller than an observation in the second sample. Gehan (1965) provides a variance estimator of this statistic under the null hypothesis, based on assuming a fixed censoring pattern and that all possible permutations of the two samples over this pattern are equally likely. Essentially, this estimator assumes that the censoring patterns are the same for both samples. When this is not the case, this variance estimator may lead to incorrect decisions.

7.4   Tests for Trend

In the previous section, we examined tests, based on a weighted comparison of the observed and expected number of deaths in each of K samples, to test the null hypothesis that the K population hazard rates are the same versus the global alternative hypothesis that, at least, one of the hazard rates is different. In this section, we shall use the statistics developed in section 7.3 to develop a test statistic with power to detect ordered alternatives, that is, we shall test

$$H_0 : h_1(t) = h_2(t) = \cdots = h_K(t), \text{ for } t \leq \tau, \qquad (7.4.1)$$

against

$$H_A : h_1(t) \leq h_2(t) \leq \cdots \leq h_K(t) \text{ for } t \leq \tau, \text{ with at least one} \\ \text{strict inequality.}$$

The alternative hypothesis is equivalent to the hypothesis that $S_1(t) \geq S_2(t) \geq \cdots \geq S_K(t)$.

The test will be based on the statistic $Z_j(\tau)$ given by (7.3.3). Any of the weight functions discussed in section 7.3 can be used in constructing the test. As discussed earlier these various weight functions give more or less weight to the time points at which the comparison between the observed and expected number of deaths in each sample is made. We let $\hat{\Sigma}$ be the full $K \times K$ covariance matrix, $\hat{\Sigma} = (\hat{\sigma}_{jg}, j, g = 1, \ldots, K)$. Here, $\hat{\sigma}_{jg}$ is given by Eqs. (7.3.4) and (7.3.5).

To construct the test, a sequence of scores $a_1 < a_2 < \cdots < a_K$ is selected. Any increasing set of scores can be used in constructing the test, and the test is invariant under linear transformations of the scores. In most cases, the scores $a_j = j$ are used, but one may take the scores to be some numerical characteristic of the jth population. The test statistic is

$$Z = \frac{\sum_{j=1}^{K} a_j Z_j(\tau)}{\sqrt{\sum_{j=1}^{K} \sum_{g=1}^{K} a_j a_g \hat{\sigma}_{jg}}}. \qquad (7.4.2)$$

When the null hypothesis is true and the sample sizes are sufficiently large, then, this statistic has a standard normal distribution. If the alternative hypothesis is true, then, the $Z_j(\tau)$ associated with larger values of a_j should tend to be large, and, thus, the null hypothesis is rejected in favor of H_A at an α Type I error rate when the test statistic is larger than the αth upper percentile of a standard normal distribution.

EXAMPLE 7.6   In section 1.8, a study of 90 patients diagnosed with cancer of the larynx in the 70s at a Dutch hospital was reported. The data consists of the times between first treatment and either death or the end of the study. Patients were classified by the stage of their disease using the American Joint Committee for Cancer Staging. We shall test the hypothesis that there is no difference in the death rates among the four stages of the disease versus the hypothesis that, the higher the stage, the higher the death rate. The data is found on our web site. The four survival curves are shown in Figure 7. 3. We shall use the scores $a_j = j$, $j = 1, \ldots, 4$ in constructing our tests.

Using the log-rank weights,

$$\mathbf{Z}(10.7) = (-7.5660, -3.0117, 2.9155, 7.6623) \text{ and}$$

$$\hat{\Sigma} = \begin{pmatrix} 12.0740 & -4.4516 & -6.2465 & -1.3759 \\ -4.4516 & 7.8730 & -2.7599 & -0.6614 \\ -6.2465 & -2.7599 & 9.9302 & -0.9238 \\ -1.3759 & -0.6614 & -0.9238 & 2.9612 \end{pmatrix}.$$

The value of the test statistic (7.4.2) is 3.72 and the p-value of the test is less than 0.0001.

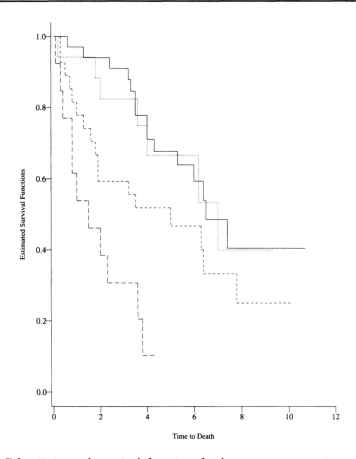

Figure 7.3 *Estimated survival function for larynx cancer patients. Stage I (————) Stage II (------) Stage III (— — —) Stage IV (—— ——)*

Using the Tarone and Ware weights, we find that the value of the test statistic is 4.06. Using Gehan's weights, the value of the test statistic is 4.22, and using the Peto and Peto weights, the value of the test statistic is 4.13. All three tests have a *p*-value of less than 0.0001, providing strong evidence that the hazard rates are in the expected order.

Practical Notes

1. The SAS procedure LIFETEST provides the statistics $Z_j(t)$ and $\hat{\Sigma}$ based on the log-rank weights and Gehan's weights.
2. This test should be applied only when there is some a priori information that the alternatives are ordered.

Theoretical Note

1. When there is no censoring, the test using Gehan or Peto–Peto weights reduces to the Jonckheere–Terpstra test.

7.5   Stratified Tests

Often in applying the tests discussed in section 7.3, an investigator is faced with the problem of adjusting for some other covariates that affect the event rates in the K populations. One approach is to imbed the problem of comparing the K populations into a regression formulation, as described in section 2.6, and perform the test to compare the populations, after an adjustment for covariates, by using one of the techniques described in Chapters 8–10. An alternative approach is to base the test on a stratified version of one of the tests discussed in section 7.3. This approach is feasible when the number of levels of the covariate is not too large or when a continuous covariate is discretized into a workable number of levels. In this section, we shall discuss how such stratified tests are constructed, and we shall see how these tests can be used to analyze the survival of matched pairs.

We assume that our test is to be stratified on M levels of a set of covariates. We shall test the hypothesis

$$H_0 : h_{1s}(t) = h_{2s}(t) = \cdots = h_{Ks}(t), \text{ for } s = 1, \ldots, M, \; t < \tau. \quad (7.5.1)$$

Based only on the data from the sth strata, let $Z_{js}(\tau)$ be the statistic (7.3.3) and $\hat{\Sigma}_s$ be the variance-covariance matrix of the $Z_{js}(\tau)$'s obtained from (7.3.4) and (7.3.5). As in the previous two sections, any choice of weight functions can be used for Z_{js}. These statistics can be used to test the hypothesis of difference in the hazard rates within stratum s by constructing the test statistic (7.3.6). A global test of (7.5.1) is constructed as follows:

$$\text{Let } Z_{j\cdot}(\tau) = \sum_{s=1}^{M} Z_{js}(\tau) \text{ and } \hat{\sigma}_{jg\cdot} = \sum_{s=1}^{M} \hat{\sigma}_{jgs}. \quad (7.5.2)$$

The test statistic, as in (7.3.6), is

$$(Z_{1\cdot}(\tau), \ldots, Z_{K-1\cdot}(\tau))\Sigma^{-1}(Z_{1\cdot}(\tau), \ldots, Z_{K-1\cdot}(\tau))^t \quad (7.5.3)$$

where Σ is the $(K-1) \times (K-1)$ matrix obtained from the $\hat{\sigma}_{jg\cdot}$'s. When the total sample size is large and the null hypothesis is true, this statistic has a chi-squared distribution with $K-1$ degrees of freedom. For the two-sample problem, the stratified test statistic is

$$\frac{\sum_{s=1}^{M} Z_{1s}(\tau)}{\sqrt{\sum_{s=1}^{M} \hat{\sigma}_{11s}}} \qquad (7.5.4)$$

which is asymptotically standard normal when the null hypothesis is true.

EXAMPLE 7.7

In section 1.10, the results of a small study comparing the effectiveness of allogeneic (allo) transplants versus autogeneic (auto) transplants for Hodgkin's disease (HOD) or non-Hodgkin's lymphoma (NHL) was presented. Of interest is a test of the null hypothesis of no difference in the leukemia-free-survival rate between patients given an allo ($j = 1$) or auto ($j = 2$) transplant, adjusting for the patient's disease state.

From only the data on Hodgkin's patients, we find $Z_{1\text{HOD}}(2144) = 3.1062$ and $\hat{\sigma}_{11\text{HOD}} = 1.5177$ using log-rank weights. For the non-Hodgkin's lymphoma patients, we find $Z_{1\text{NHL}}(2144) = -2.3056$ and $\hat{\sigma}_{11\text{NHL}} = 3.3556$. The stratified log-rank test is

$$Z = \frac{3.1062 + (-2.3056)}{\sqrt{1.5177 + 3.3556}} = 0.568,$$

which has a p-value of 0.5699.

In this example, if we perform the test only on the Hodgkin's disease patients, we find that the test statistic has a value of 2.89 ($p = 0.004$), whereas using only non-Hodgkin's patients, we find that the test statistic is -1.26 ($p = 0.2082$). The small value of the combined statistic is due, in part, to the reversal of the relationship between transplant procedure and disease-free survival in the Hodgkin's group from that in the non-Hodgkin's group.

EXAMPLE 7.4

(continued) In Example 7.4, we found that there was evidence of a difference in disease-free survival rates between bone marrow patients with ALL, low-risk AML and high-risk AML. A proper comparison of these three disease groups should account for other factors which may influence disease-free survival. One such factor is the use of a graft-versus-host prophylactic combining methotretexate (MTX) with some other agent. We shall perform a stratified Gehan weighted test for differences in the hazard rates of the three disease states. Using (7.3.2)–(7.3.5), for the no MTX strata,

$$Z_{1\text{NOMTX}} = -103, \ Z_{2\text{NOMTX}} = -892, \ Z_{3\text{NOMTX}} = 995,$$

$$\hat{\Sigma}_{\text{NOMTX}} = \begin{pmatrix} 49366.6 & -32120.6 & -17246.0 \\ -32120.6 & 69388.9 & -37268.2 \\ -17246.0 & -37268.2 & 54514.2 \end{pmatrix},$$

and, for the MTX strata,

$$Z_{1\text{MTX}} = 20, \quad Z_{2\text{MTX}} = -45, \quad Z_{3\text{MTX}} = 25,$$

and

$$\hat{\Sigma}_{\text{MTX}} = \begin{pmatrix} 5137.1 & -2685.6 & -2451.6 \\ -2685.6 & 4397.5 & -1711.9 \\ -2451.6 & -1711.9 & 4163.5 \end{pmatrix}.$$

Pooling the results in the two strata,

$$Z_{1.} = -83, \quad Z_{2.} = -937, \quad Z_{3.} = 1020, \quad \text{and}$$

$$\hat{\Sigma}_. = \begin{pmatrix} 54503.7 & -34806.2 & -19697.6 \\ -34806.2 & 73786.1 & -38980.1 \\ -19697.6 & -38980.1 & 58677.7 \end{pmatrix}.$$

The stratified Gehan test statistic is

$$(-83, \quad -937) \begin{pmatrix} 54503.7 & -34806.2 \\ -34806.2 & 73786.1 \end{pmatrix}^{-1} \begin{pmatrix} -83 \\ -937 \end{pmatrix} = 19.14$$

which has a p-value of less than 0.0001 when compared to a chi-square with two degrees of freedom. The tests on the individual strata give test statistics of $\chi^2 = 19.1822$ ($p = 0.0001$) for the no MTX group and $\chi^2 = 0.4765$ ($p = 0.7880$) in the MTX arm. The global test, ignoring MTX, found a test statistic of 16.2407 with a p-value of 0.0003.

Another use for the stratified test is for matched pairs, censored, survival data. Here we have paired event times (T_{1i}, T_{2i}) and their corresponding event indicators $(\delta_{1i}, \delta_{2i})$, for $i = 1, \ldots, M$. We wish to test the hypothesis $H_0 : h_{1i}(t) = h_{2i}(t)$, $i = 1, \ldots, M$. Computing the statistics (7.3.3) and (7.3.4),

$$Z_{1i}(\tau) = \begin{cases} W(T_{1i})(1 - 1/2) = W(T_{1i})/2 & \text{if } T_{1i} < T_{2i}, \ \delta_{1i} = 1 \\ & \text{or } T_{1i} = T_{2i}, \ \delta_{1i} = 1, \ \delta_{2i} = 0 \\ W(T_{2i})(0 - 1/2) = -W(T_{2i})/2 & \text{if } T_{2i} < T_{1i}, \ \delta_{2i} = 1 \\ & \text{or } T_{2i} = T_{1i}, \ \delta_{2i} = 1, \ \delta_{1i} = 0 \\ 0 & \text{otherwise} \end{cases},$$

(7.5.5)

and

$$\hat{\sigma}_{11i} = \begin{cases} W(T_{1i})^2(1/2)(1 - 1/2) = W(T_{1i})^2/4 & \text{if } T_{1i} < T_{2i}, \ \delta_{1i} = 1 \\ & \text{or } T_{1i} = T_{2i}, \ \delta_{1i} = 1, \ \delta_{2i} = 0 \\ W(T_{2i})^2(1/2)(1 - 1/2) = W(T_{2i})^2/4 & \text{if } T_{2i} < T_{1i}, \ \delta_{2i} = 1 \\ & \text{or } T_{2i} = T_{1i}, \ \delta_{2i} = 1, \ \delta_{1i} = 0 \\ 0 & \text{otherwise} \end{cases}.$$

For any of the weight functions we have discussed,

$$Z_{1.}(\tau) = w \frac{D_1 - D_2}{2}$$

(7.5.6)

and

$$\hat{\sigma}_{11.} = w^2 \frac{D_1 + D_2}{4},$$

where D_1 is the number of matched pairs in which the individual from sample 1 experiences the event first and D_2 is the number in which the individual from sample 2 experiences the event first. Here w is the value of the weight function at the time when the smaller of the pair fails. Because these weights do not depend on which group this failure came from, the test statistic is

$$\frac{D_1 - D_2}{\sqrt{D_1 + D_2}}, \tag{7.5.7}$$

which has a standard normal distribution when the number of pairs is large and the null hypothesis is true. Note that matched pairs, where the smaller of the two times corresponds to a censored observation, make no contribution to Z_1 or $\hat{\sigma}_{11.}$.

EXAMPLE 7.8   In section 1.2, the results of a clinical trial of the drug 6-mercaptopurine (6-MP) versus a placebo in 42 children with acute leukemia was described. The trial was conducted by matching pairs of patients at a given hospital by remission status (complete or partial) and randomizing within the pair to either a 6-MP or placebo maintenance therapy. Patients were followed until their leukemia returned (relapse) or until the end of the study.

Survival curves for the two groups were computed in Example 4.1. We shall now test the hypothesis that there is no difference in the rate of recurrence of leukemia in the two groups. From Table 1.1, we find $D_{\text{Placebo}} = 18$ and $D_{\text{6-MP}} = 3$, so that the test statistic is $(18 - 3)/(18 + 3)^{1/2} = 3.27$. The p-value of the test is $2Pr[Z \geq 3.27] = 0.001$, so that there is sufficient evidence that the relapse rates are different in the two groups.

Practical Notes

1. The test for matched pairs uses only information from those pairs where the smaller of the two times corresponds to an event time. The effective sample size of the test is the number of such pairs.
2. The test for matched pairs is the censored-data version of the sign test.
3. The stratified tests will have good power against alternatives that are in the same direction in each stratum. When this is not the case, these statistics may have very low power, and separate tests for each stratum are indicated. (See Example 7.5.)

Theoretical Notes

1. The test for matched pairs relies only on intrapair comparisons. Other tests for matched pairs have been suggested which assume a bivariate model for the paired times, but make interpair comparisons.
2. The asymptotic chi-squared distribution of the stratified tests discussed in this section is valid when either the number of strata is fixed and the number within each stratum is large or when the number in each stratum is fixed and the number of strata is large. See Andersen et al. (1993) for details on the asymptotics of these tests.

7.6   Renyi Type Tests

In section 7.3, a series of tests to compare the survival experience of two or more samples were presented. All of these tests were based on the weighted integral of estimated differences between the cumulative hazard rates in the samples. When these tests are applied to samples from populations where the hazard rates cross, these tests have little power because early differences in favor of one group are canceled out by late differences in favor of the other treatment. In this section, we present a class of tests with greater power to detect crossing hazards. We will focus on the two sample versions of the test.

The test statistics to be used are called Renyi statistics and are censored-data analogs of the Kolmogorov–Smirnov statistic for comparing two uncensored samples. To construct the tests, we will find the value of the test statistic (7.3.3) for some weight function at each death time. When the hazard rates cross, the absolute value of these sequential evaluations of the test statistic will have a maximum value at some time point prior to the largest death time. When this value is too large, then, the null hypothesis of interest $H_0 : h_1(t) = h_2(t)$, $t < \tau$, is rejected in favor of $H_A : h_1(t) \neq h_2(t)$, for some t. To adjust for the fact that we are constructing multiple test statistics on the same set of data, a correction is made to the critical value of the test.

To construct the test, suppose that we have two independent samples of size n_1 and n_2, respectively. Let $n = n_1 + n_2$. Let $t_1 < t_2 < \cdots < t_D$ be the distinct death times in the pooled sample. In sample j let d_{ij} be the number of deaths and Y_{ij} the number at risk at time t_i, $i = 1, \ldots, D$, $j = 1, 2$. Let $Y_i = Y_{i1} + Y_{i2}$ be the total number at risk in both samples and $d_i = d_{i1} + d_{i2}$ be the total number of deaths in the combined sample at time t_i. Let $W(t)$ be a weight function. For example, for the "log-rank" version of this test $W(t) = 1$ and, for the "Gehan–Wilcoxon" version, $W(t_i) = Y_{i1} + Y_{i2}$. For each value of t_i we compute, $Z(t_i)$, which is the value of the numerator of the statistic (7.3.7) using only the death times observed up to time t_i, that is,

$$Z(t_i) = \sum_{t_k \le t_i} W(t_k) \left[d_{k1} - Y_{k1} \left(\frac{d_k}{Y_k} \right) \right], \quad i = 1, \ldots, D. \qquad (7.6.1)$$

Let $\sigma(\tau)$ be the standard error of $Z(\tau)$ which, from (7.3.7), is given by

$$\sigma^2(\tau) = \sum_{t_k \le \tau} W(t_k)^2 \left(\frac{Y_{k1}}{Y_k} \right) \left(\frac{Y_{k2}}{Y_k} \right) \left(\frac{Y_k - d_k}{Y_k - 1} \right) (d_k); \qquad (7.6.2)$$

where τ is the largest t_k with $Y_{k1}, Y_{k2} > 0$.

The test statistic for a two-sided alternative is given by

$$Q = \sup\{|Z(t)|, \ t \le \tau\}/\sigma(\tau). \qquad (7.6.3)$$

When the null hypothesis is true, then, the distribution of Q can be approximated by the distribution of the $\sup(|B(x)|, \ 0 \le x \le 1)$ where B is a standard Brownian motion process. Critical values of Q are found in Table C.5 in Appendix C.

The usual weighted log rank test statistic is $Z(\tau)/\sigma(\tau)$. For this test, when the two hazard rates cross, early positive differences between the two hazard rates are canceled out by later differences in the rates, with opposite signs. The supremum version of the statistic should have greater power to detect such differences between the hazard rates.

EXAMPLE 7.9   A clinical trial of chemotherapy against chemotherapy combined with radiotherapy in the treatment of locally unresectable gastric cancer was conducted by the Gastrointestinal Tumor Study Group (1982). In this trial, forty-five patients were randomized to each of the two arms and followed for about eight years. The data, found in Stablein and Koutrouvelis (1985), is as follows:

Chemotherapy Only											
1	63	105	129	182	216	250	262	301	301	342	354
356	358	380	383	383	388	394	408	460	489	499	523
524	535	562	569	675	676	748	778	786	797	955	968
1000	1245	1271	1420	1551	1694	2363	2754*	2950*			

Chemotherapy Plus Radiotherapy											
17	42	44	48	60	72	74	95	103	108	122	144
167	170	183	185	193	195	197	208	234	235	254	307
315	401	445	464	484	528	542	547	577	580	795	855
1366	1577	2060	2412*	2486*	2796*	2802*	2934*	2988*			

*Denotes censored observation.

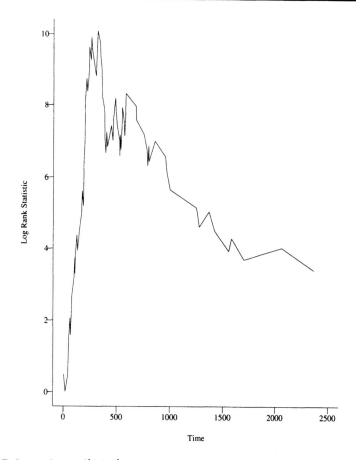

Figure 7.4   *Values of $|Z(t_i)|$ for the gastrointestinal tumor study*

We wish to test the hypothesis that the survival rate of the two groups is the same by using the log rank version ($W(t_i) = 1$) of the Renyi statistics. Figure 7.4 shows the value of $|Z(t_i)|$. Here, the maximum occurs at $t_i = 315$ with a value of 9.80. The value of $\sigma(2363) = 4.46$, so $Q = 2.20$. From Table C.5 in Appendix C we find that the p-value of this test is 0.053 so the null hypothesis of no difference in survival rates between the two treatment groups is not rejected at the 5% level. If we had used the nonsequential log-rank test, we have $Z(2363) = -2.15$, yielding a p-value of 0.6295 which is not significant. From Figure 7.5, which plots the Kaplan–Meier curves for the two samples, we see that the usual log rank statistic has a small value because early differences in favor of the chemotherapy only group are negated by a late survival advantage for the chemotherapy plus radiotherapy group.

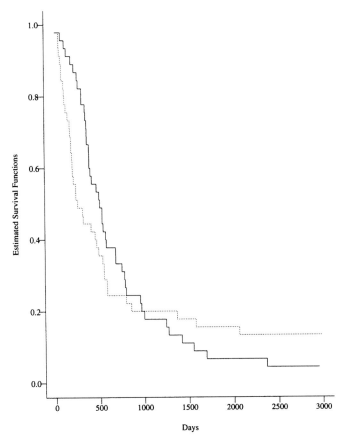

Figure 7.5 *Estimated survival functions for the gastrointestinal tumor study. Chemotherapy only (————) Chemotherapy plus radiation (------)*

Practical Notes

1. A one-sided test of the hypothesis $H_0 : S_1(t) = S_2(t)$ against $H_A : S_1(t) < S_2(t)$ can be based on $Q^* = \sup[Z(t), t \leq \tau]/\sigma(\tau)$. When H_0 is true, Q^* converges in distribution to the supremum of a Brownian motion process $B(t)$ (see Theoretical Note 3 above). The p-value of a one-sided test is given by $Pr[\sup B(t) > Q^*] = 2[1 - \Phi(Q^*)]$, where $\Phi()$ is the standard normal cumulative distribution function.

Theoretical Notes

1. The supremum versions of the weighted, log-rank tests were proposed by Gill (1980). He calls the statistic (7.6.3) a "Renyi-type"

statistic. Further development of the statistical properties of the test can be found in Fleming and Harrington (1991) and Fleming et al. (1980). Schumacher (1984) presents a comparison of this class of statistics to tests based on the complete test statistic and other versions of the Kolmogorov–Smirnov statistic.

2. For a standard Brownian motion process $B(t)$, Billingsly (1968) shows that

$$Pr[\sup |B(t)| > y] = 1 - \frac{4}{\pi} \sum_{k=0}^{\infty} \frac{(-1)^k}{2k + 1} \exp[-\pi^2(2k + 1)^2/8y^2].$$

3. Using the counting process methods introduced in section 3.6, one can show, under H_0, that, if $\sigma^2(t)$ converges in probability to a limiting variance $\sigma_0^2(t)$ on $[0, \infty)$ then, $Z(t)$ converges weakly to the process $B[\sigma_0^2(t)]$ on the interval $[0, \infty]$. This implies that $\sup[Z(t)/\sigma(\infty) : 0 < t < \infty]$ converges in distribution to $[\sup B(t) : t \in A]$ where $A = \{\sigma_0^2(t)/\sigma_0^2(\infty), 0 \leq t \leq \infty\}$. When the underlying common survival function is continuous, then, A is the full unit interval, so the asymptotic p-values are exact. When the underlying common survival function is discrete, then, A is a subset of the interval $(0, 1)$, and the test is a bit conservative. See Fleming et al. (1987) for details of the asymptotics.

4. Other extensions of the Kolmogorov–Smirnov test have been suggested in the literature. Schumacher (1984) provides details of tests based on the maximum value of either $\log[\hat{H}_1(t)] - \log[\hat{H}_2(t)]$ or $\hat{H}_1(t) - \hat{H}_2(t)$. In a Monte Carlo study, he shows that the performance of these statistics is quite poor, and they are not recommended for routine use.

5. Both Schumacher (1984) and Fleming et al. (1987) have conducted simulation studies comparing the Renyi statistic of this section to the complete test statistic of section 7.3. For the log-rank test, they conclude there is relatively little loss of power for the Renyi statistics when the hazard rates are proportional and there is little censoring. For nonproportional or crossing hazards the Renyi test seems to perform much better than the usual log-rank test for light censoring. The apparent advantage of the Renyi statistic for light censoring diminishes as the censoring fraction increases.

7.7 Other Two-Sample Tests

In this section, we present three other two-sample tests which have been suggested in the literature. These tests are constructed to have greater power than the tests in section 7.3 to detect crossing hazard rates. All three tests are analogs of common nonparametric tests for uncensored data.

The first test is a censored-data version of the Cramer–von Mises test. For uncensored data, the Cramer–Von Mises test is based on the integrated, squared difference between the two empirical survival functions. For right-censored data, it is more appropriate to base a test on the integrated, squared difference between the two estimated hazard rates. This is done to obtain a limiting distribution which does not depend on the relationship between the death and censoring times and because such tests arise naturally from counting process theory. We shall present two versions of the test.

To construct the test, recall, from Chapter 4, that the Nelson–Aalen estimator of the cumulative hazard function in the jth sample is given by

$$\tilde{H}_j(t) = \sum_{t_i \leq t} \frac{d_{ij}}{Y_{ij}}, \quad j = 1, 2. \tag{7.7.1}$$

An estimator of the variance of $\tilde{H}_j(t)$ is given by

$$\sigma_j^2(t) = \sum_{t_j \leq t} \frac{d_{ij}}{Y_{ij}(Y_{ij} - 1)}, \quad j = 1, 2. \tag{7.7.2}$$

Our test is based on the difference between $\tilde{H}_1(t)$ and $\tilde{H}_2(t)$, so that we need to compute $\sigma^2(t) = \sigma_1^2(t) + \sigma_2^2(t)$, which is the estimated variance of $\tilde{H}_1(t) - \tilde{H}_2(t)$. Also let $A(t) = n\sigma^2(t)/[1 + n\sigma^2(t)]$.

The first version of the Cramer-von Mises statistic is given by

$$Q_1 = \left(\frac{1}{\sigma^2(\tau)} \right)^2 \int_0^\tau [\tilde{H}_1(t) - \tilde{H}_2(t)]^2 \, d\sigma^2(t)$$

which can be computed as

$$Q_1 = \left(\frac{1}{\sigma^2(\tau)} \right)^2 \sum_{t_i \leq \tau} [\tilde{H}_1(t_i) - \tilde{H}_2(t_i)]^2 [\sigma^2(t_i) - \sigma^2(t_{i-1})], \tag{7.7.3}$$

where $t_0 = 0$, and the sum is over the distinct death times less than τ. When the null hypothesis is true, one can show that the large sample distribution of Q_1 is the same as that of $R_1 = \int_0^1 [B(x)]^2 \, dx$, where $B(x)$ is a standard Brownian motion process. The survival function of R_1 is found in Table C.6 of Appendix C.

An alternate version of the Cramer–von Mises statistic is given by

$$Q_2 = n \int_0^\tau \left[\frac{\tilde{H}_1(t) - \tilde{H}_2(t)}{1 + n\sigma^2(t)} \right]^2 \, dA(t)$$

which is computed as

$$Q_2 = n \sum_{t_i \leq \tau} \left[\frac{\tilde{H}_1(t_i) - \tilde{H}_2(t_i)}{1 + n\sigma^2(t_i)} \right]^2 [A(t_i) - A(t_{i-1})]. \qquad (7.7.4)$$

When the null hypothesis is true, the large sample distribution of Q_2 is the same as that of $R_2 = \int_0^{A(\tau)} [B^0(x)]^2 dx$, where $B^0()$ is a Brownian bridge process. Table C.7 of Appendix C provides critical values for the test based on Q_2.

EXAMPLE 7.2   *(continued)*   We shall apply the two Cramer–von Mises tests to the comparison of the rate of cutaneous exit-site infections for kidney dialysis patients whose catheters were placed surgically (group 1) as compared to patients who had percutaneous placement of their catheters (group 2). Routine calculations yield $Q_1 = 1.8061$ which, from Table C.6 of Appendix C, has a p-value of 0.0399. For the second version of the Cramer–von Mises test, $Q_2 = 0.2667$ and $A(\tau) = 0.99$. From Table C.7 of Appendix C, we find that this test has a p-value of 0.195.

A common test for uncensored data is the two-sample t-test, based on the difference in sample means between the two populations. The second test we present is a censored-data version of this test based on the Kaplan–Meier estimators in the two samples, $\hat{S}_1(t)$ and $\hat{S}_2(t)$. In section 4.5, we saw that the population mean can be estimated by the area under the Kaplan–Meier curve $\hat{S}(t)$. This suggests that a test based on the area under the curve $\hat{S}_1(t) - \hat{S}_2(t)$, over the range where both of the two samples still have individuals at risk, will provide a censored data analog of the two-sample t-test. For censored data, we have seen that the estimate of the survival function may be unstable in later time intervals when there is heavy censoring, so that relatively small differences in the Kaplan–Meier estimates in the tail may have too much weight when comparing the two survival curves. To handle this problem, the area under a weighted difference in the two survival functions is used. The weight function, which downweights differences late in time when there is heavy censoring, is based on the distribution of the censoring times.

To construct the test, we pool the observed times in the two samples. Let $t_1 < t_2 < \cdots < t_n$ denote the ordered times. Notice that, here, as opposed to the other procedures where only event times are considered, the t_i's consist of both event and censoring times. Let d_{ij}, c_{ij}, Y_{ij} be, respectively, the number of events, the number of censored observations, and the number at risk at time t_i in the jth sample, $j = 1, 2$. Let $\hat{S}_j(t)$ be the Kaplan–Meier estimator of the event distribution using data in the jth sample and let $\hat{G}_j(t)$ be the Kaplan–Meier estimator of the time to censoring in the jth sample, that is, $\hat{G}_j(t) = \prod_{t_i \leq t} [1 - c_{ij}/Y_{ij}]$.

Finally, let $\hat{S}_p(t)$ be the Kaplan–Meier estimator based on the combined sample.

To construct the test statistic, we define a weight function by

$$w(t) = \frac{n\hat{G}_1(t)\hat{G}_2(t)}{n_1\hat{G}_1(t) + n_2\hat{G}_2(t)}, \quad 0 \le t \le t_D \qquad (7.7.5)$$

where n_1 and n_2 are the two sample sizes and $n = n_1 + n_2$. Notice that $w(t)$ is constant between successive censored observations and, when there is heavy censoring in either sample, $w(t)$ is close to zero. When there is no censoring, $w(t)$ is equal to 1. The test statistic is given by

$$W_{\text{KM}} = \sqrt{\frac{n_1 n_2}{n}} \int_0^{t_D} w(t)[\hat{S}_1(t) - \hat{S}_2(t)]dt$$

which can be computed by

$$W_{\text{KM}} = \sqrt{\frac{n_1 n_2}{n}} \sum_{i=1}^{D-1} [t_{i+1} - t_i]w(t_i)[\hat{S}_1(t_i) - \hat{S}_2(t_i)]. \qquad (7.7.6)$$

To find the variance of W_{KM}, first, compute

$$A_i = \int_{t_i}^{t_D} w(u)\hat{S}_p(u)du = \sum_{k=i}^{D-1}(t_{k+1} - t_k)w(t_k)\hat{S}_p(t_k). \qquad (7.7.7)$$

The estimated variance of $W_{\text{KM}} = \hat{\sigma}_p^2$ is given by

$$\hat{\sigma}_p^2 = \sum_{i=1}^{D-1} \frac{A_i^2}{\hat{S}_p(t_i)\hat{S}_p(t_{i-1})} \frac{n_1\hat{G}_1(t_{i-1}) + n_2\hat{G}_2(t_{i-1})}{n\hat{G}_1(t_{i-1})\hat{G}_2(t_{i-1})}[\hat{S}_p(t_{i-1}) - \hat{S}_p(t_i)]. \quad (7.7.8)$$

Note that the sum in (7.7.8) has only nonzero contributions when t_i is a death time, because, at censored observations, $\hat{S}_p(t_{i-1}) - \hat{S}_p(t_i) = 0$. When there is no censoring, $\hat{\sigma}_p^2$ reduces to the usual sample variance on the data from the combined sample.

To test the null hypothesis that $S_1(t) = S_2(t)$, the test statistic used is $Z = W_{\text{KM}}/\hat{\sigma}_p$ which has an approximate standard normal distribution when the null hypothesis is true. If the alternative hypothesis is that $S_1(t) > S_2(t)$, then, the null hypothesis is rejected when Z is larger than the αth upper percentage point of a standard normal, whereas, for a two-sided alternative, the null hypothesis is rejected when the absolute value of Z is larger than $\alpha/2$ upper percentage point of a standard normal.

EXAMPLE 7.5   *(continued)* We shall calculate the weighted difference of Kaplan–Meier estimators statistic for the comparison of auto versus allo transplants. The calculations yield a value of 5.1789 for W_{KM} and 141.5430 for $\hat{\sigma}_p^2$, so $Z = 0.4353$. The p-value of the two-sided test of equality of the two survival curves is $2Pr[Z \ge 0.4553] = 0.6634$.

The final test we shall present is a censored-data version of the two-sample median test proposed by Brookmeyer and Crowley (1982b). This test is useful when one is interested in comparing the median survival times of the two samples rather than in comparing the difference in the hazard rates or the survival functions over time. The test has reasonable power to detect differences between samples when the hazard rates of the two populations cross. It is sensitive to differences in median survival for any shaped survival function.

To construct the test, we have two, independent, censored samples of sizes n_1 and n_2, respectively. Let $n = n_1 + n_2$ be the total sample size. For each sample, we construct the Kaplan–Meier estimator (4.2.1), $\hat{S}_j(t)$, $j = 1, 2$. When the null hypothesis of no difference in survival between the two groups is true, an estimator of the common survival function is the weighted Kaplan–Meier estimator,

$$\hat{S}_W(t) = \frac{n_1 \hat{S}_1(t) + n_2 \hat{S}_2(t)}{n}. \tag{7.7.9}$$

This weighted Kaplan–Meier estimator represents the survival function of an average individual on study and is a function only of the survival experiences in the two samples. It does not depend on the censoring patterns in the two samples, as would the Kaplan–Meier estimate based on the combined sample.

Using the weighted Kaplan–Meier estimator, an estimate of the pooled sample median $\hat{M}$ is found as follows. Let $t_1 < t_2 < \cdots < t_D$ be the event times in the pooled sample. If $\hat{S}_W(t_i) = 0.5$ for some death time, set $\hat{M} = t_i$. If no event time gives a value of $\hat{S}_W$ equal to 0.5, set M_L as the largest event time with $\hat{S}_W(M_L) > 0.5$ and M_U as the smallest event time with $\hat{S}_W(M_U) < 0.5$. The pooled median must lie in the interval (M_L, M_U) and is found by linear interpolation, that is,

$$\hat{M} = M_L + \frac{(0.5 - \hat{S}_W(M_L))(M_U - M_L)}{\hat{S}_W(M_U) - \hat{S}_W(M_L)}. \tag{7.7.10}$$

To compute this median, we are using a version of the weighted Kaplan–Meier estimator, which connects the values of the estimator at death times by a straight line, rather than the usual estimator which is a step function.

Once the pooled sample median is found, the estimated probability that a randomly selected individual in the jth sample exceeds this value is computed from each sample's Kaplan–Meier estimator. Again, a smoothed version of the Kaplan–Meier estimator, which connects the values of the estimator at each death time, is used in each sample. We find the two death times in the jth sample that bracket $\hat{M}$, $T_{Lj} \leq \hat{M} < T_{Uj}$. The estimated probability of survival beyond $\hat{M}$ in the jth sample, found by linear interpolation, is given by

$$\hat{S}_j(\hat{M}) = \hat{S}_j(T_{Lj}) + \frac{(\hat{S}_j(T_{Uj}) - \hat{S}_j(T_{Lj}))(\hat{M} - T_{Lj})}{(T_{Uj} - T_{Lj})}, \, j = 1, 2. \quad (7.7.11)$$

The test is based on comparing this value to 0.5, the expected survival if the null hypothesis of no difference in median survival between the two groups is true, that is, the test is based on the statistic $n^{1/2}[S_1(\hat{M}) - 0.5]$. For sufficiently large samples, this statistic has an approximate normal distribution with a variance found as follows. As usual, let t_{ij} denote the distinct death times in the jth sample, d_{ij} the number of deaths at time t_{ij} and Y_{ij} the number at risk at time t_{ij}. For $j = 1, 2$, define

$$V_j = \left[\hat{S}_j(T_{Uj}) \left(\frac{\hat{M} - T_{Lj}}{T_{Uj} - T_{Lj}} \right) \right]^2 \sum_{t_{ij} \leq T_{Uj}} \frac{d_{ij}}{Y_{ij}(Y_{ij} - d_{ij})} \quad (7.7.12)$$

$$+ \left\{ \left[\hat{S}_j(T_{Lj}) \left(\frac{T_{Uj} - \hat{M}}{T_{Uj} - T_{Lj}} \right) \right]^2 + \frac{2(\hat{M} - T_{Lj})(T_{Uj} - \hat{M})}{(T_{Uj} - T_{Lj})^2} \hat{S}(T_{Uj})\hat{S}(T_{Lj}) \right\}$$

$$\times \sum_{t_{ij} \leq T_{Lj}} \frac{d_{ij}}{Y_{ij}(Y_{ij} - d_{ij})}.$$

Then, the variance of $n^{1/2}[S_1(\hat{M}) - 0.5]$ is estimated consistently by

$$\sigma^2 = \frac{n_2^2}{n}\{V_1 + V_2\}, \quad (7.7.13)$$

and the (two-sided) test statistic is

$$\chi^2 = n\frac{[S_1(\hat{M}) - 0.5]^2}{\sigma^2}, \quad (7.7.14)$$

which has a chi-squared distribution with one degree of freedom when the null hypothesis is true.

EXAMPLE 7.5

(continued)   We shall illustrate the median test on the data comparing allo and auto transplants. Here the estimated median from the pooled sample $\hat{M} = 17.9225$. Using the data from the first sample, we find $\hat{S}_1(\hat{M}) = 0.5409$ and $\hat{S}_2(\hat{M}) = 0.4395$. Routine calculations find that $V_1 = 0.0122$ and $V_2 = 0.0140$, so $\sigma^2 = 0.6754$. Thus, $\chi^2 = 101(0.5409 - 0.5)^2/0.6754 = 0.2496$. The p-value of the test is the probability that a chi-squared, random variable with one degree of freedom will exceed this value, which is 0.6173.

Practical Notes

1. Schumacher (1984) has studied the small-sample properties of the two Cramer–von Mises type tests. He concludes that there is some loss of power using these tests, compared to the log-rank test of section 7.3, when the hazard rates in the two samples are proportional. However the test based on Q_1 seems to perform quite well in this case. Tests based on Q_2 perform extremely well compared to other tests, when there is a large early difference in the hazard rates or when the hazard rates cross.

2. An alternative Cramer–von Mises test was proposed by Koziol (1978). This test is based on the statistic

$$-\frac{n_1 n_2}{n_1 + n_2} \int_0^\tau [\hat{S}_1(t) - \hat{S}_2(t)]^2 d[\hat{S}_p(t)],$$

   where $\hat{S}_1(t)$, $\hat{S}_2(t)$, and $\hat{S}_p(t)$ are the Kaplan–Meier estimates of the survival function from the first, second, and pooled samples, respectively. This statistic reduces to the usual Cramer–von Mises test when there is no censoring. The asymptotic distribution of the statistic, however, can only be derived under the additional assumption that the two censoring distributions are equal and that the hazard rate of the censoring distribution is proportional to the hazard rate of the death distribution. In his Monte Carlo study, Schumacher (1984) shows that the performance of this test is similar to that of Q_2.

3. Any weight function with the property that $|w(t)| \leq \gamma \hat{G}_j^{(1/2)+\delta}$, for $\gamma, \delta > 0$ can be used in the calculation of W_{KM}. Another choice of $w(t)$, suggested by Pepe and Fleming (1989), is the square root of equation (7.7.5). If one wishes to compare the two survival curves over some range, say $t \geq t_0$, the weight function $w(t)I[t \geq t_0]$ may be appropriate. Other choices of weights could be motivated by some measure of quality of life or the cost of treatment.

4. When there is no censoring, the estimated variance based on (7.7.8) is equal to the sample variance based on the pooled sample. This is a different estimator of the common variance in the two samples than the usual pooled sample variance constructed as a weighted average of the individual sample variances found in most elementary text books.

5. An alternate estimator of the variance of W_{KM} can be constructed by using an unpooled variance estimator. Here, let

$$A_{ij} = \int_{t_i}^{t_D} w(u)\hat{S}_j(u)du, \quad j = 1, 2, \ i = 1, \ldots, D-1.$$

   The unpooled estimator of the variance is

$$\hat{\sigma}_{up}^2 = \frac{n_1 n_2}{n} \left\{ \sum_{j=1}^{2} \frac{1}{n_j - 1} \sum_{i=1}^{D-1} \frac{A_{ij}^2}{\hat{S}_j(t_i)\hat{S}_j(t_{i-1})\hat{G}_j(t_{i-1})} [\hat{S}_j(t_{i-1}) - \hat{S}_j(t_i)] \right\},$$

which, in the uncensored-data case, reduces to $(n_2/n)S_1^2+(n_1/n)S_2^2$, with S_j^2 the usual sample variance. Monte Carlo studies by Pepe and Fleming (1989) show that this variance estimator performs poorer than the pooled estimator (7.7.8) and that its performance is poor when the censoring patterns are different in the two samples.

6. For n_j moderately large (> 50), one can approximate V_j in (7.7.12) by the simple expression

$$\hat{S}_j(\hat{M})^2 \sum_{t_{ij}\leq\hat{M}} \frac{d_{ij}}{Y_{ij}(Y_{ij} - d_{ij})}.$$

7. Brookmeyer and Crowley (1982b) present an extension of this two-sample median test to the K-sample problem.

Theoretical Notes

1. The weighted Kaplan–Meier test was proposed by Pepe and Fleming (1989) who developed its properties. This statistic can not be derived using counting process techniques. Pepe and Fleming (1991) give details of the asymptotic properties of the test.
2. A small-sample Monte Carlo study reported in Pepe and Fleming (1989) shows that, when the hazard rates in the two populations are proportional, the power of W_{KM} is slightly less than that of the log-rank test. The test performs substantially better when the two hazard rates cross. This observation is confirmed in Pepe and Fleming (1991) who base these observations on the asymptotic relative efficiency of the two tests.
3. Brookmeyer and Crowley (1982b) discuss the asymptotic relative efficiency of the median test as compared to the log-rank and Wilcoxon tests. They show that the asymptotic relative efficiency is about half of that of the log-rank test for proportional hazards alternatives, but about twice that of the log-rank test for a translation alternative. The performance of these tests is also presented for small-sample sizes based on a Monte Carlo study.

7.8   Test Based on Differences in Outcome at a Fixed Point in Time

Up to this point we have considered tests that compare hazard rates or survival functions over a range of time points. Occasionally we are interested in comparing K survival curves or K cumulative incidence curves at a predetermined fixed point in time, t_0. It is important to emphasize that the fixed point, t_0, must be selected before the data is examined. It would make the p-values invalid if the curves were compared for a variety of points. For example, one may wish to compare the survival curves at 1 year or the cumulative incidence curves at 3 years. We have available to us Kaplan–Meier estimators of the survival function or estimated cumulative incidence functions as well as estimates of the variances of these statistics (See Chapter 4 for the calculation of these quantities).

The tests statistics we use are special cases of tests for contrasts between a set of parameters. If we let $\Theta' = (\theta_1, \ldots, \theta_p)$ be a p-parameter λ_j vector, then a contrast is a linear combination of the covariates. A contrast is a set of coefficients $\mathbf{c} = (c_1 \ldots c_p)$ which define a new parameter $\theta^c = \mathbf{c}\Theta = c_1\theta_1 + \cdots + c_p\theta_p$. For example, if $p = 3$, then the vector $\mathbf{c} = (1, -1, 0)$ yields $\theta^c = \theta_1 - \theta_2$ and a test of the hypothesis that $\theta^C = 0$ is a test that $\theta_1 = \theta_2$.

Suppose that we have q contrasts $\mathbf{c}_k = (c_{k1}, \ldots, c_{kp}), k = 1, \ldots, q$, and we wish to test the hypothesis that $\mathbf{c}_k\Theta = 0$ for all k, then the test statistic will be a *quadratic form* constructed using the estimates of $\theta_1, \ldots, \theta_p$. To construct the quadratic form we define a $q \times p$ contrast matrix

$$\mathbf{C} = \begin{pmatrix} \mathbf{c}_1 \\ \mathbf{c}_2 \\ \vdots \\ \mathbf{c}_q \end{pmatrix}. \tag{7.8.1}$$

We compute an estimate of θ_j, $\hat{\theta}_j$ and the variance matrix, $\mathbf{V}$, with elements, $\hat{\text{Var}}[\hat{\theta}_j, \hat{\theta}_k]$. To test the hypothesis $H_0 : \mathbf{C}\Theta' = \mathbf{0}$, the test statistic is

$$\chi^2 = [\mathbf{C}\hat{\Theta}]^t[\mathbf{C}\mathbf{V}\mathbf{C}^t]^{-1}\mathbf{C}\hat{\Theta}. \tag{7.8.2}$$

When the estimators are approximately normally distributed, this form has an asymptotically chi-squared with q degrees of freedom.

In a survival analysis context we wish to test

$$H_0 : S_1(t_0) = S_2(t_0) = \cdots = S_K(t_0) \quad \text{versus} \tag{7.8.3}$$

H_A : at least one of the $S_j(t_0)$'s is different, for predetermined t_0,

or

$$H_0 : \text{CI}_1(t_0) = \text{CI}_2(t_0) = \cdots = \text{CI}_K(t_0) \quad \text{versus} \tag{7.8.4}$$

H_A : at least one of the $\text{CI}_j(t_0)$'s is different, for predetermined t_0.

Notation similar to that used in sections 4.2, 4.7, and 7.3 will be used and the groups will be assumed to be independent. Let $\hat{\theta}_j$ be the Kaplan–Meier estimate of the jth survival curve or the estimate of the

jth cumulative incidence curve at the predetermined time point t_0. $\mathbf{C}$ will be taken to be the $p - 1 \times p$ matrix

$$
\mathbf{C} = \begin{pmatrix}
1 & 0 & 0 & \cdots & 0 & -1 \\
0 & 1 & 0 & \cdots & 0 & -1 \\
 & & & \cdot & & \\
 & & & \cdot & & \\
 & & & \cdot & & \\
0 & 0 & 0 & \cdots & 1 & -1
\end{pmatrix}
$$

Here $\mathbf{V}$ is a diagonal matrix with elements $V_k = \hat{V}(\hat{\theta}_k(t_0))$, $k = 1, \ldots, p$.

The quadratic form (7.8.2) is

$$
\chi^2 = \begin{pmatrix}
\hat{\theta}_1 - \hat{\theta}_p \\
\hat{\theta}_2 - \hat{\theta}_p \\
\cdot \\
\cdot \\
\cdot \\
\hat{\theta}_{p-1} - \hat{\theta}_p
\end{pmatrix}^t
\begin{pmatrix}
V_1 + V_p & V_p & \cdots & V_p \\
V_p & V_2 + V_p & \cdots & V_p \\
\cdot & \cdot & \cdots & \\
\cdot & \cdot & \cdots & \\
\cdot & \cdot & \cdots & \\
V_p & V_p & \cdots & V_{p-1} + V_p
\end{pmatrix}^{-1}
\begin{pmatrix}
\hat{\theta}_1 - \hat{\theta}_p \\
\hat{\theta}_2 - \hat{\theta}_p \\
\cdot \\
\cdot \\
\cdot \\
\hat{\theta}_{p-1} - \hat{\theta}_p
\end{pmatrix}
$$
(7.8.5)

EXAMPLE 7.2   *(continued)* In example 7.2, data from a clinical trial of the effectiveness of two methods for placing catheters in kidney dialysis patients was used to illustrate various two-sample tests over the entire range of time. The estimated survival functions for the two groups are given in Figure 7.1. Suppose an investigator is interested in comparing the survival functions at 3 months (short duration of time to infection). Thus, using the Kaplan–Meier estimates of the survival functions from (4.2.1) and the estimated variances of these survival functions from (4.2.2) for the jth goup, we obtain the Z test statistic as the square root of the chi-squared quadratic form with one degree of freedom from (7.8.5) to be

$$
Z = \frac{\hat{S}_1(t_0) - \hat{S}_2(t_0)}{\sqrt{\hat{V}[\hat{S}_1(t_0)] + \hat{V}[\hat{S}_2(t_0)]}}
$$
(7.8.6)

which, when H_0 is true, has a standard normal ditribution for large samples. Using this statistic, an α level test of the alternative hypothesis $H_A : S_1(t_0) > S_2(t_0)$ is rejected when $Z \geq Z_\alpha$, the αth upper percentage point of a standard normal distribution. The test of $H_A : S_1(t_0) \neq S_2(t_0)$ rejects when $|Z| > Z_{a/2}$.

The estimates of the survival functions are

$$
\hat{S}_1(3) = 0.9767 \quad \text{and} \quad \hat{S}_2(3) = 0.882,
$$

and estimates of the variances are

$$\hat{V}[\hat{S}_1(3)] = 0.00053 \quad \text{and} \quad \hat{V}[\hat{S}_2(3)] = 0.00141.$$

This leads to a test statistic of

$$Z = \frac{0.9767 - 0.8882}{\sqrt{0.00053 + 0.00141}} = 2.01,$$

which leads to a two-sided p-value of 0.044. This difference is due to the first group's (surgical placement of catheters) having a smaller probability of infection at three months than the second group (percutaneous placement of catheters).

It should be noted that another investigator comparing the survival function of the different placement of the catheters at a large time period would get the opposite conclusion. This again emphasizes the need to preselect t_0 before examining the data.

This example is also an illustration of what can occur when the hazards are not proportional (an assumption formally tested in Chapter 9, Example 9.2).

EXAMPLE 7.4   *(continued)*   In the example of section 4.7 the relapse cumulative incidence curve for 39 ALL patients was calculated as shown in Table 4.8. At one year the estimated cumulative incidence was 0.2380 (variance = 0.0048). In this data set there are two additional groups, AML low-risk and AML high-risk, whose relapse cumulative incidences are 0.0556 (variance = 0.0010) and 0.3556 (variance = 0.0054), respectively. A test of the hypothesis of no difference in three-year cumulative incidence between the three disease groups at one year has a $\chi^2 = 17.32$, which has a p-value of 0.0002 when compared to a chi-square distribution with 2 degrees of freedom.

If one is interested in comparing K groups in a pairwise simultaneous manner then an adjustment for multiple tests must be made. One such method that can be used is the Bonferroni method of multiple comparisons. Here if $K(K-1)/2$ pairwise comparisons are made and one still wants to maintain an overall α-level test, then each individual test is carried out at an $\alpha^* = \alpha/K(K-1)/2$ (or $\alpha/2K(K-1)/2 = \alpha/K(K-1)$ for two-sided tests) level of significance and if all null hypotheses are actually true, then the probability is at least $1 - \alpha$ that none of the null hypotheses will be wrongly rejected. This method is somewhat conservative and becomes more conserative as the number of comparisons increases.

EXAMPLE 7.4   *(continued)*   For our previous example of the three groups (ALL, AML low-risk, AML high-risk) when the Bonferroni method of multiple comparisons (in our case, $K = 3$) is used to make pairwise comparisons of the cumulative incidence curves, each test needs to be carried out at the $0.05/3 = 0.017$ level of significance. The contrasts $(1, -1, 0)$, $(1, 0, -1)$, and $(0, 1, -1)$ may be used to test each of the individual pairwise comparisons. Using the appropriate variances in (7.8.5), we get

$$\text{for} \quad H_0 : CI_1(t_0) = CI_2(t_0) \text{ at } t_0 = 1$$

we have

$$Z = 2.41, \ p\text{-value} = 0.016,$$

$$\text{for} \quad H_0 : CI_1(t_0) = CI_3(t_0) \text{ at } t_0 = 1$$

we have

$$Z = -1.17, \ p\text{-value} = 0.242,$$

and

$$\text{for} \quad H_0 : CI_2(t_0) = CI_3(t_0) \text{ at } t_0 = 1$$

we have

$$Z = -3.76, \ p\text{-value} = 0.0002.$$

Thus we conclude that the AML high-risk group is statistically different from the other two groups and that the ALL and AML low-risk groups are not statistically different from each other.

Practical Notes

1. One may test a hypothesis for any linear combination of several groups. For example, if one wants to test whether the cumulative incidence curves for the ALL patients are different than those for the AML (both high-risk and low-risk) patients, then one may select the linear contrast $(2, -1, -1)$ and use the quadratic form (7.8.5).

7.9   Exercises

7.1   In a study of the effectiveness of a combined drug regimen for the treatment of rheumatoid arthritis, 40 white patients were followed for a period ranging from 1 to 18 years. During the course of the study, 9 patients died. The ages at entry into the study and at death for these 9 patients were as follows:

Female deaths: (66, 74), (60, 76), (70, 77), (71, 81)
Male deaths: (50, 59), (60, 66), (51, 69), (69, 71), (58, 71)

For the 31 patients still alive at the end of the study their ages at entry and last follow-up were as follows:

Female Survivors: (50, 68), (55, 72), (56, 60), (45, 55), (48, 51), (44, 55), (33, 51), (44, 50), (60, 70), (55, 60), (60, 72), (77, 80), (70, 75), (66, 70), (59, 63), (62, 63)
Male Survivors: (53, 68), (55, 62), (56, 63), (45, 51), (48, 61), (49, 55), (43, 51), (44, 54), (61, 70), (45, 60), (63, 72), (74, 80), (70, 76), (66, 72), (54, 70)

Using the all-cause U.S. mortality table for 1989 (Table 2.1) test the hypothesis that the death rate of these rheumatoid arthritis patients is not different from that in the general population using the log-rank test.

7.2　In Exercise 5 of Chapter 6, the survival experience of patients given an autologous transplant was compared to a postulated exponential survival rate with a hazard rate of 0.045. Using the data in Table 1.4 of Chapter 1, test the hypothesis that the hazard rate of these auto transplant patients is equal to 0.045 against the alternative that it is larger than 0.045 using the one-sample, log-rank test. Repeat this test using a weight function which gives heavier weight to departures early in time from this hazard rate.

7.3　Consider the data reported in section 1.6 on the times until staphylococcus infection of burn patients (see our web page).

(a) Using the log-rank test, test the hypothesis of no difference in the rate of staphylococcus infection between patients whose burns were cared for with a routine bathing care method versus those whose body cleansing was initially performed using 4% chlorhexidine gluconate. Use a two-sided test and a 0.05 significance level.

(b) Repeat the test using Gehan's test.

(c) Repeat the test using the Tarone and Ware weights.

7.4　In section 1.11, data from a study of the effect of ploidy on survival for patients with tumors of the tongue was reported.

(a) Test the hypothesis that the survival rates of patients with cancer of the tongue are the same for patients with aneuploid and diploid tumors using the log-rank test.

(b) If primary interest is in detecting differences in survival rates between the two types of cancers which occur soon after the diagnosis of the cancer, repeat part a using a more appropriate test statistic.

7.5　Using the data on laryngeal cancers in Example 7.6, test, by the log-rank statistic, the null hypothesis of no difference in death rates among the four stages of cancer against the global alternative that at least one of the death rates differs from the others. Compare your results to those found in Example 7.6.

7.6   One of the goals of recent research is to explore the efficacy of triple-drug combinations of antiretroviral therapy for treatment of HIV-infected patients. Because of limitations on potency and the continuing emergence of drug resistance seen with the use of currently available antiretroviral agents in monotherapy and two-drug regimens, triple-combination regimens should represent a more promising approach to maximize antiviral activity, maintain long-term efficacy, and reduce the incidence of drug resistance. Towards this end, investigators performed a randomized study comparing AZT + zalcitabine (ddC) versus AZT + zalcitabine (ddC) + saquinavir. The data, time from administration of treatment (in days) until the CD4 count reached a prespecified level, is given below for the two groups.

AZT + zalcitabine (ddC): 85, 32, 38+, 45, 4+, 84, 49, 180+, 87, 75, 102, 39, 12, 11, 80, 35, 6

AZT + zalcitabine (ddC) + saquinavir: 22, 2, 48, 85, 160, 238, 56+, 94+, 51+, 12, 171, 80, 180, 4, 90, 180+, 3

Use the log rank statistic to test if there is a difference in the distribution of the times at which patient's CD4 reaches the prespecified level for the two treatments.

7.7   A study was performed to determine the efficacy of boron neutron capture therapy (BNCT) in treating the therapeutically refractory F98 glioma, using boronophenylalanine (BPA) as the capture agent. F98 glioma cells were implanted into the brains of rats. Three groups of rats were studied. One group went untreated, another was treated only with radiation, and the third group received radiation plus an appropriate concentration of BPA. The data for the three groups lists the death times (in days) and is given below:

Untreated	Radiated	Radiated + BPA
20	26	31
21	28	32
23	29	34
24	29	35
24	30	36
26	30	38
26	31	38
27	31	39
28	32	42$^+$
30	35$^+$	42$^+$

$^+$Censored observation

(a) Compare the survival curves for the three groups.

(b) Perform pairwise tests to determine if there is any difference in survival between pairs of groups.

(c) There is a priori evidence that, if there is a difference in survival, there should be a natural ordering, namely, untreated animals will have the worst survival, radiated rats will have slightly improved survival, and the radiated rats + BPA should have the best survival. Perform the test for trend which would test this ordered hypothesis.

7.8   In Example 7.4, we compared the disease-free survival rates of ALL patients with those of high-risk and low risk AML patients. Because acute graft-versus-host (aGVHD) disease is considered to have an antileukemic effect, one would expect lower relapse rates for patients who have developed aGVHD than for those that do not develop aGVHD. Using the data on out web page, examine the validity of this finding by

(a) testing if the hazard rate for the occurrence of aGVHD is the same for the three groups,

(b) testing if the hazard rate for relapse is the same in all three groups, and

(c) testing if the hazard rate for relapse in the three disease groups is the same for patients who have developed aGVHD. (Hint: For this test, the data is left-truncated at the time of aGVHD).

7.9   On our web page, data is reported on the death times of 863 kidney transplant patients (see section 1.7). Here, patients can be classified by race and sex into one of four groups.

(a) Test the hypothesis that there is no difference in survival between the four groups.

(b) Provide individual tests, for each sex, of the hypothesis of no racial differences in survival rates. Also, adjusting by stratification for the sex of the patient, test the hypothesis that blacks have a higher mortality rate than whites.

7.10   In Example 7.6 we found that the four populations of cancer patients had ordered hazard rates. Of interest is knowing which pairs of the hazard rates are different. Using the log-rank test, perform the three pairwise tests of the hypothesis $H_{0j} : h_j(t) = h_{j+1}(t)$ versus $H_{Aj} : h_j(t) < h_{j+1}(t)$, for $j = 1, 2, 3$. For each test, use only those individuals with stage j or $j + 1$ of the disease. Make an adjustment to your critical value for multiple testing to give an approximate 0.05 level test.

One method to making the pairwise comparisons is to base the pairwise tests on the full $\mathbf{Z}(\tau)$ vector. To perform this test, recall that this vector has an asymptotic K variate normal distribution with mean 0 and covariance matrix $\hat{\Sigma}$ under the null hypothesis. Thus, the statistic $Z_j(\tau) - Z_{j+1}(\tau)$ has a normal distribution with mean 0 and variance $\hat{\sigma}_{jj} + \hat{\sigma}_{j+1j+1} - 2\hat{\sigma}_{jj+1}$ when the null hypothesis is true. Large negative values of this test statistic will suggest that the hazard rate in sample j is smaller than in sample $j + 1$, so the hypothesis H_{0j} : $h_j(t) = h_{j+1}(t)$ is rejected in favor of H_{Aj} : $h_j(t) < h_{j+1}(t)$ when $[Z_j(\tau) - Z_{j+1}(\tau)]/[\hat{\sigma}_{jj} + \hat{\sigma}_{j+1j+1} - 2\hat{\sigma}_{jj+1}]^{1/2}$ is smaller than the αth lower percentile of a standard normal. Use the information in Example 7.6 and this statistic to make the multiple comparisons.

7.11   The data on laryngeal cancer patients was collected over the period 1970–1978. It is possible that the therapy used to treat laryngeal cancer may have changed over this nine year period. To adjust for this possible confounding fact, test the hypothesis of no difference in survival between patients with different stages of disease against a global alternative using a test which stratifies on the cancer being diagnosed prior to 1975 or not. Also perform a separate test of the hypothesis of interest in each stratum.

7.12   (a) Repeat Exercise 3 using the log-rank version of the Renyi statistic.

   (b) Repeat Exercise 4 using the Gehan version of the Renyi statistic.

7.13   In Table 1.3 of section 1.5, the data on time to death for breast cancer patients who where classed as lymph node negative by standard light microscopy (SLM) or by immunohistochemical (IH) examination of their lymph nodes is reported. Test the hypothesis that there is no difference in survival between theses two groups using

   (a) the log-rank test,

   (b) the Renyi statistic based on the log-rank test,

   (c) the Cramer-von Mises statistic, and

   (d) the weighted difference in the Kaplan–Meier statistic W_{KM}.

7.14   Repeat Exercise 7 using

   (a) the Renyi statistic based on the log-rank test,

   (b) the Cramer-von Mises statistic, and

   (c) the weighted difference in the Kaplan–Meier statistic W_{KM}.

7.15   Using the data of section 1.3,

   (a) compare the three survival functions for ALL, AML low-risk, and AML high-risk at one year;

   (b) perform pairwise multiple comparisons for the three groups employing the Bonferroni correction for multiple tests.

8
Semiparametric Proportional Hazards Regression with Fixed Covariates

8.1 Introduction

Often one is interested in comparing two or more groups of times-to-event. If the groups are similar, except for the treatment under study, then, the nonparametric methods of Chapter 7 may be used directly. More often than not, the subjects in the groups have some additional characteristics that may affect their outcome. For example, subjects may have demographic variables recorded, such as age, gender, socio-economic status, or education; behavioral variables, such as dietary habits, smoking history, physical activity level, or alcohol consumption; or physiological variables, such as blood pressure, blood glucose levels, hemoglobin levels, or heart rate. Such variables may be used as covariates (explanatory variables, confounders, risk factors, independent variables) in explaining the response (dependent) variable. After adjustment for these potential explanatory variables, the comparison of survival times between groups should be less biased and more precise than a simple comparison.

Another important problem is to predict the distribution of the time to some event from a set of explanatory variables. Here, the interest is in predicting risk factors for the event of interest. Statistical strategies for prediction are similar to those utilized in ordinary regression. However, the details for regression techniques in survival studies are unique.

In section 2.6, we introduced models which allow us to quantify the relationship between the time to event and a set of explanatory variables. In this chapter, we will consider in more detail the widely used multiplicative hazards model due to Cox (1972), often called the proportional hazards model.

As before, let X denote the time to some event. Our data, based on a sample of size n, consists of the triple $(T_j, \delta_j, \mathbf{Z}_j(t))$, $j = 1, \ldots, n$ where T_j is the time on study for the jth patient, δ_j is the event indicator for the jth patient ($\delta_j = 1$ if the event has occurred and $\delta_j = 0$ if the lifetime is right-censored) and $\mathbf{Z}_j(t) = (Z_{j1}(t), \ldots, Z_{jp}(t))^t$ is the vector of covariates or risk factors for the jth individual at time t which may affect the survival distribution of X. Here the $Z_{jk}(t)$'s, $k = 1, \ldots, p$, may be time-dependent covariates whose value changes over time, such as current disease status, serial blood pressure measurements, etc., or they may be constant (or fixed) values known at time 0, such as sex, treatment group, race, initial disease state, etc. In this chapter, we shall consider the fixed-covariate case where $\mathbf{Z}_j(t) = \mathbf{Z}_j = (Z_{j1}, \ldots, Z_{jp})^t$, and the former situation involving time-dependent covariates will be treated in Chapter 9.

Let $h(t \mid \mathbf{Z})$ be the hazard rate at time t for an individual with risk vector $\mathbf{Z}$. The basic model due to Cox (1972) is as follows:

$$h(t \mid \mathbf{Z}) = h_0(t)c(\boldsymbol{\beta}^t\mathbf{Z}) \qquad (8.1.1)$$

where $h_0(t)$ is an arbitrary baseline hazard rate, $\boldsymbol{\beta} = (\beta_1, \ldots, \beta_p)^t$ is a parameter vector, and $c(\boldsymbol{\beta}^t\mathbf{Z})$ is a known function. This is called a semiparametric model because a parametric form is assumed only for the covariate effect. The baseline hazard rate is treated nonparametrically. Because $h(t \mid \mathbf{Z})$ must be positive, a common model for $c(\boldsymbol{\beta}^t\mathbf{Z})$ is

$$c(\boldsymbol{\beta}^t\mathbf{Z}) = \exp(\boldsymbol{\beta}^t\mathbf{Z}) = \exp\left(\sum_{k=1}^{p} \beta_k Z_k\right)$$

yielding

$$h(t \mid \mathbf{Z}) = h_0(t)\exp(\boldsymbol{\beta}^t\mathbf{Z}) = h_0(t)\exp\left(\sum_{k=1}^{p} \beta_k Z_k\right) \qquad (8.1.2)$$

and, thus, the logarithm of $h(t \mid \mathbf{Z})/h_0(t)$ is $\sum_{k=1}^{p} \beta_k Z_k$ in the spirit of the usual linear models formulation for the effects of covariates. The coding of factors and their interaction effects follows the usual rules for linear models. For example, if a factor has four levels, three indicator (or dummy) variables may be constructed to model the effect of the factor. An interaction between two or more factors may be examined by constructing new variables which are the product of the variables associated with the individual factors as is commonly done in other (least squares or logistic) regression contexts. One needs to take care in interpreting coefficients so constructed.

The Cox model is often called a proportional hazards model because, if we look at two individuals with covariate values $\mathbf{Z}$ and $\mathbf{Z}^*$, the ratio of their hazard rates is

$$\frac{h(t \mid \mathbf{Z})}{h(t \mid \mathbf{Z}^*)} = \frac{h_0(t) \exp\left[\sum_{k=1}^{p} \beta_k Z_k\right]}{h_0(t) \exp\left[\sum_{k=1}^{p} \beta_k Z_k^*\right]} = \exp\left[\sum_{k=1}^{p} \beta_k (Z_k - Z_k^*)\right] \quad (8.1.3)$$

which is a constant. So, the hazard rates are proportional. The quantity (8.1.3) is called the relative risk (hazard ratio) of an individual with risk factor $\mathbf{Z}$ having the event as compared to an individual with risk factor $\mathbf{Z}^*$. In particular, if Z_1 indicates the treatment effect ($Z_1 = 1$ if treatment and $Z_1 = 0$ if placebo) and all other covariates have the same value, then, $h(t \mid \mathbf{Z})/h(t \mid \mathbf{Z}^*) = \exp(\beta_1)$, is the risk of having the event if the individual received the treatment relative to the risk of having the event should the individual have received the placebo.

In section 8.2 coding of both quantitative and qualitative covariates and a discussion of their interpretation is presented. Typically the goal of an investigation is to make an inference about β in a global sense, as discussed in sections 8.3 (for distinct event time data) and 8.4 (when ties are present), or, more often than not, to make an inference about a subset of β (called a local test) as discussed in section 8.5. Sometimes an investigator would like to treat a continuous covariate as binary. An example of such a covariate might be blood pressure, which is, in theory, a continuous variate; but a researcher might want to classify a patient as being normotensive or hypertensive. The rationale and details of the methodology of discretizing a continuous covariate are provided in section 8.6.

In section 8.7 these techniques are used to build the most appropriate model for survival. Inference for β in these sections is based on a partial or conditional likelihood rather than a full likelihood approach. In these analyses, the baseline hazard, $h_0(t)$, is treated as a nuisance parameter function. Sometimes, however, one is interested in estimating the survival function for a patient with a certain set of conditions and characteristics. This is accomplished by utilizing the results described in section 8.8.

8.2   Coding Covariates

In general regression analyses one may have either quantitative or qualitative independent variables. The dependent variable, in the context of this book, is a quantitative variable, the time-to-event, along with an indicator variable which indicates whether or not the event of interest occurred. As indicated in section 8.1, the independent variables may be quantitative—such as blood pressure, blood glucose levels, age, heart rate, or waiting time until a transplant—or they may be qualitative—such as gender, smoking behavior, stage of disease, or the presence or absence of any particular characteristic which the researcher wishes to investigate. Qualitative variables can be used in a regression analysis, just as quantitative variables can be used; however, more care needs to be taken in the manner they are coded and interpreted. Usually, independent variables are known at the start of the study. They are called fixed time covariates. Occasionally independent·variables may change values after the study starts and are known as time-dependent covariates. It is extremely important to make this distinction since the methods of analyses differ substantially for time-dependent covariates. First, we shall discuss fixed time covariates. Time-dependent covariates are discussed in Chapter 9.

There are many ways of coding qualitative variables. For dichotomous variables, like gender, the obvious way is to code one of the genders as 1, the other as 0. Which way we do this coding is arbitrary and immaterial. The interpretation of the results, of course, will depend on the way the coding is actually done. For example, if we code the gender variable as $Z_1 = 1$ if male, 0 if female, the hazard rate for males will be $h(t \mid Z) = h_0(t)\exp(\beta_1)$, and for females will be $h(t \mid Z) = h_0(t)\exp(0) = h_0(t)$. Here the natural logarithm of the ratio of the hazard function for males relative to the hazard function for females is β_1, and the ratio of the hazard functions for males relative to females (the relative risk) will be $\exp(\beta_1)$. The variable Z_1 is called an indicator (or dummy) variable since it indicates to which group the subject under consideration belongs. If we had coded another variable as $Z_2 = 1$ if female, 0 if male, then the hazard rate for females would have been $h(t \mid Z) = h_0(t)\exp(\beta_2)$ and for males will be $h(t \mid Z) = h_0(t)\exp(0) = h_0(t)$. Here the natural logarithm of the ratio of the hazard function for females relative to the hazard function for males is β_2, and the ratio of the hazard functions for females relative to males (the relative risk) will be $\exp(\beta_2) = 1/\exp(\beta_1) = \exp(-\beta_1)$. Either way the coding is performed, the interpretation will lead to the same conclusion. Consider the coding for an example which will be used in a subsequent section.

EXAMPLE 8.1   In section 1.5 we introduced a study designed to determine if female breast cancer patients, originally classified as lymph-node-negative by standard light microscopy (SLM), could be more accurately classified by immunohistochemical (IH) examination of their lymph nodes with an anticytokeratin monoclonal antibody cocktail. The data for 45 female breast cancer patients with negative axillary lymph nodes and a minimum 10-year follow-up were selected from The Ohio State University Hospitals Cancer Registry. Of the 45 patients, 9 were immunoperoxidase positive and the remaining 36 still remained negative.

In this example we wish to perform a proportional hazards regression with immunoperoxidase status as the single covariate in the model. We adopt the usual regression formulation of a dichotomous independent variable and construct a dummy (or indicator) variable as follows.

Let $Z = 1$ if the patient is immunoperoxidase positive, 0 otherwise. The model is $h(t \mid Z) = h_0(t)\exp(\beta Z)$, where $h_0(t)$ is an arbitrary baseline hazard rate and β is the regression coefficient for Z. The ratio of the hazard functions for patient being immunoperoxidase positive relative to the patient being immunoperoxidase negative (the relative risk) will be $\exp(\beta)$. In a later example in section 8.3, the estimate of β, denoted by b, is determined to be 0.9802. Thus the relative risk of dying for an immunoperoxidase-positive patient relative to an immunoperoxidase-negative patient is $\exp(0.9802)=2.67$. That is, a patient who is immunoperoxidase positive is 2.67 times more likely to die than an immunoperoxidase-negative patient.

When the qualitative variables (sometimes called factors) have more than two categories, there is more choice in the coding schemes. For example, when coding a factor, sometimes termed a "risk group," which has three categories, we utilize a simple extension of the indicator variable coding scheme described above. In particular, we code two indicator variables as

$$Z_1 = 1 \text{ if the subject is in category 1, 0 otherwise,}$$

$$Z_2 = 1 \text{ it the subject is in category 2, 0 otherwise.} \qquad (8.2.1)$$

One might be tempted to make a third category $Z_3 = 1$ if the subject is in category 3, 0 otherwise; but to do this would make the three variables $Z_i(i = 1, 2, 3)$ dependent upon each other. This can be seen because if you know the value of Z_1 and the value of Z_2, then you would know the value of Z_3. This is contrary to the principle of multiple regression, where you wish to introduce independent variables into the model. The independent variables may be correlated but they should not be completely dependent, since this introduces an undesirable complexity in analysis and interpretation.

There are many other ways to code the categories so as to obtain two "independent" variables which perform the same test of $\beta_1 = \beta_2 = 0$, but we shall not dwell on them, except to say that the interpretation must be carefully understood. Instead we will elaborate a bit more on the coding scheme discussed above. Consider a three level factor, such as race (black, white, Hispanic), using the coding as in (8.2.1)

$$Z_1 = 1 \quad \text{if the subject is black, 0 if otherwise,}$$

$$Z_2 = 1 \quad \text{if the subject if white, 0 otherwise,}$$

The hazard rate, in general, is $h(t \mid Z) = h_0(t)\exp\{\sum_{k=1}^{2} \beta_k Z_k\}$ and, in particular, the hazard rates for blacks, whites, and Hispanics, respectively, is as follows:

$$h(t \mid Z_1 = 1, Z_2 = 0) = h_0(t)\exp(\beta_1),$$

$$h(t \mid Z_1 = 0, Z_2 = 1) = h_0(t)\exp(\beta_2),$$

$$h(t \mid Z_1 = 0, Z_2 = 0) = h_0(t). \tag{8.2.2}$$

Here we can see that the risk of the events occurring among blacks relative to the risk of the events occurring among Hispanics is $\exp(\beta_1)$, the risk of the events occurring among whites relative to the risk of the events occurring among Hispanics is $\exp(\beta_2)$, and the risk of the events occurring among blacks relative to the risk of the events occurring among whites is $\exp(\beta_1 - \beta_2)$.

A note of caution is in order here. If the independent variable is strictly categorical with more than two groups, then it would be inappropriate to code the variable as a single covariate. Suppose we have a categorical covariate with $k(> 2)$ categories and we define a single covariate $Z = i$, if the individual belongs to category $i, i = 1, \ldots, k$. The proportional hazards model assumes that the relative risk of an event for an individual in category i as compared to an individual in category $i - 1$ is e^{β} for any $i = 2, \ldots, k$.

For example, suppose we code the patient's race as 1 if black, 2 if white, and 3 if Hispanic. A consequence of this model is that the following relationships between the relative risks must hold:

$$RR(\text{White/Black}) = RR(\text{Hispanic/White}) = e^{\beta}$$

and

$$RR(\text{Hispanic/Black}) = e^{2\beta}$$

relationships which are not likely to be true.

EXAMPLE 8.2   In section 1.8, a study of 90 males diagnosed with cancer of the larynx was described. In addition to the outcome variable, time from first treatment until either death or the end of the study, the independent variables, patient's age (in years) at the time of diagnosis and the stage of the patient's cancer, were recorded. A basic test for trend on stage was performed in section 7.4.

Here we wish to illustrate the coding of the variable stage of disease in preparation for performing a proportional hazards regression test with only stage in the model. Since stage has four levels, we adopt the usual indicator variable coding methodology as in (8.2.1) and construct the dummy (or indicator) variables as follows.

$$\text{Let} \quad Z_1 = 1 \quad \text{if the patient is in stage II, 0 otherwise,}$$

$$Z_2 = 1 \quad \text{if the patient is in Stage III, 0 otherwise,}$$

and

$$Z_3 = 1 \quad \text{if the patient is in Stage IV, 0 otherwise.} \quad (8.2.3)$$

This places the patient with Stage I cancer in the referent group; i.e., such a patient will have $Z_1 = Z_2 = Z_3 = 0$. Usually the coding is accomplished so that the referent group is expected to be at either extreme of risk from a subject matter point of view.

In section 8.4 we shall see that $b_1 = 0.06576$, $b_2 = 0.61206$, $b_3 = 1.172284$. The full model for this situation is

$$h(t \mid Z) = h_0(t)\exp\{\beta'Z\} = h_0(t)\exp\{\beta_1 Z_1 + \beta_2 Z_2 + \beta_3 Z_3\}.$$

Thus the estimated relative risks of dying for patients with Stage II, III, and IV disease relative to Stage I disease is $\exp(0.06576) = 1.068$, $\exp(0.61206) = 1.844$, and $\exp(1.72284) = 5.600$, respectively. One may also calculate the relative risk of dying for patients for Stage III disease relative to patients for Stage II disease as $\exp(0.61206 - 0.06576) = 1.727$.

A basic test for trend was performed on the data of section 1.8 in Example 7.6 of section 7.4. Since the scores test in the proportional hazards model is identical to the log rank test, when there are no ties (see Practical Note 3 in section 8.3), one could approximate the test for trend in Example 7.6 by taking the stage variable as a continuous variable (stage = 1, 2, 3, 4). The scores test in this proportional hazards model has a chi-squared of 13.64 with 1 degree of freedom, a result consistent with what we found in Example 7.6. As discussed earlier, the estimate of β must be interpreted with caution since it assumes equal relative risk between adjacent stages of disease.

On the other hand, if an independent variable is continuous, such as age, then it would be appropriate to code the variable as a single covariate. In this case, the exponentiated coefficient, e^β, for the variable

$$Z = \text{age (in years)}$$

would be the relative risk of an event for an individual of age i years compared to an individual of age $i - 1$ years. Sometimes we wish to speak of the relative risk of the event for an individual 10 years older than another individual. In this case, the ratio of the hazard (or risk) of the individual 10 years older compared to the hazard (or risk) of the referent individual would be relative risk $= \exp(10\beta)$. There may be other covariates in the model, in which case, the coefficient is termed a partial regression coefficient. Such partial regression coefficients relate the relationship of that variable, say, age, to the outcome variable, time-to-event, controlling for all other variables in the model. Tests based on partial regression coefficients utilize local tests as described in section 8.5. The results from a data set analyzed in section 8.5 are used to illustrate the interpretation of such parameters below.

EXAMPLE 8.2   *(continued)*   Continuing the examination of the data set in section 1.8, we will introduce the age covariate, $Z_4 =$ age of the patient, in addition to the stage indicator variables defined in (8.2.3). The model then is

$$h(t \mid \mathbf{Z}) = h_0(t)\exp\{\beta'\mathbf{Z}\}$$
$$= h_0(t)\exp\{\beta_1 Z_1 + \beta_2 Z_2 + \beta_3 Z_3 + \beta_4 Z_4\}. \qquad (8.2.4)$$

Here the natural logarithm of the ratio of the hazard function for a 50-year-old individual with Stage IV disease relative to the hazard function for a 40-year-old individual with Stage IV disease is $10\beta_4$; i.e., the relative risk for a 50-year-old patient compared to a 40-year-old (both with Stage IV disease) is $\exp(10\beta_4)$, since the stage of disease parameter will cancel out in the proportional hazards model.

The estimates of the parameters are obtained in section 8.5 as

$$b_1 = 0.1386, \; b_2 = 0.6383, \; b_3 = 1.6931, \quad \text{and} \quad b_4 = 0.0189. \qquad (8.2.5)$$

Thus the relative risk for a 50-year-old patient compared to a 40-year-old (both with Stage IV disease) is $\exp(10b_4) = 1.2$. Another way of stating the interpretation of a partial relative risk is that a 50-year-old patient has a probability of dying 1.2 times greater than the probability of dying for a 40-year-old patient with the same stage of disease.

Factors such as gender, age, race, or stage of disease taken individually are often referred to as main effects, i.e., their relationship with the time-to-event outcome is tested for statistical significance as if their relationship does not depend on other factors. An important concept in regression is the consideration of the effect of one factor in the presence of another factor. This concept is termed interaction.

As in other (least squares or logistic) regression contexts, interaction effects between variables may exist and these effects may be very important. An interaction effect exists if the relative risk for two levels of one factor differs for distinct levels of a second factor. Consider modeling a clinical trial of two treatments based on a proportional hazards model with covariate coded as $Z_1 = 1$ for treatment 1, 0 for treatment 2. Here $\exp(\beta_1)$ is the risk of the first treatment relative to the second. Suppose there is the potential for males and females to respond differently to the two treatments so the relative risk of treatment 1 compared to treatment 2 may depend on sex. As usual we code sex as $Z_2 = 1$ if male and 0 if female. Interactions are formed by multiplying the independent variables of the two individual factors, termed main effects, together. That is, a third variable $Z_3 = Z_1 \times Z_2$ will be created. Here, the exponential of the coefficient of Z_3, the product of the treatment and gender covariate, is the excess relative risk of treatment 1 compared to treatment 2 for males compared to females. Now the full model will be

$$h(t \mid Z) = h_0(t)\exp\{\beta'\mathbf{Z}\} = h_0(t)\exp\{\beta_1 Z_1 + \beta_2 Z_2 + \beta_3 Z_3\}. \quad (8.2.6)$$

The relative risk of treatment 1 compared to treatment 2 for males is $\exp\{\beta_1 + \beta_3\}$, while for females it is $\exp\{\beta_1\}$. If $\beta_3 = 0$, then the relative risk of treatment 1 compared to treatment 2 will be identical for the two sexes.

The following example illustrates the construction of the interaction of two categorical variables.

EXAMPLE 8.3

In section 1.7 a data set of 863 kidney transplant patients with data on race (white, black) and gender is described. In this study there were 432 white males, 92 black males, 280 white females, and 59 black females. Again, there are various coding options. First, one may treat this study as a four-group problem as we have done in Example 8.2. The three indicator variables may be defined in any desirable way but usually one wants either the best or the worst survival group as the referent group. For example, we may code

$Z_1 = 1$ if the subject is a black male, 0 otherwise,

$Z_2 = 1$ if the subject is a white male, 0 otherwise,

and

$Z_3 = 1$ if the subject is a black female, 0 otherwise.

Here the referent group is being a white female. Again the full model will be

$$h(t \mid Z) = h_0(t)\exp\{\beta'\mathbf{Z}\} = h_0(t)\exp\{\beta_1 Z_1 + \beta_2 Z_2 + \beta_3 Z_3\}.$$

The estimates of the parameters are obtained in section 8.5 as

$$b_1 = 0.1596,\ b_2 = 0.2484,\ b_3 = 0.6567.$$

Thus the relative risks for black male, white male, and black female relative to white female are 1.17, 1.28, 1.93, respectively.

Alternatively, we may code the variables as two main effect terms, race and gender, and an interaction term. For example,

$Z_1 = 1$ if the subject is a female, 0 otherwise,

$Z_2 = 1$ if the subject is black, 0 otherwise,

and

$Z_3 = Z_1 \times Z_2 = 1$ if the subject is a black female, 0 otherwise.

Again the full model will be

$$h(t \mid Z) = h_0(t)\exp\{\beta'\mathbf{Z}\} = h_0(t)\exp\{\beta_1 Z_1 + \beta_2 Z_2 + \beta_3 Z_3\}.$$

Note that the parameters β_i will have a different interpretation. The estimates of the parameters are obtained in section 8.5 as $b_1 = -0.2484$, $b_2 = -0.0888$, $b_3 = 0.7455$. Here the interest will center on the interaction term β_3, which will be tested in section 8.5. Here, the exponential of the coefficient of Z_3, the product of the race and gender covariate, is the excess relative risk of being black for females compared to males, $\exp(0.7455) = 2.11$. It is also instructive to see that the relative risks for black male, white male, and black female relative to white female are $\exp(-0.0888 - (-0.2484)) = 1.17$, $\exp(0 - (-0.2484)) = 1.28$, $\exp(-0.2484 - 0.0888 + 0.7455 - (-0.2484)) = 1.93$, respectively, just as we obtained for the earlier coding. These are two different coding schemes; the first treats the samples as four groups and the second treats the samples as a 2×2 factorial, where interest may center on the interaction between gender and race. The interpretation of the two coding schemes are equivalent in that they lead to the same relative risks and the same likelihood.

The following example illustrates the construction of the interaction of a continuous variable and a categorical variable.

EXAMPLE 8.2

(continued) Consider two of the factors, namely age and stage of disease, in the data introduced in section 1.8. As usual, Z_i, $i = 1, 2, 3$ are defined as before in (8.2.3) and Z_4 will be the age of the patient. The interaction between age and stage will involve three product terms, namely, $Z_5 = Z_1 Z_4$; $Z_6 = Z_2 Z_4$ and $Z_7 = Z_3 Z_4$. Thus, for a 50-year-old man with Stage II cancer, the three interaction variables will take on the following values: $Z_5 = Z_1 Z_4 = (1)(50) = 50$; $Z_6 = Z_2 Z_4 = (0)(50) = 0$ and $Z_7 = Z_3 Z_4 = (0)(50) = 0$. Other combinations of age and stage can be appropriately formed. Now the full model will be

$$h(t \mid \mathbf{Z}) = h_0(t)\exp\{\beta'\mathbf{Z}\}$$

$$= h_0(t)\exp\{\beta_1 Z_1 + \beta_2 Z_2 + \beta_3 Z_3 + \beta_4 Z_4 + \beta_5 Z_5 + \beta_6 Z_6 + \beta_7 Z_7\}.$$

$$(8.2.7)$$

The null hypothesis of no interaction between age and stage may be written as $H_0 : \beta_5 = \beta_6 = \beta_7 = 0$ vs. the alternate hypothesis, which will be the negation of the null. This example will be considered in much more detail in section 8.5. The estimates of the parameters and their interpretation will be delayed until that discussion.

8.3   Partial Likelihoods for Distinct-Event Time Data

As indicated earlier, our data is based on a sample of size n consisting of the triple $(T_j, \delta_j, \mathbf{Z}_j)$, $j = 1, \ldots, n$. We assume that censoring is noninformative in that, given $\mathbf{Z}_j$, the event and censoring time for the jth patient are independent. Suppose that there are no ties between the event times. Let $t_1 < t_2 < \cdots < t_D$ denote the ordered event times and $Z_{(i)k}$ be the kth covariate associated with the individual whose failure time is t_i. Define the risk set at time t_i, $R(t_i)$, as the set of all individuals who are still under study at a time just prior to t_i. The partial likelihood (see Theoretical Notes 1 and 2), based on the hazard function as specified by (8.1.2), is expressed by

$$L(\boldsymbol{\beta}) = \prod_{i=1}^{D} \frac{\exp\left[\sum_{k=1}^{p} \beta_k Z_{(i)k}\right]}{\sum_{j \in R(t_i)} \exp\left[\sum_{k=1}^{p} \beta_k Z_{jk}\right]}. \qquad (8.3.1)$$

This is treated as a usual likelihood, and inference is carried out by usual means. It is of interest to note that the numerator of the likelihood depends only on information from the individual who experiences the event, whereas the denominator utilizes information about all individuals who have not yet experienced the event (including some individuals who will be censored later).

Let $LL(\boldsymbol{\beta}) = \ln[L(\boldsymbol{\beta})]$. Then, after a bit of algebra, we can write $LL(\boldsymbol{\beta})$ as

$$LL(\boldsymbol{\beta}) = \sum_{i=1}^{D} \sum_{k=1}^{p} \beta_k Z_{(i)k} - \sum_{i=1}^{D} \ln\left[\sum_{j \in R(t_i)} \exp\left(\sum_{k=1}^{p} \beta_k Z_{jk}\right)\right]. \qquad (8.3.2)$$

The (partial) maximum likelihood estimates are found by maximizing (8.3.1), or, equivalently, (8.3.2). The efficient score equations are found by taking partial derivatives of (8.3.2) with respect to the β's as follows. Let $U_h(\boldsymbol{\beta}) = \delta LL(\boldsymbol{\beta})/\delta \beta_h$, $h = 1, \ldots, p$.

Then,

$$U_b(\beta) = \sum_{i=1}^{D} Z_{(i)b} - \sum_{i=1}^{D} \frac{\sum_{j \in R(t_i)} Z_{jb} \exp\left[\sum_{k=1}^{p} \beta_k Z_{jk}\right]}{\sum_{j \in R(t_i)} \exp\left[\sum_{k=1}^{p} \beta_k Z_{jk}\right]}. \tag{8.3.3}$$

The information matrix is the negative of the matrix of second derivatives of the log likelihood and is given by $\mathbf{I}(\beta) = [I_{gb}(\beta)]_{p \times p}$ with the (g, b)th element given by

$$I_{gb}(\beta) = \sum_{i=1}^{D} \frac{\sum_{j \in R(t_i)} Z_{jg} Z_{jb} \exp\left[\sum_{k=1}^{p} \beta_k Z_{jk}\right]}{\sum_{j \in R(t_i)} \exp\left[\sum_{k=1}^{p} \beta_k Z_{jk}\right]}$$

$$- \sum_{i=1}^{D} \left[\frac{\sum_{j \in R(t_i)} Z_{jg} \exp\left(\sum_{k=1}^{p} \beta_k Z_{jk}\right)}{\sum_{j \in R(t_i)} \exp\left(\sum_{k=1}^{p} \beta_k Z_{jk}\right)} \right]$$

$$\left[\frac{\sum_{j \in R(t_i)} Z_{jb} \exp\left(\sum_{k=1}^{p} \beta_k Z_{jk}\right)}{\sum_{j \in R(t_i)} \exp\left(\sum_{k=1}^{p} \beta_k Z_{jk}\right)} \right].$$

$$\tag{8.3.4}$$

The (partial) maximum likelihood estimates are found by solving the set of p nonlinear equations $U_b(\beta) = 0$, $b = 1, \ldots, p$. This can be done numerically, as shown in Appendix A, using a Newton–Raphson technique (or some other iterative method), with (8.3.3) and (8.3.4). Most major software packages will perform this iterative maximization. Note that (8.3.2) does not depend upon the baseline hazard rate $h_0(x)$, so that inferences may be made on the effects of the explanatory variables without knowing $h_0(x)$.

There are three main tests (described in more detail in Appendix B) for hypotheses about regression parameters β. Let $\mathbf{b} = (b_1, \ldots, b_p)'$ denote the (partial) maximum likelihood estimates of β and let $\mathbf{I}(\beta)$ be the $p \times p$ information matrix evaluated at β. The first test is the usual test based on the asymptotic normality of the (partial) maximum likelihood estimates, referred to as Wald's test. It is based on the result that, for large samples, $\mathbf{b}$ has a p-variate normal distribution with mean β and variance-covariance estimated by $I^{-1}(\mathbf{b})$. A test of the global hypothesis of $H_0 : \beta = \beta_0$ is

$$X_W^2 = (\mathbf{b} - \beta_0)' \mathbf{I}(\mathbf{b})(\mathbf{b} - \beta_0) \tag{8.3.5}$$

which has a chi-squared distribution with p degrees of freedom if H_0 is true for large samples.

The second test is the likelihood ratio test of the hypothesis of $H_0 : \beta = \beta_0$ and uses

$$X_{LR}^2 = 2[LL(\mathbf{b}) - LL(\beta_0)] \tag{8.3.6}$$

which has a chi-squared distribution with p degrees of freedom under H_0 for large n.

The third test is the scores test. It is based on the efficient scores, $\mathbf{U}(\boldsymbol{\beta}) = (U_1(\boldsymbol{\beta}), \ldots, U_p(\boldsymbol{\beta}))'$ where $U_h(\boldsymbol{\beta})$ is defined by (8.3.3). For large samples, $\mathbf{U}(\boldsymbol{\beta})$ is asymptotically p-variate normal with mean $\mathbf{0}$ and covariance $\mathbf{I}(\boldsymbol{\beta})$ when H_0 is true. Thus a test of $H_0 : \boldsymbol{\beta} = \boldsymbol{\beta}_0$ is

$$X_{SC}^2 = \mathbf{U}(\boldsymbol{\beta}_0)'\mathbf{I}^{-1}(\boldsymbol{\beta}_0)\mathbf{U}(\boldsymbol{\beta}_0) \tag{8.3.7}$$

which has a large-sample chi-squared distribution with p degrees of freedom under H_0.

EXAMPLE 8.1

(continued) In section 1.5, we introduced a study designed to determine if female breast-cancer patients, originally classified as lymph node negative by standard light microscopy (SLM), could be more accurately classified by immunohistochemical (IH) examination of their lymph nodes with an anticytokeratin, monoclonal antibody cocktail.

In this example, we wish to perform a proportional hazards regression with immunoperoxidase status as the single covariate in the model. We adopt the usual regression formulation of a dichotomous independent variable and construct a dummy (or indicator) variable as follows.

Let $Z = 1$ if the patient is immunoperoxidase positive, 0 otherwise. The model is $h(t \mid Z) = h_0(t)\exp(\beta Z)$, where $h_0(t)$ is an arbitrary baseline hazard rate and β is the regression coefficient.

For this model, $\sum_{i=1}^{D} Z_{(i)} = d_1$, the number of deaths in the immunoperoxide positive sample, and $\sum_{j \in R(t_i)} \exp(\beta Z_j) = Y_{0i} + Y_{1i}e^{\beta}$, where Y_{0i} (Y_{1i}) is the number of individuals at risk in the immunoperoxidase negative (positive) sample at time t_i. From (8.3.2)–(8.3.4),

$$LL(\beta) = \beta d_1 - \sum_{i=1}^{D} \ln[Y_{0i} + Y_{1i}e^{\beta}],$$

$$U(\beta) = d_1 - \sum_{i=1}^{D} \frac{Y_{1i}e^{\beta}}{Y_{0i} + Y_{1i}e^{\beta}},$$

and

$$I(\beta) = \sum_{i=1}^{D} \left[\frac{Y_{1i}e^{\beta}}{(Y_{0i} + Y_{1i}e^{\beta})} - \frac{Y_{1i}^2 e^{2\beta}}{(Y_{0i} + Y_{1i}e^{\beta})^2} \right].$$

The simplest test of the hypothesis that $\beta = 0$ is the score test. In this case,

$$U(0) = d_1 - \sum_{i=1}^{D} \frac{Y_{1i}}{Y_{0i} + Y_{1i}},$$

and

$$I(0) = \sum_{i=1}^{D} \frac{Y_{1i}}{(Y_{0i} + Y_{1i})} - \frac{Y_{1i}^2}{(Y_{0i} + Y_{1i})^2} = \sum_{i=1}^{D} \frac{Y_{1i} Y_{0i}}{(Y_{0i} + Y_{1i})^2}.$$

Note that, in this case, where there are no ties between the death times, the score statistic $X_{Sc}^2 = U(0)^2/I(0)$ is precisely the two-sample, log-rank test presented in section 7.3. In this example, $U(0) = 4.19$ and $I(0) = 3.19$ so the value of $X_{Sc}^2 = 5.49$ which has a p-value of 0.019, when compared to the chi-squared distribution, with one degree of freedom.

To obtain the estimate of β, the likelihood is maximized by a numerical technique. Routines for this maximization are available in most statistical packages. Using the Newton–Raphson algorithm described in Appendix A, we start with an initial guess at β of $b_0 = 0$ and compute an updated estimate at the mth stage by $b_m = b_{m-1} + U(b_{m-1})/I(b_{m-1})$. The iterative process is declared converged when the relative change in log likelihoods between successive steps is less than 0.0001. Here we have the results of three iterations:

m	b_{m-1}	$LL(b_{m-1})$	$U(b_{m-1})$	$I(b_{m-1})$	$b_m = b_{m-1}+$ $\dfrac{U(b_{m-1})}{I(b_{m-1})}$	$LL(b_m)$	$\dfrac{LL(b_m) - LL(b_{m-1})}{\lvert LL(b_{m-1})\rvert}$
1	0	−83.7438	4.1873	3.1912	1.3121	−81.8205	0.0230
2	1.3121	−81.8205	−1.8382	5.7494	0.9924	−81.5210	0.0037
3	0.9924	−81.5210	−0.0646	5.3084	0.9802	−81.5206	<0.0001

The Newton-Raphson algorithm converges after three steps.

To test the hypothesis $H_0 : \beta = 0$ using the likelihood ratio test, $X_{LR}^2 = 2\{LL(0.9802) - LL(0)\} = 2[-81.52 - (-83.74)] = 4.44$ which has a p-value of 0.035. To perform the Wald test we first estimate the standard error of our estimate of β as $SE(b) = 1/I(0.9802)^{1/2} = 1/5.2871^{1/2} = 0.4349$. The Wald test is $(0.9802 - 0)^2/(0.4349)^2 = 5.08$, which has a p-value of 0.024.

The exponential of b gives the estimated relative risk, which in this example is $e^{0.9802} = 2.67$. This number tells us that a patient, who is immunoperoxidase positive, is 2.67 times more likely to die than an immunoperoxidase negative patient. Using the asymptotic normality of b, a 95% confidence interval for the relative risk is $\exp(0.9802 \pm 1.96 \times 0.4349) = (1.14, 6.25)$.

Practical Notes

1. Algorithms for the estimation of β in the Cox regression model are available in many statistical packages. The procedure PHREG in SAS and coxph in S-Plus provide estimates of β, its standard error and the Wald, score and likelihood ratio tests of the global hypothesis of no covariate effect. A Newton–Raphson algorithm is used to estimate β with 0 as an initial value.

2. If a covariate is perfectly correlated with the event times, that is, the covariates are ordered with $Z_{(1)k} \leq Z_{(2)k} \leq \cdots \leq Z_{(D)k}$ (or $Z_{(1)k} \geq Z_{(2)k} \geq \cdots \geq Z_{(D)k}$) the (partial) maximum likelihood estimate of β_k will be ∞ (or $-\infty$). When declaring convergence of a numerical maximization routine based on differences in likelihoods at successive iterations, one should carefully check that successive values of the estimates are close to each other as well to avoid this problem.

3. If there are no ties between the event times, the scores test in the proportional hazards model is identical to the log-rank test.

4. Empirical studies have shown that the convergence rate of the likelihood ratio and Wald tests are similar. The score test converges less rapidly to the limiting chi-squared distribution.

5. The tests performed in this section have assumed that the hazard rates are proportional. They, indeed, are but we shall present tools for checking this assumption in Chapters 9 and 11.

Theoretical Notes

1. The probability that an individual dies at time t_i with covariates $\mathbf{Z}_{(i)}$, given one of the individuals in $R(t_i)$ dies at this time, is given by

$$P[\text{individual dies at } t_i \mid \text{one death at } t_i]$$

$$= \frac{P[\text{individual dies at } t_i \mid \text{survival to } t_i]}{P[\text{one death at } t_i \mid \text{survival to } t_i]}$$

$$= \frac{h[t_i \mid \mathbf{Z}_{(i)}]}{\sum_{j \in R(t_i)} h[t_i \mid \mathbf{Z}_j]} = \frac{h_0(t_i) \exp[\boldsymbol{\beta}'\mathbf{Z}_{(i)}]}{\sum_{j \in R(t_i)} h_0(t_i) \exp[\boldsymbol{\beta}'\mathbf{Z}_j]}$$

$$= \frac{\exp[\boldsymbol{\beta}'\mathbf{Z}_{(i)}]}{\sum_{j \in R(t_i)} \exp[\boldsymbol{\beta}'\mathbf{Z}_j]}.$$

The partial likelihood is formed by multiplying these conditional probabilities over all deaths, so we have the likelihood function

$$L(\boldsymbol{\beta}) = \prod_{i=1}^{D} \frac{\exp[\boldsymbol{\beta}'\mathbf{Z}_{(i)}]}{\sum_{j \in R(t_i)} \exp\{\boldsymbol{\beta}'\mathbf{Z}_j\}} \qquad \text{as in (8.3.1).}$$

2. The Cox partial likelihood can be derived as a profile likelihood from the full censored-data likelihood, as discussed by Johansen (1983). Here, we start with the complete censored-data likelihood, which, by the discussion in section 3.5, is given by

$$L[\boldsymbol{\beta}, h_0(t)] = \prod_{j=1}^{n} h(T_j \mid \mathbf{Z_j})^{\delta_j} S(T_j \mid \mathbf{Z_j})$$

$$= \prod_{j=1}^{n} h_0(T_j)^{\delta_j} [\exp(\boldsymbol{\beta}'\mathbf{Z_j})]^{\delta_j} \exp(-H_0(T_j) \exp(\boldsymbol{\beta}'\mathbf{Z_j})).$$

(8.3.8)

Now, fix $\boldsymbol{\beta}$ and consider maximizing this likelihood as a function of $h_0(t)$ only. The function to be maximized is

$$L_{\boldsymbol{\beta}}(h_0(t)) = \left[\prod_{i=1}^{D} h_0(t_i) \exp(\boldsymbol{\beta}'\mathbf{Z}_{(i)}) \right] \exp \left[-\sum_{j=1}^{n} H_0(T_j) \exp(\boldsymbol{\beta}'\mathbf{Z_j}) \right].$$

(8.3.9)

This function is maximal when $h_0(t) = 0$ except for times at which the events occurs. Let $h_{0i} = h_0(t_i)$, $i = 1, \ldots, D$ so $H_0(T_j) = \sum_{t_i \le T_j} h_{0i}$. Thus, after some simplification, (8.3.9) can be written as

$$L_{\boldsymbol{\beta}}(h_{01}, \ldots, h_{0D}) \propto \prod_{i=1}^{D} h_{0i} \exp \left[-h_{0i} \sum_{j \in R(t_i)} \exp(\boldsymbol{\beta}'\mathbf{Z_j}) \right]$$

and the profile maximum likelihood estimator of h_{0i} is given by

$$\hat{h}_{0i} = \frac{1}{\sum_{j \in R(t_i)} \exp(\boldsymbol{\beta}'\mathbf{Z_j})}.$$

Combining these estimates yields an estimate of $H_0(t)$ given by

$$\hat{H}_0(t) = \sum_{t_i \le t} \frac{1}{\sum_{j \in R(t_i)} \exp(\boldsymbol{\beta}'\mathbf{Z_j})}.$$

This is Breslow's estimator of the baseline cumulative hazard rate in the case of, at most, one death at any time and is discussed in more detail in section 8.8. Substituting $\hat{H}_0(t)$ in (8.3.8) and simplifying yields a profile likelihood proportional to the partial likelihood of Eq. (8.3.1).

8.4   Partial Likelihoods When Ties Are Present

In section 8.3, we presented the partial likelihood for the proportional hazards regression problem when there are no ties between the event times. Often, due to the way times are recorded, ties between event times are found in the data. Alternate partial likelihoods have been provided by a variety of authors when there are ties between event times.

Let $t_1 < t_2 < \cdots < t_D$ denote the D distinct, ordered, event times. Let d_i be the number of deaths at t_i and $\mathbb{D}_i$ the set of all individuals who die at time t_i. Let $\mathbf{s}_i$ be the sum of the vectors $\mathbf{Z}_j$ over all individuals who die at t_i. That is $\mathbf{s}_i = \sum_{j \in \mathbb{D}_i} \mathbf{Z}_j$. Let R_i be the set of all individuals at risk just prior to t_i.

There are several suggestions for constructing the partial likelihood when there are ties among the event times. The first, due to Breslow (1974), arises naturally from the profile likelihood construction discussed in Theoretical Note 2 of the previous section. The partial likelihood is expressed as

$$L_1(\boldsymbol{\beta}) = \prod_{i=1}^{D} \frac{\exp(\boldsymbol{\beta}'\mathbf{s}_i)}{\left[\sum_{j \in R_i} \exp(\boldsymbol{\beta}'\mathbf{Z}_j)\right]^{d_i}}. \qquad (8.4.1)$$

This likelihood considers each of the d_i events at a given time as distinct, constructs their contribution to the likelihood function, and obtains the contribution to the likelihood by multiplying over all events at time t_i. When there are few ties, this approximation works quite well, and this likelihood is implemented in most statistical packages.

Efron (1977) suggests a partial likelihood of

$$L_2(\boldsymbol{\beta}) = \prod_{i=1}^{D} \frac{\exp(\boldsymbol{\beta}'\mathbf{s}_i)}{\prod_{j=1}^{d_i}\left[\sum_{k \in R_i} \exp(\boldsymbol{\beta}'\mathbf{Z}_k) - \frac{j-1}{d_i}\sum_{k \in \mathbb{D}_i} \exp(\boldsymbol{\beta}'\mathbf{Z}_k)\right]}, \qquad (8.4.2)$$

which is closer to the correct partial likelihood based on a discrete hazard model than Breslow's likelihood. When the number of ties is small, Efron's and Breslow's likelihoods are quite close.

The third partial likelihood due to Cox (1972) is based on a discrete-time, hazard-rate model. This likelihood is constructed by assuming a logistic model for the hazard rate, that is, if we let $h(t \mid \mathbf{Z})$ denote the conditional death probability in the interval $(t, t+1)$ given survival to the start of the interval and if we assume

$$\frac{h(t \mid \mathbf{Z})}{1 - h(t \mid \mathbf{Z})} = \frac{h_0(t)}{1 - h_0(t)} \exp(\boldsymbol{\beta}'\mathbf{Z}),$$

then, this likelihood is the proper partial likelihood. To construct the likelihood, let Q_i denote the set of all subsets of d_i individuals who could be selected from the risk set R_i. Each element of Q_i is a d_i-tuple of individuals who could have been one of the d_i failures at time t_i. Let $q = (q_1, \ldots, q_{d_i})$ be one of these elements of Q_i and define $\mathbf{s}_q^* = \sum_{j=1}^{d_j} \mathbf{Z}_{qj}$. Then, the discrete log likelihood is given by

$$L_3(\boldsymbol{\beta}) = \prod_{i=1}^{D} \frac{\exp(\boldsymbol{\beta}'\mathbf{s}_i)}{\sum_{q \in Q_i} \exp(\boldsymbol{\beta}'\mathbf{s}_q^*)}. \qquad (8.4.3)$$

When there are no ties between the event times, this likelihood and Breslow's and Efron's likelihoods reduce to the partial likelihood in the previous section.

EXAMPLE 8.4   A study to assess the time to first exit-site infection (in months) in patients with renal insufficiency was introduced in section 1.4. Forty-three patients utilized a surgically placed catheter and 76 patients utilized a percutaneous placement of their catheter. Catheter failure was the primary reason for censoring. To apply a proportional hazards regression, let $Z = 1$ if the patient has a percutaneous placement of the catheter, and 0 otherwise.

To see how the three likelihoods differ, consider the contribution of the six deaths at time 0.5. All six deaths have $Z = 1$, and there are 76 patients at risk with $Z = 1$ and 43 patients at risk with $Z = 0$. The contribution to the likelihood is

Breslow:
$$\frac{\exp(6\beta)}{[43 + 76\exp(\beta)]^6},$$

Efron:
$$\frac{\exp(6\beta)}{\prod_{j=1}^{6}[76e^\beta + 43 - \frac{j-1}{6}(6e^\beta)]},$$

Discrete:
$$\frac{\exp(6\beta)}{\binom{43}{6} + \binom{43}{5}\binom{76}{1}e^\beta + \binom{43}{4}\binom{76}{2}e^{2\beta} + \binom{43}{3}\binom{76}{3}e^{3\beta} + \binom{43}{2}\binom{76}{4}e^{4\beta} + \binom{43}{1}\binom{76}{5}e^{5\beta} + \binom{76}{6}e^{6\beta}}.$$

Using the three likelihoods, we have the following results:

	Breslow's Likelihood (8.4.1)	Efron's Likelihood (8.4.2)	Discrete Likelihood (8.4.3)
Initial likelihood	−104.4533	−104.2319	−94.1869
Final likelihood	−103.2285	−103.0278	−92.9401
b	−0.6182	−0.6126	−0.6294
SE(b)	0.3981	0.3979	0.4019
Relative risk, e^b	0.539	0.542	0.553
Score test	$X^2 = 2.49$ ($p = 0.115$)	$X^2 = 2.44$ ($p = 0.117$)	$X^2 = 2.53$ ($p = 0.112$)
Wald test	$X^2 = 2.41$ ($p = 0.121$)	$X^2 = 2.37$ ($p = 0.124$)	$X^2 = 2.45$ ($p = 0.117$)
Likelihood ratio test	$X^2 = 2.45$ ($p = 0.118$)	$X^2 = 2.41$ ($p = 0.121$)	$X^2 = 2.49$ ($p = 0.114$)

Here, we see that, *assuming the proportional hazards model is appropriate*, for any of the three likelihoods the regression coefficient is not statistically significant and the relative risk is about 0.54. As we shall test formally in Chapter 9, the proportional hazards assumption is not appropriate for this data. Thus the relative risk is not constant but depends upon time, and the reported relative risk of 0.54 is not correct. Furthermore, a potentially significant result could be overlooked because the proportional hazards assumption is not satisfied. This has implications for procedures used in model building, which will be discussed in section 8.7. As a graphical check of the proportional hazards assumption, we compute the Nelson–Aalen estimator of the cumulative hazard rate for each treatment. If the proportional hazards model is correct, then, we should have $H(t \mid Z = 1) = e^{\beta} H(t \mid Z = 0)$, so that a plot of $\ln[\tilde{H}(t \mid Z = 1)] - \ln[\tilde{H}(t \mid Z = 0)]$ versus t should be roughly equal to β. The plot for this data set, shown in Figure 8.1, strongly suggests

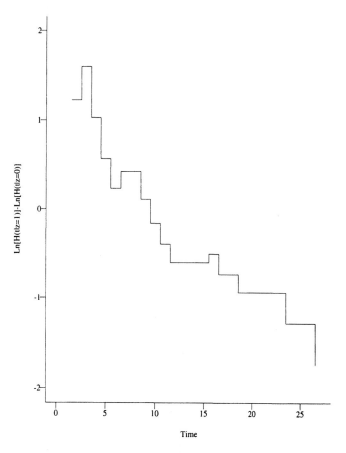

Figure 8.1 *Graphical check of the proportional hazards assumption for the renal insufficiency study.*

nonproportional hazards. Other graphical checks of the proportional hazards assumption are discussed in section 11.4.

EXAMPLE 8.2   *(continued)*   In section 1.8, a study of 90 males diagnosed with cancer of the larynx was described. In addition to the outcome variable, time from first treatment until either death or the end of the study, the independent variables, patient's age (in years) at the time of diagnosis and the stage of the patient's cancer, were recorded. A basic test for trend on stage was performed in section 7.4.

Here, we shall perform a global proportional hazards regression test with only stage in the model. Because stage has four levels, we adopt the usual indicator variable coding methodology as in (8.2.3). The maximum likelihood parameter estimates, $b_i (i = 1, \ldots, 4)$, (and their corresponding standard errors) are 0.0658 (0.4584), 0.612 (0.3552), and 1.723 (0.4197), respectively. It follows that the relative risks, RR(Stage II/Stage I) = 1.07, RR(Stage III/Stage I) = 1.84, and RR(Stage IV/Stage I) = 5.60.

The global likelihood ratio, Wald, and score chi-squared (with three degrees of freedom) statistics for stage are 16.26 (p-value = 0.001), 18.95 (p-value = 0.0003), and 22.46 (p-value = 0.0001), respectively, using Breslow's method of handling ties. All three tests suggest that the survival rates are different for, at least, one stage of cancer. In the next section, we shall consider local tests which provide information on which stages differ in survival.

The following example illustrates another example of performing a global test for different groups and will be followed up in the next section with a local test for interaction.

EXAMPLE 8.3   *(continued)*   In section 1.7 a data set of 863 kidney transplant patients with data on race (white, black) and gender is described. In this study there were 432 white males, 92 black males, 280 white females, and 59 black females. Again, there are various coding options, as described in section 8.2. First, one may treat this study as a four-group problem. The three indicator variables have been defined in the usual way as described in section 8.2 as

$Z_1 = 1$ if the subject is a black male, 0 otherwise

$Z_2 = 1$ if the subject is a white male, 0 otherwise

and

$Z_3 = 1$ if the subject is a black female, 0 otherwise.

Here the referent group is being a white female. Again, the full model will be

$$h(t \mid Z) = h_0(t)\exp\{\boldsymbol{\beta}'\mathbf{Z}\} = h_0(t)\exp\{\beta_1 Z_1 + \beta_2 Z_2 + \beta_3 Z_3\}.$$

The maximum likelihood estimates of the parameters are obtained as

$$b_1 = 0.160, \, b_2 = 0.248, \, b_3 = 0.657.$$

Thus the relative risks for black male, white male, and black female relative to white female are 1.17, 1.28, 1.93, respectively. The global likelihood ratio, Wald, and score chi-squared (with 3 degrees of freedom) statistics for groups are 4.37 (p-value = 0.22), 4.64 (p-value = 0.20), and 4.74 (p-value = 0.19), respectively, using the Breslow method of handling ties. All three tests suggest the survival rates are not different for the four groups of subjects. In the next section we shall consider local tests which provide information on testing for an interaction between race and gender.

Practical Notes

1. SAS PHREG uses Breslow's likelihood as a default and allows the user to specify that calculations be carried out using either the discrete or Efron likelihood. SAS also allows the user to specify an "exact" likelihood based on a generalized rank statistic derivation of the likelihood (see Kalbfleisch and Prentice (1980) for details). This likelihood requires a bit more computer time to implement and gives results quite close to the discrete likelihood.
2. The S-Plus function coxph uses Efron's likelihood as a default when there are ties between the event times. Breslow's likelihood and the exact likelihood are also available.

8.5   Local Tests

Often, one is interested in testing a hypothesis about a subset of the $\boldsymbol{\beta}$'s. The hypothesis is then $H_0 : \boldsymbol{\beta}_1 = \boldsymbol{\beta}_{10}$, where $\boldsymbol{\beta} = (\boldsymbol{\beta}_1', \boldsymbol{\beta}_2')'$. Here $\boldsymbol{\beta}_1$ is a $q \times 1$ vector of the $\boldsymbol{\beta}$'s of interest and $\boldsymbol{\beta}_2$ is the vector of the remaining $p - q$ $\boldsymbol{\beta}$'s.

The Wald test of $H_0 : \boldsymbol{\beta}_1 = \boldsymbol{\beta}_{10}$ is based on the maximum partial likelihood estimators of $\boldsymbol{\beta}$. Let $\mathbf{b} = (\mathbf{b}_1', \mathbf{b}_2')'$ be the maximum partial likelihood estimator of $\boldsymbol{\beta}$. Suppose we partition the information matrix $\mathbf{I}$ as

$$\mathbf{I} = \begin{pmatrix} \mathbf{I}_{11} & \mathbf{I}_{12} \\ \mathbf{I}_{21} & \mathbf{I}_{22} \end{pmatrix},$$

where $\mathbf{I}_{11}$ $(\mathbf{I}_{22})$ is the $q \times q$ $[(p-q) \times (p-q)]$ submatrix of second partial derivatives of the minus log likelihood with respect to $\boldsymbol{\beta}_1$ $(\boldsymbol{\beta}_2)$ and $\mathbf{I}_{12}$ and $\mathbf{I}_{21}$, the matrices of mixed second partial derivatives. The Wald test statistic is

$$X_W^2 = (\mathbf{b}_1 - \boldsymbol{\beta}_{10})'[\mathbf{I}^{11}(\mathbf{b})]^{-1}(\mathbf{b}_1 - \boldsymbol{\beta}_{10}) \tag{8.5.1}$$

where $\mathbf{I}^{11}(\mathbf{b})$ is the upper $q \times q$ submatrix of $\mathbf{I}^{-1}(\mathbf{b})$ (see Appendix B). For large samples, this statistic has a chi-squared distribution with q degrees of freedom under H_0.

Let $\mathbf{b}_2(\boldsymbol{\beta}_{10})$ be the (partial) maximum likelihood estimates of $\boldsymbol{\beta}_2$ based on the log likelihood (8.3.2) with the first q $\boldsymbol{\beta}$'s fixed at a value $\boldsymbol{\beta}_{10}$. The likelihood ratio test of $H_0 : \boldsymbol{\beta}_1 = \boldsymbol{\beta}_{10}$ is expressed by

$$X_{LR}^2 = 2\{LL(\mathbf{b}) - LL[\boldsymbol{\beta}_{10}, \mathbf{b}_2(\boldsymbol{\beta}_{10})]\} \tag{8.5.2}$$

which has a large sample chi-squared distribution with q degrees of freedom under H_0.

To test $H_0 : \boldsymbol{\beta}_1 = \boldsymbol{\beta}_{10}$ using the score statistic, let $\mathbf{U}_1[\boldsymbol{\beta}_{10}, \mathbf{b}_2(\boldsymbol{\beta}_{10})]$ be the $q \times 1$ vector of scores for $\boldsymbol{\beta}_1$, evaluated at the hypothesized value of $\boldsymbol{\beta}_{10}$ and at the restricted partial maximum likelihood estimator for $\boldsymbol{\beta}_2$. Then,

$$X_{SC}^2 = \mathbf{U}_1[\boldsymbol{\beta}_{10}, \mathbf{b}_2(\boldsymbol{\beta}_{10})]'[\mathbf{I}^{11}(\boldsymbol{\beta}_{10}, \mathbf{b}_2(\boldsymbol{\beta}_{10}))]\mathbf{U}_1[\boldsymbol{\beta}_{10}, \mathbf{b}_2(\boldsymbol{\beta}_{10})] \tag{8.5.3}$$

which has a large sample chi-squared distribution with q degrees of freedom under H_0. We shall illustrate these tests in the following example.

EXAMPLE 8.2 *(continued)* In section 8.4, a global test was performed on stage of cancer in a study of 90 males diagnosed with cancer of the larynx. Here we shall test the hypothesis that there is no difference in survival between patients with different stages of disease, adjusting for the age of the patient. Our test is based on the model with covariates Z_1, Z_2, and Z_3, which, as in section 8.2, are the indicators of stage II, III, and IV disease, respectively, and a covariate Z_4 which is the patient's age at diagnosis. The local hypothesis of interest is $H_0 : \beta_1 = 0$, $\beta_2 = 0$, $\beta_3 = 0$ against the alternative hypothesis that, at least, one of these β's is nonzero.

To apply the score test or the likelihood ratio test, we need to estimate the coefficient for age, β_4, in the model, with $\beta_1 = \beta_2 = \beta_3 = 0$. This involves fitting a Cox model with only the single covariate, age. Fitting this model, we find $b_4 = 0.023$ with a log partial likelihood of -195.906.

Using this value of b_4, we find that the score is

$$\mathbf{U}(0, 0, 0, 0.023) = (-2.448, 3.0583, 7.4400, 0.000)'$$

and that

$$\mathbf{I}(0,0,0,0.023) = \begin{pmatrix} 7.637 & -2.608 & -0.699 & -24.730 \\ -2.608 & 9.994 & -0.979 & -8.429 \\ -0.699 & -0.979 & 3.174 & 11.306 \\ -24.730 & -8.429 & 11.306 & 4775.716 \end{pmatrix}.$$

The inverse of this information matrix is given by

$$\mathbf{I}^{-1}(0,0,0,0.023) = \begin{pmatrix} 0.1529 & 0.0449 & 0.0448 & 0.0008 \\ 0.0449 & 0.1164 & 0.0446 & 0.0003 \\ 0.0448 & 0.0446 & 0.3404 & -0.0005 \\ 0.0008 & 0.0003 & -0.0005 & 0.0002 \end{pmatrix},$$

so the score statistic is given by

$$X_{SC}^2 = (-2.448, \quad 3.0583, \quad 7.4400) \begin{pmatrix} 0.1529 & 0.0449 & 0.0448 \\ 0.0449 & 0.1164 & 0.0446 \\ 0.0448 & 0.0446 & 0.3404 \end{pmatrix} \begin{pmatrix} -2.448 \\ 3.0583 \\ 7.4400 \end{pmatrix}$$

$$= 20.577.$$

Comparing this quantity to a chi-squared distribution with three degrees of freedom, we find that the p-value of the test of no stage effect is 0.0001.

To perform the Wald and likelihood ratio tests, we need to fit the full model with all four covariates. Here, we find

$$\mathbf{b}' = (0.1386, \quad 0.6383, \quad 1.6931, \quad 0.0189)$$

with a partial log likelihood of -188.179. The likelihood ratio test of H_0 is

$$X_{LR}^2 = 2[-188.179 - (-195.906)] = 15.454.$$

The p-value of this test is 0.0015.

To perform the Wald test, we need the information matrix based on $\mathbf{b}$. This matrix is

$$\mathbf{I}(\mathbf{b}) = \begin{pmatrix} 5.9987 & -2.3913 & -1.4565 & -22.8634 \\ -2.3913 & 10.9917 & -3.3123 & -14.0650 \\ -1.4565 & -3.3123 & 7.4979 & 25.6149 \\ -22.8634 & -14.0650 & 25.6149 & 5088.5378 \end{pmatrix}.$$

The inverse of this matrix is the covariance matrix of $\mathbf{b}$, given by

$$\mathbf{I}^{-1}(\mathbf{b}) = \begin{pmatrix} 0.2137 & 0.0683 & 0.0690 & 0.0008 \\ 0.0683 & 0.1268 & 0.0682 & 0.0003 \\ 0.0690 & 0.0682 & 0.1783 & -0.0004 \\ 0.0008 & 0.0003 & -0.0004 & 0.0002 \end{pmatrix}. \tag{8.5.4}$$

The Wald chi-squared statistic is given by

$$X_W^2 = (0.1386, \quad 0.6383, \quad 1.6931) \begin{pmatrix} 0.2137 & 0.0683 & 0.0690 \\ 0.0683 & 0.1268 & 0.0682 \\ 0.0690 & 0.0682 & 0.1783 \end{pmatrix}^{-1} \begin{pmatrix} 0.1386 \\ 0.6383 \\ 1.6931 \end{pmatrix}$$

$$= 17.63$$

which has a p-value of 0.0005.

Often it is desirable to perform tests involving one (or more) linear combination(s) of the parameters. For the Wald test, one can form a matrix of q linear combinations of parameters to test such hypotheses. Here one forms a $q \times p$ matrix of full rank ($q \le p$),

$$\mathbf{C} = \begin{pmatrix} \mathbf{c}_1^t \\ \mathbf{c}_2^t \\ \vdots \\ \mathbf{c}_q^t \end{pmatrix} \tag{8.5.5}$$

where $\mathbf{c}_k^t = (\mathbf{c}_{k1}, \mathbf{c}_{k2}, \ldots, \mathbf{c}_{kp})$ is a vector of coefficients for the kth linear combination of the betas, and the hypothesis to be tested is

$$H_0 : \mathbf{C}\boldsymbol{\beta} = \mathbf{C}\boldsymbol{\beta}_0. \tag{8.5.6}$$

From large-sample theory,

$$(\mathbf{Cb} - \mathbf{C}\boldsymbol{\beta}_0)^t [\mathbf{CI}^{-1}(\mathbf{b})\mathbf{C}^t]^{-1}(\mathbf{Cb} - \mathbf{C}\boldsymbol{\beta}_0) \tag{8.5.7}$$

will have an asymptotic chi-squared distribution with q degrees of freedom.

EXAMPLE 8.2   *(continued)*   In the previous example, we fitted a model to data on patients with cancer of the larynx. In this example, we wish to test the hypothesis $H_0 : \beta_1 = 0$. Note that the upper 1×1 submatrix of $\hat{\mathbf{V}}(\mathbf{b})$ is precisely the matrix $\mathbf{I}^{11}(\mathbf{b})$ required in (8.5.1) and the Wald chi-squared test is calculated as $(0.1386)(0.2137)^{-1}(0.1386) = 0.0898$. If we choose the linear combination approach, $\mathbf{c} = (1, 0, 0, 0)$ and $(\mathbf{Cb})^t[\mathbf{CI}^{-1}(\mathbf{b})\mathbf{C}^t]^{-1}\mathbf{Cb} = 0.0898$, the same result as above. Note that this statistic, which has a large-sample chi-squared distribution with one degree of freedom under H_0, is testing for a difference in risk of death between stage I and stage II cancer patients, adjusting for age. Here the p-value of that test is 0.7644 which suggests no difference between stage I and II patients.

Most statistics packages will produce an "Analysis of Variance" (ANOVA) table describing all such univariate Wald tests along with the estimated standard error and relative risk of the effects. Note that, in such tables, the relative risk, exp(b), is the relative risk in a differ-

TABLE 8.1

Analysis of Variance Table for Stage of the Laryngeal Cancer Patients, Utilizing the "Breslow" Method of Handling Ties

Variables	Degrees of Freedom	Parameter Estimates	Standard Errors	Wald Chi Square	p-Value	Relative risk
Z_1: Stage II	1	0.1386	0.4623	0.09	0.7644	1.15
Z_2: Stage III	1	0.6383	0.3561	3.21	0.0730	1.89
Z_3: Stage IV	1	1.6931	0.4222	16.08	<0.0001	5.44
Z_4: Age	1	0.0189	0.0143	1.76	0.1847	1.02

ence of one unit in the covariate values. So the relative risk for age in Table 8.1 reflects the excess risk of dying for each additional year of age at diagnosis. Similarly, the risks of death for a patient in Stages II, III, and IV relative to a patient in Stage I are 1.15, 1.89, and 5.44, respectively. The corresponding confidence intervals for the β_i are $[b_i - z_{1-\alpha/2}\text{SE}(b_i), b_i + z_{1-\alpha/2}\text{SE}(b_i)]$ which may be obtained from the ANOVA table and tables of the unit normal distribution. Confidence intervals for the relative risk may be found by exponentiating the lower and upper limits, respectively. For example, a 95% confidence interval for the risk of death for patients in Stage IV relative to the risk of death for patients in Stage I would be $\{\exp[1.6931 - 1.96(0.4222)], \exp[1.6931 + 1.96(0.4222)]\} = (2.38, 12.44)$. This means that, with approximately 95% confidence, $\exp(\beta_3)$ will lie between 2.38 and 12.44, so that we will reject the hypothesis that $\beta_3 = 0$ when $\alpha = 0.05$, as indicated in the table (p-value < 0.0001).

Often, one is interested in relative risks that may not appear directly in the table. For example, the risk of death for patients in Stage III relative to the risk of death for patients in Stage II is found by taking $\exp(\beta_2)/\exp(\beta_1) = \exp(\beta_2 - \beta_1)$. The point estimate of this risk is $\exp(0.6383 - 0.1386) = 1.65$ which could also have been obtained directly from the table as 1.89/1.15 (aside from round-off error). The confidence interval for this relative risk cannot be obtained directly from the table. One needs the standard error of $b_2 - b_1$ which means we need the variance-covariance matrix of the b_i's as given in (8.5.4). Calculating $\text{Var}(b_2 - b_1) = \text{Var}(b_2) + \text{Var}(b_1) - 2\text{Cov}(b_2, b_1) = 0.1268 + 0.2137 - 2(0.0683) = 0.2039$ we are led, by taking the square root, to the standard error of $(b_2 - b_1) = 0.4515$. Now, we can find a 95% confidence interval for $\beta_2 - \beta_1$ as $[b_2 - b_1 - 1.96 \text{ SE}(b_2 - b_1), b_2 - b_1 + 1.96 \text{ SE}(b_2 - b_1)] = [0.4997 - 1.96(0.4515), 0.4997 + 1.96(0.4515)] = (-0.3852, 1.3846)$. Exponentiating the lower and upper limit leads to the approximate 95% confidence interval for $\exp(\beta_2 - \beta_1)$ as $(0.68, 3.99)$. Thus, this relative risk cannot be judged to differ from one.

Both the Wald and the likelihood ratio test can be used to test the hypothesis $H_0 : \beta_1 = \beta_2, \beta_2 = \beta_3$ or, equivalently, $H_0 : \beta_1 = \beta_2 = \beta_3$. This is a test of the hypothesis that, adjusted for age of the patient, survival is the same for stage II, III and IV patients, but not necessarily the same as for stage I patients. To perform the likelihood ratio test, we fit the full model with all four covariates which has a log partial likelihood of -188.179, and we fit a model with two covariates $Z^* = Z_1 + Z_2 + Z_3$ and age (Z_4). Here Z^* is the indicator of stage II, III or IV disease. The log partial likelihood from this model is -193.137. The likelihood ratio chi-squared is $2[-188.179 - (-193.137)] = 9.916$. For large samples under H_0, this statistic has a chi-squared distribution with two degrees of freedom. (The degrees of freedom are the number of parameters in the full model minus the number of parameters in the reduced model.) The p-value of this test is 0.0070 which suggests that survival is different for at least one of the three advanced stages.

To perform the Wald test, we define the C matrix with two contrasts, namely,

$$\mathbf{C} = \begin{pmatrix} 1 & -1 & 0 & 0 \\ 0 & -1 & 1 & 0 \end{pmatrix},$$

and apply (8.5.7). The resulting statistic has a value of 10.7324 with two degrees of freedom for the large-sample chi square. The p-value of the test is 0.0047, so the conclusion is the same as for the likelihood ratio test.

Now we turn our attention to a discussion of interaction. The first example is an example of an interaction between two categorical variables.

EXAMPLE 8.3   *(continued)*   An alternative coding scheme for the data in section 1.7 discussed earlier is to code the variables as two main effect terms, race and gender, and an interaction term. For example

$$Z_1 = 1 \quad \text{if the subject is a female, 0 otherwise,}$$

$$Z_2 = 1 \quad \text{if the subject is black, 0 otherwise,}$$

and

$$Z_3 = Z_1 \times Z_2 \quad \text{if the subject is a black female, 0 otherwise.}$$

Again the full model will be

$$h(t \mid Z) = h_0(t)\exp\{\beta'\mathbf{Z}\} = h_0(t)\exp\{\beta_1 Z_1 + \beta_2 Z_2 + \beta_3 Z_3\}.$$

Note that the parameters β_i will have a different interpretation. The estimates of the parameters are obtained as

$$b_1 = -0.2484, b_2 = 0.0888, b_3 = 0.7455.$$

The complete analysis is in Table 8.2.

TABLE 8.2

Analysis of Variance Table for Race, Gender, and Interaction for the Kidney Transplant Patients Utilizing the "Breslow" Method of Handling Ties

Variables	Degrees of Freedom	Parameter Estimates	Standard Errors	Wald Chi-sqaure	p-Values	Relative Risk
Z_1: Female	1	−0.2484	0.1985	1.57	0.2108	0.78
Z_2: Black	1	−0.0888	0.2918	0.09	0.7609	0.92
Z_3: Interaction	1	0.7455	0.4271	3.05	0.0809	2.11

Here the interest will center on the interaction term β_3. However, it is instructive to see that the relative risks for black male, white male, and black female relative to white female are $\exp(-0.0888 - (-0.2484)) = 1.17$, $\exp(0 - (-0.2484)) = 1.28$, $\exp(-0.2484 - 0.0888 + 0.7455 - (-0.2484)) = 1.93$, respectively, just as we obtained for the earlier coding. These are two different coding schemes; the first treats the samples as four groups and the second treats the samples as a 2×2 factorial where interest may center on the interaction between gender and race. The interpretation of the two coding schemes are not inconsistent in that they lead to the same relative risks.

Next, we shall consider an example of an interaction between a continuous and a categorical variable.

EXAMPLE 8.2

(continued) The interaction between age and stage will involve three product terms, namely, $Z_5 = Z_1 Z_4$; $Z_6 = Z_2 Z_4$ and $Z_7 = Z_3 Z_4$, where Z_i, $i = 1, \ldots, 4$ are defined as before. Thus, for a 50-year-old man with Stage II cancer, the three interaction variables will take on the following values: $Z_5 = Z_1 Z_4 = (1)(50) = 50$; $Z_6 = Z_2 Z_4 = (0)(50) = 0$ and $Z_7 = Z_3 Z_4 = (0)(50) = 0$. Other combinations of age and stage can be appropriately formed.

For this model, the estimates of the b's are $b_1 = -7.9461$, $b_2 = -0.1225$, $b_3 = 0.8470$, $b_4 = -0.0026$, $b_5 = 0.1203$, $b_6 = 0.0114$, and $b_7 = 0.0137$. The estimated variance-covariance matrix of the estimated parameters, obtained as the inverse of the Fisher information matrix, is

$$\hat{\mathbf{V}}(\mathbf{b}) = \begin{pmatrix} 13.529 & 2.932 & 2.956 & 0.045 & -0.191 & -0.044 & -0.045 \\ 2.932 & 6.093 & 2.957 & 0.044 & -0.044 & -0.091 & -0.044 \\ 2.956 & 2.957 & 5.884 & 0.044 & -0.044 & -0.044 & -0.086 \\ 0.045 & 0.044 & 0.044 & 0.001 & -0.001 & -0.001 & -0.001 \\ -0.191 & -0.044 & -0.044 & -0.001 & 0.003 & 0.001 & 0.001 \\ -0.044 & -0.091 & -0.044 & -0.001 & 0.001 & 0.001 & 0.001 \\ -0.045 & -0.044 & -0.086 & -0.001 & 0.001 & 0.001 & 0.001 \end{pmatrix}.$$

Table 8.3 gives the analysis of variance table for this model.

TABLE 8.3

Analysis of Variance Table for Stage, Age, and the Interaction of Stage by Age for Laryngeal Cancer Patients, Utilizing the "Breslow" Method of Handling Ties

Variables	Degrees of Freedom	Parameter Estimates	Standard Errors	Wald Chi Square	p-Value
Z_1: Stage II	1	−7.946	3.6782	4.67	0.03
Z_2: Stage III	1	−0.1225	2.4683	0.003	0.96
Z_3: Stage IV	1	0.8470	2.4257	0.12	0.73
Z_4: Age	1	−0.0026	0.0261	0.01	0.92
Z_5: $Z_1 \times Z_4$	1	0.1203	0.0523	5.29	0.02
Z_6: $Z_2 \times Z_4$	1	0.0114	0.0375	0.09	0.76
Z_7: $Z_3 \times Z_4$	1	0.0137	0.0360	0.14	0.70

Table 8.3 suggests that the effect of stage II on survival may be different for different ages because a local test of $\beta_5 = 0$ may be rejected (*p*-value = 0.02). Furthermore, it is suggested by the local tests of $\beta_6 = 0$ (*p*-value = 0.76) and $\beta_7 = 0$ (*p*-value = 0.70) that the effects of stages III and IV on survival may not be different for different ages.

To test the hypothesis that $\beta_6 = \beta_7 = 0$, we need the full −2 log likelihood for all seven parameters which is 370.155 and the reduced −2 log likelihood for the first five parameters which is 370.316. The local likelihood ratio chi-squared statistic for testing that there is no interaction between age and either stage III or IV disease ($H_0 : \beta_6 = \beta_7 = 0$) is the difference between the reduced −2 log likelihood for the first five parameters minus the full −2 log likelihood for all seven parameters = 370.316 − 370.155 = 0.161 with two degrees of freedom (*p*-value = 0.92). This provides strong confirmation that the latter two interaction terms may be dropped from the model and that the risks of dying for patients with Stages III and IV relative to the risk of dying for patients with Stage I does not depend on age.

In Table 8.4 the analysis of variance table for the reduced model with only an interaction between age and stage II disease is presented.

This table suggests that there is a significant interaction between age and stage II disease, that is, the relative risk of dying for a stage II

TABLE 8.4
Analysis of Variance Table for Stage, Age, and One Interaction Term (Stage II by Age) for Laryngeal Cancer Patients, Utilizing the "Breslow" Method of Handling Ties

Variables	Degrees of Freedom	Parameter Estimates	Standard Errors	Wald Chi Square	p-Value
Z_1:Stage II	1	−7.3820	3.4027	4.71	0.03
Z_2:Stage III	1	0.6218	0.3558	3.05	0.08
Z_3:Stage IV	1	1.7534	0.4240	17.11	<0.0001
Z_4:Age	1	0.0060	0.0149	0.16	0.69
Z_5: $Z_1 \times Z_4$	1	0.1117	0.0477	5.49	0.02

patient of age Z_4 as compared to a stage I patient of the same age depends on that age. This relative risk is $\exp(\beta_1 + \beta_5 Z_4) = \exp(-7.382 + 0.1117 \text{ Age})$. For example, for a 76-year-old patient, this relative risk is 3.03 whereas for a 60-year-old it is 0.51. This linear combination of the estimated coefficients not only leads one to an estimated relative risk which depends on a patient's age at diagnosis, but also allows us to test the hypothesis that the the risk of dying for stage I and II patients is the same for a given age, that is, we wish to test that the relative risk is one or, equivalently, that $\beta_1 + \beta_5(\text{age}) = 0$. To test the hypothesis that this linear combination of the parameters is zero, one forms the quadratic form based on $C = (1, 0, 0, 0, \text{age})'$. The resulting chi-squared statistic is

$$X_{\text{w}}^2 = \frac{(b_1 + b_5 \text{ age})^2}{V(b_1) + \text{age}^2 V(b_5) + 2 \text{ age } \text{Cov}(b_1, b_5)},$$

which has a large-sample chi-squared distribution with one degree of freedom. In this example, $V(b_1) = 11.5787$, $V(b_5) = 0.00227$ and $\text{Cov}(b_1, b_5) = -0.1607$, so for a 76-year-old person, X_{w}^2 equals $(1.1072)^2/0.2638 = 4.65$ (p-value = 0.03). For a 60-year-old we have a chi-square of 0.99 with a p-value of 0.32. This suggests that for "young" ages there is little difference in survival between stage I and II patients whereas, for older patients, those with stage II disease are more likely to die.

Practical Notes

1. A Monte Carlo study (Li et al., 1996) of the small-sample properties of the likelihood ratio, Wald, and scores tests was performed with respect to inference on a dichotomous covariate effect in a Cox proportional hazards model, as assessed by size and power considerations, under a variety of censoring fractions, sample sizes, and

hazard distributions. The general conclusion of this study was that the likelihood ratio test and Wald test performed similarly (although the likelihood test ratio test slightly outperformed the Wald test for smaller sample sizes). The score test was markedly inferior and is not recommended because it tends to inflate the size of the test. These conclusions held for a variety of censoring fractions.

2. Proc PHREG provides local tests based on the Wald statistic. Tests for contrasts are available.

3. S-Plus provides building blocks of the likelihood ratio test by running a series of models. Wald tests can be constructed using the estimated covariance matrix and parameter estimates.

8.6   Discretizing a Continuous Covariate

As we saw in the previous section the Cox model can be applied when the covariates are continuous or categorical. The interpretation of the model, however, is simplest when the covariate is a binary. Here the relative risk, $\exp\{b\}$, is the ratio of the risk of the event for a patient with the characteristic versus a patient without the characteristic. Often a medical investigator would like to treat a continuous covariate, X, as a binary covariate by assigning a score of 1 to subjects with large values of X and 0 to those with small values of X. This may be done to assign patients to poor- and high-risk groups based on the value of X, to aid in making graphical plots of patients with good or bad prognosis based on the binary covariate or simply to make the resulting relative risk calculations simpler for others to understand.

In most cases a major problem is determining the value of the cut point between high- and low-risk groups. In some cases this cut point can be based on biological reasoning and this is the optimal strategy for determination of the cut point. When no *a priori* information is available a "data-oriented" method is sometimes used to choose the cut point. These methods look at the distribution of the continuous covariate and divide subjects into groups based on some predetermined statistic on the covariate. For example, quite often subjects are divided into two equal groups based on whether they are larger or smaller than the sample median. These methods tend not to perform well.

In this section we will look at the "outcome-oriented" approach to this problem. Here we seek a cut point for the covariate which gives us the largest difference between individuals in the two data-defined groups. That is, for a continuous covariate, X, we seek a binary covariate Z defined by $Z = 1$ if $X \geq C$ and 0 if $X < C$, which makes the outcomes of the groups with $Z = 1$ as different from the group with $Z = 0$ as possible based on some statistic. We would also like to test the hypothesis that this covariate in its discretized version has no effect on outcome. This test must be adjusted for the fact that we have biased the outcome of the test by considering the cut point which gives the maximum separation between the two groups.

The inference procedure we describe is due to Contal and O'Quigley (1999) and is based on the log rank test statistic discussed in section 7.3. This statistic is the score statistic from the Cox model. For the procedure we look at all possible cut points; and for each cut point, C_k, we compute the log rank statistic based on the groups defined by X being less than the cut point or greater than the cut point. That is, at each event time, t_i, we find the total number of deaths, d_i, and the total number at risk, r_i. We also find the total number of deaths with $X \geq C_k$, d_i^+ and the total number at risk with $X \geq C_k$, r_i^+. We then compute the log rank statistic,

$$S_k = \sum_{i=1}^{D} \left[d_i^+ - d_i \frac{r_i^+}{r_i} \right], \qquad (8.6.1)$$

where D is the total number of distinct death times.

The estimated cut point $\hat{C}$ is the value of C_k which yields the maximum $|S_k|$. At this cut point the Cox regression model is

$$h(t \mid X) = h_0(t)\exp\{bZ\},$$

where $Z = 1$ if $X \geq \hat{C}$, 0 otherwise. The usual tests of $H_0 : b = 0$ can not be used here since we picked the cut point $\hat{C}$, which is most favorable to rejecting H_0. To compute the proper test statistic we need first to compute the quantity s^2 defined by

$$s^2 = \frac{1}{D-1} \sum_{i=1}^{D} \left\{ 1 - \sum_{j=1}^{i} \frac{1}{D-j+1} \right\} \qquad (8.6.2)$$

The test statistic is then

$$Q = \frac{\max |S_k|}{s\sqrt{D-1}} \qquad (8.6.3)$$

which under the null hypothesis has a limiting distribution of the supremum of the absolute value of a Brownian Bridge. For $Q > 1$ the p-value of the test is approximately equal to $2\exp\{-2Q^2\}$.

EXAMPLE 8.3   *(continued)*   In section 1.7 we discussed a trial of 863 kidney transplant patients. We would like to examine categorizing the patients into high- or low-risk groups based on their age at transplant. We shall look at separate analyses by race and sex.

Consider first the sample of 92 black males. Here the transplants occurred at 43 distinct ages, which are potential candidates for a cut point. There were 14 distinct death times, which gives $s^2 = 0.8268$. The maximum value of $| S_k |$ is at age 58 and $Q = 0.8029$, which gives a p-value of at least 0.30 (see Theoretical Note 3). This suggests that age is not related to outcome for black males.

The following table gives the results for other race and sex combinations. It also gives the estimated relative risk of the high-risk (Age ≥ cut-point) group compared to the low-risk group. Also presented are the results of a model which treats age continuously. Figure 8.2 depicts the estimates $| S_k |$ for each of the four sex and race combinations that are used to find the estimated cut point. Here we find close agreement between the discrete model for age and the continuous model.

TABLE 8.5
Age Cut Points for Kidney Transplant Patients

Race/Sex	Discrete Model for Age				Continuous Model for Age	
	Cut Point	Q	p-value	RR(95%CI)	b(SE)	p
Black/male	58	0.8029	> 0.30	2.3(0.5 − 10.4)	0.036(0.024)	0.14
White/male	41	3.1232	< 0.001	2.6(1.6 − 4.1)	0.060(0.010)	< 0.001
Black/female	48	0.9445	> 0.30	2.6(0.8 − 8.4)	0.034(0.026)	< 0.20
White/female	36	1.9310	0.001	4.4(1.9 − 10.6)	0.042(0.012)	< 0.001

Theoretical Notes

1. Wu (2001) shows that if a test is based on the best cut point without some adjustment for multiple testing then this test rejects too often when the null hypothesis is true.
2. The method discussed here, based on the score statistic, is due to Contal and O'Quigley (1999). An alternative method, due to Jespersen (1986), is also based on the supremum of the absolute value of the log rank tests. His variance is slightly different than that presented here and in a Monte Carlo study. Wu (2001) shows that this statistic's performance is not quite as good as the Contral and O'Quigley statistic.
3. The limiting distribution of Q under the null hypothesis is the same as the supremum of the absolute value of a Brownian bridge. The p-value can be found by

$$P[Q \geq q] = 2 \sum_{j=1}^{\infty} (-1)^{j+1} \exp\{-2j^2 q^2\}, \qquad (8.6.4)$$

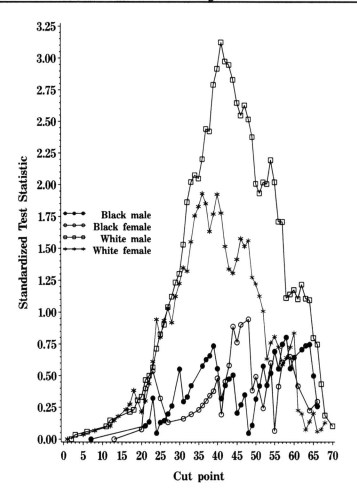

Figure 8.2

which is approximately equal to $2\exp\{-2q^2\}$ when $q > 1$. For $q \leq 1$ the p-value is at least 0.30.

Practical Notes

1. Estimation of the cut point can be performed by finding the cut point which maximizes any of the three statistics discussed in section 8.4, the Wald, Score, or likelihood ratio tests. All give approximately the correct answer.
2. Basing inference on an unadjusted Cox model with a binary covariate based on the cut-point model leads to tests which falsely reject the null hypothesis of treatment effect too often. Some correction must

be made to these tests to ensure that the overall level of inference is correct.

3. Estimation of the cut point can be performed by using existing software to identify the model which gives the maximum test statistic. The test discussed here requires additional calculations.

4. The Martingale residual plot discussed in section 11.3 can be used to check the appropriateness of discretizing a continuous covariate.

8.7   Model Building Using the Proportional Hazards Model

In an earlier example, we explored the modeling of age and stage on the survival of patients with laryngeal cancer. In many studies, a variety of explanatory factors are measured and a major question in analyzing such data sets is how to incorporate these factors in the modeling procedure.

The distinction between factors and variables is sometimes a bit vague although we shall refer to single-degree-of-freedom independent variables (such as age that is treated continuously) as either factors or variables, whereas we shall refer to multiple-degree-of-freedom independent variables (such as stage) as factors.

As mentioned at the beginning of this chapter, two distinctly different, yet important, problems in regression are i) to adjust for potential confounding (or explanatory) variables when one has a specific hypothesis in mind and the desire is to compare two or more groups with respect to survival times or ii) to predict the distribution of the time to some event from a list of explanatory variables with no particular prior hypothesis in mind. Utilizing the proportional hazards model introduced in section 2.6 and the testing procedures more fully explained in this chapter, we shall detail the approaches used for these two situations and illustrate them with two examples.

First, if one has a particular hypothesis in mind, then interest centers upon that particular hypothesis and any model building will be done to adjust that particular comparison (or comparisons) for other noncontrollable factors. Often, the other explanatory factors are simply viewed as adjusters or confounders and interest in them matters only insofar as they affect the assessment of the basic hypothesis. Examples of such possible confounders are demographic variables, such as age, gender, race, etc.; patient clinical variables at the onset of the clinical trial that may reflect the patient's condition, such as severity of disease, size of tumor, physiological variables, etc.; and, in the case of transplantation, characteristics of the donor.

The starting point of the model building process for this problem is to perform the global test of the primary hypothesis described in sections 8.3 and 8.4. This gives the investigator an impression of the simple, unadjusted relationship between the basic hypothesized factor and survival. In searching for possible confounders, it is useful to examine the simple relationship between the other explanatory factors and survival, adjusted for the factor of interest because, if there is obviously no relationship between a factor and survival, then, it is not likely to be a confounder. Thus, the next step is to consider the relationship between each of the other explanatory factors and survival, given that the factor stated in the basic hypothesis is already in the model. These local tests were described in detail in section 8.5. This process is continued by exploring the relationships between each of the remaining explanatory variables and survival, given that the factor stated in the basic hypothesis and the one next most related to survival (assuming that the basic variable is in the model) are in the model. If no significant confounders are found at any step in this process, then we stop and base our inference about the primary hypothesis on the last model. This approach is illustrated in the next example.

Another approach advocated as useful in model building, is one due to Akaike (1973) which examines the likelihood and the number of parameters included in the model. It attempts to balance the need for a model which fits the data very well to that of having a simple model with few parameters. More specifically, the Akaike information criterion (AIC), examines the statistic

$$\text{AIC} = -2 \log L + kp,$$

where p is the number of regression parameters in the model, k is some predetermined constant (which we shall take as 2), and L is the usual likelihood function. This criterion, will decrease as variables are added to the model. At some point, the criterion will increase which is a signal that the added variables are unnecessary. The AIC is reminiscent of the adjusted R^2 in least-squares regression, in that both are attempting to adjust the fit of the model by the number of parameters included. This criterion will also be recorded in the following example.

EXAMPLE 8.5   Continuing the discussion of the study of acute leukemia patients being given a bone marrow transplant, as introduced in section 1.3 and continued in Examples 4.2 and 7.5, we shall adjust the basic comparisons of the three risk groups, acute lymphoblastic leukemia (ALL), low-risk acute myeloctic leukemia (AML low-risk), and high- risk acute myeloctic leukemia (AML high-risk), so as to reduce the possible bias which may exist in making those comparisons (because this was not a randomized clinical trial). Because this chapter discusses only fixed-time covariates, we will use only the fixed-time covariates as possible confounders in making the comparisons among risk groups.

The first step in the model-building process is the global test of the hypothesis of no difference in disease-free survival. As discussed in section 8.2, we define two binary covariates ($Z_1 = 1$ if AML low-risk: $Z_2 = 1$ if AML high-risk) for the factor of interest. The global Wald chi-squared (with two degrees of freedom) statistic is 13.01 (p-value = 0.001). The AIC for this model is 737.29.

In this example, there are two sets of factors. The first set of factors is measured only on the patient. These are Z_3: waiting time from diagnosis to transplant, Z_4: indicator of FAB (French-American-British) classification M4 or M5 for AML patients, and Z_5: indicator of whether the patient was given a graft-versus-host prophylactic combining methotrexate (MTX) with cyclosporine and possibly methylprednisilone. Tests involving these factors will have one degree of freedom.

The second set of factors is based on combinations of patient and donor characteristics and cannot be described by a single covariate. These factors are sex ($Z_6 = 1$ if male donor, $Z_7 = 1$ if male recipient, and $Z_8 = Z_6 \times Z_7 = 1$ if donor and recipient are male), CMV status ($Z_9 = 1$ if donor is CMV positive, $Z_{10} = 1$ if recipient is CMV positive, and $Z_{11} = Z_9 \times Z_{10} = 1$ if donor and recipient are CMV positive), and age (Z_{12} = donor age $- 28$, Z_{13} = recipient age $- 28$, and $Z_{14} = Z_{12} \times Z_{13}$). Tests involving these factors will have three degrees of freedom.

Table 8.6 gives the local Wald tests for the six factors. Here, all models include the covariates Z_1 and Z_2 for the factor of primary interest. We find that the factor FAB classification (Z_4) has the smallest Akaike information criterion and the smallest p-value. This factor is added to the model. Table 8.7 gives the local Wald tests of all other factors not in the model with Z_1, Z_2, and Z_4 in the model. From this table, we see that the factor age (Z_{12}, Z_{13}, and Z_{14}) should be added to the model. Table 8.8 continues the model building by testing for factors not in the model, adjusted for risk group, FAB class, and age. In this table, we

TABLE 8.6

Local Tests for Possible Confounders, Adjusted for Risk Groups

Factor	Degrees of Freedom	Wald Chi-Square	p-Value	AIC
Waiting time (Z_3)	1	1.18	0.277	737.95
FAB class (Z_4)	1	8.08	0.004	731.02
MTX (Z_5)	1	2.03	0.155	737.35
Sex (Z_6, Z_7, Z_8)	3	1.91	0.591	741.44
CMV status (Z_9, Z_{10}, Z_{11})	3	0.19	0.980	743.10
Age (Z_{12}, Z_{13}, Z_{14})	3	11.98	0.007	733.18

see that all the local tests are nonsignificant and that the AIC is larger than that for the model with disease group, FAB class, and age alone. Thus, the model building process stops and the final model is given in Table 8.9. In this model, the local Wald test of the primary hypothesis of no difference between risk groups has a p-value of 0.003, which suggests that there is a difference in survival rates between at least two of the risk groups after adjustment for the patient's FAB class and for the donor and patient's age. Although we have used Wald tests in this

TABLE 8.7

Local Tests for Possible Confounders, Adjusted for Risk Groups and FAB Class

Factor	Degrees of Freedom	Wald Chi-Square	p-Value	AIC
Waiting time (Z_3)	1	1.18	0.277	731.68
MTX (Z_5)	1	2.05	0.152	731.06
Sex (Z_6, Z_7, Z_8)	3	0.92	0.820	736.11
CMV status (Z_9, Z_{10}, Z_{11})	3	0.02	0.999	737.00
Age (Z_{12}, Z_{13}, Z_{14})	3	13.05	0.004	725.98

TABLE 8.8

Local Tests for Possible Confounders, Adjusted for Risk Groups, FAB Class, and Age

Factor	Degrees of Freedom	Wald Chi-Square	p-Value	AIC
Waiting time (Z_3)	1	0.46	0.495	727.48
MTX (Z_5)	1	1.44	0.229	726.58
Sex (Z_6, Z_7, Z_8)	3	1.37	0.713	730.61
CMV status (Z_9, Z_{10}, Z_{11})	3	0.58	0.902	731.42

TABLE 8.9

Analysis of Variance Table for the Final Model for Bone Marrow Transplants

	Degrees of Freedom	b	SE(b)	Wald Chi-Square	p-Value
Z_1	1	−1.091	0.354	9.48	0.002
Z_2	1	−0.404	0.363	1.24	0.265
Z_4	1	0.837	0.279	9.03	0.003
Z_{12}	1	0.004	0.018	0.05	0.831
Z_{13}	1	0.007	0.020	0.12	0.728
Z_{14}	1	0.003	0.001	11.01	0.001

example, similar conclusions are obtained if the likelihood ratio statistic is used throughout.

The second situation, where regression techniques are useful, is in modeling the distribution of the time-to-some-event from a list of explanatory variables with no particular prior hypothesis in mind. Here, interest centers upon identifying a set of variables which will aid an investigator in modeling survival or identifying a set of variables which may be used in testing a hypothesis in some future study (hypothesis generating).

The starting point of model building for this problem is to perform separate global tests for each explanatory factor, as described in sections 8.3 and 8.4, so as to examine the simple relationship between the explanatory variables and survival. The purpose in this step is to ascertain which factor is most related to survival. The next step is to consider the relationship between each of the other explanatory factors (not the one identified as the most significant one) and survival, given that the factor identified as the most significant is already in the model. These local tests are also described in detail in section 8.5. This process is continued by exploring the relationship between each of the remaining explanatory factors and survival, assuming that the variable identified as the most significant one and the one next most related to survival (given the first variable is in the model) are already in the model. The p-value approach requires a significance level for entering variables into the model. This approach is illustrated in the next example. Furthermore, the Akaike information criterion may be used to assess the extent to which the investigator wishes to include variables into the model. This approach is especially useful for deciding how many variables to include.

EXAMPLE 8.6   In section 1.14 (see Example 5.4), a data set including times to weaning of breast-fed infants was described. In this example, we wish to find a model predictive of the distribution of time to weaning. Fixed-time factors measured by questionaire include race of mother (black, white, or other), poverty status indicator, smoking status of mother at birth of child, alcohol drinking status of mother at birth of child, age of mother at child's birth, education of mother at birth of child (less than high school, high school graduate, some college), and lack of prenatal care indicator (mother sought prenatal care after third month or never sought prenatal care).

In building a model, we are mainly interested in finding factors which contribute to the distribution of the time to weaning. Because there are many ties in this data set, we shall use the "discrete" likelihood for handling ties. Table 8.10 contains the results of the single-factor

TABLE 8.10

Global Tests for Each Factor Potentially Related to Weaning Time

Factor	Degrees of Freedom	Wald Chi-Square	p-Value	AIC
Race of mother	2	8.03	0.018	5481.67
Poverty at birth	1	0.71	0.399	5486.69
Smoking	1	10.05	0.002	5477.61
Alcohol	1	2.01	0.157	5485.48
Age	1	0.15	0.698	5487.26
Education	2	6.95	0.031	5482.36
No prenatal care	1	0.16	0.687	5487.25

TABLE 8.11

Local Tests for Each Factor Potentially Related to Weaning Time, Adjusted for Mother's Smoking Status

Factor	Degrees of Freedom	Wald Chi-Square	p-Value	AIC
Race of mother	2	12.38	0.002	5469.71
Poverty at birth	1	1.42	0.234	5478.17
Alcohol	1	1.04	0.307	5478.59
Age	1	0.01	0.954	5479.61
Education	2	3.87	0.145	5477.71
No prenatal care	1	0.02	0.888	5479.59

Wald tests. Race of mother, mother's smoking status, and education of mother are all significantly related to the time to weaning in the simple regressions. The most significant factor, mother's smoking status, is added to the model, and local tests for the remaining factors are given in Table 8.11. From this table, we add the race factor to the model and perform the local tests for the remaining factors (Table 8.12). In Table 8.12, we see that all the remaining risk factors are not significant at a 5 percent significance level. If model selection criterion is based on the p-value (< 0.05) of the local tests, we would stop at this point and take, as our final model, one with two factors, smoking status and race. The ANOVA Table for this model is given in Table 8.13A. Model building based on the AIC, however, suggests adding the poverty factor to the model because the AIC with this factor is smaller than that without the factor. Proceeding based on the AIC, we find that the AIC is increased when any other factor is added to a model with race, smoking, and poverty included as factors (table not shown). The ANOVA table for the final model, based on the AIC, is in Table 8.13B.

TABLE 8.12

Local Tests for Each Factor Potentially Related to Weaning Time, Adjusted for Mother's Smoking Status and Race

Factor	Degrees of Freedom	Wald Chi-Square	p-Value	AIC
Poverty at birth	1	2.99	0.084	5468.65
Alcohol	1	1.16	0.281	5470.58
Age	1	0.19	0.660	5471.51
Education	2	2.08	0.353	5471.60
No prenatal care	1	0.03	0.854	5471.67

TABLE 8.13A

ANOVA Table for the Time to Weaning Based on the p-Value Approach

	Degrees of Freedom	b	SE(b)	Wald Chi-Square	p-Value
Smoking	1	0.308	0.081	14.34	<0.001
Race–Black	1	0.156	0.111	1.98	0.159
Race–Other	1	0.350	0.102	11.75	<0.001

TABLE 8.13B

ANOVA Table for the Time to Weaning, Based on the AIC Approach

	Degrees of Freedom	b	SE(b)	Wald Chi-Square	p-Value
Smoking	1	0.328	0.082	15.96	<0.001
Race–Black	1	0.184	0.112	2.70	0.100
Race–Other	1	0.374	0.103	13.18	<0.001
Poverty	1	−0.163	0.094	2.99	0.084

Practical Notes

1. In the example, the stepwise model building was based on the Wald statistic. The choice of this statistic is arbitrary and either the score or likelihood ratio statistic could be used. For data sets with a large number of covariates, the score statistic may be more efficient in early steps of this process because high-dimensional models need not be fit at each step. Automated procedures which can be used, when all factors are a single covariate, are available in SAS using either the score or Wald statistic.

2. The model selection procedure discussed here is a forward selection procedure. An alternative model building procedure is a backward selection procedure which starts with the model with all factors, and, at each step, removes the least significant factor from the model. A stepwise model selection procedure combines the forward and backward procedures. All three procedures for single covariate factors are available in SAS.

3. The choice of k in the AIC reflects how conservative one wishes to be in model building (larger values of k will include fewer variables).

8.8 Estimation of the Survival Function

Once we have obtained estimates of the risk coefficients β from a proportional hazards regression model, it may be of interest to estimate the survival probability for a new patient with a given set of covariates $\mathbf{Z}_0$. The estimator of the survival function is based on Breslow's estimator of the baseline cumulative hazard rate derived in the Theoretical Notes of section 8.3.

To construct this estimator we, first, fit a proportional hazards model to the data and obtain the partial maximum likelihood estimators $\mathbf{b}$ and the estimated covariance matrix $\hat{\mathbf{V}}(\mathbf{b})$ from the inverse of the information matrix. Let $t_1 < t_2 < \cdots < t_D$ denote the distinct death times and d_i be the number of deaths at time t_i. Let

$$W(t_i; \mathbf{b}) = \sum_{j \in R(t_i)} \exp\left(\sum_{b=1}^{p} b_h Z_{jh}\right). \qquad (8.8.1)$$

The estimator of the cumulative baseline hazard rate $H_0(t) = \int_0^t h_0(u)\, du$ is given by

$$\hat{H}_0(t) = \sum_{t_i \leq t} \frac{d_i}{W(t_i; \mathbf{b})}, \qquad (8.8.2)$$

which is a step function with jumps at the observed death times. This estimator reduces to the Nelson–Aalen estimator of section 4.2, when there are no covariates present, and can be derived naturally using a profile likelihood construction (see Theoretical Note 2 of section 8.3). The estimator of the baseline survival function, $S_0(t) = \exp[-H_0(t)]$ is given by

$$\hat{S}_0(t) = \exp[-\hat{H}_0(t)]. \qquad (8.8.3)$$

This is an estimator of the survival function of an individual with a baseline set of covariate values, $\mathbf{Z} = \mathbf{0}$. To estimate the survival function for an individual with a covariate vector $\mathbf{Z} = \mathbf{Z}_0$, we use the estimator

$$\hat{S}(t \mid \mathbf{Z} = \mathbf{Z}_0) = \hat{S}_0(t)^{\exp(\mathbf{b}'\mathbf{z}_0)}. \tag{8.8.4}$$

Under rather mild regularity conditions this estimator, for fixed t, has an asymptotic normal distribution with mean $S(t \mid \mathbf{Z} = \mathbf{Z}_0)$ and a variance which can be estimated by

$$\hat{V}[\hat{S}(t \mid \mathbf{Z} = \mathbf{Z}_0)] = [\hat{S}(t \mid \mathbf{Z} = \mathbf{Z}_0)]^2[Q_1(t) + Q_2(t; \mathbf{Z}_0)]. \tag{8.8.5}$$

Here,

$$Q_1(t) = \sum_{t_i \leq t} \frac{d_i}{W(t_i, \mathbf{b})^2} \tag{8.8.6}$$

is an estimator of the variance of $\hat{H}_0(t)$ if $\mathbf{b}$ were the true value of $\boldsymbol{\beta}$. Here

$$Q_2(t; \mathbf{Z}_0) = \mathbf{Q}_3(t; \mathbf{Z}_0)' \hat{V}(\mathbf{b})\mathbf{Q}_3(t; \mathbf{Z}_0) \tag{8.8.7}$$

with $\mathbf{Q}_3$ the p-vector whose kth element is defined by

$$Q_3(t, \mathbf{Z}_0)_k = \sum_{t_i \leq t} \left[\frac{W^{(k)}(t_i; \mathbf{b})}{W(t_i; \mathbf{b})} - Z_{0k} \right] \left[\frac{d_i}{W(t_i, \mathbf{b})} \right], \quad k = 1, \ldots, p \tag{8.8.8}$$

where

$$W^{(k)}(t_i; \mathbf{b}) = \sum_{j \in R(t_i)} Z_{jk} \exp(\mathbf{b}'\mathbf{Z}_j)$$

Q_2 reflects the uncertainty in the estimation process added by estimating $\boldsymbol{\beta}$. Here, $Q_3(t, \mathbf{Z}_0)$ is large when $\mathbf{Z}_0$ is far from the average covariate in the risk set. Using this variance estimate, pointwise confidence intervals for the survival function can be constructed for $S(t \mid \mathbf{Z} = \mathbf{Z}_0)$ using the techniques discussed in section 4.3. Again, the log-transformed or arcsine-square-root-transformed intervals perform better than the naive, linear, confidence interval.

EXAMPLE 8.2   *(continued)*   We shall estimate the survival functions for survival after detection of laryngeal cancer based on the Cox regression model summarized in Table 8.1. Here, we wish to produce a curve for each of the four stages of disease. Because an adjustment for age is included in our model, we shall provide an estimate for a sixty-year-old male. The baseline survival function S_0, is estimated directly from Eqs. (8.8.2) and (8.8.3). The estimate of survival for a stage I cancer patent (of age 60 at diagnosis) is $S_0(t)^{\exp(0.0189 \times 60)}$; for a stage II patient $S_0(t)^{\exp(0.0189 \times 60 + 0.1386)}$; for a stage III patient $S_0(t)^{\exp(0.0189 \times 60 + 0.6383)}$; and for a stage IV patient $S_0(t)^{\exp(0.0189 \times 60 + 1.6931)}$. Figure 8.3 shows the estimates of the four survival curves.

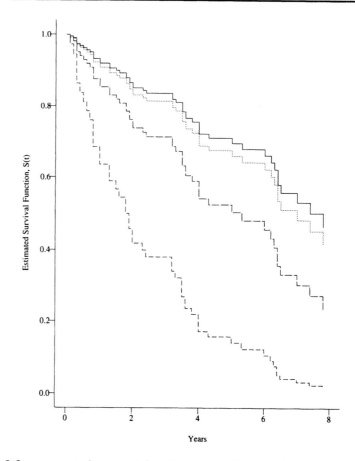

Figure 8.3 *Estimated survival functions for a 60 year old larynx cancer patient. Stage I Cancer (———) Stage II Cancer (------) Stage II Cancer (— — —) Stage IV (-- -- --)*

At five years, the estimated survival probabilities for a 60-year-old are 0.7031 for a Stage I patient, 0.6672 for a Stage II patient, 0.5132 for a Stage III patient, and 0.1473 for a Stage IV patient. Using Eqs. (8.8.5)–(8.8.8), we find that the standard errors of these estimators are 0.0737, 0.1059, 0.0949, and 0.0996, respectively. At 5 years, 95% confidence intervals for the survival function, based on the log transformation (4.3.2), are (0.5319, 0.8215), (0.4176, 0.8290), (0.3171, 0.6788), and (0.0218, 0.3834), for stages I, II, III, and IV, respectively.

Practical Notes

1. An alternative estimator of the baseline hazard rate has been proposed by Kalbfleisch and Prentice (1973). When there is at most a

single death at each time, this estimator is given by

$$\tilde{H}_0(t) = \sum_{t_i \le t} \left[1 - \left(1 - \frac{\delta_i \exp(\mathbf{b}'\mathbf{Z}_i)}{W(t_i; \mathbf{b})} \right)^{\exp(-\mathbf{b}'\mathbf{z}_i)} \right].$$

When there are ties between the death times, then, the estimator is found numerically by solving a system of equations. This statistic is used in SAS.

2. Alternative estimators of the baseline survival function are the product integral of the Breslow estimator of the cumulative hazard rate, which, in the case of no tied deaths, is given by

$$\hat{S}_2(t \mid 0) = \prod_{t_i \le t} \left(1 - \frac{\delta_i}{W(t_i; \mathbf{b})} \right),$$

or the product integral of the Kalbfleisch and Prentice estimator (see Note 1 above)

$$\hat{S}_3(t \mid 0) = \prod_{t_i \le t} \left[1 - \frac{\delta_i \exp(\mathbf{b}'\mathbf{Z}_i)}{W(t_i; \mathbf{b})} \right]^{\exp(-\mathbf{b}'\mathbf{z}_i)}.$$

Each of these can be used in Eq. (8.8.4) to obtain an estimator of the survival for an individual with a covariate vector $\mathbf{Z}_0$. An alternative estimator to $S(t \mid \mathbf{Z}_0)$ is given by first adjusting $\hat{H}_0(t)$ by the factor $\exp(\mathbf{b}'\mathbf{Z}_0)$ and, then, constructing a product-limit estimator based on $\hat{H}(t \mid \mathbf{Z}_0) = \hat{H}(t) \exp(\mathbf{b}'\mathbf{Z}_0)$ given by

$$\hat{S}_4(t \mid \mathbf{Z}_0) = \prod_{t_i \le t} \left[1 - \frac{\delta_i \exp(\mathbf{b}'\mathbf{Z}_0)}{W(t_i; \mathbf{b})} \right]$$

Under rather mild regularity conditions, each of the four estimators of $S(t \mid \mathbf{Z}_0)$ is asymptotically normal with the correct mean and a variance estimated by (8.8.5).

The estimators S_2 and S_4 can take negative values. This is only a problem when the risks sets are small in the right-hand tail of the estimates. Typically, this happens when one is attempting to predict survival for a covariate value which is extreme as compared to the covariate values of those remaining at risk when the prediction is being made. The negative value is a signal to the investigator that predictions for this covariate value should not be made in this region.

3. The SAS procedure PHREG uses the Kalbfleisch and Prentice estimator described in Note 1. A recent Monte Carlo study by Andersen and Klein (1996) shows that this estimator has a larger bias and mean-squared error than the other three estimators of survival. Breslow's estimator is also available in SAS.

4. S-Plus has both the Breslow and Kalbfleisch and Prentice estimator available in the function surv.fit.

5. As was the case for confidence intervals for the survival function discussed in Chapter 4, Andersen and Klein (1996) show that the log-transformed confidence interval for $S(t \mid \mathbf{Z}_0)$ seems to work best, and the arcsine-square-root confidence interval is a close second. The routine use of the linear confidence interval is not recommended.

6. Based on an extensive Monte Carlo study by Andersen and Klein (1996), it can be shown that the estimators S_2 and S_4 have the smallest bias and are recommended. The estimator S_3, available in SAS, seems to perform quite poorly for continuous and mixed, continuous covariate models.

8.9   Exercises

8.1   In section 1.10, times to death or relapse (in days) are given for 23 non-Hodgkin's lymphoma (NHL) patients, 11 receiving an allogenic (Allo) transplant from an HLA-matched sibling donor and 12 patients receiving an autologous (Auto) transplant. Also, data on 20 Hodgkin's lymphoma (HOD) patients, 5 receiving an allogenic (Allo) transplant from an HLA-matched sibling donor and 15 patients receiving an autologous (Auto) transplant is given.

(a) Treating NHL Allo as the baseline hazard function, state the appropriate coding which would allow the investigator to test for any difference in survival functions for the four groups, treating them as four independent groups.

(b) Treating NHL Allo as the baseline hazard function, state the appropriate coding which would allow the investigator to test for an interaction between type of transplant and disease type using main effects and interaction terms.

(c) Suppose that we have the following model for the hazard rates in the four groups:

$$h(t \mid \text{NHL Allo}) = h_0(t)$$

$$h(t \mid \text{HOD Allo}) = h_0(t)\exp(2)$$

$$h(t \mid \text{NHL Auto}) = h_0(t)\exp(1.5)$$

$$h(t \mid \text{HOD Auto}) = h_0(t)\exp(.5)$$

What are the risk coefficients, β_i, $i = 1, 2, 3$, for the interaction model in part b?

8.2   In section 1.6 a study is described which evaluates a protocol change in disinfectant practices in a large midwestern university medical center. Of primary interest in the study is a comparison of two methods of body cleansing. The first method, used exclusively from January 1983 to June 1984, consisted of a routine bathing care method (initial surface decontamination with 10% povidone-iodine followed with regular bathing with Dial soap). From June 1984 to the end of the study period in December 1985, body cleansing was initially performed using 4% chlorhexidine gluconate. Eighty-four patients were in the group who received the new bathing solution, chlorhexidine, and 70 patients served as the control group who received routine bathing care, povidone-iodine. Included in the data set is a covariate that measures the total surface area burned. The data is reported on our web site.

State the appropriate coding which would allow the investigator to test for:

(a) any difference in survival functions for the two groups.

(b) any difference in survival functions for the two groups adjusting for total area burned.

8.3   In section 1.11, a study was conducted on the effects of ploidy on the prognosis of patients with cancer of the tongue. Tissue samples were examined to determine if the tumor had a aneuploid or diploid DNA profile. Times to death for these two groups of patients are recorded in Table 1.6. To analyze this data create a single indicator variable, Z, which reflects the type of tumor.

(a) Find the p-value of a test of the hypothesis of no effect of ploidy on survival using the score test and the Breslow method of handling ties.

(b) Estimate β and its standard error using the Breslow method of handling ties. Find a 95% confidence interval for the relative risk of death of an individual with an aneuploid tumor as compared to an individual with a diploid tumor.

(c) Repeat (a) using the likelihood test. Compare your answer to that of part a.

(d) Repeat (a) using the Wald test. Compare your answer to those in parts a and c.

8.4   In Exercise 7 of Chapter 7, three different treatments were administered to rats who had F98 glioma cells implanted into their brains. The data for the three groups of rats lists the death times (in days) in that exercise. Create two dummy variables, $Z_1 = 1$ if animal is in the "radiation only" group, 0 otherwise; $Z_2 = 1$ if animal is in the "radiation plus BPA" group, 0 otherwise. Use the Breslow method of handling ties in the problems below.

(a) Estimate β_1 and β_2 and their respective standard errors. Find a 95% confidence interval for the relative risk of death of an animal radiated only compared to an untreated animal.

(b) Test the global hypothesis of no effect of either radiation or radiation plus BPA on survival. Perform the test using all the three tests (Wald, likelihood ratio, and score test).

(c) Test the hypothesis that the effect a radiated only animal has on survival is the same as the effect of radiation plus BPA (i.e., Test $H_0 : \beta_1 = \beta_2$).

(d) Find an estimate and a 95% confidence interval for the relative risk of death for a radiation plus BPA animal as compared to a radiated only animal.

(e) Test the hypothesis that any radiation given as a treatment (either radiation alone or with BPA) has a different effect on survival than no radiation. Use the likelihood ratio test.

(f) Repeat part (e) using a Wald test.

8.5 Using the data set in Exercise 1, using the Breslow method of handling ties,

(a) Analyze the data by performing a global test of no effect of group as defined in Exercise 8.1(a) on survival. Construct an ANOVA table to summarize estimates of the risk coefficients and the results of the one degree of freedom tests for each covariate in the model.

(b) Repeat part (a) using the coding as described in Exercise 8.1(b). Furthermore, test the hypothesis of disease type by transplant interaction using a likelihood ratio test based on this coding. Repeat using the Wald test.

(c) Find point estimates and 95% confidence intervals for the relative risk of death for an NHL Auto transplant patient as compared to an NHL Allo transplant patient.

(d) Find the p-value of a test of the hypothesis that the hazard rates are the same for HOD Allo transplants and NHL Allo patients, using the Wald test. Repeat a similar test for Auto patients.

(e) Test the hypothesis, using the Wald test, that the hazard rates for Auto transplant and Allo transplant patients are the same for each disease group against the alternative that the hazard rates for Auto transplant and Allo transplant patients for at least one group are different using a two-degree of freedom test of $H_0 : h(t \mid \text{NHL Allo}) = h(t \mid \text{NHL Auto})$ *and* $H_0 : h(t \mid \text{HOD Allo}) = h(t \mid \text{HOD Auto})$.

8.6 In section 1.13, data on the time to hospitalization of pneumonia in young children was discussed. The data is presented on our web site. In the sample there were 3,470 annual personal interviews. An investigator is interested in assessing race, poverty status, and their interaction on time to hospitalization of pneumonia. Use the discrete method for handling ties to answer the following questions.

(a) Estimate the parameters of your model and their standard errors. Construct and interpret an "ANOVA" table for this model.

(b) Provide point estimates and 95% confidence intervals for the relative risk of hospitalization for pneumonia for a person raised in poverty relative to a person not raised in poverty for each race.

(c) Test that blacks raised in poverty have a different hospitalization for pneumonia rate than whites not raised in poverty.

8.7   In section 1.6 a study is described which evaluates the relationship of various covariates to staphylococcus infection in a large midwestern university medical center (see Exercise 8.2). One of the covariates recorded in the data set is the total surface area burned. Use Breslow's method for handing ties to answer the following questions.

(a) Find the optimal cutpoint to categorize patients into high- or low-risk groups for staphylococcus infection based on their total surface area burned for each disinfectant practice.

(b) Test the hypothesis that there is a difference in times to infection for high- and low-risk groups using the cutpoints obtained in (a). Using the cut points obtained in (a) find the relative risk of the high-risk group compared to the low-risk group for each disinfectant practice.

(c) Analyze the data using total surface area burned as a continuous variable. Give the parameter estimate, standard error, and relative risk for total surface area burned. Compare with the answer in (b).

8.8   In section 1.3, data gathered from a multicenter trial of patients in three groups (ALL, AML low-risk, and AML high-risk) was followed after transplantation until relapse, death, or end of study. One of the covariates recorded in the data set is the waiting time to transplant (in days). Use Breslow's method for handling ties in the following.

(a) You are asked to categorize patients into high- or low-risk groups for disease-free survival based on the waiting time to transplant variable for the ALL group.

(b) Analyze the data using waiting time to transplant as a categorized variable using the cut point obtained in (a). Give the parameter estimate, standard error, and relative risk of the high-risk group compared to the low-risk group for the ALL group.

(c) Analyze the data using waiting time to transplant as a continuous variable. Give the parameter estimate, standard error, and relative risk for waiting time to transplant for the ALL group. Compare with answer in (b).

8.9   Use the Breslow method for handling ties and the Wald test in the following.

(a) Using the data set in section 1.6, test the hypothesis that the distributions of the times to staphylococcus infection are the same in the two disinfectant groups.

(b) Test the hypothesis that the distributions of the times to staphylococcus infection are the same in the two disinfectant groups adjusting for the total area burned, Z_4. Compare your results to those in part a.

(c) Also available in the data set is information on other factors that may be associated with the timing of staphylococcus infection. Some of these factors are gender, race, total surface area burned, and type of burn (chemical, scald, electrical, flame). For each factor create a set of fixed-time covariates. Test the hypothesis that the times to staphylococcus infection are the same for the two disinfectant groups using a model which adjusts for each of these factors.

(d) Since one is primarily interested in comparing the two bathing solutions, interest will center upon building a model with the view of testing that particular comparison adjusting for the other non-controllable factors in part (c). Using a forward selection approach, build such a model using the p-value approach. Based on the final model, test the hypothesis of primary interest.

8.10 In section 1.3, several event times are described for patients receiving a bone marrow transplant for leukemia. Consider the time to development of acute graft-versus-host disease (AGVHD). As a prophylactic treatment, patients at two of the hospitals were given a treatment combining methotrexate (MTX) with cyclosporine and possibly methylprednisilone. Patients at the other hospitals were not given methotrexate but rather a combination of cyclosporine and methylprednisilone. Of primary interest in studying AGVHD is a test of the effectiveness of the MTX regime to prevent AGVHD. Use Breslow's method for handling ties to answer the following exercises.

(a) Using an appropriate Cox model test the hypothesis of no difference in the rate of development of AGVHD between MTX and no MTX patients. Find a point estimate and a 95% confidence interval for the relative risk of AGVHD for patients on the MTX protocol as compared to those not given MTX.

(b) Patients were also grouped into risk categories based on their status at the time of transplantation. These categories were as follows: acute lymphoblastic leukemia (ALL) with 38 patients and acute myeloctic leukemia (AML). The latter category was further subdivided into low-risk—first remission (54 patients) and high-risk—second remission or untreated first relapse or second or greater relapse or never in remission (45 patients). Test the hypothesis of interest (no effect of MTX on development of AGVHD) adjusting for the three disease categories.

(c) Test for the possibility of an interaction effect on AGVHD between the disease categories and the use MTX.

(d) Using the factors of age, sex, CMV status, FAB class, waiting time to transplant, and disease category as defined in Example 8.5, find the best model to test the primary hypothesis of no MTX effect on the occurrence of AGVHD. Test the primary hypothesis and find an estimate of the relative risk of occurrence of AGVHD for an MTX patient as compared to a non-MTX patient.

8.11   In section 1.13, data gathered from annual personal interviews conducted for the National Longitudinal Survey of Youth (NLSY) from 1979 through 1986 was presented. This data was used to study whether or not the mother's feeding choice protected the infant against hospitalized pneumonia in the first year of life. Ages of young children at the time they were hospitalized with pneumonia were recorded as well as the observed ages of those infants that were not hospitalized with pneumonia during the study period. The data is available from our web site, which can be reached via the authors' pages at http://www.springer-ny.com. Use the discrete method for handling ties in the following.

(a) Consider the dummy variable $Z = 1$ if infants were breast fed at birth, 0 if infants were never breast fed, and test the hypothesis $H_0 : \beta = 0$, i.e., the survival functions for the two types of breast feeding are equal, using the score, likelihood ratio, and Wald tests. Find the estimate of β, b, the standard error of b, and the relative risk using the Wald test.

(b) Also available in the data set is information on other factors that may be associated with the timing of hospitalized pneumonia. These factors are age of the mother at the infant's birth, rural-urban environment of the mother, use of alcohol by the mother (no drinks, less than one drink, 1–2 drinks, 3–4 drinks, or more than 4 drinks per month), mother's cigarette use (none, less than 1 pack/day, 1 or more pack/day), region of country (northeast, north central, south, or west), birthweight of infant (less the 5.5 lbs or 5.5 lbs or more), poverty status of mother (yes/no), race of mother (white, black, or other), or number of siblings of infant. For each factor create a set of fixed-time covariates. Test the hypothesis that the times to hospitalized pneumonia are the same for the two feeding groups adjusting for each of these factors in a separate model using the Wald test.

(c) Since one is primarily interested in comparing the two types of breast feeding, interest will center upon building a model with the view of testing the particular comparison of interest adjusting for the other noncontrollable fixed covariates in part b. Build such a model using the AIC approach and the Wald test.

(d) Summarize your findings from this data set.

8.12   A major problem in certain sub-populations is the occurrence of sexually transmitted diseases (STD). Even if one ignores the lethal effects of the acquired immune deficiency syndrome, other STD's still have a significant impact on the morbidity of the community. Two of these STD's are the focus of this investigation—gonorrhea and chlamydia. Both of these diseases can be prevented and effectively treated. The purpose of the study described in section 1.12 is to identify those factors which are related to time until reinfection by either gonorrhea or chlamydia given a patient with an initial infection of gonorrhea or chlamydia. The data for this study is available from our web site.

Possible factors related to reinfection are the individual's race (black/white), marital status (divorced/separated, married, single), age at time of initial infection, years of schooling, initial infection type (gonorrhea, chlamydia, both), number of partners within the last 30 days, oral sex within the last year, rectal sex within the past year, presence of symptoms (abdominal pain, discharge, dysuria, itch, lesion, rash, lymph node involvement), and condom use. If the factors that are related to a greater risk of reinfection can be identified, then interventions could be targeted to those individuals who are at greatest risk for reinfection. Use regression techniques to find those factors which are most predictive of the distribution of the time until reinfection from this list of fixed explanatory factors with no particular prior hypothesis in mind. Build such a model using the p-value approach. Use the Breslow method for handling ties and the Wald test in the model building.

8.13 Find 95% confidence intervals for the survival functions for the two bathing solutions at 20 days for a patient with 25% of total surface area of body burned, using data in Section 1.6.

8.14 (a) Estimate the survival functions of the time to AGVHD for the MTX and no MTX treatment groups discussed in Exercise 8.10, adjusted for disease category. Provide a separate estimate for each disease group.

(b) Find 95% confidence intervals for the survival functions for the two patient treatment groups at 80 days for AML high-risk patients.

9

Refinements of the Semiparametric Proportional Hazards Model

9.1 Introduction

In Chapter 8, we modeled the hazard function for an individual as a function of fixed-time covariates. These are explanatory variables recorded at the start of the study whose values are fixed throughout the course of the study. For instance, in Example 8.5, where acute leukemia patients were given a bone marrow transplant, we considered the three risk groups, donor age, recipient age, and several other variables, as fixed-time covariates. The basic interest there was to evaluate the relationship of the risk groups to the hazard of relapse or death, controlling for possible confounding variables which might be related to relapse or death. As is typical in many survival studies, individuals are monitored during the study, and other explanatory variables are recorded whose values may change during the course of the study. Some of these variables may be instrumental in predicting survival and need to be taken into consideration in evaluating the survival distribution. Such variables which change over time are called time-dependent variables. A covariate that takes on the value 0 until some intermediate event occurs when it becomes 1 is an example of a discrete time-dependent covariate. It is also possible to include time-dependent covariates that are essentially continuous where the value of the covariate is a series of measurements of some explanatory characteristic. Examples of this type of covariate might be blood pressure, cholesterol, body mass index, size of the tumor, or rate of change in the size of the tumor recorded at different times for a patient. Section 9.2 will present methods which detail how these variables may be evaluated for their impact on survival.

As before, let X denote the time to some event and $\mathbf{Z}(t) = [Z_1(t), \ldots, Z_p(t)]'$ denote a set of covariates or risk factors at time t which may effect the survival distribution of X. Here the $Z_k(t)$'s may be time-dependent covariates, whose value changes over time or they may be constant (or fixed) values known at time 0, as we have discussed in Chapter 8. For time-dependent covariates, we assume that their value is predictable in the sense that the value of the covariate is known at an instant just prior to time t. The basic model due to Cox (1972) is as in (8.1.2) with $\mathbf{Z}$ replaced by $\mathbf{Z}(t)$ and, for the commonly used model,

$$h[t \mid \mathbf{Z}(t)] = h_o(t) \exp[\boldsymbol{\beta}^t \mathbf{Z}(t)] = h_o(t) \exp \left[\sum_{k=1}^{p} \beta_k Z_k(t) \right]. \qquad (9.1.1)$$

A common use of time-dependent covariates is for testing the proportional hazards assumption. Here a new covariate is created which incorporates a time variable into the relative risk formulation. Section 9.2 discusses details of this application of time-dependent covariates.

If the proportional hazard assumption is violated for a variable, then, one approach to dealing with this problem is to stratify on this variable. Stratification fits a different baseline hazard function for each stratum, so that the form of the hazard function for different levels of this variable is not constrained by their hazards being proportional. It is assumed, however, that the proportional hazards model is appropriate within strata for the other covariates. Usually one assumes the same β's for the other variables in each stratum. Details of this approach are given in section 9.3.

The basic proportional hazards model can be extended quite easily to allow for left-truncated survival data. These extensions are discussed in section 9.4. In section 9.5 we see how these methods can be used to analyze multistate survival data. By combining the notions of time-dependent covariates along with left-truncated regression models, it is possible to develop predicted survival probabilities for a patient, given the patient's history at some time. This prediction changes as more and more of the patient's history is observed. This approach is illustrated by the bone marrow transplant experiment first presented in section 1.3.

9.2   Time-Dependent Covariates

In this section, our data, based on a sample of size n, consists of the triple $[T_j, \delta_j, [\mathbf{Z}_j(t), 0 \leq t \leq T_j]]$, $j = 1, \ldots, n$ where T_j is the time on study for the jth patient, δ_j is the event indicator for the jth patient ($\delta_j = 1$ if event has occurred, 0 if the lifetime is right-censored) and $\mathbf{Z}_j(t) = [Z_{j1}(t), \ldots, Z_{jp}(t)]^t$ is the vector of covariates for the jth individual. For the covariate process, we assume that the value of $\mathbf{Z}_j(t)$ is known for any time at which the subject is under observation. As in Chapter 8, we assume that censoring is noninformative in that, given $\mathbf{Z}_j(t)$, the event and censoring time for the jth patient are independent. If the event times are distinct and $t_1 < t_2 < \cdots < t_D$ denotes the ordered event times, $\mathbf{Z}_{(i)}(t_i)$ is the covariate associated with the individual whose failure time is t_i and $R(t_i)$ is the risk set at time t_i (that is, $R(t_i)$ is the set of all individuals who were still under study at a time just prior to t_i), then, the partial likelihood as described by (8.2.1) is given by

$$L(\beta) = \prod_{i=1}^{D} \frac{\exp\left[\sum_{b=1}^{p} \beta_b Z_{(i)b}(t_i)\right]}{\sum_{j \in R(t_i)} \exp\left[\sum_{b=1}^{p} \beta_b Z_{jb}(t_i)\right]} \tag{9.2.1}$$

based on the hazard formulation (9.1.1). Estimation and testing may proceed as in Chapter 8 with the appropriate alterations of $\mathbf{Z}$ to $\mathbf{Z}(t)$. If ties are present, then, generalizations of the partial likelihoods described in section 8.4 may be used.

We shall illustrate the use of time-dependent covariates in the following example which is a continuation of Example 8.5.

EXAMPLE 9.1   In Chapter 8, we examined the relationship between disease-free survival and a set of fixed-time factors for patients given a bone marrow transplant. In addition to the covariates fixed at the time of transplant, there are three intermediate events that occur during the transplant recovery process which may be related to the disease-free survival time of a patient. These are the development of acute graft-versus-host disease (aGVHD), the development of chronic graft-versus-host disease (cGVHD) and the return of the patient's platelet count to a self-sustaining level (platelet recovery). The timing of these events, if they occur, is random. In this example, we shall examine their relationship to the disease-free survival time and see how the effects of the fixed covariates change when these intermediate events occur. As in the case of fixed factors, we shall make adjustments for these factors in the light of the primary comparison of interest, the potential differences in leukemia-free survival among the risk groups.

Each of these time-dependent variables may be coded as an indicator variable whose value changes from 0 to 1 at the time of the occurrence of the intermediate event. We define the covariates as follows:

$$Z_A(t) = \begin{cases} 0 & \text{if } t < \text{time at which acute graft-versus-host disease occurs} \\ 1 & \text{if } t \geq \text{time at which acute graft-versus-host disease occurs} \end{cases}$$

$$Z_P(t) = \begin{cases} 0 & \text{if } t < \text{time at which the platelets recovered} \\ 1 & \text{if } t \geq \text{time at which the platelets recovered} \end{cases}$$

and

$$Z_C(t) = \begin{cases} 0 & \text{if } t < \text{time at which chronic graft-versus-host disease occurs} \\ 1 & \text{if } t \geq \text{time at which chronic graft-versus-host disease occurs} \end{cases}$$

Because the interest in this example is in eliminating possible bias in comparing survival for the three risk groups, local tests may be performed to assess the significance for each time-dependent covariate in a model that already has covariates for the two risk groups included. As in Chapter 8, we define $Z_1 = 1$ if AML low-risk; $Z_2 = 1$ if AML high-risk, and we fit a separate Cox model for each of the three intermediate events which include the disease factor (Z_1, Z_2). The likelihood ratio chi-squared statistics (and the associated p-values) of the local tests that the risk coefficient β is zero for the time-dependent covariate are $X^2 = 1.17$ ($p = 0.28$) for $Z_A(t)$, 0.46 ($p = 0.50$) for $Z_C(t)$, and 9.64 ($p = 0.002$) for $Z_P(t)$. A summary of the coefficients, standard errors, Wald chi-square statistics and Wald p-values appears in Table 9.1 for each of the three regressions.

Here, we see that only the return to a self-sustaining level of the platelets has a significant impact on disease-free survival. The negative value of b_P suggests that a patient whose platelets have recovered at a given time has a better chance of survival than a patient who, at that time, has yet to have platelets recover. The relative risk of $\exp(-1.1297) = 0.323$ suggests that the rate at which patients are relapsing or dying after their platelets recover is about one-third the rate prior to the time at which their platelets recover.

TABLE 9.1

Time Dependent Variables and the Results of Univariate Proportional Hazards Regression in Comparing Risk Groups in Bone Marrow Transplant Study

	Degrees of Freedom	b	SE(b)	Wald Chi Square	p-Value
Z_1	1	-0.5516	0.2880	3.669	0.0554
Z_2	1	0.4338	0.2722	2.540	0.1110
$Z_A(t)$	1	0.3184	0.2851	1.247	0.2642
Z_1	1	-0.6225	0.2962	4.4163	0.0356
Z_2	1	0.3657	0.2685	1.8548	0.1732
$Z_C(t)$	1	-0.1948	0.2876	0.4588	0.4982
Z_1	1	-0.4962	0.2892	2.9435	0.0862
Z_2	1	0.3813	0.2676	2.0306	0.1542
$Z_P(t)$	1	-1.1297	0.3280	11.8657	0.0006

In the next example, we will continue the model building process, started in Example 8.5 with fixed-time covariates, by incorporating time-dependent covariates into the study of leukemia patients being given a bone marrow transplant. The basic strategy is the same as discussed in section 8.7.

EXAMPLE 9.1 *(continued):* In Example 8.5, using a forward stepwise model building procedure, we found that the factors FAB class (Z_3: AML with FAB Grade 4 or 5) and age (Z_4: Patient age -28; Z_5: Donor age -28; $Z_6 = Z_4 \times Z_5$), were key explanatory factors for disease-free survival when comparing risk groups (Z_1: AML low-risk; Z_2: AML high-risk) to explain disease-free survival after a bone marrow transplant. In the previous example, we found that the time-dependent covariate, $Z_P(t)$, which indicates whether the patient's platelets have returned to a self-sustaining level, was an important time-dependent factor in making this comparison. A natural question is whether these factors are still significantly related to disease-free survival in a model that includes both fixed and time-dependent factors. To test for this, we fitted three proportional hazards models, the first with the fixed factors of FAB class and age, the second with $Z_p(t)$, and the third, a combined model with both the fixed and time-dependent factors. The disease type factor is included in each of the models. The results of these three regressions are summarized in Table 9.2.

Using these results, we see that a local likelihood ratio test of no time-dependent covariate effect (adjusting for all fixed effects) has a chi square of $-2[-356.99 - (-353.31)] = 7.36$ with one degree of freedom ($p = 0.0067$) whereas the local likelihood ratio test of no FAB or age

TABLE 9.2
Fixed Factor Model, Time Dependent Factor Model and Combined Model for the BMT Example

	Fixed Factors Only			Time-Dependent Factor			All Factors		
	b	SE(b)	p-Value	b	SE(b)	p-Value	b	SE(b)	p-Value
Z_1	-1.091	0.354	0.002	-0.496	0.289	0.086	-1.032	0.353	0.004
Z_2	-0.404	0.363	0.265	0.381	0.267	0.154	-0.415	0.365	0.256
Z_3	0.837	0.279	0.003	–	–	–	0.813	0.283	0.004
Z_4	0.007	0.020	0.728	–	–	–	0.009	0.019	0.626
Z_5	0.004	0.018	0.831	–	–	–	0.004	0.018	0.803
Z_6	0.003	0.001	0.001	–	–	–	0.003	0.001	0.002
$Z_P(t)$	–	–	–	-1.130	0.328	0.001	-0.996	0.337	0.003
ln likelihood		-356.99			-361.82			-353.31	

factor adjustment has a chi square of $-2[-361.82 - (-353.31)] = 17.02$ with four degrees of freedom ($p = 0.0019$). Clearly, both the fixed-time and time-dependent factors should be adjusted for when comparing risk groups.

Next, we examine the relationships between the time-dependent factor and the fixed time factors. We define an additional set of time-dependent covariates that represent interactions between the timing of the return of the platelets to normal levels and the fixed-time covariates. The factors to be considered are as follows:

Fixed-Time Main Effect Factors

Risk group factor: (Z_1: AML low-risk; Z_2: AML high risk)

FAB factor: (Z_3: AML with FAB Grade 4 or 5)

Age factor (Z_4: Patient age -28; Z_5: Donor age -28; $Z_6 = Z_4 \times Z_5$)

Time-Dependent Main Effect Factor:

Platelet recovery factor [$Z_P(t)$]

Time-Dependent Interaction Factors

Risk group $\times$ Platelet recovery factor: ($Z_7(t) = Z_1 \times Z_P(t)$; $Z_8(t) = Z_2 \times Z_P(t)$)

FAB $\times$ Platelet recovery factor: ($Z_9(t) = Z_3 \times Z_P(t)$)

Age $\times$ Platelet recovery factor: ($Z_{10}(t) = Z_4 \times Z_P(t)$; $Z_{11}(t) = Z_5 \times Z_P(t)$; $Z_{12}(t) = Z_6 \times Z_P(t)$)

Note that, in this model, $\exp(\beta_1)$, for example, is the relative risk of death or relapse for an AML low-risk patient as compared to an ALL patient, and $\exp\{\beta_7\}$ is the excess relative risk between these two groups when the patient's platelets return to a normal level, that is, $\exp(\beta_1)$ is the relative risk of these two groups before platelet recovery and $\exp\{\beta_1 + \beta_7\}$ is the relative risk after platelet recovery.

To determine which of the time-dependent interaction factors should be included in the final model, we shall use a forward, stepwise selection procedure. Each model will include the three fixed-time factors and the platelet recovery factor. Here, we will base the inference on the likelihood ratio test although one would get the same final result using the Wald test. The results of this procedure are summarized in Table 9.3.

This analysis suggests that the three interaction terms between the fixed factors and the time-dependent covariate should be included in

TABLE 9.3

Likelihoods And Likelihood Ratio Tests for the Inclusion of Interactions Between Fixed Effects and the Time of Platelet Recovery

Factors in Model	Log Likelihood	Likelihood Ratio X^2	DF of X^2	p-Value
Group, FAB, age, $Z_p(t)$	−353.31			
Group, FAB, age, $Z_p(t)$, group × $Z_p(t)$	−349.86	6.89	2	0.0318
Group, FAB, age, $Z_p(t)$, FAB × $Z_p(t)$	−351.64	3.33	1	0.0680
Group, FAB, age, $Z_p(t)$, age × $Z_p(t)$	−349.36	7.90	3	0.0482

Group × $Z_p(t)$ Added to Model

Factors in Model	Log Likelihood	Likelihood Ratio X^2	DF of X^2	p-Value
Group, FAB, Age, $Z_p(t)$ Group × $Z_p(t)$,FAB × $Z_p(t)$	−347.78	4.15	1	0.0416
Group, FAB, Age, $Z_p(t)$ Group × $Z_p(t)$, Age × $Z_p(t)$	−343.79	12.14	3	0.0069

Age × $Z_p(t)$ Added to Model

Factors in Model	Log Likelihood	Likelihood Ratio X^2	DF of X^2	p-Value
Group, FAB, Age, $Z_p(t)$,Group × $Z_p(t)$ FAB × $Z_p(t)$, Age × $Z_p(t)$	−341.521	4.53	1	0.0333

FAB × $Z_p(t)$ Added To Model

the model. The ANOVA table for the final model is given in Table 9.4. Some care must be taken in interpreting these covariates. For example, here, we see that the relative risk of treatment failure (death or relapse) before platelet recovery for an AML low-risk patient compared to an ALL patient is $\exp(1.3073) = 3.696$ and a 95% confidence interval for the relative risk is $\exp(1.3073 \pm 1.96 \times 0.8186) = [0.74, 18.39]$. The risk of treatment failure after platelet recovery for an AML low-risk patient relative to an ALL patient is $\exp(1.3073 + (-3.0374)) = 0.18$. The standard error of the estimate of the risk coefficient after platelet recovery, $b_1 + b_7$ is $[V(b_1) + V(b_7) + 2\,\mathrm{Cov}(b_1, b_7)]^{1/2} = [0.6701 + 0.8570 + 2(-0.6727)]^{1/2} = 0.4262$, so a 95% confidence interval for the relative risk of treatment failure after platelet recovery for an AML low-risk patient is $\exp(-1.7301 \pm 1.96 \times 0.4262) = [0.08, 0.41]$. This suggests that the difference in outcome between the AML low-risk patients and the ALL patients is due to different survival rates after the platelets recover and that, prior to platelet recovery, the two risk groups are quite similar.

TABLE 9.4

ANOVA Table for a Model With Fixed Factors, Time to Platelet Recovery, and Their Interactions

	Degrees of Freedom	b	SE(b)	Wald Chi Square	p-Value
Z_1: AML low risk	1	1.3073	0.8186	2.550	0.1103
Z_2: AML high risk	1	1.1071	1.2242	0.818	0.3658
Z_3: AML with FAB Grade 4 or 5	1	-1.2348	1.1139	1.229	0.2676
Z_4: Patient age -28	1	-0.1538	0.0545	7.948	0.0048
Z_5: Donor age -28	1	0.1166	0.0434	7.229	0.0072
$Z_6 = Z_4 \times Z_5$	1	0.0026	0.0020	1.786	0.1814
$Z_P(t)$: Platelet Recovery	1	-0.3062	0.6936	0.195	0.6589
$Z_7(t) = Z_1 \times Z_P(t)$	1	-3.0374	0.9257	10.765	0.0010
$Z_8(t) = Z_2 \times Z_P(t)$	1	-1.8675	1.2908	2.093	0.1479
$Z_9(t) = Z_3 \times Z_P(t)$	1	2.4535	1.1609	4.467	0.0346
$Z_{10}(t) = Z_4 \times Z_P(t)$	1	0.1933	0.0588	10.821	0.0010
$Z_{11}(t) = Z_5 \times Z_P(t)$	1	-0.1470	0.0480	9.383	0.0022
$Z_{12}(t) = Z_6 \times Z_P(t)$	1	0.0001	0.0023	0.003	0.9561

A major use of time-dependent covariate methodology is to test the proportional hazards assumption. To test the proportionality assumption for a fixed-time covariate Z_1, we artificially create a time-dependent covariate, $Z_2(t)$, defined as

$$Z_2(t) = Z_1 \times g(t). \tag{9.2.2}$$

Here, $g(t)$ is a known function of the time t. In most applications, we take $g(t) = \ln t$. A proportional hazards model is fit to Z_1 and $Z_2(t)$ and the estimates of β_1 and β_2 along with the local test of the null hypothesis that $\beta_2 = 0$ is obtained. Under this proportional hazards model, the hazard rate at time t is $h(t \mid Z_1) = h_o(t) \exp[\beta_1 Z_1 + \beta_2(Z_1 \times g(t))]$, so if we compare two individuals with distinct values of Z_1, the ratio of their hazard rates is

$$\frac{h[t \mid Z_1]}{h[t \mid Z_1^*]} = \exp\{\beta_1[Z_1 - Z_1^*] + \beta_2 g(t)[Z_1 - Z_1^*]\},$$

which depends on t if β_2 is not equal to zero. (Compare this to (8.1.3) where the proportional hazards assumption holds.) Thus, a test of $H_o : \beta_2 = 0$ is a test for the proportional hazards assumption. The ability of this test to detect nonproportional hazards will depend on the choice of $g(t)$. This method will be illustrated in the following examples.

EXAMPLE 9.2

In Example 8.2, a proportional hazards model, with a single covariate Z_1 denoting the placement of a catheter either percutaneously ($Z_1 = 1$) or surgically ($Z_1 = 0$), was fit to the time to first exit-site infection (in months) in patients with renal insufficiency. In Figure 8.1, a graphical check of the proportional hazards assumption was made which casts doubt on the assumption of proportional hazards between the event times for the two types of catheters. Here, we will formally test this assumption employing the methodology of time-dependent covariates. To perform the test, we define $Z_2(t) = Z_1 \times \ln t$ and fit the Cox model with covariates Z_1 and $Z_2(t)$. Thus the relative risk of an individual with a percutaneously placed catheter compared to a surgically placed catheter is given by

$$h(t \mid Z_1 = 1)/h(t \mid Z_1 = 0) = \exp(\beta_1 + \beta_2 \ln t) = t^{\beta_2} \exp(\beta_1),$$

which is a constant only if $\beta_2 = 0$. This is the rationale for testing the local hypothesis $H_o : \beta_2 = 0$ to check the proportional hazards assumption.

The likelihood ratio statistic (and associated p-value) for this local test is 12.22 ($p = 0.0005$). The Wald chi-squared statistic for this local test is $(-1.4622)^2/0.345 = 6.19$ (p-value $= 0.013$). Thus, the evidence is strong that the hazards are not proportional, and, hence, the statistical model in Example 8.2 needs to be modified accordingly.

EXAMPLE 9.1

(continued): We shall illustrate the testing of the proportionality hazards assumption for the fixed-time factors used in Example 8.5. As in that example, we create fixed-time covariates for the patient's disease status ($Z_1 = 1$ if AML low-risk: $Z_2 = 1$ if AML high-risk); waiting time

from diagnosis to transplant (Z_3); FAB classification ($Z_4 = 1$ if M4 or M5 for AML patients); use of graft-versus-host prophylactic combining methotrexate ($Z_5 = 1$ if MTX used); and for the combined patient and donor characteristics including sex ($Z_6 = 1$ if male donor; $Z_7 = 1$ if male recipient; $Z_8 = Z_6 \times Z_7 = 1$ if donor and recipient are male); CMV status ($Z_9 = 1$ if donor is CMV positive; $Z_{10} = 1$ if recipient is CMV positive; $Z_{11} = Z_9 \times Z_{10} = 1$ if donor and recipient are CMV positive); and age ($Z_{12} =$ donor age $- 28$; $Z_{13} =$ recipient age $- 28$; $Z_{14} = Z_{12} \times Z_{13}$).

For each factor, we create a set of time-dependent covariates of the form $Z_{i+14}(t) = Z_i \times \ln t$. To check the proportional hazards assumption, we fit separate models for each factor which include the fixed values of covariates constituting the factor and the artificial time-dependent covariates created from these fixed-time covariates. A local test is then performed of the hypothesis that all β's are zero for the time-dependent covariates for this factor. The results are given in Table 9.5. Here we see that the factor MTX has nonproportional hazards whereas there is no reason to doubt the proportionality assumption for the other factors. In the next section, we will reexamine this model, adjusting for the use of MTX by fitting a stratified proportional hazards regression model.

TABLE 9.5

Tests of the Proportional Hazards Assumption for the Bone Marrow Transplant Data

Factor	Wald Chi Square	Degrees of Freedom	p-Value
Group	1.735	2	0.4200
Waiting time	0.005	1	0.9441
Fab status	0.444	1	0.5051
MTX	4.322	1	0.0376
Sex	0.220	3	0.9743
CMV status	1.687	3	0.6398
Age	4.759	3	0.1903

When the proportional hazards assumption is not satisfied, as in Example 9.2, and interest centers upon a binary covariate, Z_1, whose relative risk changes over time, one approach is to introduce a time-dependent covariate as follows. Let

$$Z_2(t) = Z_1 \times g(t) = g(t) \quad \text{if the covariate } Z_1 \text{ takes on the value 1}$$

$$= 0 \qquad\qquad\qquad \text{if the covariate } Z_1 \text{ takes on the value 0,}$$

where $g(t)$ is a known function of time. In Example 9.2, we took $g(t) = \ln t$. One difficulty with this approach is that the function $g(t)$ is usually unknown. In such cases, it may be preferable to use a procedure that would allow the function $g(t)$ to be estimated from the data.

One simple approach to this problem is to fit a model with an indicator function for $g(t)$. In the simplest approach, we define a time-dependent covariate

$$Z_2(t) = \begin{cases} Z_1 & \text{if } t > \tau \\ 0 & \text{if } t \leq \tau \end{cases}. \tag{9.2.3}$$

Here we have a proportional hazards model with hazard rate

$$h[t \mid Z(t)] = \begin{cases} h_o(t)\exp(\beta_1 Z_1) & \text{if } t \leq \tau \\ h_o(t)\exp[(\beta_1 + \beta_2)Z_1] & \text{if } t > \tau \end{cases}$$

where $h_o(t)$ is the baseline hazard rate. Note that, in this model, $\exp(\beta_1)$ is the relative risk, prior to time τ, for the group with $Z_1 = 1$ relative to the group with $Z_1 = 0$, and $\exp(\beta_1 + \beta_2)$ is the relative risk, after time τ, for the group with $Z_1 = 1$ relative to the group with $Z_1 = 0$, that is, $\exp(\beta_2)$ is the increase in relative risk after time τ and τ is sometimes referred to as the "change point" for the relative risk (Matthews and Farewell 1982 and Liang et al., 1990).

An equivalent coding for this piecewise proportional hazards model is to use a model with two time-dependent covariates, $Z_2(t)$ and $Z_3(t)$. Here, $Z_2(t)$ is as in (9.2.3),

$$Z_3(t) = \begin{cases} Z_1 & \text{if } t \leq \tau \\ 0 & \text{if } t > \tau \end{cases}, \tag{9.2.4}$$

For this coding we have

$$h(t \mid Z(t)) = \begin{cases} h_o(t)e^{\theta_3 Z_1} & \text{if } t \leq \tau \\ h_o(t)e^{\theta_2 Z_1} & \text{if } t > \tau \end{cases}.$$

The two models will have an identical log likelihood with β_1 in model 1 equal to θ_3 in the second model and $\beta_1 + \beta_2$ in the first model equal to θ_2 in the second model. Note that e^{θ_3} is the relative risk before Z and e^{θ_2} is the relative risk after Z.

To determine the optimal value of τ, either model is fit for a set of τ values, and the value of the maximized log partial likelihood is recorded. Because the likelihood will change values only at an event time, a model is fit with τ equal to each of the event times. The value of τ which yields the largest log partial likelihood is the optimal value of τ. Proportional hazards can, then, be tested for each region and if it fails, for t on either side of τ, then this process can be repeated in that region. This procedure is illustrated in the next example.

EXAMPLE 9.2   *(continued):* In Example 9.2, the proportional hazards assumption was rejected with respect to placement of the catheter. Instead of introducing a time-dependent covariate with a known function of time, a "change point" τ for the relative risk will be introduced. Because the likelihood changes only at the event times, Table 9.6 presents the log partial likelihood using the Breslow modification for ties, as a function of all τ's at the failure times.

TABLE 9.6

Log Partial Likelihood as a Function of τ at the Failure Times

Event Times	Log Partial Likelihood
0.5	-97.878
1.5	-100.224
2.5	-97.630
3.5	-97.500
4.5	-99.683
5.5	-100.493
6.5	-98.856
8.5	-100.428
9.5	-101.084
10.5	-101.668
11.5	-102.168
15.5	-100.829
16.5	-101.477
18.5	-102.059
23.5	-102.620

We see from this table that a value of τ equal to 3.5 maximizes the log partial likelihood. Using this model and the coding as in model two we have the following ANOVA table.

	Degrees of Freedom	b	$SE(b)$	Wald Chi Square	p-Value
$Z_3(t): Z_1$ if $t \le 3.5$	1	-2.089	0.7597	7.56	0.0060
$Z_2(t): Z_1$ if $t > 3.5$	1	1.081	0.7832	1.91	0.1672

Here, we see that, up to 3.5 months, patients with a percutaneously placed catheter do significantly better than patients given a surgically placed catheter (relative risk $= \exp(-2.089) = 0.124$) whereas, after 3.5 months, there is no evidence of any difference between the two groups of patients.

To check for proportional hazards within the two time intervals, we fit a model with two additional time-dependent covariates, $Z_4(t) = Z_2(t) \times \ln t$ and $Z_5(t) = Z_3(t) \times \ln t$. In this model, the test of the null hypothesis that $\beta_4 = 0$ is a test of proportional hazards in the first 3.5 months, whereas the test of the null hypothesis that $\beta_5 = 0$ is a test of the proportional hazards assumption after 3.5 months. The p-values of the local Wald tests of these hypotheses are 0.8169 and 0.2806, respectively. Thus, there is no need to further divide the subintervals.

Practical Notes

1. SAS PHREG, in the presence of ties, defaults to Breslow's likelihood and allows the user to specify either the discrete, Efron, or "exact" likelihood.
2. In S-Plus, time-dependent covariates in the proportional hazards model are handled in the routine coxph which uses Efron's likelihood as a default. Breslow's likelihood and the exact likelihood are available when there are ties between the event times.
3. To treat a covariate as a fixed-time covariate, it must be known at the onset of the study. For example, the covariate that signifies that platelets return to a self-sustaining level is not a fixed-time covariate because it is not known at the onset of the study whether a patient will experience this event or not. Such events, which occur at some intermediate time, are treated as time-dependent covariates.
4. Estimating the survival function or the cumulative hazard function is difficult for proportional hazards models with time-dependent covariates because the integral of $h_o(t) \exp[\beta' \mathbf{Z}(t)]$ depends on the random process $\mathbf{Z}(t)$. Unless this is a deterministic function, this integral requires additionally estimating the distribution of the development of $\mathbf{Z}(t)$. Christensen et al. (1986) suggest an estimator to use in this case.

Theoretical Note

1. Kalbfleisch and Prentice (1980) distinguish between two types of time-dependent covariates. The first are *external* covariates whose value at any time does not depend on the failure process. Examples of such covariates are fixed-time covariates, time-dependent covariates whose value is completely under the control of the investigator (e.g., a planned schedule of treatments under the control of the investigator), and ancillary time-dependent covariates that are the output of a stochastic process external to the failure process (e.g., daily temperature as a predictor of survival from a heart attack). Inference

for external covariates follows by the notions discussed in Chapter 8 and the survival function is estimated by the obvious changes to the estimator in section 8.6. The second type of time-dependent covariates are *internal* covariates which are time measurements taken on an individual. These covariates are measured only as long as the individual is still under observation, so that the distribution of these covariates carries information about the failure process. Examples of internal covariates are the times to acute or chronic GVHD and the time to the return of platelets to a normal level in Example 9.1. For this type of covariate, the partial likelihood construction is still valid, but it is not possible to estimate the conditional survival function because $P[X \geq t \mid \mathbf{Z}(t)] = 1$ (if $\mathbf{Z}(t)$ is known, the subject must be alive and at risk of failure).

9.3   Stratified Proportional Hazards Models

As we saw in the previous section, there are instances when the proportional hazards assumption is violated for some covariate. In such cases, it may be possible to stratify on that variable and employ the proportional hazards model within each stratum for the other covariates. Here the subjects in the jth stratum have an arbitrary baseline hazard function $h_{oj}(t)$ and the effect of other explanatory variables on the hazard function can be represented by a proportional hazards model in that stratum as

$$h_j[t \mid \mathbf{Z}(t)] = h_{oj}(t) \exp[\boldsymbol{\beta}'\mathbf{Z}(t)], \ j = 1, \ldots, s. \tag{9.3.1}$$

In this model, the regression coefficients are assumed to be the same in each stratum although the baseline hazard functions may be different and completely unrelated.

Estimation and hypothesis testing methods follow as before, where the partial log likelihood function is given by

$$LL(\boldsymbol{\beta}) = [LL_1(\boldsymbol{\beta})] + [LL_2(\boldsymbol{\beta})] + \cdots + [LL_s(\boldsymbol{\beta})], \tag{9.3.2}$$

where $LL_j(\boldsymbol{\beta})$ is the log partial likelihood (see (8.3.2)) using only the data for those individuals in the jth stratum. The derivatives for the log likelihood in (9.3.2) are found by summing the derivatives across each stratum. $LL(\boldsymbol{\beta})$ is, then, maximized with respect to $\boldsymbol{\beta}$ utilizing the methods in Chapter 8. The survival function for the jth stratum, when the covariates are all fixed at time 0, may be estimated as described in section 8.8.

EXAMPLE 9.1 *(continued):* As we saw in the previous section, the patients who where given MTX as a graft-versus-host prophylactic did not have hazard rates proportional to those patients not given MTX. One way to deal with this problem is to stratify on the use of MTX which involves fitting distinct baseline hazard rates to the two groups. Of interest, as seen in Table 9.2, is a model for the factors of disease group (Z_1, Z_2), FAB class (Z_3), Age (Z_4, Z_5, Z_6) and platelet recovery time $Z_P(t)$. Assuming that the effects of the covariates are the same for patients given MTX or not given MTX, we have the model summarized in Table 9.7.

TABLE 9.7
Anova Table for a Cox Model Stratified on the Use of MTX

Effect	Degrees of Freedom	b	SE(b)	Wald Chi Square	p-Value
Z_1: AML Low-Risk	1	−0.9903	0.3666	7.298	0.0069
Z_2: AML High-Risk	1	−0.3632	0.3714	0.957	0.3280
Z_3: AML with FAB Grade 4 or 5	1	0.8920	0.2835	9.902	0.0017
Z_4: Patient age −28	1	0.0095	0.0198	0.231	0.6305
Z_5: Donor age −28	1	−0.0014	0.0179	0.006	0.9373
$Z_6 = Z_4 \times Z_5$	1	0.0026	0.0009	7.425	0.0064
$Z_P(t)$: Platelet Recovery	1	−1.0033	0.3445	8.481	0.0036

The Wald chi square of the test of the hypothesis of no group effect ($H_0 : \beta_1 = \beta_2 = 0$) is 8.916 with a p-value of 0.0116. The results from the stratified model in this case are quite close to those obtained in the unstratified model.

A key assumption in using a stratified proportional hazards model is that the covariates are acting similarly on the baseline hazard function in each stratum. This can be tested by using either a likelihood ratio test or a Wald test. To perform the likelihood ratio test, we fit the stratified model, which assumes common β's in each stratum, and obtain the log partial likelihood, $LL(\mathbf{b})$. Using only data from the jth stratum, a Cox model is fit and the estimator $\mathbf{b}_j$ and the log partial likelihood $LL_j(\mathbf{b}_j)$ are obtained. The log likelihood under the model, with distinct covariates for each of the s strata, is $\sum_{j=1}^{s} LL_j(\mathbf{b}_j)$. The likelihood ratio chi square for the test that the β's are the same in each stratum is $-2[LL(\mathbf{b}) - \sum_{j=1}^{s} LL_j(\mathbf{b}_j)]$ which has a large-sample, chi-square distribution with $(s-1)p$ degrees of freedom under the null hypothesis.

To construct the Wald test, the model with distinct β's in each stratum is found by fitting distinct proportional hazards models to each stratum. Estimates from different strata are asymptotically independent because the information matrix of the combined model is block diagonal. The Wald test is constructed by using an appropriate contrast matrix as discussed in section 8.5. This method of testing is equivalent to testing for an interaction between a stratification variable and the covariates in a stratified proportional hazards model. These approaches are illustrated in the following continuation of the previous example.

EXAMPLE 9.1   *(continued)*   To test the hypothesis that the effects of disease group, FAB status, age, and platelet recovery are the same in both MTX strata, we fitted distinct Cox models to the two strata. The log partial likelihoods are -219.677, based on the 97 patients not given MTX, and -80.467 based on the 40 patients given MTX. The log partial likelihood from the stratified model, assuming the same β's in each stratum (Table 9.7), is -303.189. The likelihood ratio chi square is $-2\{-303.189 - [(-219.677) + (-80.467)]\} = 6.09$. The degrees of freedom of the test are 7, so the p-value of the test is 0.5292, suggesting no evidence that the covariates are operating differently on patients with or without MTX as a preventive treatment for graft-versus-host disease.

To further check the assumption of equal effects of the covariates on the two strata, we shall do a series of one-degree-of-freedom Wald tests comparing each of β's in the two strata. Here, we use the results from fitting the proportional hazards model, separately, in the two strata. For a given covariate, the estimates in the two strata are asymptotically independent, so a Wald test that $\beta_{1i} = \beta_{2i}$, where β_{ji} is the risk coefficient

TABLE 9.8

One Degree of Freedom Wald Tests Comparing Risk Coefficients in the MTX and No MTX Strata

Effect	No MTX		MTX		X^2	p-Value
	b	SE(b)	b	SE(b)		
Z_1: AML low-risk	-1.1982	0.4585	-0.5626	0.6385	0.654	0.4188
Z_2: AML high-risk	-0.2963	0.4454	-0.8596	0.9175	0.305	0.5807
Z_3: AML with FAB Grade 4 or 5	1.0888	0.3385	0.3459	0.6511	1.025	0.3114
Z_4: Patient age -28	0.0276	0.0259	0.0114	0.0391	0.120	0.7290
Z_5: Donor age -28	-0.0203	0.0253	0.0343	0.0310	1.858	0.1729
$Z_6 = Z_4 \times Z_5$	0.0022	0.0014	0.0014	0.0023	0.103	0.7489
$Z_P(t)$: Platelet recovery	-0.8829	0.4759	-1.0089	0.5511	0.030	0.8626

of the ith covariate in the jth strata, is

$$X^2 = \frac{[b_{1i} - b_{2i}]^2}{SE^2[b_{1i}] + SE^2[b_{2i}]}, \, i = 1, \ldots, 7.$$

The results, summarized in Table 9.8, confirm that, for each of the covariates, there is no reason to suspect that the β's are different in the two strata and the stratified model is appropriate.

The stratified proportional hazards model can be used to model matched pair experiments. Here, for each pair, we assume the model (9.3.1) with the strata defined by the matched pairs. When the number of pairs is large, then, the large-sample properties of the estimators from this model are valid. In this approach, the factors used in the matching are not adjusted for in the regression function, but are adjusted for by fitting distinct baseline rates for each pair. This is illustrated in the following example.

EXAMPLE 9.3 In section 1.2, the results of a clinical trial of a drug 6-mercaptopurine (6-MP) versus a placebo in 42 children with acute leukemia was described. The trial was conducted by matching pairs of patients at a given hospital by remission status (complete or partial) and randomizing within the pair to either a 6-MP or placebo maintenance therapy. Patients were followed until their leukemia returned (relapse) or until the end of the study. In Example 4.1, the survival curves for the two groups were estimated, and, in Example 7.7, using a stratified log rank test, we saw that survival was different in the two groups.

To estimate the relative risk of relapse in the 6-MP group as compared to the placebo group, we fit a Cox model stratified on the pair number. A single covariate is used with the value $Z = 1$ if the patient was given 6-MP and 0 if given a placebo. The estimate of β is -1.792 with a standard error of 0.624. The likelihood ratio chi square of the test of $\beta = 0$ is 11.887 ($p = 0.0006$), the score chi square is 10.714 ($p = 0.0011$) and the Wald chi square is 8.255 ($p = 0.0041$) suggesting a significant difference in relapse rates between the two treatment groups. Note that the score test chi square is exactly the stratified log-rank chi square found in Example 7.7. A 95% confidence interval for the relative risk is $\exp(-1.792 \pm 1.96 \times 0.6236) = [0.049, 0.566]$. Thus, patients given a placebo are between 2 to 20 times more likely to relapse than patients given 6-MP.

Practical Notes

1. When stratification is employed, the tests of hypotheses on regression coefficients will have good power only if the deviations from the null hypotheses are the same in all strata.
2. The large sample stratified tests of hypotheses on regression coefficients are appropriate when either the sample size within strata is large or when the number of strata is large.
3. Estimation of the survival function or cumulative hazard function for each stratum can be obtained using the estimators in section 8.8.

9.4   Left Truncation

In this section, we shall examine how to apply the proportional hazards regression model when the data is left-truncated. The most common situation, where left-truncated data arises, is when the event time X is the age of the subject and persons are not observed from birth but rather from some other time V corresponding to their entry into the study. This is the case for the example introduced in section 1.16 where the age, X_i, at death for the ith subject in a retirement center in California was recorded. Because an individual must survive to a sufficient age V_i to enter the retirement community, and all individuals who died prior to entering the retirement community were not included in this study, the life lengths considered in this study are left-truncated.

Another situation which gives rise to this type of data is when the event time X is measured from some landmark, but only subjects who experience some intermediate event at time V are to be included in the study. This is the case for the bone marrow transplant example where we wish to draw an inference about X, the time from transplant to death or relapse, for those patients whose platelets have recovered to a self-sustaining level. If V is the time until platelets recover for the patient, then only patients who experience this intermediate event are entered into the study. Again, life lengths in this study will be left-truncated. The times V_i are sometimes called *delayed entry times*.

To formulate a proportional hazards regression model for a set of covariates $\mathbf{Z}$, we model the conditional hazard rate of t, given $\mathbf{Z}$ and $X > V$, that is, we model

$$h(t \mid \mathbf{Z}, X > V) \cong \frac{P(X = t \mid \mathbf{Z}, X > V)}{P(X \geq t \mid \mathbf{Z}, X > V)}.$$

If the event time X and the entry time V are conditionally independent, given the covariates $\mathbf{Z}$, then a simple calculation shows that the conditional hazard $h(t \mid \mathbf{Z}(t), X > V)$ and the unconditional hazard rate, $h(t \mid \mathbf{Z})$ are equivalent (Andersen, et al., 1993).

To estimate the regression coefficients with left-truncated data, the partial likelihoods are modified to account for delayed entry into the risk set. To do this, in all of the partial likelihoods presented thus far, we define the risk set $R(t)$ at time t as the set of all individuals who are still under study at a time just prior to t. Here, $R(t) = \{j \mid V_j < t < T_j\}$. With this modification, the techniques, discussed in Chapter 8 and in earlier sections of this chapter, can be applied to left-truncated data. We shall illustrate these methods in the following two examples.

EXAMPLE 9.4   For the Channing House data set introduced in section 1.16, we look at the effect of gender on survival. To fit this model, we modify the risk set to include only those individuals at age t who entered the retirement home at an earlier age and are still under study. The size of this risk set changes with time as depicted in Figure 4.10. The estimated regression coefficient for gender is 0.3158 with a standard error of 0.1731 (Wald p-value of 0.0682). Thus, there is not a significant difference, with respect to survival, between males and females.

EXAMPLE 9.5   In the bone marrow transplant study described in section 1.3, we found, in Example 9.1, one important variable predicting that disease-free survival is the time until the platelet count returns to a self-sustaining level. It is of interest to make an inference about disease-free survival among only those patients who have had their platelets return to a self-sustaining level.

We shall fit the model stratified on the use of MTX to prevent graft-versus-host disease:

$$h(t \mid \mathbf{Z}, \text{MTX}) = h_{0j}(t) \exp(\boldsymbol{\beta}'\mathbf{Z}), \text{ for } j = \text{MTX, No MTX}.$$

The data is left-truncated because only patients whose platelets have returned to a normal level at time t are included in the risk set at that time. The resulting ANOVA table for this model is given in Table 9.9.

The Wald test of the hypothesis of no group effect has a chi square of 18.27 with two degrees of freedom. The p-value of this test is smaller than 0.0001, strongly suggesting differences among the three disease groups in disease-free survival after platelet recovery.

TABLE 9.9

Anova Table for a Cox Model for Patients Whose Platelets Have Returned to Normal Levels, Stratified on the Use of MTX

Effect	Degrees of Freedom	b	SE(b)	Wald Chi Square	p-Value
Z_1: AML low-risk	1	−1.7521	0.4376	16.03	<0.0001
Z_2: AML high-risk	1	−0.7504	0.4077	3.39	0.0657
Z_3: AML with FAB Grade 4 or 5	1	1.2775	0.3249	15.46	<0.0001
Z_4: Patient age −28	1	0.0417	0.0223	3.51	0.0611
Z_5: Donor age −28	1	−0.0346	0.0207	2.80	0.0943
$Z_6 = Z_4 \times Z_5$	1	0.0023	0.0012	3.49	0.0617

Practical Notes

1. Age is often used as a covariate when it should be used as a left-truncation point. When age is used as a left-truncation point, it is unnecessary to use it as a covariate in the model.
2. Left truncation can be performed in S-Plus and SAS.
3. The survival function for left-truncated proportional hazards regression models with fixed covariates can be estimated by using the techniques in section 8.8.

Theoretical Note

1. A key assumption for the left-truncated Cox model is that the event time X and the delayed entry time V are independent, given the covariates **Z**. If this assumption is not valid, then, the estimators of the risk coefficients are not appropriate. See Keiding (1992) for a discussion of this assumption and additional examples.

9.5   Synthesis of Time-varying Effects (Multistate Modeling)

In previous sections of this chapter, we saw how we can use time-dependent covariates or left-truncation to study time-varying effects on survival. Time-dependent covariates, in particular, provide us with important information on how changes in a subject's history effect survival. In this section, using the bone marrow transplant example, we shall illustrate how these analyses can be combined to give an investigator a complete picture of the way changes in a patient's status can affect the prediction of patient outcome.

The basis of this approach is the notion of a patient's history at a given time. Intuitively, a "history" is all the information collected on a patient up to a given time t. It consists of all the patient's covariates measured at time 0 (the fixed time covariates) and the complete knowledge of all time-dependent covariates up to time t. In the bone marrow transplant example discussed in Example 9.1, there are two possible histories when we consider the effects of platelet recovery on disease-free survival. The first history, at time t, consists of all the fixed-time covariates (Z_1: AML low-risk; Z_2: AML high-risk; Z_3: AML with FAB Grade 4 or 5; Z_4: Patient age -28; Z_5: Donor age -28; $Z_6 : Z_4 \times Z_5$), the knowledge that platelets have yet to return to normal levels by time t, and the knowledge that the patient is alive and disease free. If we denote the patient's random platelet recovery time by T_p and the event time by X, then, this history can be denoted as $H_1(t) = \{\mathbf{Z}, T_p > t, X > t\}$. The second history a patient could have at time t consists of the patient's fixed-time covariates, the fact that platelets have returned to nominal levels, and the knowledge that the patient is alive and disease free. This history is denoted by $H_2(t) = \{\mathbf{Z}, T_p \leq t, X > t\}$. We shall call the process $H = [H(t), 0 \leq t < \infty]$ a "history process" for a patient. The history process reflects what happens to the patient over the course of their lifetime under study.

The goal of a survival synthesis is to make predictions of patient outcome based on their history at time t. We shall look at estimating $\pi[s \mid H(t)] = Pr[X \leq s \mid H(t)]$. This function, called a prediction process, in our example is the probability that a patient will relapse or die in the interval t to s given all the information we have observed about the patient up to time t. Notice that the prediction process depends on the patient's history H, the time t at which the history is known, and the point at which we wish to make a prediction s. By fixing t and s and varying the history, we can compare how different patient histories effect outcome. By fixing H and s and varying t, we can see how learning more and more about the patient's history affects outcome. By fixing H and t and varying s, we can obtain an analog of the survival function.

For the transplant example, the computation of π depends on three hazard rates that are functions of the fixed time covariates (see Figure 9.1). For simplicity, we will, for the moment, ignore the dependence of these rates on the fixed covariates. The first rate $h_p(t)$ is the hazard rate for the time to platelet recovery. The second hazard rate $h_1(t)$ is the rate at which individuals, whose platelets have yet to recover, either die or relapse. The third hazard rate $h_2(t)$ is the rate at which patients,

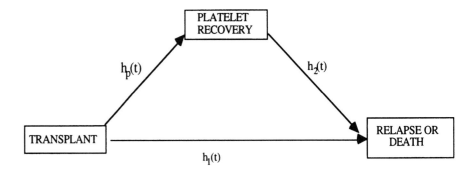

Figure 9.1   *Possible paths to relapse or death*

whose platelets have returned to normal level, die or relapse. As we shall see, these rates are directly estimable from an appropriate Cox model.

Using these rates,

$$\pi_2(s; t) = \pi(s \mid H_2(t)) = Pr(t < X \le s \mid T_p > t)$$

$$= \int_t^s h_2(r) \exp[- \int_t^r h_2(u)du]dr. \qquad (9.5.1)$$

Here the function $\exp[- \int_t^r h_2(u)du]$ is the chance that a patient will not die or relapse between t to r and $h_2(r)$ is approximately the conditional probability of treatment failure at time r, given survival to time r, so that their product is approximately the probability of failure at time r. For $H_1(t)$,

$$\pi_1(s; t) = \pi[s \mid H_1(t)] \qquad (9.5.2)$$

$$= \int_t^s \exp\left[- \int_t^r h_1(u)du - \int_t^r h_p(u)du\right] \left[h_1(r) + h_p(r)\pi_2(s, r)\right] dr.$$

Here, the exponential is the probability of not failing and not having platelet recovery between t to r, $h_1(r)$ is the conditional probability of failure at time r, and $h_p(r)\pi_1(s, r)$ is the probability of platelet recovery at time r and, then, failure in the interval r to s.

To estimate π_1 and π_2 we needed to estimate h_p, h_1, and h_2. We shall present two approaches, one based on assuming proportional hazards between h_1 and h_2 and the second based on assuming distinct baseline hazard rates. The first approach uses a Cox model with time-dependent covariates whereas the second uses a left-truncated Cox model. Both approaches require estimating the hazard rate for platelet recovery time. To estimate this rate, we fit a Cox proportional hazard rate model to the data, using platelet recovery as the event. For individuals whose platelets do not return to a nominal level, we censor their on study time at the time of treatment failure (death or relapse) or at the end of the study period if they are disease free. A careful modeling of the

TABLE 9.10

Risk Factors for Platelet Recovery

Effect	*b*	*SE(b)*	*p-Value*
Patient age -28	$+0.0360$	0.0163	0.0266
Donor age -28	-0.0262	0.0148	0.0766
Patient $\times$ donor age	-0.0027	0.0010	0.0052
MTX used	-1.0423	0.2233	>0.0001

risk factors for platelet recovery is performed using the covariates, as in Example 8.5. The best fitting model, given in Table 9.10, shows that platelet recovery depends on the use of MTX as a graft-versus-host treatment and on the patient's and donor's ages. Using these estimates, we compute Breslow's estimate of the cumulative baseline hazard rate for T_p (see 8.8.2), $\hat{H}_{op}(t)$.

The first approach to estimating h_1 and h_2 is based on assuming proportional hazards between h_1 and h_2. A time-dependent covariate approach is used, and we define 12 time-dependent covariates as follows:

Before Platelet Recovery:

$Z_1(t) = 1$ if AML low-risk and $t \le T_p$

$Z_2(t) = 1$ if AML high-risk and $t \le T_p$

$Z_3(t) = 1$ if AML FAB Grade 4 or 5 and $t \le T_p$

$Z_4(t) = $ Patient age -28 if $t \le T_p$

$Z_5(t) = $ Donor age -28 if $t \le T_p$

$Z_6(t) = Z_4(t) \times Z_5(t);$

After Platelet Recovery:

$Z_7(t) = 1$ if AML low-risk and $t > T_p$

$Z_8(t) = 1$ if AML high-risk and $t > T_p$

$Z_9(t) = 1$ if AML FAB Grade 4 or 5 and $t > T_p$

$Z_{10}(t) = $ Patient age -28 if $t > T_p$

$Z_{11}(t) = $ Donor age -28 if $t > T_p$

$Z_{12}(t) = Z_{10}(t) \times Z_{11}(t).$

Here, $Z_1(t), \dots, Z_6(t)$ are the effects of the fixed-time covariates on disease-free survival before platelet recovery, and $Z_7(t), \dots, Z_{12}(t)$ are the corresponding effects on disease-free survival after platelet recovery.

The Cox model we fit is

$$h(t \mid Z(u), 0 \le u \le t) = h_o(t) \exp\left[\sum_{j=1}^{12} \beta_j Z_j(t)\right] \quad (9.5.3)$$

$$= \begin{cases} h_o(t) \exp\left[\sum_{j=1}^{6} \beta_j Z_j(t)\right] & \text{if } t < T_p \\ h_o(t) \exp\left[\sum_{j=7}^{12} \beta_j Z_j(t)\right] & \text{if } t \ge T_p \end{cases}$$

TABLE 9.11

Estimates Of Risk Coefficients for the Two Models

Before Platelet Recovery

Effect	Proportional Hazards Model I			Left-Truncated Model II		
	b	$SE(b)$	p-Value	b	$SE(b)$	p-Value
AML low-risk	1.5353	0.6347	0.0156	1.4666	0.9117	0.1100
AML high-risk	1.3066	1.1602	0.2601	1.4478	1.3333	0.2776
AML FAB Grade 4 or 5	−1.2411	1.1155	0.2659	−1.7536	1.3214	0.1838
Patient age −28	−0.1596	0.0539	0.0031	−0.1616	0.0619	0.0091
Donor age −28	0.1194	0.0437	0.0063	0.1258	0.0475	0.0081
Patient × donor age interaction	0.0028	0.0019	0.1413	0.0032	0.0021	0.1304

After Platelet Recovery

Effect	Proportional Hazards Model I			Left-Truncated Model II		
	b	$SE(b)$	p-Value	b	$SE(b)$	p-Value
AML low-risk	−1.7622	0.4183	< 0.0001	−1.7161	0.4255	< 0.0001
AML high-risk	−0.7914	0.3991	0.0474	−0.7565	0.4075	0.0634
AML FAB Grade 4 or 5	1.2222	0.3224	< 0.0001	1.2116	0.3222	0.0002
Patient age −28	0.0404	0.0216	0.0610	0.0387	0.0218	0.0754
Donor age −28	−0.0308	0.0203	0.1305	−0.0292	0.0205	0.1540
Patient × donor age interaction	0.0027	0.0012	0.0294	0.0027	0.0012	0.0305

Fitting this model, we obtain the partial likelihood estimates $b_1, \ldots, b_{12}$ (see Table 9.11), and, using these estimates, Breslow's estimate of the cumulative baseline hazard rate $\hat{H}_o(t)$ is computed. The estimates of $H_k(t) = \int_0^t h_k(u)\,du$, $k = 1, 2$ are given by

$$\hat{H}_1(t) = \hat{H}_o(t) \exp\left[\sum_{j=1}^{6} b_j Z_j(t)\right]$$

and

$$\hat{H}_2(t) = \hat{H}_o(t) \exp\left[\sum_{j=7}^{12} b_j Z_j(t)\right].$$

(9.5.4)

An alternative to the proportional hazards model is to fit a model with distinct baseline hazard rates for the time to death or relapse for patients before and after platelet recovery, that is, we fit the Cox model

$h_1(t \mid \mathbf{Z}) = h_{01}(t)\exp(\sum_{j=1}^{6}\beta_j Z_j)$ to the data before platelet recovery by censoring any individual whose platelets recover prior to death or relapse at their platelet recovery time. Using this modified data set, we obtain an estimate $\tilde{H}_1(t) = \tilde{H}_{01}(t)\exp[\sum_{j=1}^{6} b_j Z_j(t)]$, where $\tilde{H}_{01}$ is Breslow's estimator of the baseline hazard function. To estimate the hazard rate after the platelet recovery time, notice that only patients whose platelets return to nominal levels provide any information on this rate. To estimate parameters of the model $h_2(t \mid \mathbf{Z}) = h_{02}(t)\exp(\sum_{j=1}^{6} \alpha_j Z_j)$, we use a left-truncated likelihood with patients entering the risk set at time T_p. Using the estimates of $\boldsymbol{\alpha}$ obtained from maximization of this partial likelihood, an estimate of $H_{02}(t)$ is obtained using Breslow's estimator (8.8.2) where $W(t; a)$ is based on the left-truncated risk set at time t. The estimate of $H_2(t)$ is $\tilde{H}_2(t) = \tilde{H}_{02}(t)\exp[\sum_{j=1}^{6} a_j Z_j(t)]$.

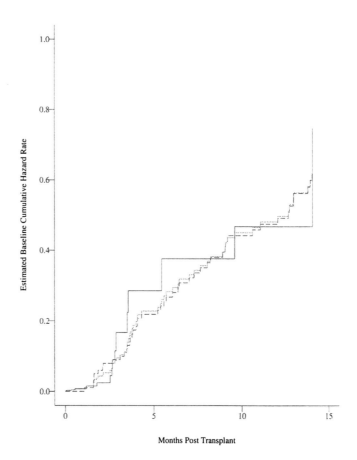

Figure 9.2   *Estimated baseline cumulative hazard rates under the two models. Model 1 (------) Model 2: pre platelet recovery (———) post platelet recovery (— — —)*

Having estimated the basic cumulative hazard rates H_p, H_1, and H_2, estimating π_1 and π_2 proceeds by substituting these values in Eq. (9.5.1) and (9.5.2). Thus, we have the following estimates:

$$\hat{\pi}_2(s, t) = \sum_{i:(t < r_i \leq s)} \exp\{-[\hat{H}_2(r_i) - \hat{H}_2(t)]\}\Delta\hat{H}_2(r_i), \qquad (9.5.5)$$

and

$$\hat{\pi}_1(s, t) = \sum_{i:(t < r_i \leq s)} \exp\{-[\hat{H}_1(r_i) - \hat{H}_1(t)] - [\hat{H}_p(r_i) - \hat{H}_p(t)]\} \qquad (9.5.6)$$

$$\cdot \{\Delta\hat{H}_1(r_i) + \Delta\hat{H}_p(r_i)\hat{\pi}_2(s, r_i)\}.$$

Here, the times r_i are when an individual either has platelets recover or when they experience an event. The values $\Delta\hat{H}(r_i)$ are the jump sizes of the estimate $\hat{H}(r_i)$ at the time r_i.

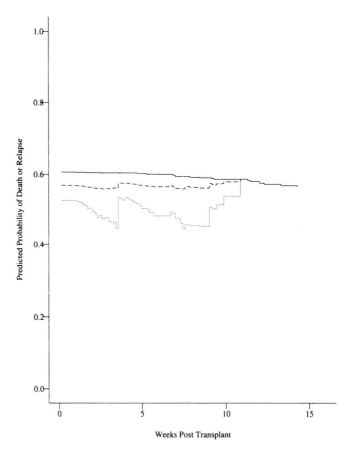

Figure 9.3 *Comparison of predicted probability of death or relapse in the first two years after transplant for an ALL patient. Platelets recovered (———) Platelets not recovered No MTX (------) Platelets not recovered MTX (— — —)*

In the current example, the two models give quite similar pictures of the effects of fixed covariates and of platelet recovery on disease-free survival. Figure 9.2 is a plot of the cumulative baseline hazards for model I, $\hat{H}_{op}(t)$, and the before and after platelet recovery rates, $\hat{H}_{01}(t)$ and $\hat{H}_{02}(t)$, respectively. From these plots, we see that the baseline rates from the two models are quite similar. In the remainder of this section, we shall base our discussion on Model I, because this model, which requires estimating a single baseline hazard rate, has a higher statistical precision.

First, we consider the effects of platelet recovery for a fixed time period. Here, we look at a comparison of $\pi_1(2 \text{ years}, t)$ and $\pi_2(2 \text{ years}, t)$ as a function of the number of weeks post transplant at which the prediction is to be made. Because these probabilities depend on the

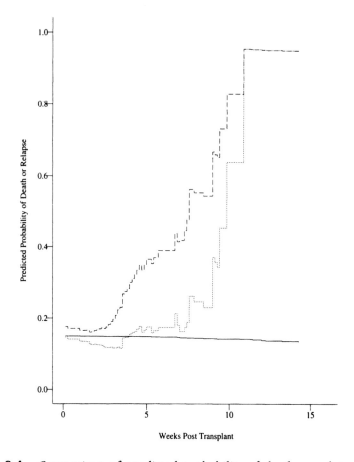

Weeks Post Transplant

Figure 9.4 *Comparison of predicted probability of death or relapse in the first two years after transplant for an AML low risk patient. Platelets recovered (——) Platelets not recovered No MTX (------) Platelets not recovered MTX (— — —)*

fixed-time covariates, we fix the patient FAB status at not being M4 or M5 ($Z_3 = 0 = Z_9$) and patient and donor age at 28 years ($Z_4 = Z_5 = Z_6 = Z_{10} = Z_{11} = Z_{12} = 0$). In Figure 9.3, we present results for ALL patients ($Z_1 = Z_2 = Z_7 = Z_8 = 0$), in Figure 9.4 for AML low-risk patients ($Z_1 = Z_7 = 1; Z_2 = Z_8 = 0$), and, in Figure 9.5, for AML high-risk patients ($Z_1 = Z_7 = 0; Z_2 = Z_8 = 1$). A single curve (the solid line) is given for the probability of death or relapse within the first two years after transplant for a patient who at t weeks has had platelets recover. Two curves are presented for the corresponding probability for a patient who has yet to have platelets recover. The first (short dashed line) is for patients not given MTX and the second (long dashed line) for those that did receive MTX. Note that, because this covariate affects only the platelet recovery rate, there is a single curve for individuals

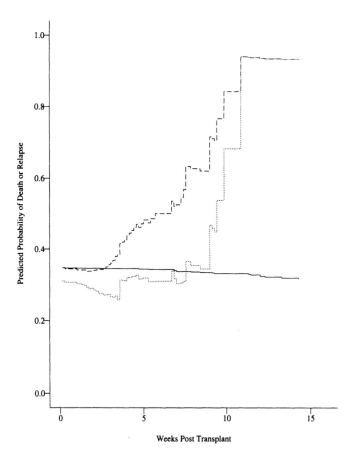

Figure 9.5   *Comparison of predicted probability of death or relapse in the first two years after transplant for an AML high risk patient. Platelets recovered (———) Platelets not recovered No MTX (------) Platelets not recovered MTX (— — —)*

whose platelets have recovered. A careful examination of these figures shows that, for ALL patients (with this set of other covariates), delayed platelet recovery seems to have only a small effect. For AML patients, it seems clear that delayed platelet recovery beyond about 4 weeks seems to predict a much greater chance of death or relapse than individuals who have had platelets return to normal prior to this time. Clearly, for AML patients, if the platelets do not recover by week 10–11, the patient has a very poor prognosis, and some therapeutic measures are indicated.

Figures 9.6–9.8 provide an alternate approach to looking at the effect of platelet recovery on disease-free survival. Here, we fix the time, when the history is known, at either 3, 7, or 10 weeks and look at

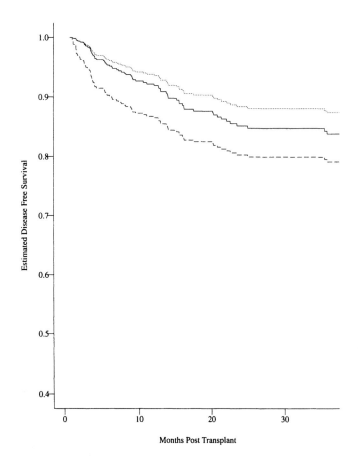

Figure 9.6 *Disease free survival probabilities for an ALL patient given their history at 3 weeks. Platelets recovered (———) Platelets not recovered No MTX (------) Platelets not recovered MTX (— — —)*

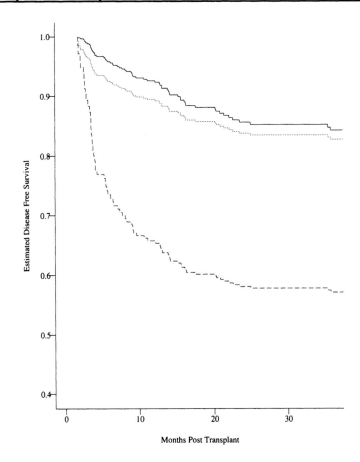

Figure 9.7 *Disease free survival probabilities for an ALL patient given their history at 7 weeks. Platelets recovered (————) Platelets not recovered No MTX (------) Platelets not recovered MTX (— — —)*

the disease-free survival curves for an ALL patient with one of the two histories at that time, that is, we compute $1 - \pi_1(s, t_o)$ and $1 - \pi_2(s, t_o)$, for $t_o = 3$, 7, or 10 weeks. Again, the fixed-time covariates for FAB status and age are set at 0, and separate curves are fitted for patients with or without the MTX treatment. From Figure 9.6, again, we see only a small effect of platelet recovery if we make estimates based on the history at week 3. At week 7 we see that patients who were given MTX at transplant at this time, and whose platelets have yet to return to normal do much worse than other patients. At week 10, this pattern is dramatically enhanced and, here, patients, who were not given MTX and whose platelets have yet to recover, also do poorly.

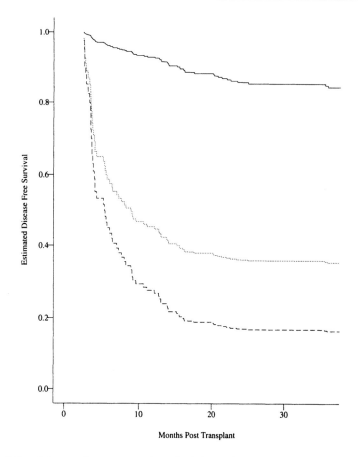

Figure 9.8 *Disease free survival probabilities for an ALL patient given their history at 10 weeks. Platelets recovered (————) Platelets not recovered No MTX (------) Platelets not recovered MTX (— — —)*

Practical Notes

1. A more detailed example, which extends these techniques to multiple intermediate events and end points using Model I, can be found in Klein, Keiding and Copelan (1994).
2. Extensions of Model II to more complex situations can be found in Andersen et al. (1993).
3. Qian (1995) provides standard error estimates for these estimators.

9.6   Exercises

9.1   In Exercise 8.3, a proportional hazards model was fit to data from a study of the effects of ploidy on survival for patients with cancer of the tongue. A single binary covariate was used. Using an appropriate time-dependent covariate, test the hypothesis that the hazard rates for the two groups are proportional.

9.2   In Exercise 8.4, a proportional hazards model was fit to data from a study of the survival of rats implanted with F98 glioma cells in their brains. Three groups of rats were considered: control rats, rats given radiation only, and rats given radiation plus BPA. Using an appropriate set of time-dependent covariates, test that the hazard rates of the three groups are proportional.

9.3   In Example 7.9, data from a clinical trial of chemotherapy and chemotherapy combined with radiotherapy in treating locally unresectable gastric cancer is given. Of interest in this study is a comparison of the efficacy of the two treatments on overall survival.

(a) Using an appropriate proportional hazards model, test the hypothesis of difference in survival between the two treatment regimes. Find a 95% confidence interval for the relative risk of death for patients treated only with chemotherapy compared to patients treated with chemotherapy plus radiation.

(b) Confirm that the hazard rates for the two treatment groups have nonproportional hazards using a time-dependent covariate in the proportional hazards model.

(c) Because the hazard rates for the two treatment groups are not proportional, consider a model with two time-dependent covariates:

$$Z_1(t) = \begin{cases} 1 & \text{if chemotherapy only and } t \leq \tau, \text{ and} \\ 0 & \text{otherwise} \end{cases}$$

$$Z_2(t) = \begin{cases} 1 & \text{if chemotherapy only and } t > \tau, \\ 0 & \text{otherwise} \end{cases}.$$

Find the value of τ which maximizes the partial likelihood for this model.

(d) Using the model constructed in part c discuss the relationship between the two treatments for locally unresectable gastric cancer and survival. Compare the relative risks obtained from this model with the relative risks obtained in part a. Explain how a physician should present this model to a patient.

9.4   Consider the data on bone marrow transplantation for acute leukemia patients discussed in section 1.3. As noted in Exercise 7.8, graft-versus-host (GVHD) disease is considered to have an antileukemic effect. To test this hypothesis, a Cox regression model will be fit to the times to relapse of these leukemia patients. Here patients who die prior to relapse are considered as censored observations.

Fit a proportional hazards model with appropriate time-dependent covariates which can be used to determine which of four time-varying GVHD groups (patient's yet to develop any GVHD, patient's who have had acute GVHD, chronic GVHD, or both acute and chronic GVHD) has the lowest relapse risk. Estimate model parameters, and make the appropriate hypothesis tests. Provide point estimates and 95% confidence intervals for the relative risk of relapse for the GVHD groups as compared to the group with no GVHD at time t.

9.5 In Exercise 11 of Chapter 7, a stratified test of the equality of the four stages of laryngeal cancer was conducted. In that problem, the test was stratified on the cancer being diagnosed prior to 1975 or not. The data for this comparison is found on our web site.

(a) Fit a proportional hazards model, stratified on the cancer being diagnosed either prior to 1975 or not. Include, in the model, indicator variables for stage of disease and a continuous covariate for patient age, as in Example 8.3. Produce an ANOVA table for the fitted model, and compare this to the results for the unstratified model found in Example 8.3.

(b) Using a likelihood ratio test, test the hypothesis that the effects of the stage and age factors are the same in the two strata.

(c) Repeat part b using a Wald test.

9.6 In Exercise 13 of Chapter 7, data was presented on a litter-matched study of the tumorigenesis of a drug. The data is found in that exercise.

(a) Ignoring the fact that this was a litter-matched study, fit a proportional hazards model to this data to estimate the relative risk of tumorigenesis of the drugged rats as compared to the control rats. Find a 95% confidence interval for this relative risk.

(b) Repeat part a using a proportional hazards model stratified on litter. Compare your results.

9.7 In Example 8.5, a proportional hazards model was built to the data on disease-free survival for bone marrow transplantation patients. Of primary interest in that example was the comparison of disease states, and possible factors to be adjusted for were the patients' FAB status, their waiting time to transplant, donor and recipient gender, CMV status, and age. Because patients who developed acute graft-versus-host disease may have different risk factors for disease-free survival, find the best fitting model for these factors for those patients who have experienced acute graft-versus-host disease. Compare your final model to that found in Example 8.5.

9.8 In the burn study described in section 1.6 and as a follow-up to Exercises 8.2 and 8.9—

(a) Introduce a time-dependent covariate that reflects the time at which a wound was excised. Investigate the effects of the timing of wound excision on the time until infection occurs.

(b) Introduce another time-dependent covariate that reflects the time when a prophylactic antibiotic treatment was administered. Investigate the effect of having a prophylactic antibiotic treatment on the time until infection occurs.

(c) Fit a full model, adjusting for all other explanatory covariates as needed to the time until infection occurs. Test for proportional hazards and deal with any variables with nonproportional hazards, as you deem appropriate.

(d) Make an inference about the time until infection among those individuals who had a prophylactic antibiotic treatment administered. Adjust for all other explanatory covariates, as needed. Test for proportional hazards, and deal with any variables with nonproportional hazards, as you deem appropriate.

10
Additive Hazards
Regression Models

10.1 Introduction

In the last two chapters, we examined regression models for survival data based on a proportional hazards model. In this model, the effect of the covariates was to act multiplicatively on some unknown baseline hazard rate. Covariates which do not act on the baseline hazard rate in this fashion were modeled either by the inclusion of a time-dependent covariate or by stratification.

In this chapter, we shall consider an alternative to the semiparametric multiplicative hazard model, namely, the additive hazard model. As in the multiplicative hazards model, we have an event time X whose distribution depends on a vector of, possibly, time-dependent covariates, $\mathbf{Z}(t) = [Z_1(t), \ldots, Z_p(t)]$. We assume that the hazard rate at time t, for an individual with covariate vector $\mathbf{Z}(t)$, is a linear combination of the $Z_k(t)$'s, that is,

$$h[t \mid \mathbf{Z}(t)] = \beta_o(t) + \sum_{k=1}^{p} \beta_k(t) Z_k(t),$$

where the $\beta_k(t)$'s are covariate functions to be estimated from the data.

Two additive models are presented in this chapter. The first, due to Aalen (1989), allows the regression coefficients, $b_k(t)$, to be functions whose values may change over time. This model is flexible enough to allow the effects of a fixed time covariate to change over time. For this model, discussed in Section 10.2, a "least-squares" approach is used to estimate the cumulative regression functions, $B_k(t) = \int_0^t b_k(u)du$ and the standard errors of these functions. These estimators can then be smoothed to obtain estimators of $b_k(t)$.

The second model presented in section 10.3, due to Lin and Ying (1995), replaces the regression functions, $b_k(t)$, by constants, b_k. The model is

$$h(t \mid \mathbf{Z}(t)) = b_0(t) + \sum_{k=1}^p b_k Z_k(t).$$

For this model estimation is based on an estimating equation which is obtained from the score equation.

10.2   Aalen's Nonparametric, Additive Hazard Model

The proportional hazards model, discussed in the previous two chapters, assumes that the effects of the covariates are to act multiplicatively on an unknown baseline hazard function. Estimation of the risk coefficients was based on the partial likelihood. In the proportional hazards model, these risk coefficients were unknown constants whose value did not change over time. In this section, we present an alternative model based on assuming that the covariates act in an additive manner on an unknown baseline hazard rate. The unknown risk coefficients in this model are allowed to be functions of time so that the effect of a covariate may vary over time. As opposed to the proportional hazards model where likelihood based estimation techniques are used, estimators of the risk coefficients are based on a least-squares technique. The derivation of these estimators, which is outlined in the technical notes, is based on the counting process approach to survival analysis and is similar to the derivation of the Nelson–Aalen estimator of the cumulative hazard rate presented in section 3.6. Tests of hypotheses are based on stochastic integrals of the resulting estimators, as in Chapter 7.

The data consists of a sample $[T_j, \delta_j, \mathbf{Z}_j(t)]$, $j = 1, \ldots, n$ where, as in Chapters 8 and 9, T_j is the on study time, δ_j the event indicator, and $\mathbf{Z}_j(t) = [Z_{j1}(t), \ldots, Z_{jp}(t)]$ is a p-vector of, possibly, time-dependent covariates. For the jth individual we define

$$Y_j(t) = \begin{cases} 1 & \text{if individual } j \text{ is under observation (at risk) at time } t, \\ 0 & \text{if individual } j \text{ is not under observation (not at risk) at time } t. \end{cases}$$

If the data is left-truncated, then, note that $Y_j(t)$ is 1 only between an individual's entry time into the study and exit time from the study. For right-censored data $Y_j(t)$ is 1 if $t \leq T_j$.

For individual j, we shall model the conditional hazard rate at time t, given $\mathbf{Z}_j(t)$, by

$$h[t \mid \mathbf{Z}_j(t)] = \beta_o(t) + \sum_{k=1}^{p} \beta_k(t)Z_{jk}(t) \tag{10.2.1}$$

where $\beta_k(t), k = 1, \ldots, p$, are unknown parametric functions to be estimated. Direct estimation of the $\beta(t)$'s is difficult in practice. Analogous to the estimation of the hazard rate in Chapters 4 and 6, we directly estimate the cumulative risk function $B_k(t)$, defined by

$$B_k(t) = \int_0^t \beta_k(u)du, k = 0, 1, \ldots, p. \tag{10.2.2}$$

Crude estimates of $\beta_k(t)$ are given by the slope of our estimate of $B_k(t)$. Better estimates of $\beta_k(t)$ can be obtained by using a kernel-smoothing technique, as discussed in section 6.2 (see Example 10.2).

To find the estimates of $B_k(t)$ a least-squares technique is used. We need to define an $n \times (p + 1)$ design matrix, $\mathbf{X}(t)$, as follows: For the ith row of $\mathbf{X}(t)$, we set $\mathbf{X}_i(t) = Y_i(t)(1, \mathbf{Z}_j(t))$. That is, $\mathbf{X}_i(t) = (1, Z_{j1}(t), \ldots, Z_{jp}(t))$ if the ith subject is at risk at time t, and a $p + 1$ vector of zeros if this subject is not at risk. Let $\mathbf{I}(t)$ be the $n \times 1$ vector with ith element equal to 1 if subject i dies at t and 0 otherwise. The least-squares estimate of the vector $\mathbf{B}(t) = (B_0(t), B_1(t), \ldots, B_p(t))^t$ is

$$\hat{\mathbf{B}}(t) = \sum_{T_i \leq t}[\mathbf{X}^t(T_i)\mathbf{X}(T_i)]^{-1}\mathbf{X}^t(T_i)\mathbf{I}(T_i). \tag{10.2.3}$$

The variance-covariance matrix of $\mathbf{B}(t)$ is

$$\widehat{\mathrm{Var}}(\hat{\mathbf{B}}(t)) = \sum_{T_i \leq t}[\mathbf{X}^t(T_i)\mathbf{X}(T_i)]^{-1}\mathbf{X}^t(T_i)\mathbf{I}^{\mathbf{D}}(T_i)\mathbf{X}(T_i)\{[\mathbf{X}^t(T_i)\mathbf{X}(T_i)]^{-1}\}^t.$$

$$\tag{10.2.4}$$

Here the matrix, $\mathbf{I}^{\mathbf{D}}(t)$ is the diagonal matrix with diagonal elements equal to $\mathbf{I}(t)$. The estimator $\mathbf{B}(t)$ only exists up to the time, t, which is the smallest time at which $\mathbf{X}^t(T_i)\mathbf{X}(T_i)$ becomes singular.

The estimators $\hat{B}_k(t)$ estimate the integral of the regression functions b_k in the same fashion as the Nelson–Aalen estimator discussed in Chapter 4 estimates the integral of the hazard rate in the univariate case. Confidence intervals and confidence bands, based on the linear formulation, for the integrated regression functions, $\hat{B}_k(t)$, are constructed in exactly the same fashion as were confidence intervals or confidence bands for the cumulative hazard functions discussed in sections 4.3 and 4.4 (see Practical Notes 4.3.1 and 4.4.1). Here, one replaces $\hat{H}(t)$ and

$T_H(t)$ in (4.3.4) by $\hat{B}_k(t)$ and $\widehat{\text{Var}}[\hat{B}_k(t)]^{1/2}$, respectively. Crude estimates of the regression functions can be found by examining the slope of the fitted $\hat{B}_k(t)$'s. Better estimates of the regression function can be found by using the kernel smoothing techniques discussed in section 6.2. To obtain the smoothed estimates one simply replaces $D\tilde{H}(t)$ and $D\hat{V}[\hat{H}(t)]$ by $D\hat{B}_k(t)$ and $D\widehat{\text{Var}}[\hat{B}_k(t)]$, respectively, in (6.2.4–6.2.5).

To illustrate these calculations we will derive the estimates in Aalen's model for the two sample problems. Here we have a single covariate Z_{j1} equal to 1 if the jth observation is from sample 1 and 0 otherwise. The design matrix is

$$\mathbf{X}(t) = \begin{pmatrix} Y_1(t) & Y_1(t)Z_{11} \\ & \vdots & \\ Y_n(t) & Y_n(t)Z_{1n} \end{pmatrix}$$

and

$$\mathbf{X}'(t)\mathbf{X}(t) = \begin{pmatrix} N_1(t) + N_2(t) & N_1(t) \\ N_1(t) & N_1(t) \end{pmatrix}.$$

Here $N_k(t)$ is the number at risk in the kth group at time t. The $\mathbf{X}'(t)\mathbf{X}(t)$ matrix is nonsingular as long as there is at least one subject still at risk in each group. The inverse of the $\mathbf{X}'(t)\mathbf{X}(t)$ is

$$[\mathbf{X}'(t)\mathbf{X}(t)]^{-1} = \begin{pmatrix} \dfrac{1}{N_2(t)} & -\dfrac{1}{N_2(t)} \\ -\dfrac{1}{N_2(t)} & \dfrac{1}{N_1(t)} + \dfrac{1}{N_2(t)} \end{pmatrix}.$$

From (10.2.3) we find that

$$\hat{B}_0(t) = \sum_{T_i \leq t} d_i \left\{ \frac{1}{N_2(T_i)} - \frac{Z_{i1}}{N_2(T_i)} \right\}$$

$$= \sum_{\substack{T_i \leq t \\ i \in \text{ sample 2}}} \frac{d_i}{N_2(T_i)}, \, t \leq t, \qquad (10.2.5)$$

the Nelson–Aalen estimator of the hazard rate using the data in sample 2 only. Also

$$\hat{B}_1(t) = \sum_{T_i \leq t} d_i \left\{ \frac{-1}{N_2(T_i)} + Z_{i1} \left[\frac{1}{N_1(T_i)} + \frac{1}{N_2(T_i)} \right] \right\}$$

$$= \sum_{\substack{T_i \leq t \\ i \in \text{ sample 1}}} \frac{d_i}{Y_1(T_i)} - \sum_{\substack{T_i \leq t \\ i \in \text{ sample 2}}} \frac{d_i}{Y_2(T_i)}, \, t \leq t, \qquad (10.2.6)$$

the difference of the Nelson–Aalen estimators of the hazard rate in the two samples. The estimates of the variance-covariance functions of

$\hat{B}_0(t)$ and $\hat{B}_1(t)$ are

$$\widehat{\mathrm{Var}}[\hat{B}_0(t)] = \sum_{\substack{T_i \leq t \\ i \in \text{ sample 2}}} \frac{d_i}{N_2(T_i)^2}; \qquad (10.2.7)$$

$$\hat{\mathrm{Var}}[\hat{B}_1(t)] = \sum_{\substack{T_i \leq t \\ i \in \text{ sample 1}}} \frac{d_i}{N_1(T_i)^2} + \sum_{\substack{T_i \leq t \\ i \in \text{ sample 2}}} \frac{d_i}{N_2(T_i)^2}; \quad (10.2.8)$$

and

$$\widehat{\mathrm{Cov}}[\hat{B}_0(t), \hat{B}_1(t)] = - \sum_{\substack{T_i \leq t \\ i \in \text{ sample 2}}} \frac{d_i}{N_2(t)^2}, \, t \leq t. \qquad (10.2.9)$$

To illustrate these estimation procedures we present two examples.

EXAMPLE 10.1 To illustrate the additive hazard model in the two-sample problem, we shall use the data on death times of breast cancer patients, presented in section 1.5 and analyzed by the use of a proportional hazards model in Example 8.1. We shall denote the immunoperoxidase negative group as sample 1 and the immunoperoxidase positive group as sample 2. $B_0(t)$, $B_1(t)$ and their respective variances are estimated by Eqs. (10.2.5)–(10.2.9). Here, $\tau = 144$, which is the maximal time when, at least, one patient is at risk in both samples. When all subjects remaining at risk are in one of the samples the $\mathbf{X}'(t)\mathbf{X}(t)$ matrix has a zero determinant.

Figures 10.1 and 10.2 show the estimated cumulative regression functions, $\hat{B}_0(t)$ and $\hat{B}_1(t)$, and 95% pointwise naive confidence intervals. Pointwise confidence intervals were found here as

$$\hat{B}_k(t) \pm 1.96\sqrt{\widehat{\mathrm{Var}}[\hat{B}_k(t)]}, \, k = 0, 1. \qquad (10.2.10)$$

Figure 10.1, which shows the plot of $B_0(t)$ in this two-sample case, is a plot of the cumulative hazard rate of the immunoperoxidase positive sample. The slope of this line is an estimate of $\beta_0(t)$, in this case the hazard rate of the immunoperoxidase positive sample. Here, it appears that $\beta_0(t)$ is roughly constant over the range 20–90 months with $\beta_0(t) = 0.009$ deaths/month. $\hat{B}_1(t)$ (see Figure 10.2) is an estimate of the cumulative excess risk of death due to being immunoperoxidase negative. The slope of the line is an estimate of $\beta_1(t)$, the excess mortality due to being immunoperoxidase negative.

EXAMPLE 10.2 Next, we consider the use of an additive model to compare the four stages of laryngeal cancer, adjusting for age in the study of 90 male

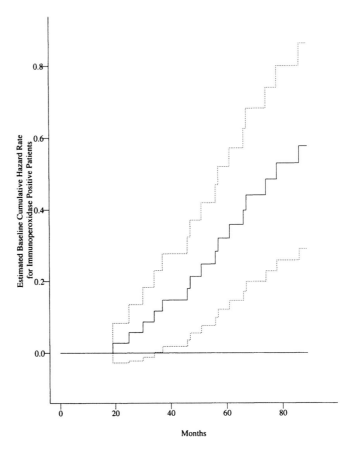

Figure 10.1 *Estimate of $B_0(t)$ and 95% pointwise confidence intervals for breast cancer patients.*

laryngeal cancer patients. The data is described in section 1.8, and this data was examined by using the Cox proportional hazards model in Example 8.3. We define four fixed time covariates:

$$Z_1 = \begin{cases} 1 & \text{if Stage II disease} \\ 0 & \text{if Stage I, III or IV disease;} \end{cases}$$

$$Z_2 = \begin{cases} 1 & \text{if Stage III disease} \\ 0 & \text{if Stage I, II or IV disease} \end{cases}$$

$$Z_3 = \begin{cases} 1 & \text{if Stage IV disease} \\ 0 & \text{if Stage I, II or III disease;} \end{cases}$$

$$Z_4 = \text{Age at diagnosis} - 64.11.$$

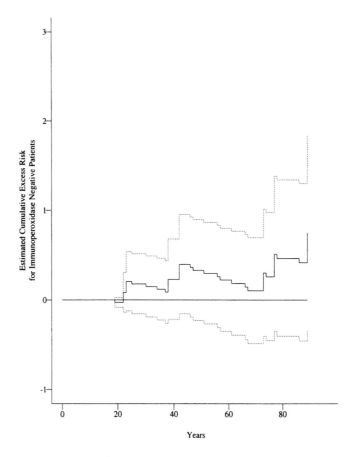

Figure 10.2 *Estimate of the cumulative effect of being Immunoperoxidase negative ($B_1(t)$) and a 95% pointwise confidence interval for breast cancer patients.*

Here, we have centered the age covariate at its mean. In this example, $\tau = 4.4$ which is the largest T_j when, at least, one patient is still at risk in each of the four disease stages.

Figures 10.3–10.7 show the estimates of $B_k(t), k = 0, \ldots, 4$ and 95% pointwise confidence intervals constructed by using (10.2.10). Here, the estimated cumulative baseline hazard $\hat{B}_0(t)$ (Figure 10.3) is an estimate of the cumulative hazard rate of a stage I patient aged 64.11. $\hat{B}_1(t)$, $\hat{B}_2(t)$, and $\hat{B}_3(t)$ (Figures 10.4–10.6) show the cumulative excess risk due to stage II, III or IV patients of a given age compared to stage I patients with a similar age. Here, it appears there is little excess risk due to being a stage II patient, whereas stage III and stage IV patients have an elevated risk in the first two years following diagnosis where the slopes of the two cumulative hazards are nonzero. Figure 10.7 shows the excess risk due to age.

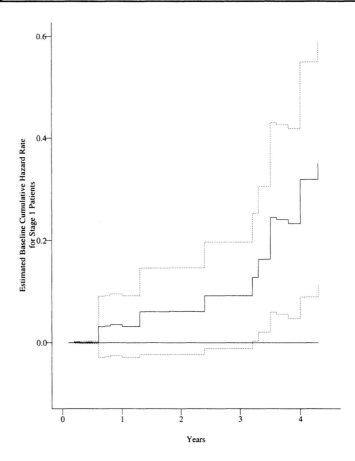

Figure 10.3 *Estimate of the cumulative baseline hazard rate (B₀) and a 95% pointwise confidence interval for laryngeal cancer patients.*

To better understand the differences in survival due to the various stages of disease, Figure 10.8 is a plot of the kernel-smoothed estimate of $\beta_k(t)$, $k = 1, 2, 3$. Here, a biweight kernel (6.2.3, 6.2.8) was used with a bandwidth of one year. Equation (6.2.4) was used to obtain the estimate, with $\Delta \hat{B}_k(t)$ substituted for $\Delta \tilde{H}(t)$. This figure suggests that there is little excess risk associated with stage II disease, that the excess risk associated with stage III disease is negated after about two and a half years and that stage IV disease is the worst case with an excess risk of about 0.4 deaths per year over the first two years after transplant.

We now consider the problem of testing the hypothesis of no regression effect for one or more the covariates. That is, we wish to test the hypothesis $H_0 : \beta_k(t) = 0$ for all $t \leq \tau$ and all k in some set K. The test, described in Aalen (1993), is an analog of tests discussed in Chapter 7 that are based on a weighted stochastic integral of the estimated

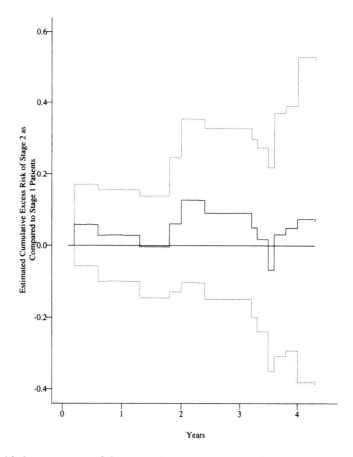

Figure 10.4   *Estimate of the cumulative excess risk of Stage II cancer as compared to Stage I cancer (B_1) and a 95% pointwise confidence interval for laryngeal cancer patients.*

value of $\beta_k(t)$ as compared to its expected value, zero, under the null hypothesis.

To perform the test we need a matrix of weights to use in constructing the test. This weight matrix is a diagonal matrix $\mathbf{W}(t)$ with diagonal elements, $\mathbf{W}_j(t)$, $j = 1, \ldots, p + 1$. Using these weights we can construct the test statistic vector $\mathbf{U}$ as

$$\mathbf{U} = \sum_{T_i} \mathbf{W}(T_i)[\mathbf{X}^t(T_i)\mathbf{X}(T_i)]^{-1}\mathbf{X}^t(T_i)\mathbf{I}(T_i). \qquad (10.2.11)$$

The $(j + 1)$st element of $\mathbf{U}$ is the test statistic we use for testing the hypothesis $H_j : \beta_j(t) = 0$. The covariance matrix $\mathbf{U}$ is given by

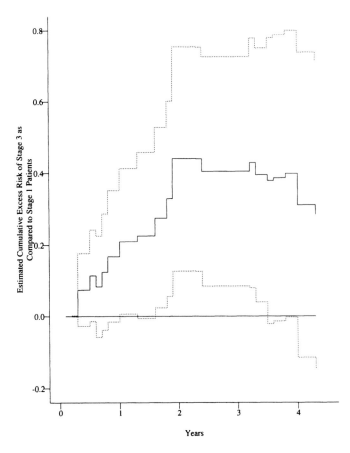

Figure 10.5 *Estimate of the cumulative excess risk of Stage III cancer as compared to Stage I cancer (B_2) and a 95% pointwise confidence interval for laryngeal cancer patients.*

$$\mathbf{V} = \sum_{T_i} \mathbf{W}(T_i)[\mathbf{X}^t(T_i)\mathbf{X}(T_i)]^{-1}\mathbf{X}^t(T_i)\mathbf{I}^D(T_i)\mathbf{X}(T_i)\{[\mathbf{X}^t(T_i)\mathbf{X}(T_i)]^{-1}\}^t\mathbf{W}(T_i)$$

$$(10.2.12)$$

Note that elements of $\mathbf{U}$ are weighted sums of the increments of $\hat{B}_k(t)$ and elements of $\mathbf{V}$ are also obtained from elements of $\widehat{\mathrm{Var}}(\hat{B}(t))$. Using these statistics a simultaneous test of the hypothesis that $\beta_j(t) = 0$ for all $j \in J$ where J is a subset of $\{0, 1, \ldots, p + 1\}$ is

$$X = \mathbf{U}_J^t\mathbf{V}_J^{-1}\mathbf{U}_J \qquad (10.2.13)$$

Here $\mathbf{U}_J$ is the subvector of $\mathbf{U}$ corresponding to elements in $\mathbf{J}$ and $\mathbf{V}_J$ the corresponding subcovariance matrix.

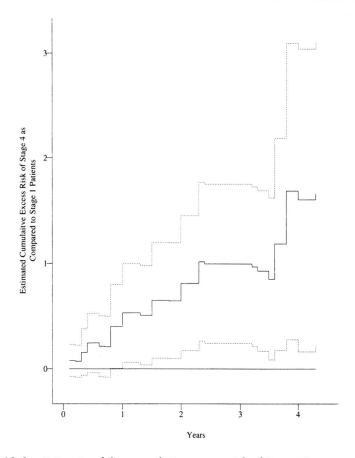

Figure 10.6   *Estimate of the cumulative excess risk of Stage IV cancer as compared to Stage I cancer (B_3) and a 95% pointwise confidence interval for laryngeal cancer patients.*

Any weight function can be used in the calculation of the test statistics. Aalen suggests one use $\mathbf{W}(t) = \{\text{diag}[[\mathbf{X}^t(t)\mathbf{X}(t)]^{-1}]\}^{-1}$. This weight function, as shown below, is useful in the two-sample problem where we are interested in testing the equality of two treatments, but in other cases may lead to inconsistent results. The problem is that this weight function is not the same for all subhypotheses. We prefer a weight function which is common to all subhypotheses as was the case for the weighted log rank tests. Possible candidates are $W_j(t)$ = number at risk at time t or $W_j(t)$ = constant. We will use the former in the numerical examples.

To show how these calculations are carried out we again consider the two-sample problem in Example 10.1. Here the $\mathbf{U}$ vector is

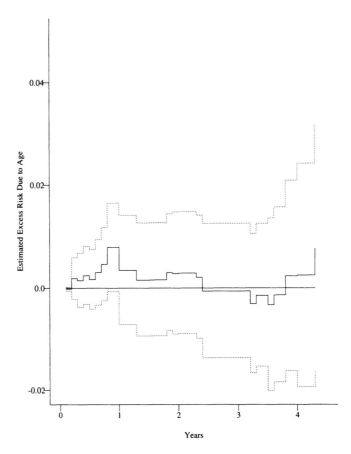

Figure 10.7 *Estimate of the cumulative effect of age (B_4) and a 95% pointwise confidence interval for laryngeal cancer patients.*

$$
\mathbf{U} = \left(
\begin{array}{c}
\sum_{T_i \leq \tau} \delta_i \, W_{11}(T_i) \left\{ \dfrac{1}{N_2(T_i)} - \dfrac{Z_{i1}}{N_2(T_i)} \right\} \\[2ex]
\sum_{T_i \leq \tau} W_{22}(T_i) \, \delta_i \left\{ \dfrac{-1}{N_2(T_i)} + Z_{i1} \left[\dfrac{1}{N_1(T_i)} + \dfrac{1}{N_2(T_i)} \right] \right\}
\end{array}
\right).
$$

If we use Aalen's suggestion of $\mathbf{W}(t) = \{\mathrm{diag}[[\mathbf{X}^t(T_i)\mathbf{X}(T_i)]^{-1}]\}^{-1}$, we have

$$
W_{22}(t) = \left(\frac{1}{N_1(T_i)} + \frac{1}{N_2(T_i)} \right)^{-1} = \left(\frac{N_1(T_i)N_2(T_i)}{N_1(T_i) + N_2(T_i)} \right),
$$

which yields a test statistic for testing the hypothesis the $\beta_1(t) = 0$ of

$$
U_1 = \sum_{i=1}^{n} \delta_i \left(Z_{i1} - \frac{N_1(T_i)}{N_1(T_i) + N_2(T_i)} \right) = D_1 - \sum_{i=1}^{n} \frac{\delta_i N_1(T_i)}{N_1(T_i) + N_2(T_i)}
$$

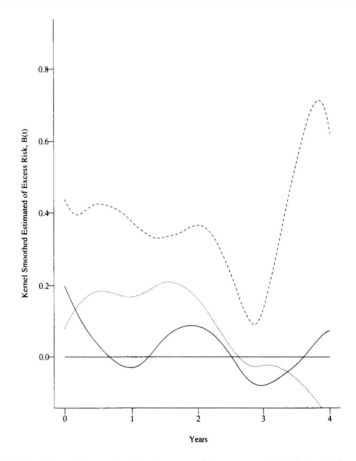

Figure 10.8 *Kernel smoothed estimates of the excess risk of death for Stage II, III, or IV cancer as compared to Stage I cancer. Stage II (———) Stage III (------) Stage IV (— — —).*

Note that U_1 is the difference between the observed and expected number of events in the first sample and is exactly the numerator of the log rank statistic (see 7.3.3). We find after some simplification

$$V_{11} = \sum_{\text{sample } 2} \delta_i \frac{N_1(T_i)^2}{(N_1(T_i) + N_2(T_i))^2} + \sum_{\text{sample } 1} \delta_i \frac{N_2(T_i)^2}{(N_1(T_i) + N_2(T_i))^2},$$

which is distinct from the log rank variance. The test statistic is given by $U_1/\sqrt{V_{11}}$, which has a standard normal distribution when H_0 is true.

When the weight function $W_{ii}(t) = N_1(T_i) + N_2(T_i) = N(T_i)$ is used, we have

$$U_1 = \sum_{i=1}^{n} [N_1(T_i) + N_2(t)] \left\{ \frac{\delta_{i1}}{N_1(T_i)} - \frac{\delta_{i2}}{N_2(T_i)} \right\},$$

where $\delta_{ik}, k = 1, 2, i = 1, \ldots, n$ is 1 if the ith patient died at time T_i and was in the kth treatment group. Note that this is the weighted difference of the Nelson–Aalen estimators for the two samples. Here the variance is

$$V_{11} = \sum_{\text{sample 2}} \frac{(N_1(T_i) + N_2(T_i))^2}{N_1(T_i)^2} + \sum_{\text{sample 1}} \frac{(N_1(T_i) + N_2(T_i))^2}{N_2(T_i)^2}$$

EXAMPLE 10.1   *(continued)*   We shall apply the two-sample test based on the additive hazard model to the comparison of immunoperoxidase-negative and -positive breast cancer patients. Here, using Aalen's weight function, $[N_1(t)^{-1} + N_2(t)^{-1}]^{-1}$, we have $U_1 = -4.187$ and $\sigma_{11} = 6.04$. The test statistic is $Z = -4.187/\sqrt{6.040} = -1.704$ and the p-value is 0.088, which suggests no evidence of a difference in survival between the two types of patients. Using the weight $N_1(t) + N_2(t)$, we have $Z = -33.62/\sqrt{396.26} = -1.69$ and a p-value of 0.0912. When the log rank test is used, we have a test statistic of $-4.187/\sqrt{3.1912} = -2.344$, which has a p-value of 0.019. The difference between the log rank test and the test based on Aalen's weights is due to different estimators of the variability of U_1.

EXAMPLE 10.2   *(continued)*   We now apply these methods to test the hypothesis of no difference in survival between the four stages of larynx cancer, adjusting for age at diagnosis in the study of 90 male larynx cancer patients. We have using the weight function $W_j(t) = $ number at risk at time t,

$$\mathbf{U} = (4.93 \quad 26.17 \quad 106.83 \quad 0.38)^t$$

and

$$\mathbf{V} = \begin{pmatrix} 177.69 & 55.20 & 92.14 & 4.27 \\ 55.20 & 216.41 & 61.35 & 2.16 \\ 92.14 & 61.35 & 1727.30 & 5.73 \\ 4.27 & 2.13 & 5.78 & 0.51 \end{pmatrix}$$

A test of the hypothesis that $\beta_1 = \beta_2 = \beta_3$ is based on the first three elements of $\mathbf{U}, \mathbf{U}_q$, and the upper 3×3 submatrix of $\mathbf{V}, \mathbf{V}_q$. That is,

$$\chi^2 = \mathbf{U}_q^t \mathbf{V}_q^{-1} \mathbf{U}_q = 9.18,$$

which has a p-value of 0.027 when compared to a chi-squared random variable with 3 degrees of freedom.

To examine further the relationship between stage of disease and patient survival, we construct an ANOVA table by using the one degree of freedom tests for $\beta_k(t)$.

Effect	Chi-Square	df	p-Value
Z_1: Stage II disease	0.14	1	0.7111
Z_2: Stage III disease	3.16	1	0.0753
Z_3: Stage IV disease	6.61	1	0.0102
Z_4: Age	0.28	1	0.5913

This table suggests that, adjusted for age, there is little difference between the survival rates of Stage I and Stage II patients, but that Stage IV patients have a significantly different survival rate than Stage I patients.

Practical Notes

1. The additive model measures additional risk due to the effect of a covariate in *absolute* terms whereas the proportional hazards model, discussed in Chapters 8–9, measures this excess risk in *relative* terms. It is quite possible that the relative risk is constant over time (as assumed in a Cox model), but that the additive risk varies with time. In fact, changes in absolute risk with time give no information on changes in relative risk with time.

2. The baseline hazard rate provides an estimate of the hazard rate for an individual with a zero value for all covariates. For a continuous covariate, the most easily interpretable baseline hazard is obtained when the covariate is centered at its mean. In this case, $\beta_0(t)$ is the baseline hazard rate for an individual with an average value of the covariate.

3. The estimates of the baseline hazard rate are not constrained to be nonnegative by this least-squares estimation procedure. In fact, if continuous covariates are not centered at their mean values, the estimator of $\beta_0(t)$ may be negative.

4. Aalen (1989) shows that if a covariate is independent of all the other covariates in the model, then, the regression model with this covariate eliminated is the same as the regression model with this variable included. Only the baseline function $B_0(t)$ is changed. Note that this is not true for the Cox proportional hazard model.

5. If a covariate in the linear model is measured with an additive, normally distributed, random error, then, the linear model is preserved with the regression functions reduced by a constant factor. This is not true for the Cox proportional hazard model.

6. When all covariates are fixed at time zero, an estimate of the cumulative hazard function for $t \leq \tau$, for an individual with a covariate vector $\mathbf{Z} = (Z_1, \ldots, Z_p)$ is given by $\hat{H}(t \mid \mathbf{Z}) =$

$\hat{B}_0(t) + \sum_{k=1}^{p} \hat{B}_k(t)Z_k$ with an estimated variance of $\widehat{\text{Var}}[\hat{H}(t \mid \mathbf{Z})] = \sum_{g=0}^{p} \sum_{k=0}^{p} Z_g Z_k \widehat{\text{Cov}}[\hat{B}_g(t), \hat{B}_k(t)]$. Two asymptotically equivalent estimators of the survival function for this individual are given by $\hat{S}(t \mid \mathbf{Z}) = \exp[-\hat{H}(t \mid \mathbf{Z})]$, and $\hat{S}^*(t \mid \mathbf{Z}) = \prod_{t_j \leq t}[1 - \Delta \hat{H}(t_j \mid \mathbf{Z})]$. The estimated variances of these estimators are $\widehat{\text{Var}}[\hat{H}(t \mid \mathbf{Z})]\hat{S}(t \mid \mathbf{Z})^2$ and $\widehat{\text{Var}}[\hat{H}(t \mid \mathbf{Z})]\hat{S}^*(t \mid \mathbf{Z})^2$, respectively. These two estimators correspond to the survival function constructed from the Nelson–Aalen estimator and the product-limit estimator, discussed in Chapter 4. Some care is needed in interpreting either estimate because $\hat{H}(t \mid \mathbf{Z})$ need not be monotone over the time interval. Also, note that these estimates are defined only over the range where the matrix $\mathbf{A}(t)$ is full rank.

7. The model for excess mortality, discussed in section 6.3, is a special case of the additive model discussed in this section. If we let $Z_j(t) = \theta_j(t)$, the value of the reference hazard rate at time t for the jth individual, then, $\hat{B}_1(t) = \hat{A}(t)$, where $\hat{A}(t)$ is given by Eq. (6.3.6). To adjust for other prognostic factors in an excess mortality model, this coding of the known risk covariate can be used.

8. A SAS macro to fit the additive model is available at our worldwide Web site. Details of its use are found in Howell (1996).

Theoretical Notes

1. The derivation of the least-squares estimators and their variances follows by using the theory of counting processes, discussed in section 3.6. We shall sketch that development in this note. We start by defining a counting process $N_j(t)$ for each individual which has the value 1, if $T_j \leq t$, $\delta_i = 1$, 0, otherwise. As above, let $Y_j(t)$ denote a stochastic process which indicates whether individual j is at risk at time t. Let $\mathbf{N}(t) = [N_1(t), \ldots, N_n(t)]^t$ be the column vector of the n individual counting processes. Let $\mathbf{X}(t)$ be the $n \times (p + 1)$ matrix whose jth row is given by

$$[Y_j(t), Y_j(t)Z_{j1}(t), \ldots, Y_j(t)Z_{jp}], j = 1, \ldots, n.$$

The intensity function for $N_j(t)$, given $\mathbf{Z}_j(t)$ under the assumed additive hazard model, is $[Y_j(t), Y_j(t)Z_{j1}(t), \ldots, Y_j(t)Z_{jp}]\boldsymbol{\beta}(t)$, where $\boldsymbol{\beta}(t)$ is the $(p + 1) \times 1$ vector $[\beta_0(t), \beta_1(t), \ldots, \beta_p(t)]^t$. This implies that

$$\mathbf{M}(t) = \mathbf{N}(t) - \int_0^t \mathbf{X}(u)\boldsymbol{\beta}(u)du$$

is an $n \times 1$ vector of martingales. Thus

$$d\mathbf{N}(t) = \mathbf{X}(t)\boldsymbol{\beta}(t) + d\mathbf{M}(t), \tag{10.2.14}$$

analogous to the relationship used in section 3.6 to derive the Nelson–Aalen estimator. The martingale can be regarded as statistical noise, so set $d\mathbf{M}(t)$ to zero in (10.2.14) and solve for the vector $\mathbf{B}(t) = \int_0^t \boldsymbol{\beta}(u)du$. A solution is possible only when $\mathbf{X}(t)$ has full rank $(p + 1)$. We let $\mathbf{X}^-(t)$ be a generalized inverse of $\mathbf{X}(t)$. That is, $\mathbf{X}^-(t)$ is a $(p + 1) \times n$ matrix with the property that $\mathbf{X}^-(t)\mathbf{X}(t) = \mathbf{I}$, the $(p + 1) \times (p + 1)$ identity matrix. The solution to (10.2.14) is

$$\hat{\mathbf{B}}(t) = \int_0^t \mathbf{X}^-(u)d\mathbf{N}(u) \tag{10.2.15}$$

$$\hat{\mathbf{B}}(t) = \sum_{T_j \le t} \delta_j \mathbf{X}^-(T_j), \text{ for } t \le \tau.$$

Any generalized inverse can be used. In deriving the estimator in this section, the generalized inverse $\mathbf{X}^-(t) = [\mathbf{X}'(t)\mathbf{X}(t)]^{-1}\mathbf{X}'(t)$ is used.

To derive the variance estimator of $\hat{\mathbf{B}}(t)$, for $t \le \tau$, note that, in this range, we can write $(\hat{\mathbf{B}} - \mathbf{B})(t) = \int_0^t \mathbf{X}^-(u)d\mathbf{M}(u)$, the stochastic integral of a predictable process $\mathbf{X}^-(t)$ with respect to a martingale. The predictable variation process $\langle \hat{\mathbf{B}} - \mathbf{B} \rangle(t)$ provides the mean-squared error of the process $(\hat{\mathbf{B}} - \mathbf{B})(t)$, which, using a matrix version of (3.6.4), is given by

$$\langle \hat{\mathbf{B}} - \mathbf{B} \rangle(t) = \int_0^t \mathbf{X}^-(u) < d\mathbf{M}(u) > \mathbf{X}^-(u)^t$$

$$= \int_0^t \mathbf{X}^-(u)[\text{diag } \mathbf{h}(u)]\mathbf{X}^-(u)^t,$$

where $\text{diag } \mathbf{h}(u)$ is the $n \times n$ diagonal matrix with (j, j)th element $h[t \mid \mathbf{Z}_j(u)]Y_j(u)$. The variance is estimated by the matrix $\hat{\boldsymbol{\Sigma}} = \int_0^t \mathbf{X}^-(u)[\text{diag } d\mathbf{N}(u)]\mathbf{X}^-(u)^t$, where $\text{diag } d\mathbf{N}(u)$ is the $n \times n$ diagonal matrix with (j, j)th element $dN_j(u)$. When the generalized inverse $\mathbf{X}^-(t) = [\mathbf{X}'(t)\mathbf{X}(t)]^{-1}\mathbf{X}'(t)$ is used, this reduces to the estimated covariances given by (10.2.4).

A more rigorous derivation of these results is found in Aalen (1980), McKeague(1988) or Huffer and McKeague (1991).

2. As seen in the previous theoretical note, the estimators discussed in this section are based on the generalized inverse of $\mathbf{X}(t)$ defined by $\mathbf{X}^-(t) = [\mathbf{X}'(t)\mathbf{X}(t)]^{-1}\mathbf{X}'(t)$. This choice of an inverse does not account for the possibility that the individual martingales $M_j(t)$ may have distinct variances. Because any generalized inverse can be used in developing these estimators, there is no guarantee that this choice of generalized inverse is optimal. Huffer and McKeague (1991) and MacKeague (1988) suggest using a weighted, least-squares, generalized inverse

$$\mathbf{X}^-(t) = [\mathbf{X}'(t)\mathbf{W}(t)\mathbf{X}(t)]^{-1}\mathbf{X}'(t)\mathbf{W}(t). \tag{10.2.16}$$

Here $\mathbf{W}(t)$ is an $n \times n$ diagonal matrix taken to have the (j, j)th element proportional to the inverse of the variance of $dM_j(t)$. Because the variance of $dM_j(t)$ is given by $h[t \mid \mathbf{Z}_j(t)] = \beta_0(t) + \sum_{k=1}^{p} \beta_k(t) Z_{jk}(t)$, a two-stage estimation procedure is suggested.

In the first stage of the weighted, least-squares procedure, the unweighted, least-squares estimator of $B_k(t), k = 0, \ldots, p$ is computed via (10.2.3). Using this estimator, a kernel-smoothed estimator of $\beta_k(t)$, $\hat{\beta}_k(t)$, is obtained. These estimators are used to estimate the weight function as $W_{jj} = [\hat{\beta}_0(t) + \sum_{k=1}^{p} \hat{\beta}_k(t) Z_{jk}(t)]^{-1}$. These estimated weights are used in (10.2.16) to compute the generalized inverse which is used in (10.2.15) to obtain the second-stage, weighted, least-squares estimator.

3. Aalen (1993) discusses a method to check for the goodness of fit of this model.

4. The difference between the two-sample test presented in this section and the log rank test presented in section 7.3 is the estimated variance of U. For the log rank test the variance of U is estimated under the null hypothesis of no difference between the two survival curves, while for the test presented here the variance is estimated in the general mode, which does not assume equal survival.

5. Other test statistics for the additive hazard model have been suggested by Huffer and McKeague (1991). They consider Renyi tests that are in the same spirit of those presented in Chapter 7.

6. Andersen et al. (1993) present a detailed outline of the large sample theory needed to prove the asymptotic chi-squared distribution of the test statistics presented here.

10.3   Lin and Ying's Additive Hazards Model

In the previous section we studied an additive model for the conditional hazard rate of an individual given a set of covariates. In that model we allowed the regression coefficients to be functions of time. Lin and Ying (1994, 1995, 1997) propose an alternate additive hazards regression model. For their model the possibly time-varying regression coefficients in the Aalen model are replaced by constants. That is, the Lin and Ying additive model for the conditional hazard rate for individual j with covariate vector $\mathbf{Z}_j(t)$ is

$$h(t \mid \mathbf{Z}_j(t)) = \alpha_0(t) + \sum_{k=1}^{p} \alpha_k Z_{jk}(t) \qquad (10.3.1)$$

where $\alpha_k, k = 1, \ldots, p$ are unknown parameters and $\alpha_0(t)$ is an arbitrary baseline function. When all the covariate values are fixed at time

0 it is easy to estimate the regression coefficient, $\alpha_k, k = 1, \ldots, p$. In fact, as opposed to the estimates in the Cox model an explicit formula is available for the estimates and their variances, and as opposed to the Aalen model we can directly estimate the regression coefficients. In this section we will focus on the case where all the covariates are fixed at time 0 and refer readers to the papers by Lin and Ying (1994, 1997) for details of the calculations when time-varying covariates are present.

As usual our data consists of a sample $(T_j, \delta_j, \mathbf{Z}_j), j = 1, \ldots, n$ where T_j is the on study time, δ_j the event indicator, and $\mathbf{Z}_j = \{Z_{j1}, \ldots, Z_{jp}\}$ is a p-vector of possibly fixed time-dependent covariates. We assume that the T_j are ordered from smallest to largest with $0 = T_0 \leq T_1 \leq T_2 \leq \cdots \leq T_n$. For the jth individual we define

$$
Y_j(t) = \begin{cases} 1 & \text{if individual } j \text{ is under observation} \\ & \quad (\text{at risk}) \text{ at time } t \\ 0 & \text{if individual } j \text{ is not under observation} \\ & \quad (\text{not at risk}) \text{ at time } t. \end{cases}
$$

Note that if the data is left truncated then $Y_j(t)$ is 1 only between an individual's entry time into the study and their exit time from the study. For right-censored data $Y_j(t)$ is 1 if $t \leq T_j$.

To construct the estimates of $\alpha_k, k = 1, \ldots, p$, we need to construct the vector $\bar{\mathbf{Z}}(t)$, which is the average value of the covariates at time t. That is,

$$
\bar{\mathbf{Z}}(t) = \frac{\sum_{i=1}^{n} \mathbf{Z}_i Y_i(t)}{\sum_{i=1}^{n} Y_i(t)}. \tag{10.3.2}
$$

Note the numerator is the sum of the covariates for all individuals at risk at time t and the denominator is the number at risk at time t. We next construct the $p \times p$ matrix $\mathbf{A}$ given by

$$
\mathbf{A} = \sum_{i=1}^{n} \sum_{j=1}^{i} (T_j - T_{j-1})[\mathbf{Z}_i - \bar{\mathbf{Z}}(T_j)]'[\mathbf{Z}_i - \bar{\mathbf{Z}}(T_j)]; \tag{10.3.3}
$$

the p-vector $\mathbf{B}$ given by

$$
\mathbf{B}^t = \sum_{i=1}^{n} \delta_i[\mathbf{Z}_i - \bar{\mathbf{Z}}(T_i)]; \tag{10.3.4}
$$

and the $p \times p$ matrix $\mathbf{C}$ by

$$
\mathbf{C} = \sum_{i=1}^{n} \delta_i[\mathbf{Z}_i - \bar{\mathbf{Z}}(T_j)]^t[\mathbf{Z}_i - \bar{\mathbf{Z}}(T_j)]. \tag{10.3.5}
$$

The estimate of $\alpha = (\alpha_1, \ldots, \alpha_p)$ is

$$
\hat{\alpha} = \mathbf{A}^{-1}\mathbf{B}^t \tag{10.3.6}
$$

and the estimated variance of $\hat{\alpha}$ is given by

$$\hat{\mathbf{V}} = \widehat{\text{Var}}(\hat{\alpha}) = \mathbf{A}^{-1}\mathbf{C}\mathbf{A}^{-1}. \tag{10.3.7}$$

To test the hypothesis $H_j : \alpha_j = 0$, we can use the statistic

$$\frac{\hat{\alpha}_j}{\sqrt{\hat{\mathbf{V}}_{jj}}}, \tag{10.3.8}$$

which has a standard normal distribution for large n under the null hypothesis. The test of the hypothesis that $\alpha_j = 0$ for all $j \in \mathbf{J}$ is based on the quadratic form

$$\chi^2 = [\hat{\alpha}_j - \mathbf{0}]'\hat{\mathbf{V}}_j^{-1}[\hat{\alpha}_j - \mathbf{0}], \tag{10.3.9}$$

where $\hat{\alpha}_j$ is the subvector of estimates with subscripts in the set $\mathbf{J}$ and $\hat{\mathbf{V}}_j$ is the corresponding part of the covariance matrix. Under the null hypothesis the statistic has a chi squared distribution with degrees of freedom equal to the dimension of $\mathbf{J}$.

EXAMPLE 10.1

(continued) We shall apply this technique to the data comparing immunoperoxidase-negative and -positive breast cancer patients discussed in the previous section. Here we have a single covariate, Z, with value of 1 if the patient was in the immunoperoxidase-negative arm. The model is $h(t \mid Z) = \alpha_0(t) + \alpha Z$. For the two-sample problem we have

$$\bar{\mathbf{Z}}(t) = \frac{N_1(t)}{N_1(t) + N_2(t)},$$

where $N_k(t)$ is the number at risk at time t in group $k, 2 = 1, 2$. Note that this is an estimate of the chance a subject alive at time t will be in group 1. Applying (10.3.3) to (10.3.7), we find $\hat{\alpha} = 0.00803$ with a standard error of 0.00471. The test of the hypotheses that $\alpha = 0$ from (10.3.8) has a value of 1.704 with a p-value of 0.0880 in close agreement with the more complicated Aalen model.

EXAMPLE 10.2

(continued) We now apply these methods to the larynx cancer data. We have four covariates, Z_1, the indicator of Stage II disease; Z_2, the indicator of Stage III disease; Z_3, the indicator of Stage IV disease; and Z_4, the patient's age at diagnosis. The model is $h(t \mid Z_1, \ldots, Z_4) = \alpha_0(t) + \sum_{j=1}^{4} Z_j\alpha_j$. Applying (10.3.3)–(10.3.7), we have

$$\hat{\alpha} = (0.01325 \quad 0.07654 \quad 0.37529 \quad 0.00256)'$$

and

$$
\hat{\mathbf{V}} = \begin{pmatrix}
1.90 \times 10^{-3} & 5.38 \times 10^{-4} & 4.51 \times 10^{-4} & 3.80 \times 10^{-5} \\
5.38 \times 10^{-4} & 1.95 \times 10^{-3} & 3.07 \times 10^{-4} & 1.17 \times 10^{-5} \\
4.51 \times 10^{-4} & 3.07 \times 10^{-4} & 1.97 \times 10^{-2} & 9.88 \times 10^{-6} \\
3.80 \times 10^{-5} & 1.17 \times 10^{-5} & 9.88 \times 10^{-6} & 4.74 \times 10^{-6}
\end{pmatrix}.
$$

To examine the relationship between stage of disease and patient survival we construct an ANOVA table by using the one degree of freedom test for $\alpha_k(t)$.

Effect	α	Standard Error	Chi-Square	df	p-value
Z_1: Stage II disease	0.01325	0.0435	0.09	1	0.7609
Z_2: Stage III disease	0.07654	0.0442	3.00	1	0.0833
Z_3: Stage IV disease	0.37529	0.1403	7.16	1	0.0075
Z_4: Age	0.00256	0.0022	1.38	1	0.2399

This table suggests that, adjusted for age, there is little difference between the survival rates of Stage I and Stage II patients, but that Stage IV patients have a significantly different survival rate than Stage I patients. Results are quite similar to the result for Aalen's model in the previous section.

The test statistic of the hypothesis of no effect of stage is

$$
\chi^2 = (0.01325 \quad 0.07654 \quad 0.37529)
$$
$$
\times \begin{pmatrix}
1.90 \times 10^{-3} & 5.38 \times 10^{-4} & 4.51 \times 10^{-4} \\
5.38 \times 10^{-4} & 1.95 \times 10^{-3} & 3.07 \times 10^{-4} \\
4.51 \times 10^{-4} & 3.07 \times 10^{-4} & 1.97 \times 10^{-2}
\end{pmatrix}
\begin{pmatrix}
0.01325 \\
0.07654 \\
0.37529
\end{pmatrix}
$$

which is equal to 9.84 and gives a p-value based on the chi-square distribution with three degrees of freedom of 0.0200.

Theoretical Notes

1. This estimator is derived based on counting process theory. As usual we start with a counting process $N_j(t)$ for each individual which has the value 0 if $T_j > t$ and 1 if $T_j \leq t, \delta_i = 1$. We let $Y_j(t)$ denote stochastic process which indicates whether individual j is at risk at time t. For this model we have that $h(t \mid \mathbf{Z}_i(t)) = \alpha_0(t) + \alpha' \mathbf{Z}_i(t)$, where $\mathbf{Z}_i(t) = (Z_{i1}(t), \ldots, Z_{ip}(t))'$ is a vector of possible time-dependent covariates and $\alpha = (\alpha_1, \ldots, \alpha_p)^t$ are the regression coefficients. The basic counting process, $N_i(t)$, can be decomposed

into

$$N_i(t) = M_i(t) = \int_0^t Y_i(u)[\alpha_0(t) + \boldsymbol{\alpha}'\mathbf{Z}_i(t)]\,dt, \qquad (10.3.10)$$

where $M_i(t)$ is a martingale. If $\boldsymbol{\alpha}$ were known then as in the construction of the Nelson–Aalen estimator (see section 3.6), we have an estimator of $A_0(t) = \int_0^t \alpha_0(u)\,du$ of

$$\hat{A}_0(t) = \int_0^t \frac{\sum_{i=1}^n \{dN_i(u) - Y_i(u)\boldsymbol{\alpha}'\mathbf{Z}_i(u)\,du\}}{\sum_{i=1}^n Y_i(u)}. \qquad (10.3.11)$$

Recall the partial likelihood function for the Cox regression model was based on a profile likelihood construction where we substituted our estimate of the baseline hazard rate, as a function of the regression coefficients, into the likelihood (see Theoretical Note 2 in section 8.2). This leads to a partial score equation for the Cox model of

$$\mathbf{U}(\boldsymbol{\beta}) = \sum_{i=1}^n \int_0^\infty \mathbf{Z}_i(t)[dN_i(t) - Y_i(t)\exp\{\boldsymbol{\beta}'\mathbf{Z}_i(t)\}d\hat{\Lambda}_0(t,\boldsymbol{\beta})$$

where $\hat{\Lambda}_0(t,\boldsymbol{\beta})$ is Breslow's estimator of the baseline hazard in the Cox model. Lin and Ying mimic this score equation for the additive model as

$$\mathbf{U}(\boldsymbol{\alpha}) = \sum_{i=1}^n \int_0^\infty \mathbf{Z}_i(t)[dN_i(t) - Y_i(t)\,d\hat{A}_0(t) - Y_i(t)\boldsymbol{\alpha}'\mathbf{Z}_i(t)\,dt], \quad (10.3.12)$$

which upon substitution of (10.3.11) for $\hat{A}_0(t)$ gives a score equation of

$$\mathbf{U}(\boldsymbol{\alpha}) = \sum_{i=1}^n \int_0^\infty [\mathbf{Z}_i(t) - \bar{\mathbf{Z}}(t)][dN_i(t) - Y_i(t)\boldsymbol{\alpha}'\mathbf{Z}_i(t)\,dt]. \qquad (10.3.13)$$

The value of $\boldsymbol{\alpha}$ which maximizes (10.3.13) is

$$\hat{\boldsymbol{\alpha}} = \left[\sum_{i=1}^n \int_0^\infty Y_i(t)[\mathbf{Z}_i(t) - \bar{\mathbf{Z}}(T_j)]'[\mathbf{Z}_i(t) - \bar{\mathbf{Z}}(T_j)]\,dt\right]^{-1}$$

$$\times \left[\sum_{i=1}^n \int_0^\infty [\mathbf{Z}_i(t) - \bar{\mathbf{Z}}(T_i)]\,dN_i(t)\right], \qquad (10.3.14)$$

which reduces to (10.3.6) when all the covariates are fixed at time zero.

Practical Note

1. The estimate of the cumulative baseline hazard rate $\alpha_0(t)$ is given by (10.3.11). When all the covariates are fixed this reduces to

$$\hat{A}_0(t) = \sum_{i=1}^{k} \frac{\delta_i}{Y.(T_i)} - \sum_{j=1}^{k} \sum_{i=j}^{n} \alpha' \mathbf{z}_i \frac{T_j - T_{j-1}}{Y.(T_j)} - \sum_{i=k+1}^{n} \alpha' \mathbf{z}_i \frac{t - T_k}{Y.(T_k)},$$

for $T_k \leq t \leq T_{k+1}$. Here $Y.(t)$ is the total number at risk at time t. Note that the first term here is the usual Nelson–Aalen estimator of the hazard rate as presented in Chapter 4. This function is a piecewise linear function with jumps at the observed deaths.

10.4   Exercises

10.1   In section 1.10 we presented a study comparing two types of transplants for two types of lymphoma. In the study patients were given either an allogenic (Allo) or autologous (Auto) transplant for Hodgkin's lymphoma or non-Hodgkin's lymphoma (NHL). Using Aalen's additive model and this data set, answer the following questions:

(a) Ignoring disease status, estimate the effect of type of transplant. Provide an estimate of the cumulative regression coefficient and its standard error.

(b) Test the hypothesis that there is no difference between Allo and Auto transplants based on the model in part a using the number at risk as weights.

(c) Consider fitting the model with a term for disease ($Z_1 = 1$ if NHL, 0 otherwise), a term for type of transplant ($Z_2 = 1$ if Allo, 0 if Auto), and an interaction term, $Z_3 = Z_1 \times Z_2$. Find the estimates of the cumulative regression coefficients for this model. Provide a test of the hypothesis that there is no interaction between disease and type of transplant.

10.2   In section 1.6 a study which evaluated disinfectant practices at a large midwestern medical center was presented. The effect of two methods of burn cleansing (routine and chlorhexidine) on the distribution of the times to staphylococcus infections is of primary interest. A covariate to be adjusted for in the analysis is the total surface area burned.

(a) Fit an additive hazard model for the times to staphylococcus infections using an indicator variable for the type of cleansing used and a continuous covariate, centered at its mean, for the total surface area burned. Plot your estimates of the cumulative regression coefficients.

    (b) Provide a plot of the estimated survival function of time to staphylococcus infection for both types of cleansing for an individual with burns over 50% of their total surface area.

    (c) Test the hypothesis of no difference in the rates of staphylococcus infection in the two bathing groups.

    (d) Using a kernel-smoothed estimator with a bandwidth of 10 days and a biweight kernel, estimate the regression function for the type of cleansing. Interpret this estimate.

10.3   Using the data in section 1.10 on transplants for patients with lymphoma—

    (a) Estimate the effect of type of transplant, ignoring the disease using the Lin–Ying model.

    (b) Consider fitting a model with an indicator of disease type and type of transplant. Estimate the regression coefficients in the Lin and Ying model and test the hypothesis that there is no interaction between disease and type of transplant.

11
—— Regression Diagnostics

11.1   Introduction

In Chapters 8 and 9, methods for analyzing semiparametric proportional hazards models were presented. In these chapters, the focus was on estimating and testing covariate effects assuming that the model was correctly chosen. Only limited checks of the assumptions were made. A check of the proportional hazards assumption in the Cox model with a constructed time-dependent covariate was used. In this chapter, we shall look at a series of regression diagnostics for the Cox models, based on residual plots.

We are interested in examining four aspects of the proportional hazards model. First, for a given covariate, we would like to see the best functional form to explain the influence of the covariate on survival, adjusting for other covariates. For example, for a given covariate Z, is its influence on survival best modeled by $h_o(t) \exp(\beta Z)$, by $h_o(t) \exp(\beta \log Z)$, or, perhaps, by a binary covariate defined by 1 if $Z \geq Z_o$; 0 if $Z < Z_o$? In the last case, the choice of Z_o was discussed in section 8.6.

The second aspect of the model to be checked is the adequacy of the proportional hazards assumption. If this assumption is not valid, then, one may be appreciably misled by the results of the analyses. While we have looked at the use of a time-dependent covariate to check this assumption, a graphical check may provide some additional insight into any departure from proportionality.

The third aspect of the model to be checked is its accuracy for predicting the survival of a given subject. Here, we are interested in patients who died either too early or too late compared to what the fitted model predicts. This will tell us which patients are potential outliers and, perhaps, should be excluded from the analysis.

The final aspect of the model to be examined is the influence or leverage each subject has on the model fit. This will also give us some information on possible outliers.

In the usual linear regression setup, it is quite easy to define a residual for the fitted regression model. In the regression model presented in Chapters 8 and 9, the definition of the residual is not as clear-cut. A number of residuals have been proposed for the Cox model. Different residuals are useful for examining different aspects of the model. In section 11.2, we present the first residual proposed for the Cox model, the so-called *Cox–Snell (1968) residuals*. These residuals are useful for checking the overall fit of the final model.

In section 11.3, the notion of a *martingale residual* is presented. These residuals are useful for determining the functional form of a covariate to be included in a proportional hazards regression model.

In section 11.4, graphical checks of the proportional hazards assumption in the Cox model are presented. These checks include the use of *score residuals, Arjas plots*, and plots based on estimates of the cumulative hazard from a stratified model.

In section 11.5, the problem of examining model accuracy for each individual is addressed. Here, the use of the *deviance residual* for this purpose is illustrated.

In section 11.6, the problem of determining leverage points is addressed. Here, we wish to estimate, efficiently, the difference between an estimate of β based on a full sample and one based on a sample with the ith observation omitted. Approaches to this problem are based on the *partial residual* or *score residual*.

11.2 Cox–Snell Residuals for Assessing the Fit of a Cox Model

The Cox and Snell (1968) residuals can be used to assess the fit of a model based on the Cox proportional hazards model. Suppose a Cox model was fitted to the data $(T_j, \delta_j, \mathbf{Z}_j)$, $j = 1, \ldots, n$. For simplicity we assume that $\mathbf{Z}_j = (Z_{j1}, \ldots, Z_{jp})^t$ are all fixed-time covariates. Suppose that the proportional hazards model $h(t \mid \mathbf{Z}_j) = h_o(t) \exp(\Sigma \beta_k Z_{jk})$ has been fitted to the model. If the model is correct, then, it is well known that, if we make the probability integral transformation on the true death time X, the resulting random variable has a uniform distribution on the unit interval or that the random variable $U = H(X_j \mid \mathbf{Z}_j)$ has an exponential distribution with hazard rate 1. Here, $H(x \mid \mathbf{Z}_j)$ is the true cumulative hazard rate for an individual with covariate vector $\mathbf{Z}_j$.

If the estimates of the β's from the postulated model are $\mathbf{b} = (b_1, \ldots, b_p)^t$, then, the Cox–Snell residuals are defined as

$$r_j = \hat{H}_o(T_j) \exp \left(\sum_{k=1}^{p} Z_{jk} b_k \right), j = 1, \ldots, n. \qquad (11.2.1)$$

Here, $\hat{H}_o(t)$ is Breslow's estimator of the baseline hazard rate defined by Eq. (8.8.2) in section 8.8. If the model is correct and the b's are close to the true values of β then, the r_j's should look like a censored sample from a unit exponential distribution.

To check whether the r_j's behave as a sample from a unit exponential, we compute the Nelson–Aalen estimator of the cumulative hazard rate of the r_j's, discussed in section 4.2 (see Eq. (4.2.3)). If the unit exponential distribution fits the data, then, this estimator should be approximately equal to the cumulative hazard rate of the unit exponential $H_E(t) = t$. Thus, a plot of the estimated cumulative hazard rate of the r_j's, $\hat{H}_r(r_j)$, versus r_j should be a straight line through the origin with a slope of 1.

The Cox–Snell residual can be defined for more complex proportional hazards models. For example, for a model with time-dependent covariates and/or a model where the baseline hazard is stratified on some risk factor, if we let $\hat{H}_{oj}(t)$ be the appropriate baseline cumulative hazard estimator for the jth individual, then, the residual is given by

$$r_j = \hat{H}_{oj}(T_j) \exp \left[\sum_{k=1}^{p} Z_{jk}(T_j) b_k \right], j = 1, \ldots, n. \qquad (11.2.2)$$

EXAMPLE 11.1   We shall use the Cox–Snell residual plots to check the fit of a model for disease-free survival following a bone marrow transplantation. The data is presented in section 1.3, and this study was previously analyzed using the proportional hazards model in Examples 8.5 and 9.1. Here shall focus on the effects of the following covariates on disease-free survival:

Z_1 = Patient age at transplant (centered at 28 years)

Z_2 = Donor age at transplant (centered at 28 years)

$Z_3 = Z_1 \times Z_2$ (Patient–Donor age interaction)

$Z_4 =$ "Low-Risk" acute leukemia

$Z_5 =$ "High-Risk" acute leukemia

$Z_6 =$ FAB score of 4 or 5 and acute leukemia

$Z_7 =$ Waiting time to transplant in months (centered at 9 months)

$Z_8 =$ MTX used as an aGVHD prophylactic

A Cox regression model is fit to the data with these eight covariates, and the residuals r_j are computed using (11.2.1). The data (r_j, δ_j), $j = 1, \ldots, 137$ is, then, used to obtain the Nelson–Aalen estimate of the cumulative hazard rate of r_j. Figure 11.1 is a plot of the residuals versus the estimated cumulative hazard of the residuals. If the Cox model fits the data, the plot should follow the 45° line. The plot suggests that this model does not fit too badly.

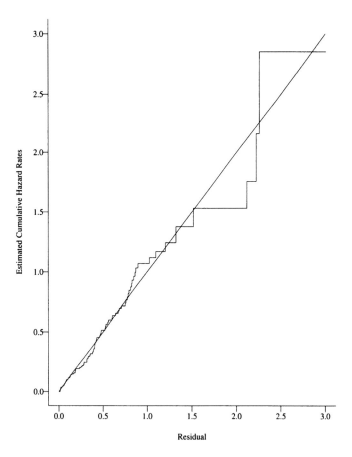

Figure 11.1 *Cox–Snell residual plot treating MTX as a fixed time covariate*

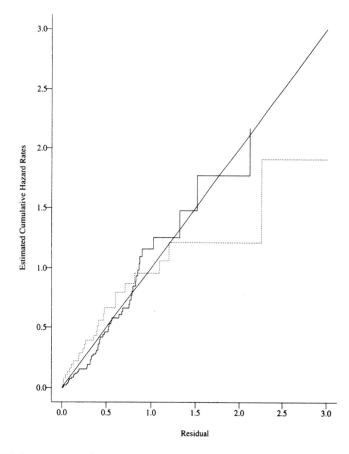

Figure 11.2   *Cox–Snell residual plots for MTX and no MTX patients separately treating MTX as a fixed covariate in the model. MTX patients (------) No MTX patients (———)*

To further examine the model, we construct the Nelson–Aalen estimator of the cumulative hazard rate of the residuals, based on the complete data set, separately, for patients given MTX and not given MTX. Figure 11.2 shows the resulting hazard plot for each of the groups. The estimated cumulative hazard rate for the MTX group (dotted line) lies above the 45° line, except in the tail where the variability in the estimate of the cumulative hazard rate is large. This suggests that a model stratified on the use of MTX may be more appropriate. A formal test of the hypothesis of proportional hazards for MTX, with a time-dependent covariate $Z_9(t) = Z_8 \ln t$, yields a p-value of 0.0320 from the Wald test. Figure 11.3 plots the estimates of the cumulative hazard rates of the residuals from a Cox model stratified on the use of MTX (see (11.2.2)). Here, both estimates seem to be close to the 45° line, except in the

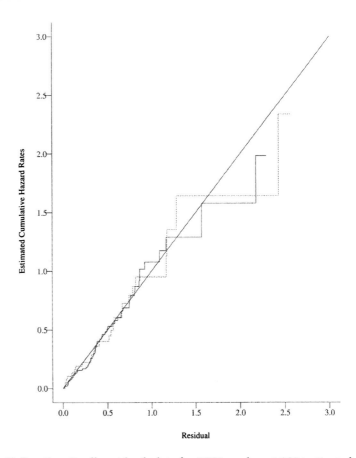

Figure 11.3 *Cox–Snell residual plots for MTX and no MTX patients based on a model stratified on MTX usage. MTX patients (------) No MTX patients (————)*

tail where the estimates are quite variable. This analysis suggests that the stratified model fits better than the unstratified model. In section 11.4, we shall look at other methods for checking the proportionality assumption.

Practical Notes

1. The Cox–Snell residuals are most useful for examining the overall fit of a model. A drawback of their use is they do not indicate the type of departure from the model detected when the estimated cumulative hazard plot is not linear (see Crowley and Storer 1983).
2. Cox–Snell residuals are available in SAS Proc PHREG and S-Plus routine coxph.

3. The closeness of the distribution of the r_j's to the unit exponential depends heavily on the assumption that, when β and $H_o(t)$ are replaced by their estimates, the probability integral transform still yields uniform $[0, 1]$ distributed variates. This approximation is somewhat suspect for small samples (see Lagakos, 1981).

4. Kay (1984) surveys other plots one may make based on these residuals. For example, one may plot the residual from a model with a covariate omitted versus the omitted covariate to assess how the omitted covariate should be modeled. For this purpose, we prefer the use of the related martingale residual discussed in the following section.

5. The Cox–Snell residuals should be used with some caution. The exponential distribution for the residuals holds only when the actual parameter values are used in (11.2.1). When the estimates of these quantities are used to compute the residuals, as we do here, departures from the exponential distribution may be partly due to the uncertainty in estimating β and H. This uncertainty is the largest in the right-hand tail of the distribution and for small samples.

11.3   Determining the Functional Form of a Covariate: Martingale Residuals

In this section, we shall examine the problem of determining the functional form to be used for a given covariate to best explain its effect on survival through a Cox proportional hazards model. The best functional form may be a transform of the covariates, such as $\log Z$, Z^2, or $Z \log Z$, or it may be a discretized version of the covariate. As discussed in section 8.4, it is common practice in many medical studies to discretize continuous covariates, and the residuals presented here are useful for determining cut points for the covariates.

The residual we shall use here, called a *martingale residual*, is a slight modification of the Cox–Snell residual discussed in the previous section. To define the martingale residual in the most general sense, suppose that, for the jth individual in the sample, we have a vector $\mathbf{Z}_j(t)$ of possible time-dependent covariates. Let $N_j(t)$ have a value 1 at time t if this individual has experienced the event of interest and 0 if the individual has yet to experience the event of interest. Let $Y_j(t)$ be the indicator that individual j is under study at a time just prior to time t. Finally, let $\mathbf{b}$ be the vector of regression coefficients and $\hat{H}_o(t)$ the Breslow estimator of the cumulative baseline hazard rate.

The martingale residual is defined as

$$\hat{M}_j = N_j(\infty) - \int_0^\infty Y_j(t) \exp[\mathbf{b'Z}_j(t)] d\hat{H}_o(t), \ j = 1, \ldots, n. \quad (11.3.1)$$

When the data is right-censored and all the covariates are fixed at the start of the study, then, the martingale residual reduces to

$$\hat{M}_j = \delta_j - \hat{H}_o(T_j) \exp\left(\sum_{k=1}^p Z_{jk} b_k\right) = \delta_j - r_j, \ j = 1, \ldots, n. \quad (11.3.2)$$

The residuals have the property $\sum_{j=1}^n \hat{M}_j = 0$. Also, for large samples the $\hat{M}_j$'s are an uncorrelated sample from a population with a zero mean.

The martingale residuals are motivated by the property that, if the true value of $\boldsymbol{\beta}$ and $H_o()$, rather than the sample values, were used in (11.3.1), then, the functions M_j would be martingales (see section 3.6 and the theoretical notes in this section). The residuals can be interpreted as the difference over time of the observed number of events minus the expected number of events under the assumed Cox model, that is, the martingale residuals are an estimate of the excess number of events seen in the data but not predicted by the model. In this section, we shall use these residuals to examine the best functional form for a given covariate using an assumed Cox model for the remaining covariates. For computational simplicity, we will focus on fixed-time covariates. Suppose that the covariate vector $\mathbf{Z}$ is partitioned into a vector $\mathbf{Z}^*$, for which we know the proper functional form of the Cox model, and a single covariate Z_1 for which we are unsure of what functional form of Z_1 to use. We assume that Z_1 is independent of $\mathbf{Z}^*$. Let $f(Z_1)$ be the best function of Z_1 to explain its effect on survival. Our optimal Cox model is, then,

$$H(t \mid \mathbf{Z}^*, Z_1) = H_o(t) \exp(\boldsymbol{\beta}^* \mathbf{Z}^*) \exp[f(Z_1)]. \quad (11.3.3)$$

To find f, we fit a Cox model to the data based on $\mathbf{Z}^*$ and compute the martingale residuals, $\hat{M}_j, \ j = 1, \ldots, n$. These residuals are plotted against the value of Z_1 for the jth observation. A smoothed fit of the scatter diagram is, typically, used. The smoothed-fitted curve gives an indication of the function f. If the plot is linear, then, no transformation of Z_1 is needed. If there appears to be a threshold, then, a discretized version of the covariate is indicated. This process is illustrated in the following example.

EXAMPLE 11.2   To illustrate the use of martingale residuals in model selection, consider the data comparing allogeneic (allo) bone marrow transplants from an HLA-matched sibling or an autogeneic (auto) bone marrow transplant for patients with either Hodgkin's disease (HL) or non-Hodgkin's lymphoma (NHL) (see section 1.10) first analyzed in Example 7.6. Two additional covariates to be adjusted for in the analysis are the patient's initial Karnofsky score, a measure of the patient's condition at transplant, and Z_1 the waiting time from diagnosis to transplant in months. We shall examine the best function of the variable Z_1 to be used in an analysis. Covariates included in the Cox Model are the type of transplant (1-allo, 0-auto), Disease (1-HL, 0-NHL), a disease-transplant type interaction (1 if allo and HL, 0 otherwise), and the initial Karnofsky score. A Cox model is fitted with these four covariates and the martingale residuals are plotted along with a LOWESS smooth (Cleveland (1979)). The LOWESS smoothing function is available in both SAS and S-plus. Figure 11.4 shows the results. The smoothed curve is roughly

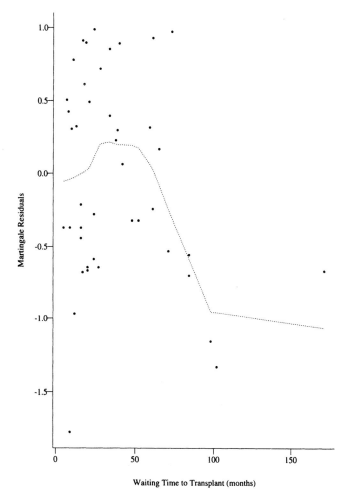

Figure 11.4 *Plot of martingale residual verus waiting time to transplant and LOWESS smooth*

zero up to about 50 months, decreases linearly up to about 100 months, and, then, levels off. This suggests that waiting time from diagnosis to transplant can be coded in the Cox model as a indicator variable.

To find the best break point for the variable Z_1, we apply the techniques in section 8.4. Here we find the best cut point is at 84 months. The value of the scaled score statistic is 0.697, which is not significant.

Practical Notes

1. Martingale residuals are available in SAS PROC PHREG and in S-plus.
2. Some authors have suggested plots of the martingale residuals against survival times or against the ranks of the survival times as an indication of model fit. We prefer to use other residuals for assessing the general lack of fit of models because the martingale residuals are skewed.

Theoretical Notes

1. The martingale residual is based on the fact that the process

$$M_j(t) = N_j(t) - \int_0^t Y_j(u) \exp[\boldsymbol{\beta}'\mathbf{Z}_j(u)]dH_o(u) \qquad (11.3.4)$$

is a martingale when the proportional hazards model is correctly specified. The martingale residual, defined in (11.3.1), is obtained by substituting the estimates of $\boldsymbol{\beta}$ and $H_o()$ in this expression and evaluating the estimated martingale at time $t = \infty$. To see why the martingale residuals can be used for model identification, suppose that the model (11.3.3) holds. Let $g(Z_1) = \exp[f(Z_1)]$. Then, the expected value of $M_j(t)$, given Z_{1j}, is approximately equal to $[1 - g^*/g(Z_{1j})]E[N_j(t) \mid Z_{1j}]$, where g^* is the average value of $g(Z_1)$ over both time and the expected distribution of the risk set at a given time. If one expands this expectation in a Taylor series, we see that

$$E(M_j \mid Z_{1j}) \approx [f(Z_{1j}) - \log(g^*)]w,$$

where w is the total number of events divided by the total sample size. Thus, a smoothed plot of $\hat{M}_j$ versus a covariate should reveal the correct functional form for including Z_1 in a Cox model. A detailed derivation of these results is found in Fleming and Harrington (1991) or Therneau et al. (1990).

11.4 Graphical Checks of the Proportional Hazards Assumption

The validity of Cox's regression analysis relies heavily on the assumption of proportionality of the hazard rates of individuals with distinct values of a covariate. In Chapter 9, we examined techniques for testing this assumption using time-dependent covariates. In this section, we shall look at several graphical techniques for checking this assumption. We are interested in checking for proportional hazards for a given covariate Z_1 after adjusting for all other relevant covariates in the model, that is, we write the full covariate vector as $\mathbf{Z} = (Z_1, \mathbf{Z}_2')'$ where $\mathbf{Z}_2$ is the vector of the remaining $p - 1$ covariates in the model. We assume that there is no term in the model for interaction between Z_1 and any of the remaining covariates.

The first series of plots requires that the covariate Z_1 has only K possible values. For a continuous covariate, we stratify the covariate into K disjoint strata, $G_1, G_2, \ldots, G_K$, whereas, for a discrete covariate, we assume that Z_1 takes only the values $1, 2, \ldots, K$. We, then, fit a Cox model stratified on the discrete values of Z_1, and we let $\hat{H}_{go}(t)$ be the estimated cumulative baseline hazard rate in the gth stratum. If the proportional hazards model holds, then, the baseline cumulative hazard rates in each of the strata should be a constant multiple of each other. This serves as the basis of the first graphical check of the proportional hazards assumption.

To check the proportionality assumption one could plot $\ln[\hat{H}_{1o}(t)]$, $\ldots, \ln[\hat{H}_{Ko}(t)]$ versus t. If the assumption holds, then, these should be approximately parallel and the constant vertical separation between $\ln[\hat{H}_{go}(t)]$ and $\ln[\hat{H}_{ho}(t)]$ should give a crude estimate of the factor needed to obtain $\hat{H}_{ho}(t)$ from $\hat{H}_{go}(t)$. An alternative approach is to plot $\ln[\hat{H}_{go}(t)] - \ln[\hat{H}_{1o}(t)]$ versus t for $g = 2, \ldots, K$. If the proportional hazards model holds, each curve should be roughly constant. This method has the advantage that we are seeking horizontal lines for each curve rather than comparing parallel curves. Note that both plots give us only the information that the baseline hazards for each stratum are not proportional. The plots do not give detailed information about the type of departure from proportionality we are seeing.

Another graphical method based on $\hat{H}_{go}(t)$ is the so-called Andersen (1982) plots. Here, we plot, for all t (or for a selected subset of t), $\hat{H}_{go}(t)$ versus $\hat{H}_{1o}(t)$ for $g = 2, \ldots, K$. If the proportional hazards model holds, these curves should be straight lines through the origin. If $H_{go}(t) = \exp(\gamma_g)H_{1o}(t)$, then, the slope of these lines is a crude estimate of $\exp(\gamma_g)$. Gill and Schumacher (1987) have shown that, if the plot of

$\hat{H}_{go}(t)$ versus $\hat{H}_{10}(t)$ is a convex (concave) function, then, the ratio $h_{go}(t)/h_{10}(t)$ is an increasing (decreasing) function of t. If the plot is piecewise linear, then, this ratio is piecewise constant. All three plots should be interpreted with some care because the variances of the curves are not constant over time.

EXAMPLE 11.3

Here, we shall check the proportional hazards assumption graphically for the data set comparing the difference in disease-free survival between patients given an autologous (auto) and allogeneic (allo) bone marrow transplant for acute leukemia (see Section 1.9).

Because there is a single covariate Z taking a value 1 if the patient has an auto transplant and 0 if the patient had an allo transplant for this data set, the estimated baseline hazard rates in each group reduce to the Nelson–Aalen estimators. Figure 11.5 shows a plot of the logarithm of

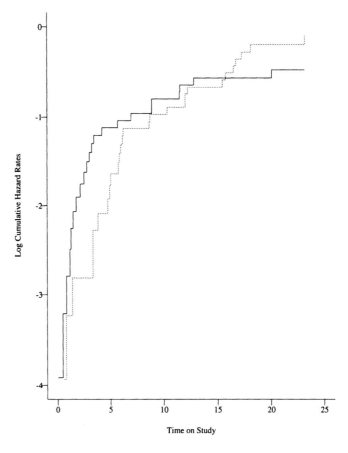

Figure 11.5 *Plot of log cumulative baseline hazard rates versus time on study for the allo transplant (——) and the auto transplant (------) groups.*

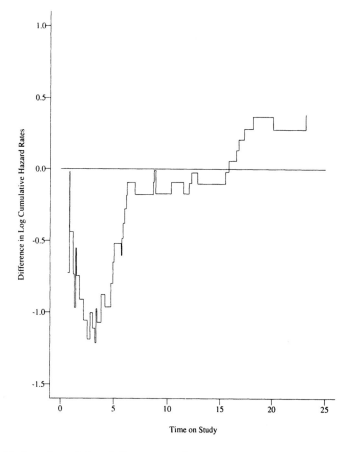

Figure 11.6   *Plot of the difference in log cumulative baseline hazard rates (Auto-Allo) versus time on study*

the two Nelson–Aalen estimators. We see that the two baseline hazard rates are not proportional and, in fact, the baseline hazard rates cross, clearly suggesting nonproportional hazards. Figure 11.6 shows the difference in log baseline cumulative hazard rates (auto-allo) for the two samples. Again, the proportional hazards assumption is rejected because the plotted curve is not roughly constant over time. This figure shows an early advantage for autologous transplants due, in part, to less early transplant-related mortality. After this early period, we see an advantage for the allogeneic transplants, primarily, due to a decreased relapse rate in these patients.

Figure 11.7 gives the Andersen plot for these data. If the model held, we would have expected a linear plot through the origin, but this is not the case in this plot. The plot appears to be roughly concave, suggesting that the ratio $h(t \mid \text{allo})/h(t \mid \text{auto})$ is a function decreasing with time.

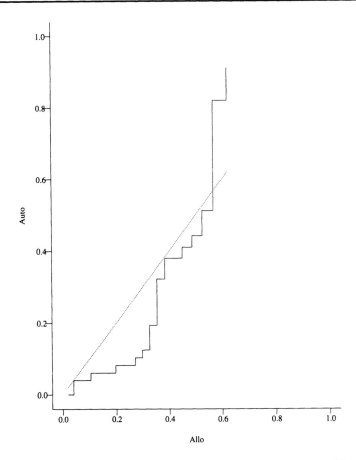

Figure 11.7   *Andersen plot to check the assumption of proportional hazards of the type of transplant covariate*

EXAMPLE 11.1   *(continued):*   We shall use these techniques to examine the proportional hazards assumption for the binary covariate Z_8, the use of MTX as a GVHD prophylactic, and, separately, for the covariate Z_7, the waiting time from diagnosis to transplant, for the bone marrow transplant data. To examine the proportionality of Z_8, we first fit a Cox model with the covariates Z_1, Z_2, Z_3, Z_4, Z_5, Z_6, and Z_7 stratified on the use of MTX. The baseline cumulative hazards are estimated using Breslow's estimator (8.8.2) for each stratum. Figures 11.8–11.10 show the log cumulative baseline hazards, difference in log cumulative baseline hazards, and the Andersen plot, based on these estimates. These figures are suggestive of nonproportional hazards. In particular, Figure 11.9 suggests that the proportional hazards model is questionable about 80 days after transplant.

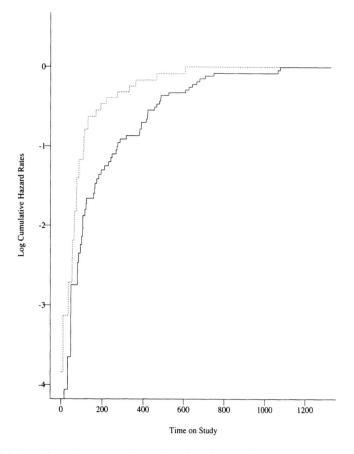

Figure 11.8 *Plot of log cumulative baseline hazard rates versus time on study for the no MTX (————) and the MTX (------) groups.*

To examine the proportionality assumption for Z_7, we stratify this continuous covariate as

$$Z_7^* = \begin{cases} 1 & \text{if } Z_7 \leq -5 \\ 2 & \text{if } -5 < Z_7 \leq -3.06 \\ 3 & \text{if } -3.06 < Z_7 \leq 0 \\ 4 & \text{if } Z_7 > 0 \end{cases}.$$

These categories were chosen to give about 25% of the data in each of the strata. We, then, fit a Cox model with the covariates $Z_1, Z_2, Z_3, Z_4, Z_5, Z_6$, and Z_8 stratified on Z_7^*, and the four baseline cumulative hazard rates were estimated. Figure 11.11 shows a plot of the log cumulative baseline rates, Figure 11.12 shows a plot of $\ln[\hat{H}_{g o}(t)] - \ln[\hat{H}_{1 o}(t)]$ versus t for $g = 2, 3, 4$, and Figure 11.13 shows

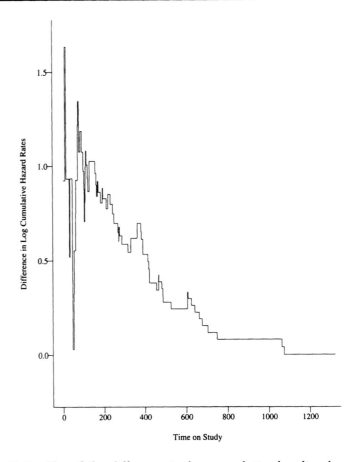

Figure 11.9 *Plot of the difference in log cumulative baseline hazard rates (MTX – No MTX) versus time on study*

the Andersen plot of $\hat{H}_{g0}(t)$ versus $\hat{H}_{10}(t)$, for $g = 2, 3, 4$. In Figure 11.11, we see no gross departure from the hypothesis of parallel curves, and, in Figure 11.12, although there is some early random variation from a constant curve, we see no gross departure from the proportional hazards assumption. The Andersen curves also appear to be approximately linear. All the figures suggest that there is little difference in hazard rates for individuals who are in strata 1–3 whereas individuals in stratum 4 are at higher risk of death or relapse. This suggests that, although the proportional hazards model may be reasonable for the waiting time to transplant, the assumption that waiting time has a linear effect on the event may be suspect.

A second method of checking the proportional hazards assumption is the use of Arjas (1988) plots. These plots can also be used to check the

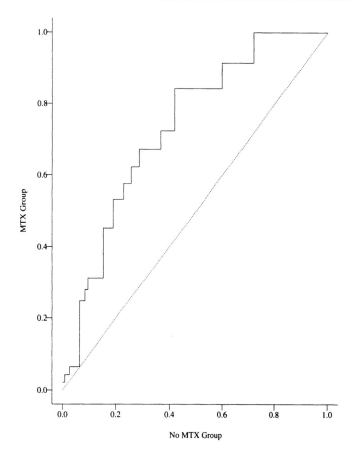

Figure 11.10   *Andersen plot to check the assumption of proportional hazards of the MTX risk factor*

overall fit of the proportional hazards regression model. Suppose that a Cox model has been fitted with a covariate vector $\mathbf{Z}^*$ of p variables and we wish to check if an additional covariate Z should be included in the model or if the new covariate has proportional hazards after adjustment for covariates $\mathbf{Z}^*$. We fit the proportional hazards model with the covariates $\mathbf{Z}^*$ and let $\hat{H}(t \mid \mathbf{Z}_j^*)$ be the estimated cumulative hazard rate for the jth individual in the sample at time t. Note that, in this approach, the fitted model ignores Z_1 and does not stratify on Z_1 as in the Andersen plots. As before, we group values of Z_1 into K classes if the covariate is continuous. At each event time for each level of Z_1, we compute the "total time on test" of the estimated cumulative hazard rates up to this time and the observed number of events that

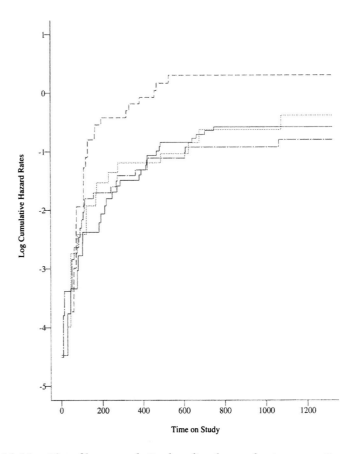

Figure 11.11   *Plot of log cumulative baseline hazard rates versus time on study for the four waiting time to transplant groups. $Z_7 \leq -5$ (———) $-5 < Z_7 \leq -3.06$ (------) $-3.06 < Z_7 \leq 0$ (— · —) $Z_7 > 0$ (— — —)*

have occurred up to this time. That is, at each event time t_i, we compute

$$\text{TOT}_g(t_i) = \sum_{Z_{1j=g}} \hat{H}(\min(t_i, T_j) \mid \mathbf{Z}_j^*), \qquad (11.4.1)$$

and

$$N_g(t_i) = \sum_{Z_{1j=g}} \delta_j I(T_j \leq t_i). \qquad (11.4.2)$$

If the covariate Z_1 does not need to be in the model, then, for each level of Z_1, the quantity $N_g(t_i) - \text{TOT}_g(t_i)$ is a martingale and a plot of $N_g(t_i)$ versus $\text{TOT}_g(t_i)$ should roughly be a 45° line through the origin. Departures from this pattern provide evidence of a lack of fit

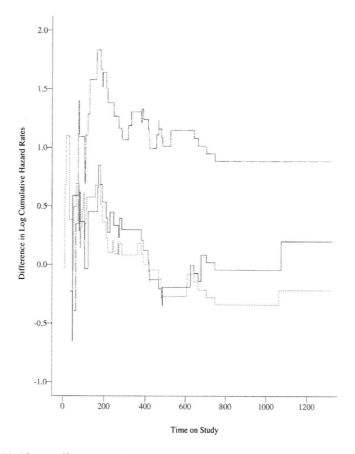

Figure 11.12   *Difference in log cumulative baseline hazard rates of the waiting time to transplant strata. Strata 2-strata 1 (———) Strata 3-strata 1 (------) Strata 4 -strata 1 (— · —)*

of the model. If covariate Z_1 should be included in the model, so that the correct model is $h(t \mid Z_1 = g, \mathbf{Z}^*) = h_0(t) \exp\{\gamma_g\} \exp(\boldsymbol{\beta}'\mathbf{Z}^*)$, then, plots will gives curves which are approximately linear but with slopes differing from 1. If the omitted covariate Z_1 has a nonproportional hazards effect on the hazard rate, then, the curves will differ nonlinearly from the 45° line. Particular departures from linearity give some clue to the relationship between hazards for individuals with different levels of Z_1. For example, if the true model is $h(t \mid Z_1 = g, \mathbf{Z}^*) = h_{og}(t) \exp(\boldsymbol{\beta}'\mathbf{Z}^*)$, then, if the ratio $h_{go}(t)/h_{g'o}(t)$ is increasing in t for $g < g'$, the curve for $Z_1 = 1$ should be concave whereas the curve with $Z_1 = K$ should be convex.

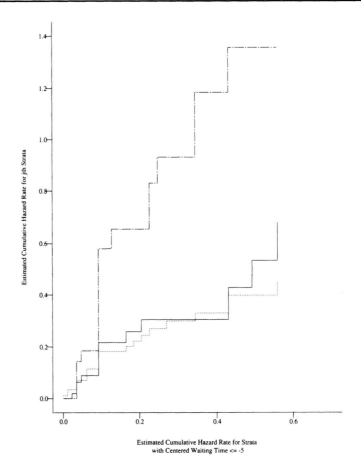

Figure 11.13   *Andersen plot to check the assumption of proportional hazards of the waiting time to transplant risk factor. Strata 2 (———) Strata 3 (------) Strata 4 (— · —)*

EXAMPLE 11.3

(continued):   Using an Arjas plot, we shall examine the proportionality assumption for comparing the disease-free survival rates of allo and auto transplant patients. In this case, no Cox model is fitted, but, rather, the residuals are the Nelson–Aalen estimates of the cumulative hazard rate, ignoring the covariate Z. These estimates are stratified on the binary covariate Z and the statistics N_g and $\text{TOT}_g, g = 1, 2$ are computed. Figure 11.14 shows the diagnostic plot. Because neither curve closely follows the 45° line, the type of transplant should be adjusted for in the model. The nonlinear appearance of the two curves strongly suggests a nonproportional hazards correction.

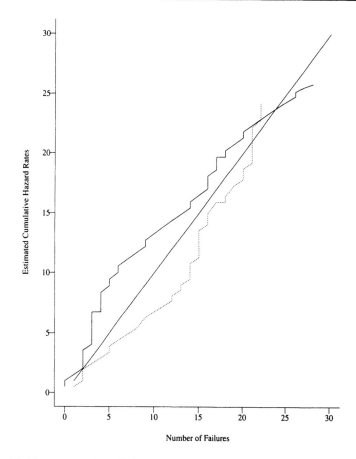

Figure 11.14 *Arjas plot of the estimated cumulative hazard rates for auto transplant (———) and allo transplant (------) patients versus the number of events*

EXAMPLE 11.1   *(continued):*   Now, we will use the Arjas plots to examine the effect of MTX, as a prophylatic treatment, and, separately, the role of time from diagnosis to transplant in disease-free survival. Figure 11.15 shows the Arjas plot for the MTX and no MTX groups. Here, the fitted Cox model included the factors $Z_1, Z_2, Z_3, Z_4, Z_5, Z_6$, and Z_7. The two curves vary appreciably from the 45° line, the MTX curve appears to be concave, and the no MTX curve convex. This suggests that the ratio of the MTX baseline hazard to the no MTX baseline hazard is increasing with time, and a stratified model is appropriate.

Figure 11.16 shows the Arjas plot for the covariate Z_7^*, the grouped values of the waiting time to transplant. Here, all of the curves seem to

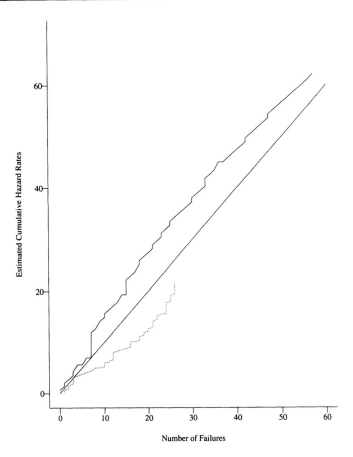

Figure 11.15   *Arjas plot of the estimated cumulative hazard rates for no MTX patients (————) and MTX patients (------) versus the number of events*

fit closely to the 45° line, suggesting that this covariate has little effect on disease-free survival and can be dropped from the model.

A third approach to examining a model for departures from proportional hazards for a given covariate is the use of plots based on the score residuals. To construct these plots, we first fit the Cox model with all p covariates. Here, no transformation is made to discretize the continuous covariates. Let **b** be the estimate of the risk coefficient β and $\hat{H}_o(t)$ be the estimated baseline hazard rate. As in the definition of the martingale residual, let $N_j(t)$ indicate whether the jth individual has experienced the event of interest and let $Y_j(t)$ indicate that individual j is under study at a time just prior to time t. For the kth covariate,

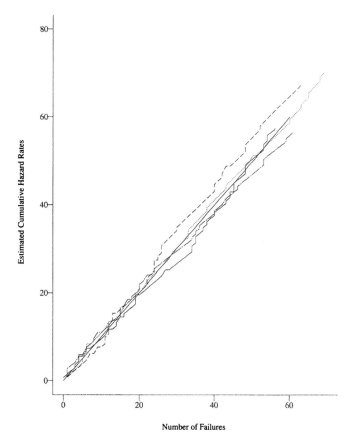

Number of Failures

Figure 11.16 *Arjas plot of the estimated cumulative hazard rates to assess the proportionality of the waiting time to transplant risk factor. $Z_7 \le -5$ (————) $-5 < Z_7 \le -3.06$ (------) $-3.06 < Z_7 \le 0$ (— · —) $Z_7 > 0$ (— — —)*

$k = 1, \ldots, p$, let

$$\bar{Z}_k(t) = \frac{\sum\limits_{j=1}^{n} Y_j(t) Z_{jk}(t) \exp[\mathbf{b}'\mathbf{Z}_j(t)]}{\sum\limits_{j=1}^{n} Y_j(t) \exp[\mathbf{b}'\mathbf{Z}_j(t)]} \qquad (11.4.3)$$

Note that, in (11.4.3), only individuals at risk at time t contribute to the sums. Finally let $\hat{M}_j(t)$ be the martingale residual at time t for individual j defined by

$$\hat{M}_j(t) = N_j(t) - \int_0^t Y_j(u) \exp[\mathbf{b}'\mathbf{Z}_j(u)]d\hat{H}_o(u), \, j = 1, \ldots, n. \quad (11.4.4)$$

The score residual for the kth covariate and the jth individual at time t is defined by

$$S_{jk}(t) = \int_0^t \{Z_{jk}(u) - \bar{Z}_k(u)\}d\hat{M}_j(u). \qquad (11.4.5)$$

Using the scores for each of the n individuals, we define a score process for the kth covariate as

$$U_k(t) = \sum_{j=1}^n S_{jk}(t). \qquad (11.4.6)$$

When all the covariates are fixed at time 0 the score process for the kth covariate is

$$U_k(t) = \sum_{\text{deaths} \le t} (Z_{jk} - \bar{Z}_k(T_j)). \qquad (11.4.7)$$

The terms $[Z_{jk} - \bar{Z}_k(T_j)]$ at death times are the Schoenfeld residuals available in SAS or S-plus. The scores process is the first partial derivative of the partial likelihood function for the fitted Cox model, using only the information accumulated up to time t. Clearly $U_k(0) = 0$ and $U_k(\infty) = 0$ because the value of $\boldsymbol{\beta}$, used in constructing the score residuals, is the solution to the vector equation $U_k(\infty) = 0$, $k = 1, \ldots, p$. If the model fits properly, then, the process, $W_k(t) = U_k(t) \times$ Standard Error of b_k, converges to a tied down Brownian bridge (provided $\text{Cov}(b_k, b'_k) = 0$, for $k \ne k'$). Thus, a plot of $W_k(t)$ versus time should look like a tied down random walk. If the hazard rates for different levels of a covariate are nonproportional, then, the plots will have a maximum value that is too large in absolute value at some time.

The use of score residuals to check for nonproportional hazards has several advantages over the other two approaches. First, continuous covariates are treated naturally and need not be discretized. Second, only a single Cox model needs to be fit to check for proportional hazards for all covariates in the model. However, because the power of the score process to detect nonproportional hazards has not been compared to the Andersen or Arjas plots, we recommend that all three approaches be used to check for proportional hazards.

EXAMPLE 11.3   *(continued):* We shall use the score residuals to check the assumption of proportional hazards for type of transplant on the disease-free survival time of patients given a bone marrow transplant. The score residuals are computed by, first, fitting a Cox model with the single covariate Z and, then, computing the score process $U(t)$ in (11.4.7). This process is standardized by multiplying by the estimated standard

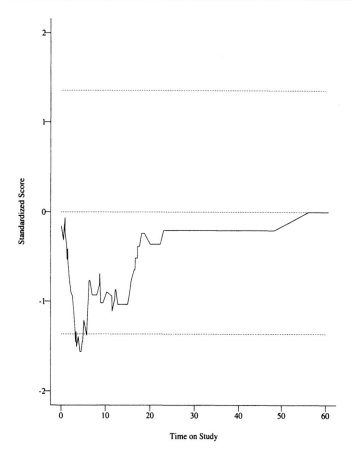

Figure 11.17   *Standardized score residuals to check the proportionality of the type of transplant factor on disease-free survival*

deviation of *b*. Figure 11.17 shows the score process. The dotted lines at ±1.3581 are chosen so that the probability that the supremum of a Brownian bridge is beyond these values is 0.05 at most. Should the plot exceed these boundaries at any point, the assumption of proportional hazards can be rejected at a 5% level of significance. Here again, we find evidence of nonproportionality of the hazards for patients given different types of transplants.

These findings suggest that the tests discussed in section 7.3, which are based on a weighted difference between the hazard rates in the two groups, should be used with caution and the tests based on the Renyi statistic in section 7.6 will have higher power to detect differences between the two groups.

EXAMPLE 11.1   *(continued):* Now, we will use the score process plots to examine the effect of MTX as a prophylactic treatment and the role of time from diagnosis to transplant on disease-free survival. To compute the score residuals, we now fit a Cox model with all eight covariates. Note that, here, the waiting time from diagnosis to transplant is treated as a continuous covariate rather then a categorical variable, as used in the Andersen and Arjas plots. Figures 11.18 and 11.19 show the standardized score process for the two covariates. As in the previous example, the dotted lines are at ±1.3581 for reference to a 5% level test of the hypothesis of proportional hazards. In Figure 11.18, we see some evidence of nonproportional hazards based on the use of MTX whereas, in Figure 11.19, we see no evidence of nonproportional hazards for the covariate, waiting time to diagnosis.

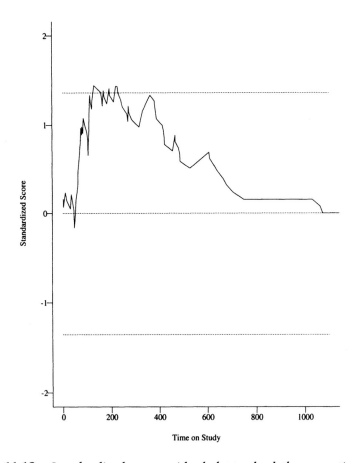

Figure 11.18 *Standardized score residual plot to check the proportionality of the MTX risk factor*

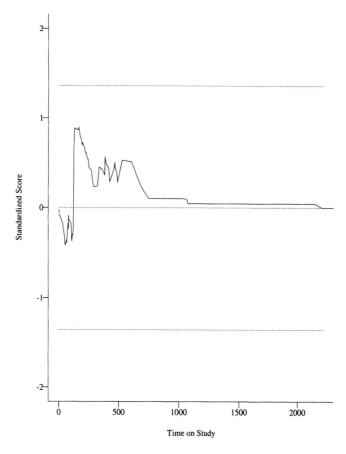

Figure 11.19 *Standardized score residual plot to check the proportionality of the waiting time to transplant risk factor*

Practical Notes

1. The advantage of checking proportionality with the score process is that the covariate need not be discretized to make the plot. A different stratification of a continuous covariate in the other plots described here may lead to conflicting results.
2. We prefer either the Andersen plots or plots based on the difference in baseline hazard rates over simple plots of the baseline hazard rates. In many cases, it is quite difficult to judge proportionality between curves whereas the Andersen and difference in baseline rate curves ask the eye to distinguish the estimated curves from a line.
3. The Arjas plots have an advantage over the other plots based on the baseline hazard rate because they allow the investigator to change

the way the covariate under study is stratified without refitting the basic Cox model.

4. Departures from the assumed model discovered in the right-hand tail of the plots, based on the estimated baseline hazard rates, should be interpreted with caution because these baseline rates are most variable for large time values.

5. Formal tests of the proportionality assumption are available. A good survey of these tests is found in Andersen et al. (1993). We prefer graphical checks to formal goodness-of-fit tests because we believe that the proportional hazards assumption only approximates the correct model for a covariate and any formal test, based on a large enough sample, will reject the null hypothesis of proportionality.

Theoretical Notes

1. The Arjas plots were introduced by Arjas (1988) and are based on the following development. If the proportional hazards model fits the data without inclusion of the covariate Z_1, then, the martingale residuals from the fitted Cox model, based on the remaining covariates $\mathbf{Z}^* = (Z_2, \ldots, Z_p)'$,

$$M_j(t) = N_j(t) - \int_0^t Y_j(u) \exp[\boldsymbol{\beta}' \mathbf{Z}_j^*(u)] dH_o(u)$$

are martingales. This fact that $E[M_j(t)] = 0$ implies that $E[N_j(t)] = E \int_0^t Y_j(u) \exp[\boldsymbol{\beta}' \mathbf{Z}_j(u)] dH_o(u)$. This suggests that, if the model fits, $\sum_{Z_{1j}=g} E[N_j(t)]$ should be equal to

$$\sum_{Z_{1j}=g} E \int_0^t Y_j(u) \exp[\boldsymbol{\beta}' \mathbf{Z}_j(u)] dH_o(u)$$

for each stratum. The Arjas plot is a plot of the estimated values of these quantities.

The Arjas plot can be motivated by an alternative formulation. Recall that if the model with Z_1 omitted fits, the Cox–Snell residual defined by (11.2.2) should be a censored sample from a unit exponential distribution. The Arjas plot is a Total Time on Test plot of these residuals for each stratum (see Barlow and Campo, 1975 for a discussion of Total Time on Test Plots).

Formal hypothesis tests for the assumption of proportional hazards, based on these plots, are found in Arjas and Haara (1988).

2. The Andersen plots are based on an estimate of the "relative trend function" $y = H_g^{-1}[H_1(t)]$, where $H_g(t)$ are the estimated cumulative hazard rates, adjusted for other covariates, for different levels of a covariate Z. If $H_g(t) = e^\gamma H_1(t)$, then the plots should be linear with slope e^γ.

Dabrowska et al. (1989) study these plots, based on a relative change model $(H_g - H_1)/H_1$ and give formal tests of proportionality in the two-sample problem.

3. The score residual $S_{jk}(t)$ is a martingale because it is the integral of a predictable process with respect to a martingale, when the model holds true. This residual is the first partial derivative with respect to β_k of the contribution to the log likelihood of the jth individual using only the information on this person accumulated up to time t. The score process $U_k(t)$ is the first partial derivative with respect to β_k of the log likelihood if the study were stopped at time t. Because the partial likelihood estimator maximizes the final log likelihood it follows that $U_k(\infty)$ is 0. Therneau et al. (1990) show that the normalized $U_k(t)$ process converges weakly to a tied down Brownian bridge process, as long as the asymptotic covariance of b_k and b_h is 0 for $k \neq h$.

11.5   Deviance Residuals

In this section, we shall consider the problem of examining a model for outliers, after a final proportional hazards model has been fit to the data. As in the usual regression formulation, we would like a plot which shows us which, if any, of the observations has a response not well predicted by the fitted model.

In section 11.3, we saw that the martingale residual $\hat{M}_j$ is a candidate for the desired residual. This quantity give us a measure of the difference between the indicator of whether a given individual experiences the event of interest and the expected number of events the individual would experience, based on the model. The problem with basing a search for outliers on the martingale residual is that the residuals are highly skewed. The maximum value of the residual is +1 but the minimum possible value is $-\infty$.

The deviance residual is used to obtain a residual which has a distribution more normally shaped than the martingale residual. The deviance residual is defined by

$$D_j = \text{sign}[\hat{M}_j]\{-2[\hat{M}_j + \delta_j \log(\delta_j - \hat{M}_j)]\}^{1/2}. \quad (11.5.1)$$

This residual has a value of 0 when $\hat{M}_j$ is zero. The logarithm tends to inflate the value of the residual when $\hat{M}_j$ is close to 1 and to shrink large negative values of $\hat{M}_j$.

To assess the effect of a given individual on the model, we suggest constructing a plot of the deviance residuals D_j versus the risk scores

$\sum_{k=1}^{p} b_k Z_{jk}$. When there is light to moderate censoring, the D_j should look like a sample of normally distributed noise. When there is heavy censoring, a large collection of points near zero will distort the normal approximation. In either case, potential outliers will have deviance residuals whose absolute values are too large.

EXAMPLE 11.2

(continued): We shall examine the use of deviance residuals to check for outliers in the model constructed to compare allogeneic (allo) bone marrow transplants from an HLA-matched sibling with an autogeneic (auto) bone marrow transplant for patients with either Hodgkin's lymphoma (HL) or non-Hodgkin's lymphoma (NHL). The covariates used in the final model and the estimates of their risk coefficients are as follows:

Risk Factor: Z	b	se(b)
Z_1: Type of Transplant (1-allo, 0-auto)	−1.3339	0.6512
Z_2: Disease (1-HL, 0-NHL)	−2.3087	0.7306
Z_3: Interaction between disease and type of transplant ($Z_1 \times Z_2$)	2.1096	0.9062
Z_4: Karnofsky score	−0.0564	0.0122
Z_5: Waiting time to transplant (1 if time≥ 84, 0 otherwise)	−2.1001	1.0501

Figures 11.20 and 11.21 show a plot of the martingale residuals and deviance residuals, respectively, against the risk score $-1.3339Z_{j1} - 2.3087Z_{j2} + 2.1096Z_{j3} - 0.0564Z_{j4} - 2.1001Z_{j5}$. An examination of Figure 11.20 does not suggest any potential outliers with the possible exception of the individual with a risk score of about −3.22 who had a martingale residual of −1.88. Upon examining the deviance residual plot, we see that this individual has a deviance residual of −1.28, well within the acceptable range for the deviance residual. Figure 11.21 does suggest two outliers with risk scores of −6.40 ($D_i = 2.26$) and −7.384 ($D_i = 2.78$). The first patient had a covariate vector of $(1, 0, 0, 90, 0)$ and died at 1 month whereas the second patient had a covariate vector of $(0, 1, 0, 90, 0)$ and died at 0.9333 months. Based on their risk profiles, these patients should have had relatively long survival times but they were in fact, two of the shortest lived patients.

Practical Notes

1. Both martingale and deviance residuals are available in SAS and S-plus.

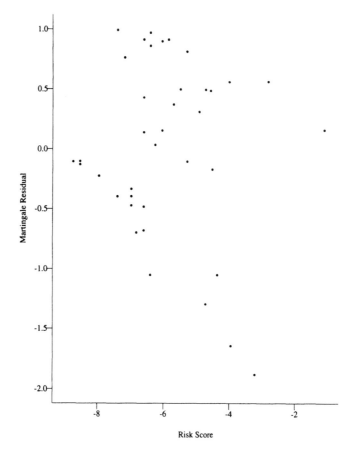

Figure 11.20 *Plot of the martingale residuals versus risk scores for the bone marrow transplant example*

2. Therneau et al. (1990) report that Monte Carlo studies have shown that both types of residuals detect outliers who lived longer than expected by the model. Outliers, who die sooner than expected, are detected only by the deviance residual.

Theoretical Notes

1. The deviance residual is motivated by an examination of the deviance as used in the theory of general linear models (see McCullagh and Nelder, 1989). The deviance of a model is defined as $d = 2$ [Log likelihood (saturated model) − log likelihood (fitted model)] the saturated model is one in which each observation is allowed to have its own coefficient vector $\boldsymbol{\theta}_j$. In computing the

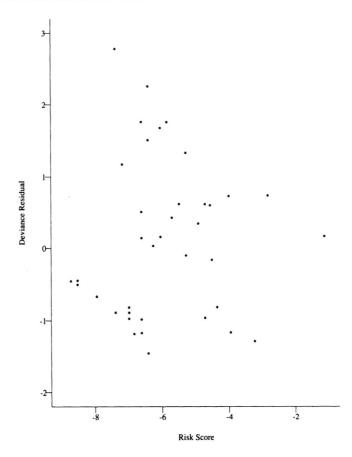

Figure 11.21   *Plot of the deviance residuals versus risk scores for the bone marrow transplant example*

deviance, nuisance parameters (here $H_o(t)$) are held fixed between the saturated and fitted model. For the Cox model, the deviance is

$$d = 2\sup \sum_{j=1}^{n} \int_0^{\infty} [\boldsymbol{\theta}_j' \mathbf{Z}_j - \mathbf{b}' \mathbf{Z}_j] dN_j(s) - \int_0^{\infty} Y_j(s) [\exp(\boldsymbol{\theta}_j' \mathbf{Z}_j)$$

$$- \exp(\mathbf{b}' \mathbf{Z}_j)] dH_o(s).$$

where the supremum is over the vector $\boldsymbol{\theta} = (\boldsymbol{\theta}_1, \ldots, \boldsymbol{\theta}_n)$. Using a Lagrange multiplier argument, one can show that, at the maximum, $\int_0^{\infty} Y_j(s) [\exp(\boldsymbol{\theta}_j' \mathbf{Z}_j)] dH_o(s) = \int_0^{\infty} dN_j(s)$. If we let $\hat{M}_j(t) = N_j(t) - \int_0^t \exp(\mathbf{b}' \mathbf{Z}_j) dH_o(s)$, the martingale residual (11.3.4) with $\boldsymbol{\beta}$ replaced by $\mathbf{b}$, then, after some factorization which is valid if all the covariates are fixed at time 0,

$$d = -2 \sum_{j=1}^{n} \left\{ \hat{M}_j(t_j) + N_j(\infty) \log \left[\frac{N_j(\infty) - \hat{M}_j(t_j)}{N_j(\infty)} \right] \right\}.$$

The deviance residual is the signed square root of the summands with $\hat{M}_j(t_j)$ replaced by the martingale residual. These residuals are similar to the deviance residuals for a Poisson model suggested by McCullagh and Nelder (1989).

11.6   Checking the Influence of Individual Observations

In section 11.5, the deviance residual was introduced as a means of checking for outliers following a Cox regression analysis. In this section, we shall examine graphical techniques to check the influence of each observation on the estimate of $\boldsymbol{\beta}$. The optimal means of checking the influence of a given observation on the estimation process is to compare the estimate $\mathbf{b}$ one obtains by estimating $\boldsymbol{\beta}$ from all the data, to the estimate $\mathbf{b}_{(j)}$ obtained from the data with the given observation deleted from the sample. If $\mathbf{b} - \mathbf{b}_{(j)}$ is close to zero, the jth observation has little influence on the estimate, whereas large deviations suggest a large influence. To compute $\mathbf{b} - \mathbf{b}_{(j)}$ directly requires fitting $n + 1$ Cox regression models, one with the complete data and n with a single observation eliminated. Although this may be feasible in some small-sample-size problems, it is not feasible in larger problems. An approximation based on the Cox model fitted from the complete data can be used to circumvent this computational problem.

The approximation is based on the score residuals defined in section 11.4. Let $S_{jk} = S_{jk}(\infty)$, where $S_{jk}(t)$ is defined by (11.4.5), that is, when all of the covariates are fixed at time 0,

$$S_{jk} = \delta_j[Z_{jk} - \bar{Z}_k(T_j)] - \sum_{t_b \leq T_j} [Z_{jk} - \bar{Z}_k(t_b)] \exp(\mathbf{b}'\mathbf{z}_j)[\hat{H}_o(t_b) - \hat{H}_o(t_{b-1})],$$

$$(11.6.1)$$

for $j = 1, \ldots, n$ and $k = 1, \ldots, p$. The first term $\delta_j\{Z_{jk} - \bar{Z}_k(T_j)\}$, Schoenfeld's (1982) *partial residual*, is the difference between the covariate Z_{jk} at the failure time and the expected value of the covariate at this time. It can be shown that $\mathbf{b} - \mathbf{b}_{(j)}$ is approximated by $\Delta = \mathbf{I}(\mathbf{b})^{-1}(S_{j1}, \ldots, S_{jp})'$, where $\mathbf{I}(\mathbf{b})$ is the observed Fisher information (see Theoretical Note 1). Plots of these quantities against the case number or against the covariate Z_{jk} for each covariate are used to gage the influence of the jth observation on the kth covariate.

EXAMPLE 11.2

(continued):   We shall examine the influence of each observation on the risk coefficients for type of transplant Z_1, disease type Z_2, disease by type of transplant interaction Z_3, and Karnofsky score Z_4. A model is fit with the five covariates $Z_1, \ldots, Z_5$, where Z_5 is the indicator that the waiting time to transplant is at least 84 months, and the score residuals (11.6.1) are computed. To compute $\mathbf{I(b)}^{-1}$, we use the fact that this is the covariance matrix of $\mathbf{b}$. To assess how well standardized scores work, we directly estimate $\mathbf{b} - \mathbf{b}_{(j)}$ by fitting the n Cox models with the jth observation omitted. These values are represented by "+" in the following figures and the estimated values from the score residual by a "o". Figures 11.22–11.24 show the results. Here, we have plotted $b_k - b_{(j)k}$ versus the observation number j. In all figures, we note remarkably good agreement between the exact and approximate estimate of $\mathbf{b} - \mathbf{b}_{(j)}$. In all cases, the signs are the same and the influence is well approximated when $b_k - b_{(j)k}$ is close to zero. The approximations will improve as the sample size increases. Table 11.1 shows the risk vector $\mathbf{Z}_j$, the on study times T_j, the event indicators δ_j, and the values of $\Delta_1, \ldots, \Delta_4$, for each observation.

Examining Figures 11.22–11.24, we see that observation 17 has the largest effect on the estimates of $Z_1, Z_2,$ and Z_3. This is an individual with the risk vector $Z_{17} = (0, 0, 0, 70, 0)$ who dies at 2.633 months and has $\Delta_1 = 0.326$, $\Delta_2 = 0.334$ and $\Delta_3 = -0.322$. This is the longest lived patient with non-Hodgkin's lymphoma who had an auto transplant, all unfavorable risk factors. Based on the full sample analysis, he/she should have died earlier.

Turning now to the estimate of the Karnofsky score coefficient, we see that observation 12 has the most influence on the estimate. This is a patient with a relatively low Karnofsky score, who lives a relatively long time. Other influential observations are patient 3 who has a good Karnofsky score, but dies quite early, and patient 11 who also has a good Karnofsky score, but dies quite early.

We note that, in this example, the sample size is relatively small and, in small samples, each observation will have a relatively strong influence on the estimated value of $\boldsymbol{\beta}$. When an influential observation is found, it must be a cooperative decision between the statistician and the clinician whether this observation should be deleted from the data set. At the least, detection of an outlier calls for reexamining the original data for possible recording errors.

TABLE 11.1
Standardized Score Residuals for Transplant Data

j	T	δ	Z_1	Z_2	Z_3	Z_4	Z_5	Δ_1	Δ_2	Δ_3	Δ_4
1	0.1	1	0	0	0	20	0	−0.037	−0.051	0.037	−0.001
2	0.1	1	0	0	0	50	0	−0.139	−0.166	0.146	−0.001
3	0.9	1	0	1	0	90	0	0.039	0.290	−0.244	0.005
4	1.0	1	1	0	0	90	0	0.167	0.048	−0.163	0.004
5	1.1	1	0	1	0	30	0	0.009	0.108	−0.149	−0.002
6	1.2	1	1	0	0	80	0	0.161	0.052	−0.186	0.002
7	1.4	1	1	0	0	70	0	0.145	0.035	−0.173	0.001
8	1.4	1	1	1	1	80	0	0.021	0.057	0.070	0.004
9	1.6	1	0	1	0	40	0	0.022	0.123	−0.158	−0.001
10	1.7	1	1	0	0	60	0	0.096	0.031	−0.136	−0.001
11	1.8	1	1	1	1	90	0	0.039	0.093	0.038	0.005
12	1.9	1	1	1	1	30	0	−0.029	0.023	−0.136	0.005
13	2.1	1	1	0	0	90	0	0.191	0.091	−0.206	0.002
14	2.1	1	1	1	1	60	0	0.032	0.031	0.002	−0.001
15	2.4	1	0	0	0	80	0	0.049	0.043	−0.040	−0.001
16	2.6	1	0	0	0	60	1	−0.091	−0.061	0.085	−0.001
17	2.6	1	0	0	0	70	0	0.326	0.334	−0.322	−0.001
18	2.7	1	1	1	1	50	0	−0.108	−0.107	0.029	0.000
19	2.8	1	0	1	0	60	0	−0.086	0.003	0.028	−0.002
20	3.6	1	1	0	0	70	0	−0.095	−0.075	0.112	−0.001
21	4.4	1	1	0	0	60	0	−0.222	−0.057	0.255	0.001
22	4.7	1	1	1	1	100	0	−0.051	−0.066	0.168	0.001
23	5.9	1	1	1	1	80	0	−0.036	−0.082	0.079	−0.002
24	6.0	0	1	0	0	100	0	−0.041	−0.005	0.039	−0.001
25	7.0	0	1	1	1	90	0	0.013	−0.001	−0.073	−0.001
26	8.4	1	1	1	1	90	0	−0.049	−0.094	0.156	0.000
27	10.2	0	1	0	0	100	0	−0.047	0.001	0.041	−0.001
28	11.9	1	0	1	0	70	0	−0.029	−0.070	0.063	−0.001
29	13.5	0	1	0	0	100	0	−0.054	0.008	0.042	−0.001
30	14.9	0	1	0	0	100	0	−0.054	0.008	0.042	−0.001
31	15.9	0	1	1	1	90	0	0.026	0.019	−0.122	−0.002
32	16.1	0	1	0	0	90	1	−0.017	−0.011	0.018	0.000
33	17.5	1	1	1	1	90	0	−0.027	−0.096	0.119	−0.001
34	24.9	0	1	0	0	90	1	−0.021	−0.011	0.020	0.000
35	31.1	0	0	1	0	90	0	0.007	−0.069	0.058	−0.001
36	34.6	0	1	1	1	90	1	−0.002	−0.009	−0.012	0.000
37	35.9	0	0	1	0	100	0	0.003	−0.043	0.032	−0.001
38	39.4	0	0	1	0	90	0	0.007	−0.069	0.058	−0.001
39	43.0	0	1	0	0	90	0	−0.113	0.054	0.077	0.000
40	44.8	0	1	0	0	80	1	−0.037	−0.016	0.037	0.000
41	52.0	0	0	1	0	80	0	0.013	−0.112	0.104	0.000
42	70.5	0	0	1	0	80	0	0.013	−0.112	0.104	0.000
43	71.5	0	0	1	0	90	0	0.007	−0.069	0.058	−0.001

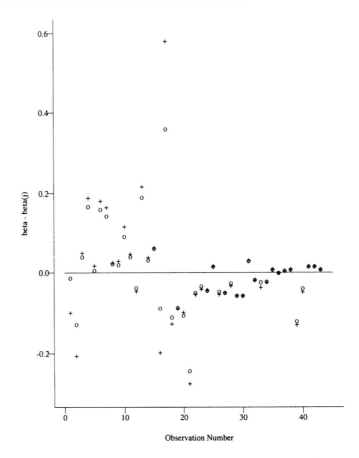

Figure 11.22   *Score residuals estimate b − b$_{(j)}$ for the type of transplant co-variate. Exact values (O) Estimated values (+)*

Practical Note

1. Score residuals are available in the packages S-Plus and SAS.
2. Schoenfeld residuals are available in S-Plus and SAS.

Theoretical Note

1. To see that the normalized score residual approximates $\mathbf{b} - \mathbf{b}_{(j)}$, let $U(\boldsymbol{\beta})$ be the score vector based on the complete data evaluated at the point $\boldsymbol{\beta}$. If we expand $U(\boldsymbol{\beta})$ about the point $\boldsymbol{\beta}_o$,

$$U_k(\boldsymbol{\beta}) = U_k(\boldsymbol{\beta}_o) - \mathbf{I}_k(\boldsymbol{\beta}_k^*)(\boldsymbol{\beta} - \boldsymbol{\beta}_o), k = 1, \ldots, p. \qquad (11.6.2)$$

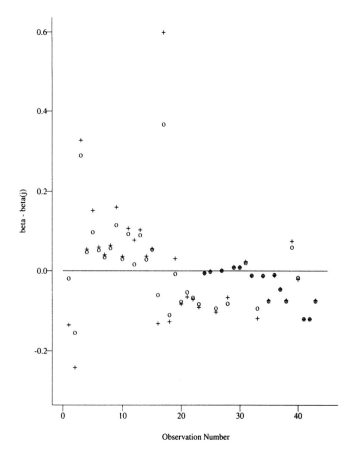

Figure 11.23   *Score residuals estimate* $b - b_{(j)}$ *for the type of disease covariate. Exact values (O) Estimated values (+)*

Here, $\mathbf{I}_k(\boldsymbol{\beta}^*)$ is the row vector with elements

$$-\frac{\partial U_k(\boldsymbol{\beta})}{\partial \boldsymbol{\beta}_b}, \quad b = 1, \ldots, p,$$

and, for each k, $\boldsymbol{\beta}_k^*$ is on the line segment between $\boldsymbol{\beta}$ and $\boldsymbol{\beta}_o$. When $\boldsymbol{\beta} = \mathbf{b}$, then, $\mathbf{U}(\mathbf{b})$ is zero so that solving equation (11.6.2) yields the fact that

$$(\mathbf{b} - \boldsymbol{\beta}_o) = \mathbf{I}^{-1}(\boldsymbol{\beta}_o)\mathbf{U}(\boldsymbol{\beta}_o). \tag{11.6.3}$$

We can write (11.6.3) as

$$\mathbf{I}^{-1}(\boldsymbol{\beta}_o)\mathbf{U}(\boldsymbol{\beta}_o) = \mathbf{I}^{-1}(\boldsymbol{\beta}_o)\left\{\sum_{j=1}^{n} \delta_j[\mathbf{Z}_j - \bar{\mathbf{Z}}(T_j)]\right\}$$

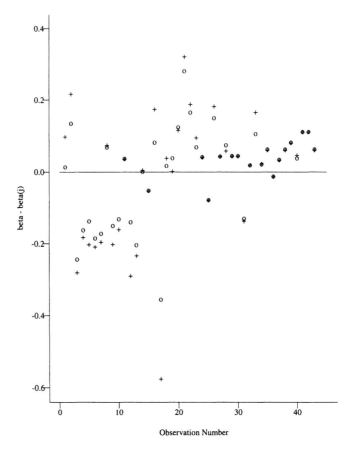

Figure 11.24 *Score residuals estimate* $b - b_{(j)}$ *for the interaction between disease and transplant covariate. Exact values (O) Estimated values (+)*

where $\bar{\mathbf{Z}}(t)$ is the vector $(\bar{Z}_k(t), k = 1, \ldots, p)$. For $\mathbf{b}_{(j)}$, we have a similar expression with the jth observation deleted. If we assume that deletion of the jth observation does not change the value of $\mathbf{I}^{-1}(\boldsymbol{\beta}_o)$, we have a similar expression for $\boldsymbol{\beta} - \mathbf{b}_{(j)}$ with the jth observation deleted. Subtracting these two approximations gives an estimate of $(\mathbf{b} - \mathbf{b}_{(j)})$, that is,

$$(\mathbf{b} - \mathbf{b}_{(j)}) = \mathbf{I}^{-1}(\boldsymbol{\beta}_o)\left\{\sum_{b=1}^{n} \delta_b[\mathbf{Z}_b - \bar{\mathbf{Z}}(T_b)]\right\}$$

$$- \mathbf{I}^{-1}(\boldsymbol{\beta}_o)\left\{\sum_{\substack{b=1 \\ b \neq j}}^{n} \delta_b[\mathbf{Z}_b - \bar{\mathbf{Z}}_{(j)}(T_b)]\right\},$$

where $\bar{\mathbf{Z}}_{(j)}(T_j)$ is the vector $\bar{\mathbf{Z}}(t)$ defined by (11.4.3), with the jth observation omitted. Simplifying this expression yields

$$(\mathbf{b} - \mathbf{b}_{(j)}) = \mathbf{I}^{-1}(\boldsymbol{\beta}_o)\{\delta_j[\mathbf{Z}_j - \bar{\mathbf{Z}}(T_j)]\}$$

$$- \Sigma[\mathbf{Z}_j - \bar{\mathbf{Z}}_{(j)}(T_b)]\exp[\mathbf{b}'\mathbf{Z}_j(t)][\hat{H}_o(T_b) - \hat{H}_o(T_{b-1})],$$

which is approximately equal to the standardized score residual Δ.

The approximation is based on an assumption that deletion of the jth observation does not change the value of $\mathbf{I}^{-1}(\boldsymbol{\beta}_o)$ and that $\bar{\mathbf{Z}}_{(j)}$ is close to $\bar{\mathbf{Z}}$. Storer and Crowley (1985) give an alternate approximation based on the one-step application of the Newton-Raphson approximation to the estimate of $\boldsymbol{\beta}$ which attempts to remedy the first problem. In most cases, these approximations are very close to the score residuals suggested here.

11.7 Exercises

11.1   In Example 8.2, a proportional hazards model was fit to the data on the death times of 90 males diagnosed with cancer of the larynx (see section 1.8). A model with three covariates for stage of disease was considered.

(a) Determine if adding the patient's age into the model is appropriate using a martingale residual plot based on a Cox model adjusted for disease stage. If age should not enter the model as a linear term suggest a functional form for age.

(b) Repeat part a for the covariate year of transplant.

(c) Fit a model with the factor stage of disease and a linear term for age. Perform a general examination of this model using a Cox–Snell residual.

11.2   In section 1.14 a study of the times to weaning of breast-fed newborns was presented. This data is available on our web site. Categorical variables which could explain the difference in weaning times are the mother's race (white, black, other), smoking status, and an indicator of whether the mother was in poverty. Continuous variables which could explain outcome are the mother's age at the child's birth, mother's years of education, and the child's year of birth. Using a Cox model with appropriate terms for the mother's race, smoking status, and poverty indicator, determine if each of the three continuous covariates would enter the model as a linear function.

11.3   In section 1.8 data on the death times of patients diagnosed with cancer of the larynx was presented (see Example 8.2 and Exercise 11.1). Using this data in a model which adjusts for age, examine the proportional hazards assumption for the stage of disease by the following graphical methods.

(a) A plot of the logarithms of the cumulative baseline hazard rates for each disease stage.

(b) A plot of the difference in the log cumulative hazard rates for the disease stages.

(c) An Andersen plot.

(d) A score residual plot.

11.4   In Exercise 1 of Chapter 8 a Cox model was fit to data on the survival times of patients with an aneuploid or diploid DNA tumor profile.

(a) Check the proportional hazards assumption for this data by plotting the logarithms of the cumulative baseline hazard rates for each ploidy group.

(b) Check for proportional hazards by plotting the difference in the log cumulative hazard rates for the two groups.

(c) Check for proportional hazards by using an Andersen plot.

(d) Check for proportional hazards by using a score residual plot.

11.5   In Example 8.3 and its continuation in section 8.4 a proportional hazards model was fit to the data on the time to death of 863 kidney transplant patients. (The data is presented on our web site.) Covariates in the model were gender, race, and a gender by race interaction.

(a) Check this data for possible outliers by making an appropriate plot of the deviance residuals.

(b) For each of the three covariates in this model find the four most influential observations on the estimates of the regression coefficients. Explain why these observations are so influential.

11.6   (a) For the data on survival times of patients with an aneuploid or diploid DNA tumor profile in Exercise 4 determine which, if any, observations are outliers by making an appropriate deviance residual plot.

(b) Find the three points that have the greatest influence on the estimate of the regression effect by constructing a plot of the adjusted score residuals. Explain why these three points are so influential in light of your fitted regression model.

12

Inference for Parametric
——— Regression Models

12.1 Introduction

In previous chapters, we focused on nonparametric methods for describing the survival experience of a population and regression models for survival data which do not require any specific distributional assumptions about the shape of the survival function. In this chapter, we shall discuss the use of parametric models for estimating univariate survival and for the censored-data regression problem. When these parametric models provide a good fit to data, they tend to give more precise estimates of the quantities of interest because these estimates are based on fewer parameters. Of course, if the parametric model is chosen incorrectly, it may lead to consistent estimators of the wrong quantity.

All of the models we shall consider in this chapter have an *accelerated failure-time model* representation and a *linear model* representation in log time. Let X denote the time to the event and $\mathbf{Z}$ a vector of fixed-time explanatory covariates. The accelerated failure-time model states that the survival function of an individual with covariate $\mathbf{Z}$ at time x is the same as the survival function of an individual with a baseline survival function at a time $x \exp(\boldsymbol{\theta}^t \mathbf{Z})$, where $\boldsymbol{\theta}^t = (\theta_1, \ldots, \theta_p)$ is a vector of regression coefficients. In other words, the accelerated failure-time model is defined by the relationship

$$S(x \mid \mathbf{Z}) = S_0[\exp(\boldsymbol{\theta'}\mathbf{Z})x], \text{ for all } x. \tag{12.1.1}$$

The factor $\exp(\boldsymbol{\theta'}\mathbf{Z})$ is called an *acceleration factor* telling the investigator how a change in covariate values changes the time scale from the baseline time scale. One implication of this model is that the hazard rate for an individual with covariate $\mathbf{Z}$ is related to a baseline hazard rate by

$$h(x \mid \mathbf{Z}) = \exp(\boldsymbol{\theta'}\mathbf{Z})h_0[\exp(\boldsymbol{\theta'}\mathbf{Z})x], \text{ for all } x. \tag{12.1.2}$$

A second implication is that the median time to the event with a covariate $\mathbf{Z}$ is the baseline median time to event divided by the acceleration factor.

The second representation of the relationship between covariate values and survival is the linear relationship between log time and the covariate values. Here, we assume the usual linear model for log time, namely,

$$Y = \ln X = \mu + \boldsymbol{\gamma'}\mathbf{Z} + \sigma W, \tag{12.1.3}$$

where $\boldsymbol{\gamma'} = (\gamma_1, \ldots, \gamma_p)$ is a vector of regression coefficients and W is the error distribution. The regression coefficients have an interpretation similar to those in standard normal theory regression.

The two representations are closely related. If we let $S_0(x)$ be the survival function of the random variable $\exp(\mu + \sigma W)$, then, the linear log-time model is equivalent to the accelerated failure-time model with $\boldsymbol{\theta} = -\boldsymbol{\gamma}$.

A variety of models can be used for W or, equivalently, for S_0 (see Table 2.2). In section 12.2, we focus on estimation for the Weibull distribution for which W has a standard extreme value distribution. This model is very flexible because it has a hazard rate that can be either increasing, decreasing, or constant. It has the unique property that, along with the accelerated failure-time representation, it is the only parametric distribution that also has a proportional hazards representation.

In section 12.3, we focus on the log logistic distribution for which W has a standard logistic distribution. This model has a hazard rate that is hump-shaped (see Chapter 2). This model is the only accelerated failure-time model that also has a representation as a *proportional odds* model, that is, for the log logistic distribution, the odds of survival beyond time t are given by

$$\frac{S(x \mid \mathbf{Z})}{1 - S(x \mid \mathbf{Z})} = \exp(\boldsymbol{\beta'}\mathbf{Z})\frac{S_0(x)}{1 - S_0(x)}, \tag{12.1.4}$$

where $\boldsymbol{\beta} = -\boldsymbol{\gamma}/\sigma$.

In section 12.4, we will examine several other models popular for survival data. These include the log normal distribution and the generalized gamma distribution which can be used to discriminate between models.

In these sections, maximum likelihood estimation of the parameters is presented for each of these parametric models. The parameters of the log-time, linear model are estimated first and their variance-covariance matrix, readily available in most major software packages, is reported. From these values, the maximum likelihood estimates of functions of the parameters, along with their approximate variance-covariance matrix, may be obtained using the method of statistical differentials, also called the delta method.

In section 12.5, graphical methods for assessing the fit of these models are presented. For univariate problems, we use the hazard rates displayed in Table 2.2 and the Nelson–Aalen estimator of the cumulative hazard rate, to make hazard plots for each parametric model. A *hazard plot* is a plot of the appropriate function of the cumulative hazard function as the ordinate versus the appropriate function of time as the abscissa. Each distribution will have its own functions of the cumulative hazard and time. Such plots should be straight lines if the model is correct.

To assess the fit of regression models, we present analogs of the Cox–Snell, martingale and deviance residuals presented in Chapter 11. A quantile-quantile plot is also presented for checking that the accelerated failure-time model fits a set of data.

12.2   Weibull Distribution

The Weibull distribution, discussed in Chapter 2, is a very flexible model for lifetime data. It has a hazard rate which is either monotone increasing, decreasing, or constant. It is the only parametric regression model which has both a proportional hazards representation and an accelerated failure-time representation. In this section, we shall first examine estimating the parameters of the Weibull distribution in the univariate setting and, then, examine the regression problem for this model.

The survival function for the Weibull distribution is given by

$$S_X(x) = \exp(-\lambda x^{\alpha}),$$

and the hazard rate is expressed by

$$h_X(x) = \lambda \alpha x^{\alpha - 1}.$$

Taking the log transform of time, the univariate survival function for $Y = \ln X$ is given by

$$S_Y(y) = \exp(-\lambda e^{\alpha y}).$$

If we redefine the parameters as $\lambda = \exp(-\mu/\sigma)$ and $\sigma = 1/\alpha$, then, Y follows the form of a log linear model with

$$Y = \ln X = \mu + \sigma W, \qquad (12.2.1)$$

where W is the extreme value distribution with probability density function,

$$f_W(w) = \exp(w - e^w) \qquad (12.2.2)$$

and survival function,

$$S_W(w) = \exp(-e^w). \qquad (12.2.3)$$

Thus, the underlying probability density function and survival function, respectively, for Y, are

$$f_Y(y) = (1/\sigma)\exp[(y - \mu)/\sigma - e^{[(y-\mu)/\sigma]}] \qquad (12.2.4)$$

and

$$S_Y(y) = \exp(-e^{[(y-\mu)/\sigma]}). \qquad (12.2.5)$$

When $\alpha = 1$, or, equivalently, $\sigma = 1$, then, the Weibull distribution reduces to an exponential distribution.

The likelihood function for right-censored data, following the construction in Chapter 3, is given by

$$L = \prod_{j=1}^{n}[f_Y(y_j)]^{\delta_j}[S_Y(y_j)]^{(1-\delta_j)}$$

$$= \prod_{j=1}^{n}\left[\frac{1}{\sigma}f_W\left(\frac{y_j - \mu}{\sigma}\right)\right]^{\delta_j}\left[S_w\left(\frac{y_j - \mu}{\sigma}\right)\right]^{(1-\delta_j)}$$

where $f_Y(y)$ and $S_Y(y)$ are given in (12.2.4) and (12.2.5). Once maximum likelihood estimates of the parameters μ and σ, or equivalently, λ and α are computed (see Practical Note 1), estimates of the survival function and the cumulative hazard rate are available for the distribution of X or Y.

Estimates of μ and σ are found numerically, and routines to do so are available in most statistical packages. The variance-covariance matrix of the log linear parameters μ and σ, obtained from the observed information matrix, are also available in these packages. The invariance property of the maximum likelihood estimator provides that the maximum likelihood estimators of λ and α are given by

$$\hat{\lambda} = \exp(-\hat{\mu}/\hat{\sigma}) \text{ and } \hat{\alpha} = 1/\hat{\sigma}. \qquad (12.2.6)$$

Applying the delta method (see Theoretical Notes),

$$\text{Var}(\hat{\lambda}) = \exp(-2\hat{\mu}/\hat{\sigma})[\text{Var}(\hat{\mu})/\hat{\sigma}^2 + \hat{\mu}^2\,\text{Var}(\hat{\sigma})/\hat{\sigma}^4 \qquad (12.2.7)$$

$$- 2\hat{\mu}\,\text{Cov}(\hat{\mu}, \hat{\sigma})/\hat{\sigma}^3],$$

$$\text{Var}(\hat{\alpha}) = \text{Var}(\hat{\sigma})/\hat{\sigma}^4, \qquad (12.2.8)$$

and

$$\text{Cov}(\hat{\lambda}, \hat{\alpha}) = \exp(-\hat{\mu}/\hat{\sigma})[\text{Cov}(\hat{\mu}, \hat{\sigma})/\hat{\sigma}^3 - \hat{\mu}\,\text{Var}(\hat{\sigma})/\hat{\sigma}^4]. \qquad (12.2.9)$$

EXAMPLE 12.1   Consider the data set described in section 1.9 and studied in Chapters 7 and 11. It compares the efficacy of autologous (auto) versus allogeneic (allo) transplants for acute myelogenous leukemia. The outcome for the 101 patients was leukemia-free survival. All patients in the sample were in their first complete remission at least one year.

The Weibull maximum likelihood estimates of the log linear parameters μ and σ are $\hat{\mu}_{auto} = 3.45$, $\hat{\sigma}_{auto} = 1.11$, $\hat{\mu}_{allo} = 4.25$, and $\hat{\sigma}_{allo} = 1.94$. The corresponding maximum likelihood estimates of the parameters $\lambda = \exp(-\mu/\sigma)$ and $\alpha = 1/\sigma$ are $\hat{\lambda}_{auto} = 0.045$, $\hat{\alpha}_{auto} = 0.900$, $\hat{\lambda}_{allo} = 0.112$, and $\hat{\alpha}_{allo} = 0.514$, respectively. The variance-covariance matrix for $\hat{\mu}_{auto}$ and $\hat{\sigma}_{auto}$ is

$$\begin{pmatrix} 0.048 & 0.010 \\ 0.010 & 0.031 \end{pmatrix},$$

and the variance-covariance matrix for $\hat{\mu}_{allo}$ and $\hat{\sigma}_{allo}$ is

$$\begin{pmatrix} 0.229 & 0.088 \\ 0.088 & 0.135 \end{pmatrix}.$$

Applying (12.2.7)–(12.2.9), the variance-covariance matrix for $\hat{\lambda}_{auto}$ and $\hat{\alpha}_{auto}$ is

$$\begin{pmatrix} 0.0004 & -0.00202 \\ -0.00202 & 0.0202 \end{pmatrix}$$

and the variance-covariance matrix for $\hat{\lambda}_{allo}$ and $\hat{\alpha}_{allo}$ is

$$\begin{pmatrix} 0.0016 & -0.0032 \\ -0.0032 & 0.0095 \end{pmatrix}.$$

To test the hypothesis that the exponential model provides as good a fit to the data as the Weibull model, we shall test the hypothesis that $\sigma = 1$ (or equivalently that $\alpha = 1$). Although any of the three types of likelihood-based tests can be performed, only the likelihood ratio and score tests are invariant under the different parameterizations. We shall perform the likelihood ratio tests. For the allo transplant data, the log likelihood for the Weibull model is -72.879 whereas for the exponential model it is -81.203. For the auto transplant data, the log likelihood for the Weibull model is -68.420 whereas, for the exponential model, it is -68.653. The likelihood ratio chi square for the allo transplant data is $2[72.879 - (-81.203)] = 16.648$ which is highly significant when compared to a chi-square percentile with one degree of freedom. For auto transplants, the likelihood ratio chi square is 0.467, which is not significant. This suggests that an exponential distribution may provide as good a fit as the Weibull distribution for auto transplants, but it is not a viable model for allo transplants.

To incorporate covariates into the Weibull model, we use a linear model (12.1.3) for log time,

$$Y = \mu + \boldsymbol{\gamma}'\mathbf{z} + \sigma W, \tag{12.2.10}$$

where W has the standard extreme value distribution (12.2.2). This model leads to a proportional hazards model for X with a Weibull baseline hazard, that is,

$$h(x \mid \mathbf{Z}) = (\alpha \lambda x^{\alpha - 1}) \exp(\boldsymbol{\beta}'\mathbf{Z}), \tag{12.2.11}$$

with $\alpha = 1/\sigma$, $\lambda = \exp(-\mu/\sigma)$ and $\beta_j = -\gamma_j/\sigma$, $j = 1, \ldots, p$.

Using the accelerated failure-time representation of the Weibull regression model, the hazard rate for an individual with covariate vector $\mathbf{Z}$ is given by

$$h(x \mid \mathbf{z}) = \exp(\boldsymbol{\theta}'\mathbf{Z}) h_0[x \exp(\boldsymbol{\theta}'\mathbf{Z})] \tag{12.2.12}$$

where the baseline hazard, $h_0(x)$ is $\lambda \alpha x^{\alpha - 1}$. The factor $\exp(\boldsymbol{\theta}'\mathbf{Z})$ is called an acceleration factor. If the covariate vector is a scalar which is the indicator of treatment group ($Z = 1$ if group 1; $Z = 0$ if group 2], the acceleration factor can be interpreted naturally. Under the accelerated failure model, the survival functions between the two groups will have the following relationship:

$$S(x \mid Z = 1) = S(xe^{\theta} \mid Z = 0), \text{ for all } t.$$

For an accelerated failure time distribution with covariate $\mathbf{Z}$

$$S(x \mid \mathbf{Z}) = S_0(x \exp[\boldsymbol{\theta}'\mathbf{Z}]) \quad \text{for all } x$$

by (12.1.1). Let X_m^0 be the median of the baseline distribution. Then $S_0(X_m^0) = 1/2$. Now let X_m^z be the median, with $Z = z$, which has $S_0(X_m^z \mid z) = S_0(X_m^z \exp[\theta z]) = 1/2$ by (12.1.1). This implies that $X_m^z \exp[\theta z] = X_m^0$ or $X_m^z = X_m^0/\exp[\theta z]$. So the median of a group with $Z = z$ is the baseline median divided by $\exp[\theta z]$. This implies that the median time in the $Z = 1$ group is equal to the median time in the $Z = 0$ group divided by e^{θ}. Comparing 12.2.11 and 12.2.12, we see that $\boldsymbol{\theta} = \boldsymbol{\beta}/\alpha$ or $\boldsymbol{\theta} = -\boldsymbol{\gamma}$. The Weibull is the only continuous distribution that yields both a proportional hazards and an accelerated failure-time model.

For the Weibull regression model, estimates must be found numerically. Routines for estimation, based on (12.2.10), are found in most statistical packages. As before, the invariance property of the maximum likelihood estimators in the log linear model provides estimates of parameters in the alternative formulation (12.2.11). Using the delta method, the following is the approximate covariance matrix for these estimates based on the estimates and their covariances in the log linear model:

$$\text{Cov}(\hat{\beta}_j, \hat{\beta}_k) = \frac{\text{Cov}(\hat{\gamma}_j, \hat{\gamma}_k)}{\hat{\sigma}^2} - \frac{\hat{\gamma}_j \, \text{Cov}(\hat{\gamma}_j, \hat{\sigma})}{\hat{\sigma}^3} - \frac{\hat{\gamma}_k \, \text{Cov}(\hat{\gamma}_k, \hat{\sigma})}{\hat{\sigma}^3} \tag{12.2.13}$$

$$+ \frac{\hat{\gamma}_j \hat{\gamma}_k \, \text{Var}(\hat{\sigma})}{\hat{\sigma}^4}, \quad j, k = 1, \ldots, p;$$

$$\text{Var}(\hat{\lambda}) = \exp\left(-2\frac{\hat{\mu}}{\hat{\sigma}}\right)\left[\frac{\text{Var}(\hat{\mu})}{\hat{\sigma}^2} - 2\frac{\hat{\mu}\,\text{Cov}(\hat{\mu}, \hat{\sigma})}{\hat{\sigma}^3} + \frac{\hat{\mu}^2\,\text{Var}(\hat{\sigma})}{\hat{\sigma}^4}\right] \tag{12.2.14}$$

$$\text{Var}(\hat{\alpha}) = \frac{\text{Var}(\hat{\sigma})}{\hat{\sigma}^4} \tag{12.2.15}$$

$$\text{Cov}(\hat{\beta}_j, \hat{\lambda}) = \exp\left(-\frac{\hat{\mu}}{\hat{\sigma}}\right)\left[\frac{\text{Cov}(\hat{\gamma}_j, \hat{\mu})}{\hat{\sigma}^2} - \frac{\hat{\gamma}_j\,\text{Cov}(\hat{\gamma}_j, \hat{\sigma})}{\hat{\sigma}^3}\right. \tag{12.2.16}$$

$$\left. - \frac{\hat{\mu}\,\text{Cov}(\hat{\mu}, \hat{\sigma})}{\hat{\sigma}^3} + \frac{\hat{\gamma}_j \hat{\mu}\,\text{Var}(\hat{\sigma})}{\hat{\sigma}^4}\right], \quad j = 1;, \ldots, p;$$

$$\text{Cov}(\hat{\beta}_j, \hat{\alpha}) = \frac{\text{Cov}[\hat{\gamma}_j, \hat{\sigma}]}{\hat{\sigma}^3} - \frac{\hat{\gamma}_j\,\text{Var}[\hat{\sigma}]}{\hat{\sigma}^4} \quad j = 1, \ldots, p; \tag{12.2.17}$$

$$\text{Cov}(\hat{\lambda}, \hat{\alpha}) = \exp\left(-\frac{\hat{\mu}}{\hat{\sigma}}\right)\left[\frac{\text{Cov}(\hat{\mu}, \hat{\sigma})}{\hat{\sigma}^3} - \frac{\hat{\mu}\,\text{Var}(\hat{\sigma})}{\hat{\sigma}^4}\right]. \tag{12.2.18}$$

We shall illustrate this model on the data for times to death from laryngeal cancer.

EXAMPLE 12.2 A study of 90 males diagnosed with cancer of the larynx is described in section 1.8 and analyzed in Chapters 7 and 8. Here, we shall employ the accelerated failure-time model using the main effects of age and stage for this data. The model is given by

$$Y = \ln X = \mu + \gamma_1 Z_1 + \gamma_2 Z_2 + \gamma_3 Z_3 + \gamma_4 Z_4 + \sigma W$$

where $Z_i, i = 1, \ldots, 3$ are the indicators of stage II, III and IV disease, respectively, and Z_4 is the age of the patient. The parameter estimates, standard errors, Wald chi squares, and p-values for testing that $\gamma_i = 0$ are given in Table 12.1. Here, we see that patients with stage IV disease do significantly worse than patients with stage I disease. Note that, as opposed to the Cox model where a positive value of the risk coefficient reflects poor survival, here, a negative value of the coefficient is indicative of decreased survival.

We apply the transformation in (12.2.11)–(12.2.18) on the original time scale and obtain the parameter estimates in Table 12.2. Using these estimates and the proportional hazards property of the Weibull regression model, we find that the relative risk of death for a Stage IV patient compared to a Stage I patient is $\exp(1.745) = 5.73$. The acceleration factor for Stage IV disease compared to Stage I disease is $\exp(1.54) = 4.68$, so that the median lifetime for a Stage I patient is estimated to be 4.68 times that of a Stage IV patient.

TABLE 12.1

Analysis of Variance Table for Stage and Age for Laryngeal Cancer Patients, Utilizing the Log Linear Model, Assuming the Weibull Distribution

Variable	Parameter Estimate	Standard Error	Wald Chi Square	p-Value
Intercept $\hat{\mu}$	3.53	0.90		
Scale $\hat{\sigma}$	0.88	0.11		
Z_1: Stage II ($\hat{\gamma}_1$)	−0.15	0.41	0.13	0.717
Z_2: Stage III ($\hat{\gamma}_2$)	−0.59	0.32	3.36	0.067
Z_3: Stage IV ($\hat{\gamma}_3$)	−1.54	0.36	18.07	<0.0001
Z_4: Age ($\hat{\gamma}_4$)	−0.02	0.01	1.87	0.172

TABLE 12.2

Parameter Estimates for the Effects of Stage and Age on Survival for Laryngeal Cancer Patients, Modeling Time Directly Assuming the Weibull Distribution

Variable	Parameter Estimate	Standard Error
Intercept $\hat{\lambda}$	0.002	0.002
Scale $\hat{\alpha}$	1.13	0.14
Z_1: Stage II ($\hat{\beta}_1$)	0.17	0.46
Z_2: Stage III ($\hat{\beta}_2$)	0.66	0.36
Z_3: Stage IV ($\hat{\beta}_3$)	1.75	0.42
Z_4: Age ($\hat{\beta}_4$)	0.02	0.01

Practical Notes

1. SAS PROC LIFEREG and the S-Plus routine survreg provide maximum likelihood estimates of an intercept μ and scale parameter σ associated with the extreme value distribution, the error distribution for the Weibull model. Our parameters of the underlying Weibull distribution are the following functions of these extreme value parameters, $\lambda = \exp(-\mu/\sigma)$ and $\alpha = 1/\sigma$. SAS allows for right-, left- and interval-censored data.

2. When performing an accelerated failure time regression employing the Weibull distribution, SAS and S-Plus provide maximum likelihood estimates of an intercept μ, scale parameter σ, and regression coefficients γ_i. The parameters of the underlying Weibull distribution, when modeling time directly, are the following functions of those parameters: $\lambda = \exp(-\mu/\sigma)$, $\alpha = 1/\sigma$, and $\beta_i = -\gamma_i/\sigma$. SAS allows for right-, left- and interval-censored data.

Theoretical Notes

1. The method of statistical differentials or the delta method (Elandt–Johnson and Johnson, 1980, pp. 69–72) is based on a Taylor series expansion of a continuous function $g(.)$ of the maximum likelihood estimators of a vector of parameters. We shall illustrate how this works in the two-parameter case. Let ψ_1 and ψ_2 be the two parameters of interest, and let $\hat{\psi}_1$ and $\hat{\psi}_2$ be the maximum likelihood estimators of the parameters. Recall that, for large samples, $(\hat{\psi}_1, \hat{\psi}_2)$ has a bivariate normal distribution with mean (ψ_1, ψ_2) and a covariance matrix estimated by the inverse of the observed Fisher information observed. Let $\theta_1 = g_1(\psi_1, \psi_2)$ and $\theta_2 = g_2(\psi_1, \psi_2)$ be a reparametrization of ψ_1 and ψ_2. The invariance principle of the maximum likelihood estimator insures that the maximum likelihood estimators of θ_1 and θ_2 are $g_k(\hat{\psi}_1, \hat{\psi}_2)$, $k = 1, 2$.

   To apply the delta method, for $k = 1, 2$, we expand $g_k(\hat{\psi}_1, \hat{\psi}_2)$ in a first-order Taylor series about the true values of ψ_1 and ψ_2, that is,

   $$g_k(\hat{\psi}_1, \hat{\psi}_2) = g_k(\psi_1, \psi_2) + (\hat{\psi}_1 - \psi_1)\frac{\partial g_k(\hat{\psi}_1, \hat{\psi}_2)}{\partial \hat{\psi}_1} + (\hat{\psi}_2 - \psi_2)\frac{\partial g_k(\hat{\psi}_1, \hat{\psi}_2)}{\partial \hat{\psi}_2}$$

   where the partial derivatives are evaluated at the true values of the parameters. Thus,

   $$g_k(\hat{\psi}_1, \hat{\psi}_2) - g_k(\psi_1, \psi_2) = (\hat{\psi}_1 - \psi_1)\frac{\partial g_k(\hat{\psi}_1, \hat{\psi}_2)}{\partial \hat{\psi}_1} + (\hat{\psi}_2 - \psi_2)\frac{\partial g_k(\hat{\psi}_1, \hat{\psi}_2)}{\partial \hat{\psi}_2}$$

   If we let $g_k^b = \frac{\partial g_k(\hat{\psi}_1, \hat{\psi}_2)}{\partial \hat{\psi}_b}$, then, for large samples,

   $$\begin{aligned}
   \text{Cov}[g_k(\hat{\psi}_1, \hat{\psi}_2), g_m(\hat{\psi}_1, \hat{\psi}_2)] &= E\{g_k^1 g_m^1 (\hat{\psi}_1 - \psi_1)^2 + [g_k^1 g_m^2 + g_k^2 g_m^1] \\
   &\quad \cdot (\hat{\psi}_1 - \psi_1)(\hat{\psi}_2 - \psi_2) + g_k^2 g_m^2 (\hat{\psi}_2 - \psi_2)^2\} \\
   &= g_k^1 g_m^1 \text{Var}[\hat{\psi}_1] + [g_k^1 g_m^2 + g_k^2 g_m^1] \text{Cov}[\hat{\psi}_1, \hat{\psi}_2] \\
   &\quad + g_k^2 g_m^2 \text{Var}[\hat{\psi}_2], \quad k, m = 1, 2.
   \end{aligned}$$

12.3 Log Logistic Distribution

An alternative model to the Weibull distribution is the log logistic distribution. This distribution has a hazard rate which is hump-shaped, that is, it increases initially and, then, decreases. It has a survival function and hazard rate that has a closed form expression, as contrasted with the log normal distribution which also has a hump-shaped hazard rate.

Utilizing the notation which models time directly, as in Chapter 2, the univariate survival function and the cumulative hazard rate for X, when X follows the log logistic distribution, are given by

$$S_X(x) = \frac{1}{1 + \lambda x^\alpha} \qquad (12.3.1)$$

and

$$H_X(x) = \ln(1 + \lambda x^\alpha). \qquad (12.3.2)$$

Taking the log transform of time, the univariate survival function for $Y = \ln X$ is

$$S_Y(y) = \frac{1}{1 + \lambda e^{\alpha y}} \qquad (12.3.3)$$

This log linear model with no covariates is, from (12.1.1),

$$Y = \ln X = \mu + \sigma W, \qquad (12.3.4)$$

where W is the standard logistic distribution with probability density function,

$$f_W(w) = e^w/(1 + e^w)^2 \qquad (12.3.5)$$

and survival function,

$$S_W(w) = 1/(1 + e^w) \qquad (12.3.6)$$

Thus, the underlying probability density function and survival function, respectively, for Y, are given by

$$f_Y(y) = (1/\sigma)\exp[(y - \mu)/\sigma]/[1 + \exp[(y - \mu)/\sigma]^2 \qquad (12.3.7)$$

and

$$S_Y(y) = 1/[1 + e^{[(y-\mu)/\sigma]}]. \qquad (12.3.8)$$

Thus, one can see that the parameters of the underlying log logistic distribution for the random variable X in (12.3.1) and for the distribution of the log transformed variable Y in (12.3.3) are the following functions of the log linear parameters in (12.3.8):

$$\alpha = 1/\sigma \text{ and } \lambda = \exp(-\mu/\sigma), \qquad (12.3.9)$$

the same functions as for the Weibull model (see (12.2.6)). Thus, given estimates of μ and α, estimates of λ and α and their covariance matrix are given by (12.2.6)–(12.2.9). Estimates of μ and σ are available in most statistical packages.

EXAMPLE 12.1   *(continued):* We shall continue the example on univariate estimation for the autologous (auto) versus allogeneic (allo) transplants for acute myelogenous leukemia.

The log logistic maximum likelihood estimates of the log linear parameters μ and σ are $\hat{\mu}_{\text{auto}} = 2.944$, $\hat{\sigma}_{\text{auto}} = 0.854$, $\hat{\mu}_{\text{allo}} = 3.443$, and $\hat{\sigma}_{\text{allo}} = 1.584$, and the corresponding maximum likelihood estimates of the parameters $\lambda = \exp(-\mu/\sigma)$ and $\alpha = 1/\sigma$ are $\hat{\lambda}_{\text{auto}} = 0.032$, $\hat{\alpha}_{\text{auto}} = 1.171$, $\hat{\lambda}_{\text{allo}} = 0.114$, and $\hat{\alpha}_{\text{allo}} = 0.631$, respectively. The variance-covariance matrix for $\hat{\mu}_{\text{auto}}$ and $\hat{\sigma}_{\text{auto}}$ is

$$\begin{pmatrix} 0.0531 & 0.0085 \\ 0.0085 & 0.019 \end{pmatrix},$$

and the variance-covariance matrix for $\hat{\mu}_{\text{allo}}$ and $\hat{\sigma}_{\text{allo}}$ is

$$\begin{pmatrix} 0.2266 & 0.0581 \\ 0.0581 & 0.0855 \end{pmatrix}.$$

Inserting the maximum likelihood estimates $(\hat{\mu}, \hat{\sigma})$ and their estimated variances into (12.2.7)–(12.2.9), the variance-covariance matrix for $\hat{\lambda}_{\text{auto}}$ and $\hat{\alpha}_{\text{auto}}$ is

$$\begin{pmatrix} 3.010 \times 10^{-4} & -2.861 \times 10^{-3} \\ -2.681 \times 10^{-3} & 3.518 \times 10^{-3} \end{pmatrix}$$

and the variance-covariance matrix for $\hat{\lambda}_{\text{allo}}$ and $\hat{\alpha}_{\text{allo}}$ is

$$\begin{pmatrix} 1.951 \times 10^{-3} & -3.661 \times 10^{-3} \\ -3.661 \times 10^{-3} & 1.360 \times 10^{-2} \end{pmatrix}.$$

One of three equivalent models can be used to model the effects of covariates on survival with the log logistic distribution. The first is the linear model for log time with

$$Y = \ln X = \mu + \gamma' Z + \sigma W, \tag{12.3.10}$$

where W has the standard logistic distribution (12.3.5). The second representation is obtained by replacing λ in (12.3.3) by $\exp(\beta' Z)$. Here, the conditional survival function for the time to the event is given by

$$S_X(x \mid Z) = \frac{1}{1 + \lambda \exp(\beta' Z) x^\alpha}. \tag{12.3.11}$$

As for the Weibull distribution, the parameters are related by

$$\beta = -\gamma/\sigma, \tag{12.3.12}$$

$$\lambda = \exp[-\mu/\sigma],$$

and

$$\alpha = 1/\sigma.$$

Based on maximum likelihood estimates for μ, γ, σ, and their covariance matrix, estimates for λ, β, α, and their covariance are obtained from Eqs. (12.2.13)–(12.2.18). To interpret the factor $\exp(\beta' Z)$ for the log logistic model, note that the odds of survival beyond time t for the logistic model is given by

$$\frac{S_X(x \mid \mathbf{Z})}{1 - S_X(x \mid \mathbf{Z})} = \frac{1}{\lambda \exp[\boldsymbol{\beta}'\mathbf{Z}]x^{\alpha}} = \exp(-\boldsymbol{\beta}'\mathbf{Z})\frac{S_X(x \mid \mathbf{Z} = \mathbf{0})}{1 - S_X(x \mid \mathbf{Z} = \mathbf{0})}.$$

So, the factor $\exp(-\boldsymbol{\beta}'\mathbf{Z})$ is an estimate of how much the baseline odds of survival at any time changes when an individual has covariate $\mathbf{Z}$. Note that $\exp(\boldsymbol{\beta}'\mathbf{Z})$ is the relative odds of dying for an individual with covariate $\mathbf{Z}$ compared to an individual with the baseline characteristics.

The third representation of a log logistic regression is as an accelerated failure-time model (12.1.1) with a log logistic baseline survival function. The log logistic model is the only parametric model with both a proportional odds and an accelerated failure-time representation.

EXAMPLE 12.2   *(continued):*   Continuing the study of laryngeal cancer, we shall employ the log logistic model using the main effects of age and stage. The parameter estimates, standard errors, Wald chi squares and p-values for testing $\gamma_i = 0$, are given in Table 12.3. Here we see that Stage II is not significantly different from Stage I, Stages III and IV are significantly different from Stage I, adjusted for age, and, as in earlier analyses, age is not a significant predictor of death in these patients, adjusted for stage.

The estimates obtained by converting the parameters in the log linear model to those in the proportional odds model and calculating their standard errors using (12.2.13)–(12.2.18), are listed in Table 12.4. From Table 12.4, we see that the relative odds of survival for a Stage III patient compared to a Stage I patient are $\exp(-1.127) = 0.32$ and for a Stage IV patient are $\exp(-2.469) = 0.085$, that is, Stage IV patients have 0.085 times lesser odds of surviving than Stage I patients (or $1/0.085 = 11.81$ times greater odds of dying). Using the accelerated failure-time model for the log logistic model, we see that the acceleration factor for Stage III

TABLE 12.3
Analysis of Variance Table for Stage and Age for Laryngeal Cancer Patients, Utilizing the Log Linear Model, Assuming the Log Logistic Distribution

Parameter	Parameter Estimate	Standard Errors	Wald Chi Square	p-Value
Intercept $\hat{\mu}$	3.10	0.95		
Scale $\hat{\sigma}$	0.72	0.09		
Z_1: Stage II ($\hat{\gamma}_1$)	−0.13	0.42	0.09	0.762
Z_2: Stage III ($\hat{\gamma}_2$)	−0.81	0.35	5.18	0.023
Z_3: Stage IV ($\hat{\gamma}_3$)	−1.77	0.43	17.22	<0.0001
Z_4: Age ($\hat{\gamma}_4$)	−0.015	0.014	1.20	0.273

TABLE 12.4

Analysis of Variance Table for Stage And Age For Laryngeal Cancer Patients, Utilizing the Proportional Odds Model and the Log Logistic Distribution

$= e^{\frac{parameter}{\alpha}} \longrightarrow$

which is our odds!

Parameter	Parameter Estimate	Standard Errors	
Intercept $\hat{\lambda}$	0.013	0.018	
Scale $\hat{\alpha}$	1.398	0.168	$= 13/.72$
Z_1: Stage II ($\hat{\beta}_1$)	0.176	0.581	$=$
Z_2: Stage III ($\hat{\beta}_2$)	1.127	0.498	$= -1.77/.72$
Z_3: Stage IV ($\hat{\beta}_3$)	2.469	0.632	
Z_4: Age ($\hat{\beta}_4$)	0.021	0.019	

$e^{\beta_2} = effect$

disease compared to Stage I disease is $\exp[-(-.81)] = 2.25$ and for Stage IV disease is $\exp[-(-1.77)] = 5.87$. This suggests that the median life for Stage I patients is about 5.87 times that of Stage IV patients.

ex. $e^{-.176} = .84$. We then say odds of survival for Stage II is 84% of the odds for Stage I.

Practical Notes

1. SAS PROC LIFEREG and S-Plus routine survreg provide maximum likelihood estimates of intercept μ, and scale parameter σ, associated with the logistic distribution. The parameters of the underlying log logistic distribution are the following functions of these extreme value parameters: $\lambda = \exp(-\mu/\sigma)$ and $\alpha = 1/\sigma$. SAS allows for right-, left- and interval-censored data.

2. When performing an accelerated failure time regression employing the log logistic distribution, SAS and S-Plus provide maximum likelihood estimates of intercept μ, scale parameter σ, and regression coefficients γ_i. The parameters of the underlying log logistic distribution, when modeling time directly, are the following functions of those parameters: $\lambda = \exp(-\mu/\sigma)$, $\alpha = 1/\sigma$, and $\beta_i = -\gamma_i/\sigma$. SAS allows for right-, left- and interval-censored data.

12.4   Other Parametric Models

In section 12.2, we examined the use of the Weibull distribution as a model for survival data, and, in section 12.3, we examined the use of the log logistic model. In this section, we shall look at alternative parametric models for the survival function, focusing on the regression problem with obvious extensions to the problem of univariate estimation.

The first model to be considered is the log normal distribution. Here, given a set of covariates $\mathbf{Z} = (Z_1, \ldots, Z_p)'$, the logarithm of the time to the event follows the usual normal regression model, that is,

$$Y = \log X = \mu + \gamma'\mathbf{Z} + \sigma W, \qquad (12.4.1)$$

where W has a standard normal distribution. The general shape of the hazard rate for this model is quite similar to that of the log logistic distribution, and, in most instances, regression models based on the log normal distribution are very close to regression models based on the log logistic model.

For the log normal distribution the survival function of the time to event T is given by

$$S(x) = 1 - \Phi\{[\log(x) - (\mu + \gamma'\mathbf{Z})]/\sigma\},$$

where $\Phi\{\}$ is the standard normal cumulative distribution function.

A second model of interest is the generalized gamma distribution. This model is very useful in selecting between alternative parametric models because it includes the Weibull, exponential, and the log normal models as limiting cases. For this model, $Y = \log X$ follows the linear model (12.4.1) with W having the following probability density function:

$$f(w) = \frac{|\theta|[\exp(\theta w)/\theta^2]^{(1/\theta^2)} \exp[-\exp(\theta w)/\theta^2]}{\Gamma(1/\theta^2)}, \quad -\infty < w < \infty.$$
$$(12.4.2)$$

When θ equals 1, this model reduces to the Weibull regression model, and, when θ is 0, the model reduces to the log normal distribution. When $\theta = 1$ and $\sigma = 1$ in (12.4.1), then, (12.4.1) reduces to the exponential regression model.

The generalized gamma model is most commonly used as a tool for picking an appropriate parametric model for survival data but, rarely, as the final parametric model. Wald or likelihood ratio tests of the hypotheses that $\theta = 1$ or $\theta = 0$ provide a means of checking the assumption of a Weibull or log normal regression model, respectively.

With the exception of the Weibull and log normal distribution, it is difficult to use a formal statistical test to discriminate between parametric models because the models are not nested in a larger model which includes all the regression models discussed in this chapter. One way of selecting an appropriate parametric model is to base the decision on minimum Akaikie information criterion (AIC). For the parametric models discussed, the AIC is given by

$$\text{AIC} = -2 * \log(\text{Likelihood}) + 2(p + k), \qquad (12.4.3)$$

where $k = 1$ for the exponential model, $k = 2$ for the Weibull, log logistic, and log normal models and $k = 3$ for the generalized gamma model. We shall illustrate this on two examples considered in this chapter.

EXAMPLE 12.1

(continued): We shall reexamine the parametric models for the auto and allo transplant survival data. We will fit the exponential, Weibull, log logistic, log normal models, and generalized gamma models separately to the data on allo and auto transplants. The log likelihood and the AIC for each model are reported in Table 12.5. (Note that here $p = 0$.) Also included in this table are the estimates of θ from the generalized gamma model, their standard errors and the p-values of the Wald tests of $H_o : \theta = 0$ and $H_o : \theta = 1$. These are tests of the appropriateness of the log normal and Weibull models, respectively.

TABLE 12.5

Results of Fitting Parametric Models to the Transplant Data

		Allo Transplants	Auto Transplants
Exponential	Log likelihood	−81.203	−68.653
	AIC	164.406	139.306
Weibull	Log likelihood	−72.879	−68.420
	AIC	149.758	140.840
Log logistic	Log likelihood	−71.722	−67.146
	AIC	147.444	138.292
Log normal	Log likelihood	−71.187	−66.847
	AIC	146.374	137.694
Generalized gamma	Log likelihood	−70.892	−66.781
	AIC	147.784	139.562
	$\hat{\theta}$	−0.633	−0.261
	SE[$\hat{\theta}$]	0.826	0.725
	p-value for $H_o : \theta = 0$	0.443	0.719
	p-value for $H_o : \theta = 1$	0.048	0.082

From this table, we see that the log normal distribution provides the best fit to this data, and the log logistic distribution is a close second. The generalized gamma model, which has the smallest log likelihood, does not have a smaller AIC than these two models and the simpler models are preferred. The exponential distribution for Allo transplants has a much poorer fit than the Weibull model, and there is no evidence of an improved fit for auto transplants, using the Weibull rather than the exponential. A likelihood ratio chi-square test, with one degree of freedom, for testing the hypothesis that the Weibull shape parameter is equal to one has a value of 16.648 ($p < 0.0001$) for allo transplants and a value of 0.468 for auto transplants.

Using a log normal regression model with a single covariate Z_1 equal to 1 if the patient received an auto transplant, we have the following regression model:

Parameter	Estimate	Standard Error	Wald Chi Square	p-Value
Intercept: μ	3.177	0.355	80.036	<0.0001
Type of Transplant: γ_1	0.054	0.463	0.0133	0.9080
Scale: σ	2.084	0.230	—	—

Here, we see that there is no evidence of any difference in survival between the two types of transplants.

EXAMPLE 12.2   *(continued):*   We shall now compare the fit of the exponential, Weibull, log normal, log logistic and generalized gamma models for the data on laryngeal cancer. Recall that, here, we have four covariates:

Z_1: 1 if Stage II cancer, 0 otherwise,

Z_2: 1 if Stage III cancer, 0 otherwise,

Z_3: 1 if Stage IV cancer; 0 otherwise, and

Z_4: Patient's age at diagnosis.

We fit the log linear model

$$Y = \ln X = \mu + \sum_{k=1}^{4} \gamma_k Z_k + \sigma W,$$

TABLE 12.6
Parametric Models for the Laryngeal Cancer Study

	Exponential		Weibull		Log Logistic		Log Normal		Generalized Gamma	
	Estimate	SE	Estimate	SE	Estimate	SE	Estimate	SE	Estimate	SE
μ	3.755	0.990	3.539	0.904	3.102	0.953	3.383	0.936	3.453	0.944
α_1	−0.146	0.460	−0.148	0.408	−0.126	0.415	−0.199	0.442	−0.158	0.431
α_2	−0.648	0.355	−0.587	0.320	−0.806	0.354	−0.900	0.363	−0.758	0.394
α_3	−1.635	0.399	−1.544	0.363	−1.766	0.426	−1.857	0.443	−1.729	0.449
α_4	−0.020	0.014	−0.017	0.013	−0.015	0.014	−0.018	0.014	−0.018	0.014
σ	1.000	0.000	0.885	0.108	0.715	0.086	1.264	0.135	1.104	0.257
θ									0.458	0.584
Log L	−108.50		−108.03		−108.19		−108.00		−107.68	
AIC	227.00		228.05		228.38		227.99		229.36	

where W has the appropriate distribution for each of the models. Note that the value of σ is fixed at 1 for the exponential distribution. Table 12.6 provides the estimates of the model parameters and their standard errors, the maximized likelihoods, and the AIC criterion for all five models.

In this table, we see that all three models fit equally well. The exponential model has the smallest AIC and, in that sense, is the best fitting model. For this model,

$$Y = 3.755 - 0.146Z_1 - 0.648Z_2 - 1.635Z_3 - 0.020Z_4 + W.$$

The negative values of the coefficients of Z_1, Z_2, and Z_3 in the log linear model suggest that individuals with stages II, III, and IV cancer have shorter lifetimes than individuals with Stage I disease.

Practical Note

1. SAS PROC LIFEREG has routines for fitting the generalized gamma and log normal distributions to right-, left- and interval-censored data. The S-Plus routine survreg fits the log normal model.

12.5   Diagnostic Methods for Parametric Models

In the last three sections, we have presented a variety of models for univariate survival data and several parametric models that can be used to study the effects of covariates on survival. In this section, we shall focus on graphical checks of the appropriateness of these models. As discussed in Chapter 11, we favor graphical checks of the appropriateness rather then formal statistical tests of lack of fit because these tests tend either to have low power for small-sample sizes or they always reject a given model for large samples. The graphical checks discussed here serve as a means of rejecting clearly inappropriate models, not to "prove" that a particular parametric model is correct. In fact, in many applications, several parametric models may provide reasonable fits to the data and provide quite similar estimates of key quantities.

We shall first examine the problem of checking for the adequacy of a given model in the univariate setting. The key tool is to find a function of the cumulative hazard rate which is linear in some function of time. The basic plot is made by estimating the cumulative hazard rate by the Nelson–Aalen estimator (see section 4.2). To illustrate this technique, consider a check of the appropriateness of the log logistic distribution. Here, the cumulative hazard rate is $H(x) = \ln(1 + \lambda x^\alpha)$. This implies that, for the log logistic model,

$$\ln\{\exp[H(x)] - 1\} = \ln\lambda + \alpha\ln x, \qquad (12.5.1)$$

so, a plot of $\ln\{\exp[\hat{H}(x)] - 1\}$ versus $\ln x$ should be approximately linear. The slope of the line gives a crude estimate of α and the y intercept gives a crude estimate of $\ln\lambda$. Here, $\hat{H}$ is the Nelson–Aalen estimator. Note that, for the log logistic distribution, the quantity $\ln\{\exp[H(x)] - 1\}$ is precisely the log odds favoring survival.

For the other models discussed in this chapter, the following plots are made to check the fit of the models:

Model	Cumulative Hazard Rate	Plot	
Exponential:	λx	$\hat{H}$ versus x	(12.5.2)
Weibull:	λx^{α}	$\ln\hat{H}$ versus $\ln x$	(12.5.3)
Log normal:	$-\ln\{1 - \Phi[\ln(x) - \mu)]/\sigma\}$	$\Phi^{-1}[1 - \exp(-\hat{H})]$ versus $\ln x$	(12.5.4)

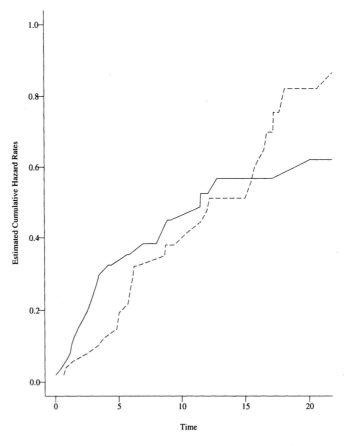

Figure 12.1   *Exponential hazard plot for the allo (solid line) and auto (dashed line) transplant groups.*

Note that the slope of the line for the Weibull hazard plot gives a crude estimate of α and, if the slope of the line is 1, then, the exponential is a reasonable model.

EXAMPLE 12.1

(continued):   To check the adequacy of the exponential, Weibull, log logistic, and log normal models for the data on auto and allo transplants, four hazard plots are presented in Figures 12.1–12.4. If the curves do not appear linear for each figure, this is evidence that the parametric model does not provide an adequate fit to the data. From Figure 12.1, the exponential plot, we see that the curves for the allo transplant group appear to be nonlinear, suggesting that the exponential is not a good model for this set of data. The curve is roughly linear for the auto transplant data, except in the tail where the estimate of H is highly variable, suggesting that the exponential may be a reasonable model. The curves for the other three models (Figures 12.2–12.4) are roughly linear, suggesting these may be appropriate models for either groups.

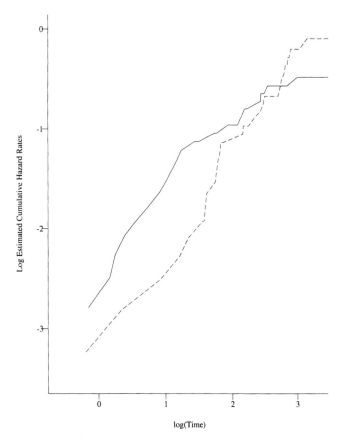

Figure 12.2   *Weibull hazard plot for the allo (solid line) and auto (dashed line) transplant groups.*

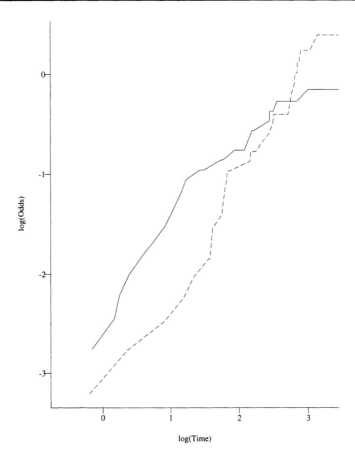

Figure 12.3 *Log logistic hazard plot for the allo (solid line) and auto (dashed line) transplant groups.*

When comparing two groups, an alternative to the proportional hazards model is the accelerated failure-time model. A *quantile-quantile* or *q-q* plot is made to check if this provides an adequate fit to the data. The plot is based on the fact that, for the accelerated failure-time model,

$$S_1(t) = S_o(\theta t), \tag{12.5.2}$$

where S_o and S_1 are the survival functions in the two groups and θ is the acceleration factor. Let t_{op} and t_{1p} be the pth percentiles of groups 0 and 1, respectively, that is

$$t_{kp} = S_k^{-1}(1 - p), k = 0, 1.$$

Using the relationship (12.5.2), we must have $S_o(t_{op}) = 1 - p = S_1(t_{1p}) = S_o(\theta t_{1p})$ for all t. If the accelerated failure time model holds, $t_{op} = \theta t_{1p}$. To check this assumption we compute the Kaplan–Meier estimators of the two groups and estimate the percentiles t_{1p}, t_{0p}, for various values of p. If we plot the estimated percentile in group 0 versus the estimated

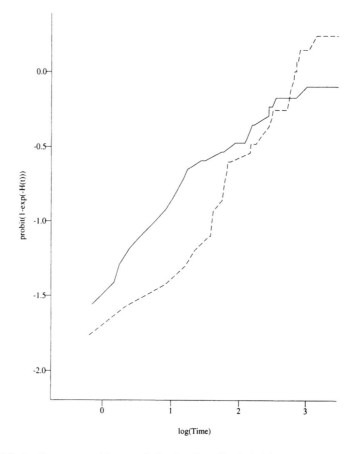

log(Time)

Figure 12.4 *Log normal hazard plot for the allo (solid line) and auto (dashed line) transplant groups.*

percentile in group 1 (i.e., plot the points t_{1p}, t_{0p} for various values of p), the graph should be a straight line through the origin, if the accelerated failure time model holds. If the curve is linear, a crude estimate of the acceleration factor q is given by the slope of the line.

EXAMPLE 12.1   *(continued):* We shall graphically check the adequacy of the accelerated failure-time model for comparing allo and auto transplants. Here, we fit the Kaplan-Meier estimator separately to each group and compute the percentiles for each group for $p = 0.05, 0.10, \ldots, 0.35$. These percentiles are in the range where the percentile could be estimated for both groups. Figure 12.5 shows the q-q plot for auto transplants (Group 1) versus allo transplants (Group 0). The figure appears to be approximately linear with a slope of about 0.6, which is a crude estimate of the acceleration factor θ.

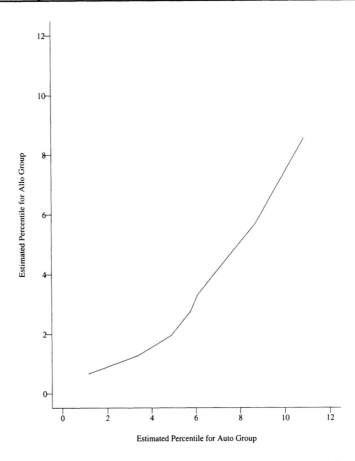

Figure 12.5 *q-q plot to check the adequacy of the accelerated failure time model for comparing Allo and Auto transplants.*

For the parametric regression problem, analogs of the residual plots described in Chapter 11 can be made with a redefinition of the various residuals to incorporate the parametric form of the baseline hazard rates. The first such residual is the Cox–Snell residual that provides a check of the overall fit of the model. The Cox–Snell residual, r_j, is defined by $r_j = \hat{H}(T_j \mid \mathbf{Z}_j)$, where $\hat{H}$ is the fitted model. If the model fits the data then the r_j's should have a standard ($\lambda = 1$) exponential distribution, so that a hazard plot of r_j versus the Nelson–Aalen estimator of the cumulative hazard of the r_j's should be a straight line with slope 1. For the four models considered in this chapter, the Cox–Snell residuals are

Exponential $r_i = \hat{\lambda} t_i \exp\{\hat{\beta}'\mathbf{Z}_i\},$

Weibull $\hat{\lambda} \exp(\hat{\beta}'\mathbf{Z}_i) t_1^{\hat{\alpha}},$

Log logistic $\ln\left[\dfrac{1}{1 + \hat{\lambda}\exp(\hat{\beta}'\mathbf{Z}_i) t_j^{\hat{\alpha}}}\right],$

and

Log normal $\ln\left[1 - \Phi\left(\dfrac{\ln T_j - \hat{\mu} - \hat{\gamma}'\mathbf{Z}_j}{\hat{\sigma}}\right)\right].$

Examination of model fit with the Cox–Snell residuals is equivalent to that done using the so-called standardized residuals based on the log linear model representation. Here, we define the standardized residuals by analogy to those used in normal theory regression as

$$s_j = \frac{\ln T_j - \hat{\mu} - \hat{\gamma}'\mathbf{Z}_j}{\hat{\sigma}}.$$

If the Weibull model holds, then, these residuals should be a censored sample from the standard extreme value distribution (12.2.2); if the log logistic distribution holds, these are a censored sample from a standard logistic distribution (12.3.1); and if the log normal distribution holds, these are a censored sample from a standard normal distribution. The hazard plot techniques discussed earlier can be used to check if the standardized residuals have the desired distribution. However, the hazard plots obtained are exactly those obtained by the exponential hazard plot for the Cox–Snell residuals.

EXAMPLE 12.2   *(continued):*   In Figures 12.6–12.9, the cumulative hazard plots for the Cox–Snell residuals are shown for the exponential, Weibull, log logistic and log normal regression models for the laryngeal cancer data. We see from these plots that all four models give reasonable fits to the data, the best being the log normal and log logistic models.

In Chapter 11, the martingale and deviance residuals were defined for Cox regression models. For a parametric model, the martingale residual is defined by $M_j = \delta_j - r_j$ and the deviance residual by

$$D_j = \mathrm{sign}[M_j]\{-2[M_j + \delta_j\ln(\delta_j - M_j)]\}^{1/2}.$$

As for the Cox model, the martingale residual is an estimate of the excess number of deaths seen in the data, but not predicted by the model. In the parametric case, note that the derivation of M_j as a martingale does not hold but, because the residuals are similar in form to those for the Cox model, the name carries through. The deviance residuals are an attempt to make the martingale residuals more symmetric about 0. If the model is correct, then, the deviance residuals should look like random noise. Plots of either the martingale or deviance residuals against time, observation number, or acceleration factor provides a check of the model's adequacy. The discussion of how to use these residuals in Chapter 11 carries over to the parametric case. We shall illustrate

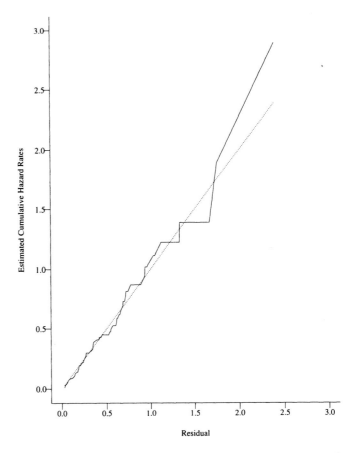

Figure 12.6 *Cox–Snell residuals to assess the fit of the exponential regression model for the laryngeal cancer data set*

the use of the deviance residuals in the following continuation of Example 12.2.

EXAMPLE 12.2   *(continued):* We shall examine the fit of the log logistic regression model to the laryngeal cancer data using the deviance residuals. Figure 12.10 is a plot of the deviance residuals versus time on study. Here, we see that the deviance residuals are quite large for small times and that they decrease with time. This suggests that the model underestimates the chance of dying for small t and overestimates this chance for large t. However, there are only a few outliers early, which may cause concern about the model. The deviance residual plots for the other three models are quite similar.

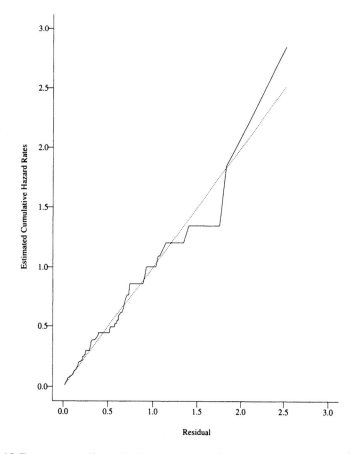

Figure 12.7 *Cox–Snell residuals to assess the fit of the Weibull regression model for the laryngeal cancer data set*

Practical Note

1. Martingale and deviance residuals for these parametric models are available in S-Plus.

Theoretical Notes

1. Further work on graphical checks for the parametric regression models can be found in Weissfeld and Schneider (1990) and Escobar and Meeker (1992).
2. It is possible to define a score residual for the various parametric models similar to that presented in section 12.6. To illustrate how

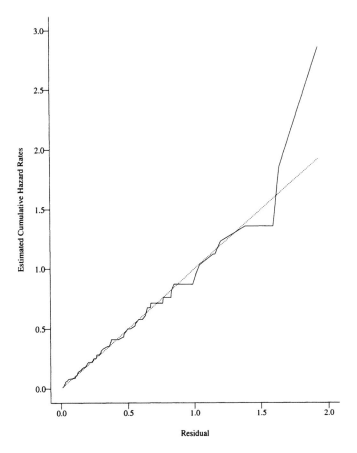

Figure 12.8 *Cox–Snell residuals to assess the fit of the log logistic regression model for the laryngeal cancer data set*

this is done, consider the Weibull regression problem with a single covariate Z. The contribution of an individual with covariate Z_j to the likelihood is given by

$$L_j = [\exp(\beta Z_j)\lambda \alpha t_j^{\alpha-1}]^{\delta_j} \exp[-\lambda \exp(\beta Z_j) T_j^{\alpha}].$$

The score residual for λ is given by

$$\frac{\partial \ln L_j}{\partial \lambda} = \frac{\delta_j}{\lambda} - \exp(\beta Z_j) T_j^{\alpha},$$

for α,

$$\frac{\partial \ln L_j}{\partial \alpha} = +\frac{\delta_j}{\alpha} + \delta_j \ln T_j - \lambda \exp(\beta Z_j) T_j^{\alpha} \ln T_j,$$

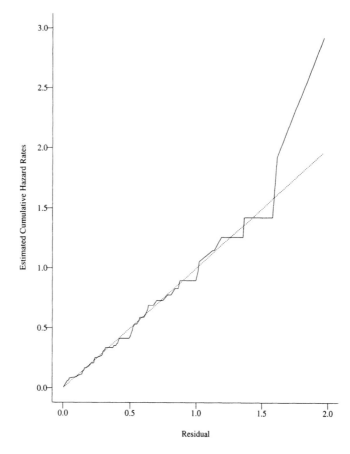

Figure 12.9 *Cox–Snell residuals to assess the fit of the log normal regression model for the laryngeal cancer data set*

and for β,

$$\frac{\partial \ln L_j}{\partial \beta} = \delta_j Z_j - \lambda Z_j \exp(\beta Z_j) T_j^{\alpha}.$$

These residuals can be used, as in section 11.6, to examine the influence of a given observation on the estimates. See Collett (1994) for additional detail. These residuals are available in S-Plus.

12.6 Exercises

12.1 In section 1.11, a study of the effects of ploidy on survival for patients with cancer of the tongue was described. In the study patients were classified as having either an aneuploid or diploid DNA profile. The data is presented in Table 1.6.

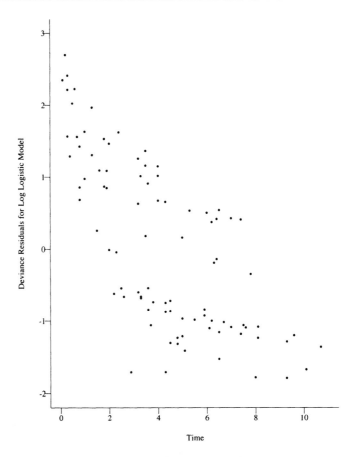

Figure 12.10 *Deviance residuals from the log logistic regression model for laryngeal cancer patients*

(a) For both the aneuploid and diploid groups fit a Weibull model to the data. Find the maximum likelihood estimates of λ and α, and their standard errors.

(b) For both groups, test the hypothesis that the shape parameter, α, is equal to 1 by both the Wald and likelihood ratio tests.

(c) Find the maximum likelihood estimates of the median survival for both groups. Use the delta method to find an estimate of the standard error of your estimates.

(d) Fit a Weibull regression model to this data with a single covariate, Z, that is equal to 1 if the patient had an aneuploid DNA profile and 0 otherwise. Test the hypothesis of no effect of ploidy on survival using the likelihood ratio test and the Wald test. Find a point estimate and 95% confidence interval for the relative risk of death for an aneuploid tumor as compared to a diploid tumor. Also find a point estimate and a 95% confidence for the acceleration factor. Provide an interpretation of this factor.

12.2 In section 1.4 the times to first exit-site infection (in months) of patients with renal insufficiency were reported. In the study 43 patients had a surgically placed catheter (Group 1) and 76 patients had a percutaneous placement of their catheter (Group 0).

(a) For both groups fit a Weibull model to the data. Find the maximum likelihood estimates of λ and α, and their standard errors.

(b) For both groups test the hypothesis that the shape parameter, α, is equal to 1 using the likelihood ratio test and the Wald test.

(c) Find the maximum likelihood estimates and 95% confidence intervals for the two survival functions at 5 months after placement of the catheter. Compare these estimates to those obtained using the product-limit estimator.

(d) Fit a Weibull regression model to this data with a single covariate, Z, that indicates group membership. Test the hypothesis of no effect of catheter placement on the time to exit site infection. Find point estimates and 95% confidence intervals for the relative risk and the acceleration factor for exit site infections. Provide an interpretation of these quantities.

12.3 In section 1.10, times to death or relapse (in days) are given for 23 non-Hodgkin's lymphoma (NHL) patients, 11 receiving an allogeneic (Allo) transplant from an HLA-matched sibling donor and 12 patients receiving an autologous (Auto) transplant. Also, data is given in Table 1.5 on 20 Hodgkin's lymphoma (HOD) patients, 5 receiving an allogeneic (Allo) transplant from an HLA-matched sibling donor and 15 patients receiving an autologous (Auto) transplant. Because there is a potential for different efficacy of the two types of transplants for the two types of lymphoma, a model with a main effect for type of transplant, a main effect for disease type and an interactive term is of interest (coding similar to 8.1b).

(a) Using a Weibull regression model, analyze this data by performing a likelihood ratio global test of no effect of transplant type and disease state on survival. Construct an ANOVA table to summarize estimates of the risk coefficients and the results of the one degree of freedom tests for each covariate in the model.

(b) Test the hypothesis of no disease–transplant type interaction using a likelihood ratio test.

(c) Find point estimates and 95% confidence intervals for the relative risk of death for an NHL Auto transplant patient as compared to an NHL Allo transplant patient.

(d) Test the hypothesis that the death rates are the same for HOD Allo transplants and NHL Allo patients. Repeat this test for Auto patients.

(e) Test the hypothesis that the death rates for Auto transplant and Allo transplant patients are the same against the alternative they are different for at least one disease group by a 2 degree of freedom test of $H_o : h(t \mid \text{NHL Allo}) = h(t \mid \text{NHL Auto})$ *and* $h(t \mid \text{HOD Allo}) = h(t \mid \text{HOD Auto})$.

(f) Compare your results to those found in Exercise 3 of Chapter 8 by using the semiparametric proportional hazards model.

12.4   Repeat Exercise 2 using the log logistic model. In part b use the Wald test and in part d provide point and interval estimates of the acceleration factor and the relative odds. Compare your results to those found in Exercise 2.

12.5   Repeat Exercise 1 using the log logistic model. In part b use the Wald test and in part d provide point and interval estimates of the acceleration factor and the relative odds. Compare your results to those found in that exercise.

12.6   Repeat Exercise 3 using the log logistic model. Compare your results to those found in that exercise. Estimate relative odds rather than relative risks in part c.

12.7   Using the ploidy data in Exercise 1, estimate the parameters and the variance-covariance matrix for the following models for each of the two groups.

(a) A log normal model.

(b) A normal model.

(c) A generalized gamma model.

(d) Using the results of part c, test the hypothesis that $\theta = 0$. Interpret your result in terms of model selection.

(e) Using the results of part c, test the hypothesis that $\theta = 1$. Interpret your result in terms of model selection.

(f) Based on your results in this exercise and in Exercises 1 and 5, which parametric model best fits the data for each of the two ploidy groups?

12.8   Using the information in Exercise 2, determine the best fitting parametric regression model to determine the effects of catheter placement on the time to first exit site infection by fitting the exponential, log normal, and generalized gamma models.

12.9   For both the aneuploid and diploid groups in Exercise 1, make an appropriate hazard plot to determine if the following models fit the data:

(a) exponential,

(b) Weibull,

(c) log normal, and

(d) log logistic.

12.10 For both catheter placement groups in Exercise 2, make an appropriate hazard plot to determine if the following models fit the data:

(a) exponential,

(b) Weibull,

(c) log normal, and

(d) log logistic.

12.11 Check the adequacy of the accelerated failure time model for describing the effects of ploidy on survival in Exercise 1 by making a quantile-quantile plot. Provide a crude estimate of the acceleration factor and compare it to the estimate you found in Exercise 1.

12.12 Check the adequacy of the accelerated failure time model for describing the effects of catheter placement on the time to first exit site infection in Exercise 2 by making a quantile-quantile plot. Provide a crude estimate of the acceleration factor and compare it to the estimate you found in Exercise 2.

12.13 In Exercise 1, you fit a Weibull regression model to explain the effect of ploidy on survival.

(a) Examine the fit of this model by making the appropriate plot of the Cox–Snell residuals.

(b) Examine the fit of this model by making the appropriate plot of the deviance residuals residuals.

(c) Repeat a and b for the log logistic regression model.

12.14 In Exercise 3 a Weibull regression model was fit to the survival times of patients given a bone marrow transplant. The model included a covariate for type of transplant, type of disease as well as an interaction term.

(a) Examine the fit of this model by making the appropriate plot of the Cox–Snell residuals.

(b) Examine the fit of this model by making the appropriate plot of the deviance residuals residuals.

(c) Repeat a and b for the log logistic regression model.

13
Multivariate Survival Analysis

13.1 Introduction

In the previous chapters of this book, we have examined a variety of techniques for analyzing survival data. With few exceptions, these techniques are based on the assumption that the survival times of distinct individuals are independent of each other. Although this assumption may be valid in many experimental settings, it may be suspect in others. For example, we may be making inferences about survival in a sample of siblings or litter mates who share a common genetic makeup, or we may be studying survival in a sample of married couples who share a common, unmeasured, environment. A third example is where we are studying the times to occurrence of different nonlethal diseases within the same individual. In each of these situations, it is quite probable that there is some association within groups of survival times in the sample.

A model that is becoming increasingly popular for modeling association between individual survival times within subgroups is the use of a *frailty model*. A frailty is an unobservable random effect shared by subjects within a subgroup. This most common model for the frailty is a common random effect that acts multiplicatively on the hazard rates of all subgroup members. In this model, families with a large value of the frailty will experience the event at earlier times then families with small values of the random effect. Thus the most "frail" individuals will die early and late survivors will tend to come from more robust families.

Frailty models are also used in making adjustments for overdispersion in univariate survival studies. Here, the frailty represents the total effect on survival of the covariates not measured when collecting information on individual subjects. If these effects are ignored, the resulting survival estimates may be misleading. Corrections for this overdispersion allow for adjustments for other unmeasured important effects.

The most common model for a frailty is the so-called shared frailty model extension of the proportional hazards regression model. Here, we assume that the hazard rate for the jth subject in the ith subgroup, given the frailty, is of the form

$$h_{ij}(t) = h_0(t)\exp(\sigma w_i + \boldsymbol{\beta}'\mathbf{Z}_{ij}), \quad i = 1, \dots, G, \, j = 1, \dots, n_i, \quad (13.1.1)$$

where $h_0(t)$ is an arbitrary baseline hazard rate, $\mathbf{Z}_{ij}$ is the vector of covariates, $\boldsymbol{\beta}$ the vector of regression coefficients, and $w_1, \dots, w_G$ the frailties. Here, we assume that the w's are an independent sample from some distribution with mean 0 and variance 1. Note that, when σ is zero, the model (13.1.1) reduces to the basic proportional hazards model discussed in Chapter 8. The model allows for the sizes n_i of the individual groups to differ from group to group and for the group sizes to be 1. In some applications, it is more convenient to write the model (13.1.1) as

$$h_{ij}(t) = h_0(t)u_i\exp(\boldsymbol{\beta}'\mathbf{Z}_{ij}), \quad i = 1, \dots, G, \, j = 1, \dots, n_i, \quad (13.1.2)$$

where the u_i's are an independent and identically distributed sample from a distribution with mean 1 and some unknown variance. We see clearly that, when nature picks a value of u_i greater than 1, individuals within a given family tend to fail at a rate faster than under an independence model where u_i is equal to 1 with probability 1. Conversely, when u_i is less than 1, individuals in a family tend to have survival longer than predicted under an independence model.

When we collect data, the w_i's are not observable, so the joint distribution of the survival times of individuals within a group, found by taking the expectation of $\exp[-\sum_{j=1}^{n_i} H_{ij}(t)]$ with respect to u_i is given by

$$S(x_{i1}, \dots, x_{in_i}) = P[X_{i1} > x_{i1}, \dots, X_{in_i} > x_{in_i}] \quad (13.1.3)$$

$$= LP\left[\sum_{j=1}^{n_i} H_0(x_{ij})\exp(\boldsymbol{\beta}'\mathbf{Z}_{ij})\right].$$

Here, $LP(v) = E_U[\exp(-Uv)]$ is the Laplace transform of the frailty U. Common models proposed in the literature for the random effect are the one-parameter gamma distribution (Clayton, 1978), the positive stable distribution (Hougaard, 1986a), and the inverse Gaussian distribution (Hougaard, 1986b). Other models for the frailty are the log normal distribution (McGilchrist and Aisbett, 1991), a power variance model (Aalen, 1988; 1992), a uniform distribution (Lee and Klein, 1988) and a threshold model (Lindley and Singpurwalla, 1986).

In this chapter, we shall examine estimation and inference procedures for random effect models. In section 13.2, we will present a score test of the hypothesis of no frailty effect ($\sigma = 0$) suggested by Commenges and Andersen (1995). This test is valid for any frailty distribution and provides an easy means of testing for a random effect in either the multivariate or univariate problem.

In sections 13.3, we present both a semiparametric and a parametric approach to estimating the risk coefficients and the frailty parameter for the most commonly used gamma frailty model. Here, estimating in the semiparametric model involves implementing an EM (Expectation-Maximization) algorithm. We give a series of references in the notes showing where these techniques have been extended to other frailty models.

In section 13.4, we present an alternative model for multivariate survival data suggested by Lee et al. (1992). For this approach, we assume that the proportional hazards model holds marginally for each individual, but individuals within groups are associated. (Note that, for the frailty approach, we assumed that the proportional hazards model is true for each individual, conditional on the frailty.) Estimators of β are the usual regression estimates, as discussed in Chapter 8. These estimators are based on an "independence working model" and are consistent for β. The usual variance estimator, based on the observed information matrix, however, is not correct when individuals are correlated. An adjusted covariance matrix is used which is found by adjusting the information matrix for the association between individuals.

13.2   Score Test for Association

To test for association between subgroups of individuals, a score test has been developed by Commenges and Andersen (1995). The test can be used to test for association, after an adjustment for covariate effects has been made using Cox's proportional hazards model, or can be applied when there are no covariates present. It can be used to test for overdispersion in the univariate Cox proportional hazards model. Although the test can be applied following a stratified Cox regression model, for simplicity, we shall focus on the nonstratified model.

The test is based on the model (13.1.1). Here we have G subgroups of subjects, and within the ith subgroup, we have n_i individuals. Note that, for the univariate model, where we are testing for overdispersion, G is the total sample size and each n_i is 1. We are interested in testing the null hypothesis that σ is equal to zero against the alternative that σ is not zero. No distributional assumptions are made about the distribution of the random effects w_i.

The data we need to construct the test statistics consists of the on study time T_{ij}, the event indicator δ_{ij}, and the covariate vector $\mathbf{Z}_{ij}$ for the jth observation in the ith group. Using this data, we can construct the "at risk" indicator $Y_{ij}(t)$ at time t for an observation. Recall that $Y_{ij}(t)$ is 1 when the ijth individual is at risk. Then, we fit the Cox proportional hazards model, $h(t \mid \mathbf{Z}_{ij}) = h_o(t)\exp(\boldsymbol{\beta}'\mathbf{Z}_{ij})$, and we obtain the partial maximum likelihood estimates $\mathbf{b}$ of $\boldsymbol{\beta}$ and the Breslow estimate $\hat{H}_o(t)$ of the cumulative baseline hazard rate (see Chapter 8). Let

$$S^{(0)}(t) = \sum_{i=1}^{G}\sum_{j=1}^{n_i} Y_{ij}(t)\exp(\mathbf{b}'\mathbf{Z}_{ij}). \tag{13.2.1}$$

Note that $S^{(0)}(t)$ is simply the sum of $\exp(\mathbf{b}'\mathbf{Z}_{ij})$ over all individuals at risk at time t. Finally, we need to compute $M_{ij} = \delta_{ij} - \hat{H}_o(T_{ij})\exp(\mathbf{b}'\mathbf{Z}_{ij})$, the martingale residual for the ijth individual (see section 11.3).

The score statistic is given by

$$T = \sum_{i=1}^{G}\left\{\sum_{j=1}^{n_i} M_{ij}\right\}^2 - D + C, \tag{13.2.2}$$

where D is the total number of deaths and

$$C = \sum_{i=1}^{G}\sum_{j=1}^{n_i}\frac{\delta_{ij}}{S^{(0)}(T_{ij})^2}\sum_{b=1}^{G}\left[\sum_{k=1}^{n_i} Y_{bk}(T_{ij})\exp(\mathbf{b}'\mathbf{Z}_{bk})\right]^2. \tag{13.2.3}$$

The test statistic can also be written as

$$T = \sum_{i=1}^{G}\sum_{j=1}^{n_i}\sum_{\substack{k=1 \\ k \neq j}}^{n_j} M_{ij}M_{ik} + \left(\sum_{i=1}^{G}\sum_{j=1}^{n_i} M_{ij}^2 - N\right) + C, \tag{13.2.4}$$

where N is the total sample size.

Here, we see that the score statistic is the sum of three terms. The first is the sum of the pairwise correlations between individuals in the subgroups, the second a measure of overdispersion, and the third a correction term C which tends to zero as N increases.

To estimate the variance of T let $\hat{\boldsymbol{\Sigma}}$ be the estimated variance-covariance matrix of $\mathbf{b}$ with elements $\hat{\sigma}_{bk}$, $b = 1, \ldots, p$, $k = 1, \ldots, p$. Let

$$p_{ij}(t) = \frac{Y_{ij}(t)\exp(\mathbf{b}'\mathbf{Z}_{ij})}{S^{(0)}(t)}$$

and

$$\bar{p}_i(t) = \sum_{j=1}^{n_i} p_{ij}(t), \ i = 1, \ldots, G. \tag{13.2.5}$$

For each distinct death time, $0 = t_o < t_1 < \ldots < t_d$, let d_k be the observed number of deaths at time t_k. Recall that the martingale residual for the jth observation in the ith group at time t_k is given by

$$M_{ij}(t_k) = \begin{cases} \delta_{ij} - H_o(T_{ij}), \exp(\mathbf{b}^t\mathbf{Z}_{ij}) & \text{if } T_{ij} \le t_k, \\ -H_o(t_k)\exp(\mathbf{b}^t\mathbf{Z}_{ij}) & \text{if } T_{ij} > t_k. \end{cases} \tag{13.2.6}$$

Let $\bar{M}_i(t_k) = \sum_{j=1}^{n_i} M_{ij}(t_k)$ and $\bar{M}_i(t_o) = 0$. For each group, compute

$$Q_i(t_k) = 2\left[\bar{M}_i(t_{k-1}) - \sum_{g=1}^{G}\bar{M}_g(t_{k-1})\bar{p}_g(t_k) - \bar{p}_i(t_k) + \sum_{g=1}^{G}\bar{p}_i(t_k)^2\right],$$

$$k = 1, \ldots, d. \tag{13.2.7}$$

Finally, for each of the p-covariates, compute

$$\theta_b = \sum_{i=1}^{G}\sum_{k=1}^{d} Q_i(t_k)d_k\left\{\sum_{j=1}^{n_i} p_{ij}(t_k)Z_{ijb}\right\}, b = 1, \ldots, p. \tag{13.2.8}$$

The estimated variance of the score statistic is given by

$$V = \sum_{i=1}^{G}\sum_{k=1}^{d} Q_i(t_k)^2\bar{p}_i(t_k)d_k + \sum_{b=1}^{p}\sum_{k=1}^{p} \theta_b\theta_k\hat{\sigma}_{bk}. \tag{13.2.9}$$

The test statistic for the test of no association is given by $T/V^{1/2}$ which has an asymptotic normal distribution.

EXAMPLE 13.1   Mantel et al. (1977) reports the results of a litter-matched study of the tumorigenesis of a drug. In the experiment, rats were taken from fifty distinct litters, and one rat of the litter was randomly selected and given the drug. For each litter, two rats were selected as controls and were given a placebo. All mice were females. Possible associations between litter mates in their times to development of tumors may be due to common genetic backgrounds shared by siblings. We shall test this hypothesis using the score test for association. The data is given in Table 13.1.

To perform the test, we first fit a Cox proportional hazards regression model to the data ignoring possible dependence between litter mates. A single covariate was used, which has the value 1 if the rat was in the treated group and 0 if the rat was in the placebo group. The estimate of β is $b = 0.8975$ with an estimated variance of 0.1007 using Breslow's likelihood for ties. The value of T computed from (13.2.4) is 8.91, and $V = 45.03$. The standardized test statistic is $8.91/\sqrt{45.03} = 1.33$ which has a p-value of 0.184 from the standard normal table. This suggests that there is no evidence of a litter effect in this experiment.

Practical Note

1. An SAS macro to compute this test is available on our web site.

TABLE 13.1
Data On 50 Litters of Rats

Group	Treated Rat	Control Rats	Group	Treated Rat	Control Rats
1	101^+	104^+, 49	26	89^+	104^+, 104^+
2	104^+	104^+, 102^+	27	78^+	104^+, 104^+
3	104^+	104^+, 104^+	28	104^+	81, 64
4	77^+	97^+, 79^+	29	86	94^+, 55
5	89^+	104^+, 104^+	30	34	104^+, 54
6	88	104^+, 96	31	76^+	87^+, 74^+
7	104	94^+, 77	32	103	84, 73
8	96	104^+, 104^+	33	102	104^+, 80^+
9	82^+	104^+, 77^+	34	80	104^+, 73^+
10	70	104^+, 77^+	35	45	104^+, 79^+
11	89	91^+, 90^+	36	94	104^+, 104^+
12	91^+	92^+, 70^+	37	104^+	104^+, 104^+
13	39	50, 45^+	38	104^+	101, 94^+
14	103	91^+, 69^+	39	76^+	84, 78
15	93^+	104^+, 103^+	40	80	80, 76^+
16	85^+	104^+, 72^+	41	72	104^+, 95^+
17	104^+	104^+, 63^+	42	73	104^+, 66
18	104^+	104^+, 74^+	43	92	104^+, 102
19	81^+	104^+, 69^+	44	104^+	98^+, 78^+
20	67	104^+, 68	45	55^+	104^+, 104^+
21	104^+	104^+, 104^+	46	49^+	83^+, 77^+
22	104^+	104^+, 104^+	47	89	104^+, 104^+
23	104^+	83^+, 40	48	88^+	99^+, 79^+
24	87^+	104^+, 104^+	49	103	104^+, 91^+
25	104^+	104^+, 104^+	50	104^+	104^+, 79

$^+$ Censored observation

13.3   Estimation for the Gamma Frailty Model

In this section we present a scheme for estimating the risk coefficients, baseline hazard rate, and frailty parameter for a frailty model based on a gamma distributed frailty. For this model, we assume that, for the jth individual in the ith group, the hazard rate given the unobservable frailty u_i is of the form (13.1.2) with the u_i's an independent and identically distributed sample of gamma random variables with density function,

$$g(u) = \frac{u^{(1/\theta-1)}\exp(-u/\theta)}{\Gamma[1/\theta]\theta^{1/\theta}}. \qquad (13.3.1)$$

With this frailty distribution, the mean of U is 1 and the variance is θ, so that large values of θ reflect a greater degree of heterogeneity among groups and a stronger association within groups. The joint survival function for the n_i individuals within the ith group is given by

$$S\{x_{i1}, \ldots, x_{in_i}\} = P[X_{i1} > x_{i1}, \ldots, X_{in_i} > x_{in_i}]$$

$$= \left[1 + \theta \sum_{j=1}^{n_i} H_0(x_{ij})\exp(\boldsymbol{\beta}'\mathbf{Z}_{ij})\right]^{-1/\theta}.$$

The association between group members as measured by Kendall's τ is $\theta/(\theta+2)$, and $\theta = 0$ corresponds to the case of independence.

Estimation for this model is based on the log likelihood function. Our data consists of the usual triple $(T_{ij}, \delta_{ij}, \mathbf{Z}_{ij})$, $i = 1, \ldots, G$, $j = 1, \ldots, n_i$. Let $D_i = \sum_{j=1}^{n_i} \delta_{ij}$ be the number of events in the ith group. Then, the observable log likelihood is given by

$$L(\theta, \boldsymbol{\beta}) = \sum_{i=1}^{G} D_i \ln \theta - \ln[\Gamma(1/\theta)] + \ln[\Gamma(1/\theta + D_i)]$$

$$- (1/\theta + D_i)\ln\left[1 + \theta \sum_{j=1}^{n_i} H_0(T_{ij})\exp(\boldsymbol{\beta}'\mathbf{Z}_{ij})\right]$$

$$+ \sum_{j=1}^{n_i} \delta_{ij}\{\boldsymbol{\beta}'\mathbf{Z}_{ij} + \ln[h_0(T_{ij})]\}. \qquad (13.3.2)$$

If one assumes a parametric form for $h_0()$, then, maximum likelihood estimates are available by directly maximizing (13.3.2). Estimates of the variability of the parameter estimates are obtained by inverting the information matrix.

If a parametric form is not assumed for $h_0()$, the semiparametric estimates are obtained by using an EM algorithm. Here, we consider the full likelihood we would have if the frailties were observed. This log likelihood is given by

$$L_{\text{FULL}} = L_1(\theta) + L_2(\boldsymbol{\beta}, H_0),$$

where

$$L_1(\theta) = -G[(1/\theta)\ln\theta + \ln\Gamma[1/\theta]] + \sum_{i=1}^{G}[1/\theta + D_i - 1]\ln u_i - u_i/\theta$$

$$(13.3.3)$$

and

$$L_2(\boldsymbol{\beta}, H_o) = \sum_{i=1}^{G} \sum_{j=1}^{n_i} \delta_{ij}[\boldsymbol{\beta}^t \mathbf{Z}_{ij} + \ln h_o(T_{ij})] - u_i H_o(T_{ij}) \exp(\boldsymbol{\beta}^t \mathbf{Z}_{ij}). \quad (13.3.4)$$

The EM algorithm provides a means of maximizing complex likelihoods. In the E (or Estimation) step of the algorithm the expected value of L_{FULL} is computed, given the current estimates of the parameters and the observable data. In the M (or maximization) step of the algorithm, estimates of the parameters which maximize the expected value of L_{FULL} from the E step are obtained. The algorithm iterates between these two steps until convergence.

To apply the E-step to our problem, we use the fact that, given the data and the current estimates of the parameters, the u_i's are independent gamma random variables with shape parameters $A_i = [1/\theta + D_i]$ and scale parameters $C_i = [1/\theta + \sum_{i=1}^{G} H_o(T_{ij}) \exp(\boldsymbol{\beta}^t \mathbf{Z}_{ij})]$. Thus,

$$E[u_i \mid \text{Data}] = \frac{A_i}{C_i} \text{ and } E[\ln u_i] = [\psi(A_i) - \ln C_i], \quad (13.3.5)$$

where $\psi(.)$ is the digamma function. Substituting these values in (13.3.3) and (13.3.4) completes the E-step of the algorithm.

For the M-step, note that $E[L_2(\boldsymbol{\beta}, H_o) \mid \text{Data}]$ is expressed as

$$E[L_2(\boldsymbol{\beta}, H_o) \mid \text{Data}] = \sum_{i=1}^{G} \sum_{j=1}^{n_i} \delta_{ij}[\boldsymbol{\beta}^t \mathbf{Z}_{ij} + \ln h_o(T_{ij})] - \frac{A_i}{C_i} H_o(T_{ij}) \exp(\boldsymbol{\beta}^t \mathbf{Z}_{ij})$$

$$(13.3.6)$$

which depends on the nuisance parameter $h_o()$. This likelihood is of the form of the complete likelihood from which the partial likelihood is constructed for the Cox model with the addition of a group specific covariate, $\ln(A_i/C_i)$, with a known risk coefficient of 1. If we let $t_{(k)}$ be the kth smallest death time, regardless of subgroup, and $m_{(k)}$ the number of deaths at time $t_{(k)}$, for $k = 1, \ldots, D$ and if we denote by $\hat{u}_h$ and $\mathbf{Z}_h$ the expected value of the frailty and the covariate vector for the hth individual, then, the partial likelihood to be maximized in the M step is

$$L_3(\boldsymbol{\beta}) = \sum_{k=1}^{D} \left\{ \mathbf{S}_{(k)} - m_{(k)} \ln \left[\sum_{h \in R(T_{(k)})} \hat{u}_h \exp(\boldsymbol{\beta}^t \mathbf{Z}_h) \right] \right\}, \quad (13.3.7)$$

where $\mathbf{S}_{(k)}$ is the sum of the covariates of individuals who died at time $t_{(k)}$. Note the similarity of this partial likelihood to the Breslow likelihood for the usual Cox regression model. An estimate of the $H_o(t)$ from this step is given by

$$\hat{H}_o(t) = \sum_{t_{(k)} \leq t} h_{ko} \quad (13.3.8)$$

where

$$h_{ko} = \frac{m_{(k)}}{\displaystyle\sum_{h \in R(t_{(k)})} \hat{u}_h \exp(\boldsymbol{\beta}^t \mathbf{Z}_h)}.$$

A full implementation of the EM algorithm is, then,

Step 0. Provide initial estimates of $\boldsymbol{\beta}$, θ and thus h_{ko}, $k = 1, .., d$.

Step 1. (E step). Compute A_i, C_i, $i = 1, \ldots, G$ and $\hat{u}_h$, $h = 1, \ldots, n$ based on the current values of the parameters.

Step 2. (M step). Update the estimate of $\boldsymbol{\beta}$ (and the h_{ko}) using the partial likelihood in (13.3.7). Update the estimate of θ based on the likelihood $L_4 = E[L_1(\theta) \,|\, \text{Data}]$ given by

$$L_4 = -G[(1/\theta)\ln\theta + \ln\Gamma(1/\theta)] + \sum_{i=1}^{G}[1/\theta + D_i - 1][\psi(A_i) - \ln C_i] - \frac{A_i}{\theta C_i}$$

Step 3. Iterate between Steps 1 and 2 until convergence.

Although this algorithm eventually converges to the maximum likelihood estimates of θ, $\boldsymbol{\beta}$, and H_o, the rate of convergence is quite slow. A quicker algorithm, suggested by Nielsen et al.(1992), is a modified profile likelihood EM algorithm. Here we use a grid of possible values for the frailty parameter θ. For each fixed value of θ, the following EM algorithm is used to obtain an estimate of $\boldsymbol{\beta}_\theta$.

Step 0. Provide an initial estimate of $\boldsymbol{\beta}$ (and hence h_{ko})

Step 1. (E step). Compute $\hat{u}_h$, $h = 1, \ldots, n$ based on the current values of the parameters.

Step 2. (M step). Update the estimate of $\boldsymbol{\beta}_\theta$ (and the h_{ko}) using the partial likelihood in (13.3.7).

Step 3. Iterate between steps 1 and 2 until convergence.

The value of the profile likelihood for this value of θ is, then, given by $L(\theta, \boldsymbol{\beta}_\theta)$ using (13.3.2). The value of θ which maximizes this quantity is, then, the maximum likelihood estimate. Both approaches will converge to the same estimates.

The standard errors of the estimates of $\boldsymbol{\beta}$, θ and h_{ko} are based on the inverse of the observed information matrix constructed by taking partial derivatives of minus the observable likelihood (13.3.2). The information matrix is a square matrix of size $D + p + 1$. The elements of this matrix are as follows:

$$\frac{-\partial^2 L}{\partial h_{ko} \partial h_{jo}} = \sum_{i=1}^{G} -(D_i + 1/\theta)\frac{Q_i^{(0)}(\boldsymbol{\beta}, t_{(k)}) Q_i^{(0)}(\boldsymbol{\beta}, t_{(j)})}{\left[1/\theta + \displaystyle\sum_{j=1}^{n_i} H_o(T_{ij}) \exp(\boldsymbol{\beta}^t \mathbf{Z}_{ij})\right]^2}$$

$$+ I(k = j)\frac{m_j}{(h_{jo})^2}, \; j, k = 1, \ldots, D,$$

$$\frac{-\partial^2 L}{\partial b_{ko}\partial \beta_v} = \sum_{i=1}^{G} \frac{D_i + 1/\theta}{\left[1/\theta + \sum_{j=1}^{n_i} H_o(T_{ij}) \exp(\boldsymbol{\beta}^{\mathrm{t}}\mathbf{Z}_{ij}) \right]^2}$$

$$\times \left\{ \left[1/\theta + \sum_{j=1}^{n_i} H_o(T_{ij}) \exp(\boldsymbol{\beta}^{\mathrm{t}}\mathbf{Z}_{ij}) \right] Q_{iv}^{(1)}(\boldsymbol{\beta}, t_{(k)}) \right.$$

$$\left. - Q_i^{(0)}(\boldsymbol{\beta}, t_{(k)}) \sum_{j=1}^{n_i} Z_{ijv} H_o(T_{ij}) \exp(\boldsymbol{\beta}^{\mathrm{t}}\mathbf{Z}_{ij}) \right\},$$

$$k = 1, \ldots, D, v = 1, \ldots, p,$$

$$\frac{-\partial^2 L}{\partial b_{ko}\partial \theta} = \sum_{i=1}^{G} Q_i^{(0)}(\boldsymbol{\beta}, t_{(k)}) \left\{ \frac{D_i - \sum_{j=1}^{n_i} H_o(T_{ij}) \exp(\boldsymbol{\beta}^{\mathrm{t}}\mathbf{Z}_{ij})}{1 + \theta \sum_{j=1}^{n_i} H_o(T_{ij}) \exp(\boldsymbol{\beta}^{\mathrm{t}}\mathbf{Z}_{ij})} \right\}, k = 1, \ldots, D,$$

$$\frac{-\partial^2 L}{\partial \beta_v \partial \beta_w} = \sum_{i=1}^{G} \frac{D_i + 1/\theta}{1/\theta + \sum_{j=1}^{n_i} H_o(T_{ij}) \exp(\boldsymbol{\beta}^{\mathrm{t}}\mathbf{Z}_{ij})}$$

$$\times \left\{ \left[1/\theta + \sum_{j=1}^{n_i} H_o(T_{ij}) \exp(\boldsymbol{\beta}^{\mathrm{t}}\mathbf{Z}_{ij}) \right] \sum_{j=1}^{n_i} H_o(T_{ij}) Z_{ijv} Z_{ijw} \exp(\boldsymbol{\beta}^{\mathrm{t}}\mathbf{Z}_{ij}) \right.$$

$$\left. - \sum_{j=1}^{n_i} H_o(T_{ij}) Z_{ijv} \exp(\boldsymbol{\beta}^{\mathrm{t}}\mathbf{Z}_{ij}) \sum_{j=1}^{n_i} H_o(T_{ij}) Z_{ijw} \exp(\boldsymbol{\beta}^{\mathrm{t}}\mathbf{Z}_{ij}) \right\},$$

$$v, w = 1, \ldots, p$$

$$\frac{-\partial^2 L}{\partial \beta_v \partial \theta} = \sum_{i=1}^{G} \frac{\left[D_i - \sum_{j=1}^{n_i} H_o(T_{ij}) \exp(\boldsymbol{\beta}^{\mathrm{t}}\mathbf{Z}_{ij}) \right] \sum_{j=1}^{n_i} H_o(T_{ij}) Z_{ijv} \exp(\boldsymbol{\beta}^{\mathrm{t}}\mathbf{Z}_{ij})}{\left[1 + \theta \sum_{j=1}^{n_i} H_o(T_{ij}) \exp(\boldsymbol{\beta}^{\mathrm{t}}\mathbf{Z}_{ij}) \right]^2},$$

$$v = 1, \ldots, p,$$

and

$$
\frac{-\partial^2 L}{\partial \theta^2} = \sum_{i=1}^{G} \left\{ \sum_{r=0}^{D_i-1} -\frac{1+2r\theta}{(r\theta^2+\theta)^2} + \frac{2\ln\theta-3}{\theta^3} + \frac{2}{\theta^3}\ln\left[1/\theta + \sum_{j=1}^{n_i} H_0(T_{ij})\exp(\boldsymbol{\beta}^{\mathrm{t}}\mathbf{Z}_{ij}) \right] \right.
$$

$$
\left. + \frac{3/\theta + D_i + 2\left[\displaystyle\sum_{j=1}^{n_i} H_0(T_{ij})\exp(\boldsymbol{\beta}^{\mathrm{t}}\mathbf{Z}_{ij})\right][2+\theta+D_i]}{\left[\theta + \theta^2 \displaystyle\sum_{j=1}^{n_i} H_0(T_{ij})\exp\{\boldsymbol{\beta}^{\mathrm{t}}\mathbf{Z}_{ij}\}\right]^2} \right\},
$$

where

$$
Q_i^{(0)}(\boldsymbol{\beta}, t) = \sum_{j=1}^{n_i}{}^* \exp(\boldsymbol{\beta}^{\mathrm{t}}\mathbf{Z}_{ij}), \quad Q_{iv}^{(1)}(\boldsymbol{\beta}, t) = \sum_{j=1}^{n_i}{}^* Z_{ijv}\exp(\boldsymbol{\beta}^{\mathrm{t}}\mathbf{Z}_{ij}) \text{ and}
$$

$$
Q_{ivw}^{(2)}(\boldsymbol{\beta}, t) = \sum_{j=1}^{n_i}{}^* Z_{ijv}Z_{ijw}\exp\{\boldsymbol{\beta}^{\mathrm{t}}\mathbf{Z}_{ij}\},
$$

with each of these starred sums only over those at risk at time t.

EXAMPLE 13.1   *(continued):* We shall fit the gamma frailty model to the litter data presented in Example 13.1. Here, we have a single covariate Z denoting whether the rat was fed the drug or placebo. Each litter defines a group. Applying the EM algorithm and after computing the information matrix, we find the following results:

Model	Treatment	Frailty
Cox Model (Frailty)	$b = 0.904, SE(b) = 0.323$	$\hat{\theta} = 0.472, SE(\hat{\theta}) = 0.462$
Cox Model (Independence)	$b = 0.897, SE(b) = 0.317$	—

Note that, when the frailty is ignored, the estimate of β is closer to zero and the estimated standard error is reduced. This is a consequence of the frailty model. In this example, we also see that there is no significant random effect because the Wald statistic $\hat{\theta}/SE(\hat{\theta}) = 1.02$ which gives a p-value of 0.31. The estimated value of Kendall's τ is $\hat{\theta}/(2+\hat{\theta}) = 0.19$ with a standard error of $2SE(\hat{\theta})/(2+\hat{\theta})^2 = 0.43$.

Practical Notes

1. A SAS macro is available at our website for implementing these techniques. They are quite computer intensive.
2. Similar estimating schemes have been developed for the positive stable frailty (Wang et al. 1995) and for the inverse gaussian frailty (Klein et al. 1992).
3. The estimating scheme can be simplified if a parametric model is assumed for the baseline hazard rate. See Costigan and Klein (1993) for a survey of this approach.
4. Additional models and examples are found in Hougaard (2001).

Theoretical Notes

1. Estimation for the semiparametric gamma frailty model was first considered by Clayton (1978) and Clayton and Cuzick (1985). Based on a suggestion of Gill (1985), the EM algorithm was developed independently by Klein (1992) and Nielsen et al. (1992).
2. Neilsen et al. (1992) give a counting process derivation of this approach to estimation.
3. The calculation of the variance-covariance matrix given here was introduced in Andersen et al. (1997). This is a nonstandard approach but there is good empirical data and some Monte Carlo work (Morsing 1994) which suggests that it is valid. It is of interest to note that, if one uses the same approach to derive a variance estimate for the usual Cox model, then, the inverse of the information matrix gives precisely the common formula for the variance.

13.4   Marginal Model for Multivariate Survival

In the previous section, we modeled multivariate survival data by a conditional proportional hazards model given the unobserved frailty. With this model, the marginal distributions no longer followed a simple Cox model. In fact, except for the positive stable frailty model, the marginal distributions do not follow a proportional hazards model.

An alternative model for multivariate survival data has been suggested by Lee et al. (1992). In this approach, a proportional hazards model is assumed marginally for each individual, that is, for the *j*th individual in the *i*th group, the marginal hazard rate given an individual's covariates

$\mathbf{Z}_{ij}$ is expressed by

$$h_{ij}(t \mid \mathbf{Z}_{ij}) = h_o(t) \exp(\boldsymbol{\beta}' \mathbf{Z}_{ij}), \, j = 1, \dots, n_i, \, i = 1, \dots, G. \quad (13.4.1)$$

As in previous sections of this chapter, we allow the individual observations within each of the G groups to be associated.

To estimate $\boldsymbol{\beta}$, we proceed with an "independence working model" for the data. We pretend that all observations are independent of each other and construct the partial likelihood function for a sample of $\sum_{i=1}^{G} n_i$ observations, as in Chapter 8. Using this partial likelihood function, $\mathbf{b}$, the estimator of $\boldsymbol{\beta}$ is found. Lee et al. (1992) show that this estimator is consistent for $\boldsymbol{\beta}$, provided the marginal model (13.4.1) is correctly specified. However, the information matrix obtained from this likelihood does not provide a valid estimator of the variance-covariance matrix of $\mathbf{b}$.

To estimate the variance of $\mathbf{b}$, a "sandwich" estimator is used. This estimator adjusts the usual covariance matrix for the possible association between the event times within groups. To construct the covariance matrix, let $\hat{\mathbf{V}}$ be the usual $p \times p$ covariance matrix for $\mathbf{b}$, based on the independence working model. We let S_{ij} be the score residual for Z_{ijk} for the jth member of the ith group, $i = 1, \dots, G, j = 1, \dots, n_i, k = 1, \dots p$. Recall, as defined in (11.6.1), the score residual for the kth covariate is

$$S_{ijk} = \delta_{ij}[Z_{ijk} - \bar{Z}_k(T_{ij})] - \sum_{t_b \leq T_{ij}} [Z_{ijk} - \bar{Z}_k(t_b)] \, d\bar{H}_0(t_b), \quad (13.4.2)$$

where $\bar{Z}_k(t)$ is the average value of the kth covariate, $t_1, \dots t_D$ are the observed death times, and $d\bar{H}_0(t_b)$ is the jump in the estimate of the baseline rate at time t_b. Summing these values, we compute the $p \times p$ matrix $\mathbf{C}$ defined by

$$C_{h,k} = \sum_{i=1}^{g} \sum_{j=1}^{n_i} \sum_{m=1}^{n_i} S_{ijh} S_{imk} \quad (13.4.3)$$

The adjusted estimator of the variance of $\mathbf{b}$ is given by

$$\tilde{\mathbf{V}} = \hat{\mathbf{V}} \mathbf{C} \hat{\mathbf{V}}.$$

The estimator $\mathbf{b}$ has a large sample p-variate normal distribution with a mean of $\boldsymbol{\beta}$ and a variance estimated by $\tilde{\mathbf{V}}$, so the global and local Wald tests discussed in Chapter 8 can be constructed. Note that, in this model, we have no estimate of the strength of association between individuals within groups. We simply have an estimator of $\boldsymbol{\beta}$ and an adjusted estimator of its standard error.

EXAMPLE 13.1   *(continued):* We shall fit the marginal model to the litter data presented in Example 13.1. Here, we have a single covariate Z denoting whether the rat was fed the drug or placebo. Each litter defines a group.

Fitting a Cox regression model, we obtain $b = 0.897$ and $\hat{V} = 0.1007$. Computing (13.4.3), we find $C = 8.893$ so $\tilde{V} = 0.1007^2(8.893) = 0.0902$ and the standard error of b is $\sqrt{0.0902} = 0.3000$. A 95% confidence interval for the relative risk of death for a rat given the drug compared to a placebo-fed rat is $\exp[0.897 \pm 1.96(0.3000)] = (1.36, 4.42)$. Note that, if the naive estimate of the standard error $\sqrt{0.1007} = 0.3173$ were used, the 95% confidence interval for the relative risk would be $(1.32, 4.57)$, which is slightly wider than that obtained using the adjusted variance.

Practical Notes

1. A FORTRAN program to compute the adjusted variance estimator can be found in Lin (1993). This variance estimator is also available in S-Plus.
2. When the event times within groups are independent, the adjusted variance estimator reduces to a robust variance estimator for the Cox regression model proposed by Lin and Wei (1989).
3. If the baseline hazard rate is assumed to be different for each group, but a common $\boldsymbol{\beta}$ acts on these rates, then, Wei et al. (1989) provide a similar approach based on an independence working model with a sandwich estimator for the variance. Details and examples are found in Lin (1993).
4. Score residuals are available in SAS.

13.5   Exercises

13.1   Batchelor and Hackett (1970) have reported the results of a study of 16 acutely burned patients treated with skin allografts. Patients received from one to four grafts. For each graft, the time in days to rejection of the graft was recorded as well as an indicator variable Z which had a value of 1 if the graft was a good match of HLA skin type and 0 if it was a poor match. The survival times of some grafts were censored by the death of the patient. The data is recorded below.

The survival time of an allograft is thought to depend on the degree of HLA matching between the patient and the donor and on the strength of the patient's immune response. Test the hypothesis of a random patient effect due to differing immune responses by applying the score test for association.

Patient	(T, Z)
1	(29, 0), (37, 1)
2	(3, 0), (19, 1)
3	(15, 0), $(57^+, 1),(57^+, 1)$
4	(26, 0), (93, 1)
5	(11, 0), (16, 1)
6	(15, 0), (21, 1)
7	(20, 1), (26, 0)
8	(18, 1), (19, 0)
9	(29, 0), (43, 0), (63, 1), (77, 1)
10	(15, 0), (18, 0), (29, 1)
11	(38, 0), $(60^+, 1)$
12	(19, 0)
13	(24, 1)
14	(18, 0), (18, 0)
15	(19, 0), (19, 0)
16	$(28^+, 0), (28^+, 0)$

$^+$ Censored observation

13.2 McGilchrist and Aisbett (1991) report data on the recurrence times of infections of 38 kidney patients using a portable dialysis machine. For each patient, two times to recurrence of an infection at the site of insertion of the catheter placement (in days), (T_1, T_2), are recorded as well the event indicators (δ_1, δ_2) for each time . Also recorded for each patient are five covariates: Z_1, the patient's age; Z_2, patient's gender (0-male, 1-female); Z_3, indicator of whether the patient had disease type GN; Z_4, indicator of whether the patient had disease type AN; and Z_5, indicator of whether the patient had disease type PKD. The data is recorded below.

Using the score test, test the hypothesis of no association between the recurrence times.

Patient	T_1	δ_1	T_2	δ_2	Z_1	Z_2	Z_3	Z_4	Z_5
1	16	1	8	1	28	0	0	0	0
2	13	0	23	1	48	1	1	0	0
3	22	1	28	1	32	0	0	0	0
4	318	1	447	1	31.5	1	0	0	0
5	30	1	12	1	10	0	0	0	0
6	24	1	245	1	16.5	1	0	0	0
7	9	1	7	1	51	0	1	0	0
8	30	1	511	1	55.5	1	1	0	0
9	53	1	196	1	69	1	0	1	0
10	154	1	15	1	51.5	0	1	0	0
11	7	1	33	1	44	1	0	1	0
12	141	1	8	0	34	1	0	0	0

Patient	T_1	δ_1	T_2	δ_2	Z_1	Z_2	Z_3	Z_4	Z_5
13	38	1	96	1	35	1	0	1	0
14	70	0	149	0	42	1	0	1	0
15	536	1	25	0	17	1	0	0	0
16	4	0	17	1	60	0	0	1	0
17	185	1	177	1	60	1	0	0	0
18	114	1	292	1	43.5	1	0	0	0
19	159	0	22	0	53	1	1	0	0
20	108	0	15	1	44	1	0	0	0
21	562	1	152	1	46.5	0	0	0	1
22	24	0	402	1	30	1	0	0	0
23	66	1	13	1	62.5	1	0	1	0
24	39	1	46	0	42.5	1	0	1	0
25	40	1	12	1	43	0	0	1	0
26	113	0	201	1	57.5	1	0	1	0
27	132	1	156	1	10	1	1	0	0
28	34	1	30	1	52	1	0	1	0
29	2	1	25	1	53	0	1	0	0
30	26	1	130	1	54	1	1	0	0
31	27	1	58	1	56	1	0	1	0
32	5	0	43	1	50.5	1	0	1	0
33	152	1	30	1	57	1	0	0	1
34	190	1	5	0	44.5	1	1	0	0
35	119	1	8	1	22	1	0	0	0
36	54	0	16	0	42	1	0	0	0
37	6	0	78	1	52	1	0	0	1
38	8	0	63	1	60	0	0	0	1

13.3 (a) Using the data on skin grafts in Exercise 1, fit a standard proportional hazards model with a single covariate reflecting the degree of matching, ignoring the information on which patient the graft was applied.

(b) Fit the semiparametric gamma frailty model to this data. As an initial guess of the frailty parameter, use a value of 0.55. Find the estimate of the regression coefficient and its standard error. Compare this value to that found in part a. Test the hypothesis of no association in this model.

13.4 (a) Using the data in Exercise 2 and ignoring any patient effect, fit a proportional hazards model to the times to infection using the five covariates $Z_1, \ldots, Z_5$. Construct an ANOVA table for your estimates.

(b) Using the marginal model with the corrected variance estimators, repeat part a. Compare the variance estimates in the two models.

13.5 (a) Using the data in Exercise 1, fit a Cox model, ignoring any patient effect.

(b) Fit the marginal model to this data and compute the corrected variance estimate. Using the corrected variance estimate, perform the Wald test of the hypothesis of no effect of degree of matching on graft survival. Compare the results of this test to a similar test based on the results of part a.

A
Numerical Techniques for Maximization

Many of the procedures discussed in this volume require maximizing the log likelihood or partial log likelihood function. For many models, it is impossible to perform this maximization analytically, so, numerical methods must be employed. In this appendix, we shall summarize some techniques which can be used in both univariate and multivariate cases. The reader is referred to a text on statistical computing, such as Thisted (1988), for a more detailed discussion of these techniques.

A.1   Univariate Methods

Suppose we wish to find the value x which maximizes a function $f(\ )$ of a single variable. Under some mild regularity conditions, x maximizes f if the score equation $f'(x)$ equals 0 and $f''(x) < 0$. We present three numerical methods which attempt to find the maximum of $f(\ )$ by solving the score equation. Some care must be taken when using these routines because they do not ensure that the second derivative of f is negative at the solution we find.

The first technique is the bisection method. Here, the algorithm starts with two initial values, x_L and x_U, which bracket the root of $f'(x) = 0$, that is, $f'(x_L) \cdot f'(x_U) < 0$. A new guess at the root is taken to be the midpoint of the interval (x_L, x_U), namely, $x_N = (x_L + x_U)/2$. If $f'(x_L)$ and $f'(x_N)$ have the same sign, x_L is replaced by x_N, otherwise x_U is replaced by x_N. In either case, the algorithm continues with the new values of x_L and x_U until the desired accuracy is achieved. At each step, the length of the interval $(x_U - x_L)$ is a measure of the largest possible difference between our updated guess at the root of $f'(\ )$ and the actual value of the root.

A second method, of use when one has good initial starting values and a complicated second derivative of $f(\ )$, is the secant method or *regula falsi*. Again, we start with two initial guesses at the root, x_o and x_1. These guesses need not bracket the root. After i steps of the algorithm, the new guess at the root of $f'(x)$ is given by

$$x_{i+1} = x_i - f'(x_i)(x_i - x_{i-1})/[f'(x_i) - f'(x_{i-1})]. \qquad (A.1)$$

Iterations continue until convergence. Typical stopping criteria are

$$|x_{i+1} - x_i| < \gamma, \ |f'(x_{i+1})| < \gamma$$

or

$$|(x_{i+1} - x_i)/x_i| < \gamma,$$

where γ is some small number.

The third method is the Newton–Raphson technique. Here, a single initial guess, x_o, of the root is made. After i steps of the algorithm, the updated guess is given by

$$x_{i+1} = x_i - f'(x_i)/f''(x_i). \qquad (A.2)$$

Again, the iterative procedure continues until the desired level of accuracy is met. Compared to the secant method, this technique has the advantage of requiring a single starting value, and convergence is quicker than the secant method when the starting values are good. Both the secant and Newton–Raphson techniques may fail to converge when the starting values are not close to the maximum.

EXAMPLE A.1

Suppose we have the following 10 uncensored observations from a Weibull model with scale parameter $\lambda = 1$ and shape parameter α, that is, $h(t) = \alpha t^{\alpha-1} e^{-t^\alpha}$.

Data: 2.57, 0.58, 0.82, 1.02, 0.78, 0.46, 1.04, 0.43, 0.69, 1.37

To find the maximum likelihood estimator of α, we need to maximize the log likelihood $f(\alpha) = \ln L(\alpha) = n \ln(\alpha) + (\alpha - 1) \sum \ln(t_j) - \sum t_j^\alpha$.

Here, $f'(\alpha) = n/\alpha + \sum \ln(t_j) - \sum t_j^\alpha \ln(t_j)$, and $f''(\alpha) = -n/\alpha^2 - \sum t_j^\alpha [\ln(t_j)]^2$.

Applying the bisection method with $\alpha_L = 1.5$ and $\alpha_U = 2$ and stopping the algorithm when $|f'(\alpha)| < 0.01$, we have the following values:

Step	α_L	α_U	α_N	$f'(\alpha_L)$	$f'(\alpha_U)$	$f'(\alpha_N)$
1	1.5	2	1.75	1.798	−2.589	−0.387
2	1.5	1.75	1.625	1.798	−0.387	0.697
3	1.625	1.75	1.6875	0.697	−0.387	0.154
4	1.6875	1.75	1.71875	0.154	−0.387	−0.116
5	1.6875	1.71875	1.70313	0.154	−0.116	0.019
6	1.70313	1.71875	1.71094	0.019	−0.116	−0.049
7	1.70313	1.71094	1.70704	0.019	−0.049	−0.015
8	1.70313	1.70704	1.70509	0.019	−0.015	0.002

So, after eight steps the algorithm stops with $\hat{\alpha} = 1.705$.

For the secant method, we shall start the algorithm with $\alpha_o = 1$ and $\alpha_1 = 1.5$. The results are in the following table:

Step	α_{i-1}	α_i	$f'(\alpha_{i-1})$	$f'(\alpha_i)$	α_{i+1}	$f'(\alpha_{i+1})$
1	1	1.5	7.065	1.798	1.671	0.300
2	1.5	1.671	1.798	0.300	1.705	0.004

Here, using the same stopping rule $|f'(\alpha)| < 0.01$, the algorithm stops after two steps with $\hat{\alpha} = 1.705$.

For the Newton–Raphson procedure, we use an initial value of $\alpha_o = 1.5$. The results of the algorithm are in the following table.

i	α_{i-1}	$f'(\alpha_{i-1})$	$f''(\alpha_{i-1})$	α_i	$f'(\alpha_i)$
1	1.5	1.798	−8.947	1.701	0.038
2	1.701	0.038	−8.655	1.705	2×10^{-6}

Again, using the same stopping rule $|f'(\alpha)| < 0.01$, the algorithm stops after two steps with $\hat{\alpha} = 1.705$. Notice the first step of the Newton–Raphson algorithm moves closer to the root than the secant method.

A.2 Multivariate Methods

We present three methods to maximize a function of more than one variable. The first is the method of steepest ascent which requires only the vector of first derivatives of the function. This method is robust to the starting values used in the iterative scheme, but may require a large number of steps to converge to the maximum. The second is the multivariate extension of the Newton–Raphson method. This method, which requires both the first and second derivatives of the function, converges quite rapidly when the starting values are close to the root, but may not converge when the starting values are poorly chosen. The third, called Marquardt's (1963) method, is a compromise between these two methods. It uses a blending constant which controls how closely the algorithm resembles either the method of steepest ascent or the Newton–Raphson method.

Some notation is needed before presenting the three methods. Let $f(\mathbf{x})$ be a function of the p-dimensional vector $\mathbf{x} = (x_1, \ldots, x_p)^t$. Let $\mathbf{u}(\mathbf{x})$ be the p-vector of first order partial derivatives of $f(\mathbf{x})$, that is,

$$\mathbf{u}(\mathbf{x}) = [u_1(\mathbf{x}), \ldots, u_p(\mathbf{x})]^t, \qquad (A.3)$$

where

$$u_j(\mathbf{x}) = \frac{\partial f(\mathbf{x})}{\partial x_j}, \, j = 1, \ldots, p.$$

Let $\mathbf{H}(\mathbf{x})$ be the $p \times p$ Hessian matrix of mixed second partial derivatives of $f(\mathbf{x})$, defined by

$$\mathbf{H}(\mathbf{x}) = (H_{ij}(\mathbf{x})), \; i, j = 1, \ldots, p \text{ where } H_{ij}(\mathbf{x}) = \frac{\partial^2 f(\mathbf{x})}{\partial x_i \partial x_j}. \qquad (A.4)$$

The method of steepest ascent starts with an initial guess, $\mathbf{x}_o$, of the point which maximizes $f(\mathbf{x})$. At any point, the gradient vector $\mathbf{u}(\mathbf{x})$ points the direction of steepest ascent of the function $f(\mathbf{x})$. The algorithm moves along this direction by an amount d to a new estimate of the maximum from the current estimate. The step size d is chosen to maximize the function in this direction, that is, we pick d to maximize $f[\mathbf{x}_k + d\mathbf{u}(\mathbf{x}_k)]$. This requires maximizing a function of a single variable, so that any of the techniques discussed earlier can be employed.

The updated guess at the point which maximizes $f(\mathbf{x})$ is given by

$$\mathbf{x}_{k+1} = \mathbf{x}_k + d\mathbf{u}(\mathbf{x}_k). \qquad (A.5)$$

The second method is the Newton–Raphson method which, like the method of steepest ascent, starts with an initial guess at the point which maximizes $f(\mathbf{x})$. After k steps of the algorithm, the updated estimate of the point which maximizes $f(\mathbf{x})$ is given by

$$\mathbf{x}_{k+1} = \mathbf{x}_k - \mathbf{H}(\mathbf{x}_k)^{-1}\mathbf{u}(\mathbf{x}_k). \qquad (A.6)$$

The Newton–Raphson algorithm converges quite rapidly when the initial guess is not too far from the maximum. When the initial guess is poor, the algorithm may move in the wrong direction or may take a step in the correct direction, but overshoot the root. The value of the function should be computed at each step to ensure that the algorithm is moving in the correct direction. If $f(\mathbf{x}_k)$ is smaller than $f(\mathbf{x}_{k+1})$, one option is to cut the step size in half and try $\mathbf{x}_{k+1} = \mathbf{x}_k - \mathbf{H}(\mathbf{x}_k)^{-1}\mathbf{u}(\mathbf{x}_k)/2$. This procedure is used in SAS and BMDP in the Cox regression procedure.

The third method is Marquardt's (1963) compromise between the method of steepest ascent and the Newton–Raphson method. This method uses a constant, γ, which blends the two methods together. When γ is zero, the method reduces to the Newton–Raphson method, and, as $\gamma \to \infty$, the method approaches the method of steepest ascent. Again, the method starts with an initial guess, $\mathbf{x}_o$. Let $\mathbf{S}_k$ be the $p \times p$ diagonal scaling matrix with diagonal element $(|\mathbf{H}_{ii}(\mathbf{x}_k)|^{-1/2})$. The updated estimate of the maximum is given by

$$\mathbf{x}_{k+1} = \mathbf{x}_k - \mathbf{S}_k(\mathbf{S}_k\mathbf{H}(\mathbf{x}_k)\mathbf{S}_k + \gamma\mathbf{I})^{-1}\mathbf{S}_k\mathbf{u}(\mathbf{x}_k),$$

where $\mathbf{I}$ is the identity matrix. Typically, the algorithm is implemented with a small value of γ for the first iteration. If $f(\mathbf{x}_1) < f(\mathbf{x}_o)$, then, we are having difficulty approaching the maximum and the value of γ is increased until $f(\mathbf{x}_1) > f(\mathbf{x}_o)$. This procedure is iterated until convergence is attained. For the final step of the algorithm, a "Newton–Raphson" step with $\gamma = 0$ is taken to ensure convergence.

In the multivariate maximization problem, there are several suggestions for declaring convergence of these algorithms. These include stopping when $f(\mathbf{x}_{k+1}) - f(\mathbf{x}_k) < \epsilon$ (or $|[f(\mathbf{x}_{k+1}) - f(\mathbf{x}_k)]/f(\mathbf{x}_k)| < \epsilon$); when $\sum \mathbf{u}_j(\mathbf{x}_{k+1})^2 < \epsilon$ (or $\max[|\mathbf{u}_1(\mathbf{x}_{k+1})|, \ldots, |\mathbf{u}_p(\mathbf{x}_{k+1})|] < \epsilon$) or when $\sum(x_{k+1,j} - x_{k,j})^2 < \epsilon$ (or $\max[|x_{k+1,1} - x_{k,1}|, \ldots, |x_{k+1,p} - x_{k,p}|] < \epsilon$).

EXAMPLE A.2 We shall fit a two-parameter Weibull model with survival function $S(t) = \exp(-\lambda t^\alpha)$ to the ten observations in Example A.2. Here the log likelihood function is given by

$$L(\lambda, \alpha) = n \ln \lambda + n \ln \alpha + (\alpha - 1)\sum \ln t_i - \lambda \sum t_i^\alpha .$$

The score vector $\mathbf{u}(\lambda, \alpha)$ is expressed by

$$u_\lambda(\lambda, \alpha) = \frac{\partial L(\lambda, \alpha)}{\partial \lambda} = \frac{n}{\lambda} - \sum t_i^\alpha$$

$$u_\alpha(\lambda, \alpha) = \frac{\partial L(\lambda, \alpha)}{\partial \alpha} = \frac{n}{\alpha} + \sum \ln t_i - \lambda \sum t_i^\alpha \ln t_i$$

and the Hessian matrix is

$$\mathbf{H}(\lambda, \alpha) = \begin{pmatrix} -\dfrac{n}{\lambda^2} & -\sum t_i^\alpha \ln t_i \\ -\sum t_i^\alpha \ln t_i & -\dfrac{n}{\alpha^2} - \lambda \sum t_i^\alpha (\ln t_i^2) \end{pmatrix}$$

To apply the method of steepest ascent, we must find the value of d_k which maximizes $L[(\lambda_k + d_k u_\lambda[\lambda_k, \alpha_k]), (\alpha_k + d_k u_\alpha[\lambda_k, \alpha_k])]$. This needs to be done numerically and this example uses a Newton–Raphson algorithm. Convergence of the algorithm is declared when the maximum of $|u_\lambda|$ and $|u_\alpha|$ is less than 0.1. Starting with an initial guess of $\alpha = 1$ and $\lambda = 10/\sum t_i = 1.024$, which leads to a log likelihood of -9.757, we have the following results:

Step k	λ_k	α_k	$L(\lambda, \alpha)$	u_λ	u_α	d_k
0	1.024	1.000	−9.757	0.001	7.035	0.098
1	1.025	1.693	−7.491	0.001	−1.80	0.089
2	0.865	1.694	−7.339	0.661	0.001	0.126
3	0.865	1.777	−7.311	0.000	−0.363	0.073
4	0.839	1.777	−7.307	0.121	0.000	0.128
5	0.839	1.792	−7.306	0.000	−0.072	0.007

Thus the method of steepest ascent yields maximum likelihood estimates of $\hat{\lambda} = 0.839$ and $\hat{\alpha} = 1.792$ after 5 iterations of the algorithm.

Applying the Newton–Raphson algorithm with the same starting values and convergence criterion yields

Step k	λ_k	α_k	u_λ	u_α	$H_{\lambda\lambda}$	$H_{\alpha\alpha}$	$H_{\alpha\lambda}$
0	1.024	1.000	0.001	7.035	−9.537	−13.449	−1.270
1	0.954	1.530	−0.471	1.684	−10.987	−8.657	−3.34
2	0.838	1.769	0.035	0.181	−14.223	−7.783	−1.220
3	0.832	1.796	−0.001	0.001	−14.431	−7.750	−4.539

This method yields maximum likelihood estimates of $\hat{\lambda} = 0.832$ and $\hat{\alpha} = 1.796$ after three iterations.

Using $\gamma = 0.5$ in Marquardt's method yields

Step k	λ_k	α_k	u_λ	u_α	$H_{\lambda\lambda}$	$H_{\alpha\alpha}$	$H_{\alpha\lambda}$
0	1.024	1.000	0.001	7.035	−9.537	−13.449	−1.270
1	0.993	1.351	−0.357	3.189	−10.136	−9.534	−2.565
2	0.930	1.585	−0.394	1.295	−11.557	−8.424	−3.599
3	0.883	1.701	−0.275	0.523	−12.813	−8.049	−4.176
4	0.858	1.753	−0.162	0.218	−13.591	−7.891	−4.453
5	0.845	1.777	−0.087	0.094	−14.013	−7.817	−4.581

Here, the algorithm converges in five steps to estimates of $\hat{\lambda} = 0.845$ and $\hat{\alpha} = 1.777$.

B
Large-Sample Tests Based on Likelihood Theory

Many of the test procedures used in survival analysis are based on the asymptotic properties of the likelihood or the partial likelihood. These test procedures are based on either the maximized likelihood itself (likelihood ratio tests), on the estimators standardized by use of the information matrix (Wald tests), or on the first derivatives of the log likelihood (score tests). In this appendix, we will review how these tests are constructed. See Chapter 9 of Cox and Hinkley (1974) for a more detailed reference.

Let $\mathbf{Y}$ denote the data and $\boldsymbol{\theta} = (\theta_1, \ldots, \theta_p)$ be the parameter vector. Let $L(\boldsymbol{\theta} : \mathbf{Y})$ denote either the likelihood or partial likelihood function. The maximum likelihood estimator of $\boldsymbol{\theta}$ is the function of the data which maximizes the likelihood, that is, $\hat{\boldsymbol{\theta}}(\mathbf{Y}) = \hat{\boldsymbol{\theta}}$ is the value of $\boldsymbol{\theta}$ which maximizes $L(\boldsymbol{\theta} : \mathbf{Y})$ or, equivalently, maximizes $\log L(\boldsymbol{\theta} : \mathbf{Y})$.

Associated with the likelihood function is the efficient score vector $\mathbf{U}(\boldsymbol{\theta}) = [U_1(\boldsymbol{\theta}), \ldots, U_p(\boldsymbol{\theta})]$ defined by

$$U_j(\boldsymbol{\theta}) = \frac{\delta}{\delta \theta_j} \ln L(\boldsymbol{\theta} : \mathbf{Y}). \tag{B.1}$$

In most regular cases, the maximum likelihood estimator is the solution to the equation $\mathbf{U}(\boldsymbol{\theta}) = \mathbf{0}$. The efficient score vector has the property that its expected value is zero when the expectation is taken with respect to the true value of θ.

A second key quantity in large-sample likelihood theory is the Fisher information matrix defined by

$$\mathbf{i}(\theta) = E_\theta[\mathbf{U}(\theta)'\mathbf{U}(\theta)] = E_\theta \left[-\frac{\delta}{\delta\theta}\mathbf{U}(\theta) \right]$$

$$= \left\{ -E_\theta \left[\frac{\delta^2}{\delta\theta_j\delta\theta_k} \ln L(\theta : \mathbf{Y}) \right] \right\}, \ j = 1, \ldots, p, \ k = 1, \ldots, p. \quad (B.2)$$

Computation of the expectation in (B.2) is very difficult in most applications of likelihood theory, so a consistent estimator of $\mathbf{i}$ is used. This estimator is the observed information, $\mathbf{I}(\theta)$, whose (j, k)th element is given by

$$I_{j,k}(\theta) = -\frac{\delta^2 \ln L(\theta : \mathbf{Y})}{\delta\theta_j\delta\theta_k}, \ j, k = 1, \ldots, p. \quad (B.3)$$

The first set of tests based on the likelihood are for the simple null hypothesis, $H_o : \theta = \theta_o$. The first test is the *likelihood ratio test* based on the statistic

$$\chi^2_{\text{LR}} = -2[\ln L(\theta_o : \mathbf{Y}) - \ln L(\hat{\theta} : \mathbf{Y})] \quad (B.4)$$

This statistic has an asymptotic chi-squared distribution with p degrees of freedom under the null hypothesis.

A second test, called the *Wald test*, is based on the large-sample distribution of the maximum likelihood estimator. For large samples, $\hat{\theta}$ has a multivariate normal distribution with mean θ and covariance matrix $\mathbf{i}^{-1}(\theta)$ so the quadratic form $(\hat{\theta} - \theta_o)\mathbf{i}(\hat{\theta})(\hat{\theta} - \theta_o)'$ has a chi-squared distribution with p degrees of freedom for large samples. Using the observed information as an estimator of the Fisher information, the Wald statistic is expressed as

$$\chi^2_W = (\hat{\theta} - \theta_o)\mathbf{I}(\hat{\theta})(\hat{\theta} - \theta_o)' \quad (B.5)$$

which has a chi-squared distribution with p degrees of freedom for large samples when H_o is true.

The third test, called the *score or Rao test*, is based on the efficient score statistics. When $\theta = \theta_o$, the score vector $\mathbf{U}(\theta_o)$ has a large-sample multivariate normal distribution with mean $\mathbf{0}$ and covariance matrix $\mathbf{i}(\theta)$. This leads to a test statistic given by

$$\chi^2_S = \mathbf{U}(\theta_o)\mathbf{i}^{-1}(\theta_o)\mathbf{U}'(\theta_o).$$

As for the Wald test, the Fisher information is replaced in most applications by the observed information, so the test statistic is given by

$$\chi^2_S = \mathbf{U}(\theta_o)\mathbf{I}^{-1}(\theta_o)\mathbf{U}'(\theta_o). \quad (B.6)$$

Again, this statistic has an asymptotic chi-squared distribution with p degrees of freedom when H_o is true. The score test has an advantage in many applications in that the maximum likelihood estimates need not be calculated.

EXAMPLE B.1   Suppose we have a censored sample of size n from an exponential population with hazard rate λ. We wish to test the hypothesis that $\lambda = 1$. Let (T_i, δ_i), $i = 1, \ldots, n$, so that the likelihood, $L(\lambda; (T_i, \delta_i), i = 1, \ldots, n)$, is given by $\prod_{i=1}^{n} \lambda^{\delta_i} e^{-\lambda T_i} = \lambda^D e^{-\lambda S}$ where $D = \sum_{i=1}^{n} \delta_i$ is the observed number of deaths and $S = \sum_{i=1}^{n} T_i$ is the total time on test (see Section 3.5). Thus,

$$\ln L(\lambda) = D \ln \lambda - \lambda S, \tag{B.7}$$

$$U(\lambda) = \frac{d}{d\lambda} \ln L(\lambda) = \frac{D}{\lambda} - S, \tag{B.8}$$

and

$$I(\lambda) = -\frac{d^2}{d\lambda^2} \ln L(\lambda) = \frac{D}{\lambda^2}. \tag{B.9}$$

Solving B.8 for λ gives us the maximum likelihood estimator, $\hat{\lambda} = D/S$. Using these statistics,

$$\chi_S^2 = \left(\frac{D}{1} - S\right)^2 \cdot \left(\frac{1^2}{D}\right) = \frac{(D - S)^2}{D},$$

$$\chi_W^2 = \left(\frac{D}{S} - 1\right)^2 \cdot \frac{D}{(D/S)^2} = \frac{(D - S)^2}{D}$$

$$\chi_{LR}^2 = -2\{(D \ln 1 - 1 \cdot S) - [D \ln(D/S) - (D/S) \cdot S]\}$$

$$= 2[S - D + D \ln(D/S)]$$

In this case, note that the Wald and Rao tests are identical. All three of these statistics have asymptotic chi-squared distributions with one degree of freedom.

All three test statistics can be used to test composite hypotheses. Suppose the parameter vector $\boldsymbol{\theta}$ is divided into two vectors $\boldsymbol{\psi}$ and $\boldsymbol{\phi}$ of lengths p_1, and p_2, respectively. We would like to test the hypothesis $H_o : \boldsymbol{\psi} = \boldsymbol{\psi}_o$. Here $\boldsymbol{\phi}$ is a nuisance parameter. Let $\hat{\boldsymbol{\phi}}(\boldsymbol{\psi}_o)$ be the maximum likelihood estimates of $\boldsymbol{\phi}$ obtained by maximizing the likelihood with respect to $\boldsymbol{\phi}$, with $\boldsymbol{\psi}$ fixed at $\boldsymbol{\psi}_o$. That is, $\hat{\boldsymbol{\phi}}(\boldsymbol{\psi}_o)$ maximizes $\ln L[(\boldsymbol{\psi}_o, \boldsymbol{\phi}) : \mathbf{Y}]$ with respect to $\boldsymbol{\phi}$. We also partition the information matrix $\mathbf{I}$ into

$$\mathbf{I} = \begin{pmatrix} \mathbf{I}_{\psi\psi} & \mathbf{I}_{\psi\phi} \\ \mathbf{I}_{\phi\psi} & \mathbf{I}_{\phi\phi} \end{pmatrix}, \tag{B.10}$$

where $\mathbf{I}_{\psi\psi}$ is of dimension $p_1 \times p_1$, $\mathbf{I}_{\phi\phi}$ is of dimension $p_2 \times p_2$, $\mathbf{I}_{\psi\phi}$ is $p_1 \times p_2$, and $\mathbf{I}_{\phi\psi}^t = \mathbf{I}_{\psi\phi}$. Notice that a partitioned information matrix has an inverse which is also a partitioned matrix with

$$\mathbf{I}^{-1} = \begin{pmatrix} \mathbf{I}^{\psi\psi} & \mathbf{I}^{\psi\phi} \\ \mathbf{I}^{\phi\psi} & \mathbf{I}^{\phi\phi} \end{pmatrix}, \tag{B.11}$$

With these refinements, the three statistics for testing $H_o : \boldsymbol{\psi} = \boldsymbol{\psi}_o$ are given by
Likelihood ratio test:

$$\chi^2_{LR} = -2\{\ln L[(\boldsymbol{\psi}_o, \hat{\boldsymbol{\phi}}(\boldsymbol{\psi}_o) : \mathbf{Y})] - \ln L(\hat{\boldsymbol{\theta}} : \mathbf{Y})\}, \tag{B.12}$$

Wald test:

$$\chi^2_W = (\hat{\boldsymbol{\psi}} - \boldsymbol{\psi}_o)[\mathbf{I}^{\psi\psi}(\hat{\boldsymbol{\psi}}, \hat{\boldsymbol{\phi}})]^{-1}(\hat{\boldsymbol{\psi}} - \boldsymbol{\psi}_o)', \tag{B.13}$$

and score test:

$$\chi^2_S = \mathbf{U}_\psi[\boldsymbol{\psi}_o, \hat{\boldsymbol{\phi}}(\boldsymbol{\psi}_o)][\mathbf{I}^{\psi\psi}(\boldsymbol{\psi}_o, \hat{\boldsymbol{\phi}}(\boldsymbol{\psi}_o))]\mathbf{U}'[\boldsymbol{\psi}_o, \hat{\boldsymbol{\phi}}(\boldsymbol{\psi}_o)]. \tag{B.14}$$

All three statistics have an asymptotic chi-squared distribution with p_1 degrees of freedom when the null hypothesis is true.

EXAMPLE B.2   Consider the problem of comparing two treatments, where the time to event in each group has an exponential distribution. For population one, we assume that the hazard rate is λ whereas for population two, we assume that the hazard rate is $\lambda\beta$. We shall test $H_o : \beta = 1$ treating λ as a nuisance parameter. The likelihood function is given by

$$L[(\lambda, \beta) : D_1, D_2, S_1, S_2] = \lambda^{D_1+D_2}\beta^{D_2}\exp(-\lambda S_1 - \lambda\beta S_2) \tag{B.15}$$

where D_i is the number of events and S_i is the total time on test in the ith sample, $i = 1, 2$. From (B.15),

$$\ln L[(\beta, \lambda)] = (D_1 + D_2)\ln\lambda + D_2\ln\beta - \lambda S_1 - \lambda\beta S_2, \tag{B.16}$$

$$U_\beta(\beta, \lambda) = \frac{\delta}{\delta\beta}\ln L(\beta, \lambda) = \frac{D_2}{\beta} - \lambda S_2, \tag{B.17}$$

$$U_\lambda(\beta, \lambda) = \frac{\delta}{\delta\lambda}\ln L(\beta, \lambda) = \frac{D_1 + D_2}{\lambda} - S_1 - \beta S_2, \tag{B.18}$$

$$I_{\beta\beta}(\beta, \lambda) = -\frac{\delta^2\ln L(\beta, \lambda)}{\delta\beta^2} = \frac{D_2}{\beta^2}, \tag{B.19}$$

$$I_{\lambda\lambda}(\beta, \lambda) = -\frac{\delta^2\ln L(\beta, \lambda)}{\delta\lambda^2} = \frac{D_1 + D_2}{\lambda^2}, \tag{B.20}$$

and

$$I_{\beta\lambda} = -\frac{\delta^2\ln L(\beta, \lambda)}{\delta\lambda\delta\beta} = S_2. \tag{B.21}$$

Solving the system of equations $U_\beta(\beta, \lambda) = 0$, $U_\lambda(\beta, \lambda) = 0$ yields the global maximum likelihood estimators $\hat{\beta} = S_1 D_2/(S_2 D_1)$ and $\hat{\lambda} = D_1/S_1$.

Solving $U_\lambda(\beta, \lambda) = 0$, for β fixed at its value under H_o, yields $\hat{\lambda}(\beta = 1)$, denoted by $\hat{\lambda}(1), = (D_1 + D_2)/(S_1 + S_2)$. Thus, we have from B.12 a likelihood ratio test statistic of

$$\chi^2_{LR} = -2\{[(D_1 + D_2)\ln[\hat{\lambda}(1)] - \hat{\lambda}(1)(S_1 + S_2)]$$

$$- [(D_1 + D_2)\ln[\hat{\lambda}] + D_2\ln[\hat{\beta}] - \hat{\lambda}S_1 - \hat{\lambda}\hat{\beta}S_2]\}.$$

$$= 2D_1 \ln\left[\frac{D_1(S_1 + S_2)}{S_1(D_1 + D_2)}\right] + 2D_2 \ln\left[\frac{D_2(S_1 + S_2)}{S_2(D_1 + D_2)}\right].$$

From B.19–B.21,

$$I^{\beta\beta}(\beta, \lambda) = \frac{\beta^2(D_1 + D_2)}{[D_2(D_1 + D_2) - (\lambda\beta S_2)^2]}$$

so, the Wald test is given by

$$\chi^2_W = (\hat{\beta} - 1)^2 \left\{\frac{\hat{\beta}^2(D_1 + D_2)}{[D_2(D_1 + D_2) - (\hat{\lambda}\hat{\beta}S_2)^2]}\right\}^{-1}$$

$$= \frac{D_1^2(S_1 D_2 - S_2 D_1)^2}{D_2 S_1^2(D_1 + D_2)}$$

The score test is given by

$$\chi^2_S = (D_2 - \hat{\lambda}(1)S_2)^2 \frac{(D_1 + D_2)}{[D_2(D_1 + D_2) - (\hat{\lambda}(1)S_2)^2]}$$

$$= \frac{[D_2(S_1 + S_2) - (D_1 + D_2)S_2]^2}{D_2(S_1 + S_2)^2 - (D_1 + D_2)S_1^2}$$

If, for example, $D_1 = 10$, $D_2 = 12$, $S_1 = 25$, and $S_2 = 27$, then $\chi^2_{LR} = 0.0607$, $\chi^2_W = 0.0545$ and $\chi^2_S = 0.0448$, all nonsignificant when compared to a chi-square with one degree of freedom.

C

Statistical Tables

TABLE C.1

Standard Normal Survival Function P[Z ≥ z]

z	0	0.01	0.02	0.03	0.04	0.05	0.06	0.07	0.08	0.09
0.0	0.5000	0.4960	0.4920	0.4880	0.4840	0.4801	0.4761	0.4721	0.4681	0.4641
0.1	0.4602	0.4562	0.4522	0.4483	0.4443	0.4404	0.4364	0.4325	0.4286	0.4247
0.2	0.4207	0.4168	0.4129	0.4090	0.4052	0.4013	0.3974	0.3936	0.3897	0.3859
0.3	0.3821	0.3783	0.3745	0.3707	0.3669	0.3632	0.3594	0.3557	0.3520	0.3483
0.4	0.3446	0.3409	0.3372	0.3336	0.3300	0.3264	0.3228	0.3192	0.3156	0.3121
0.5	0.3085	0.3050	0.3015	0.2981	0.2946	0.2912	0.2877	0.2843	0.2810	0.2776
0.6	0.2743	0.2709	0.2676	0.2643	0.2611	0.2578	0.2546	0.2514	0.2483	0.2451
0.7	0.2420	0.2389	0.2358	0.2327	0.2296	0.2266	0.2236	0.2206	0.2177	0.2148
0.8	0.2119	0.2090	0.2061	0.2033	0.2005	0.1977	0.1949	0.1922	0.1894	0.1867
0.9	0.1841	0.1814	0.1788	0.1762	0.1736	0.1711	0.1685	0.1660	0.1635	0.1611
1.0	0.1587	0.1562	0.1539	0.1515	0.1492	0.1469	0.1446	0.1423	0.1401	0.1379
1.1	0.1357	0.1335	0.1314	0.1292	0.1271	0.1251	0.1230	0.1210	0.1190	0.1170
1.2	0.1151	0.1131	0.1112	0.1093	0.1075	0.1056	0.1038	0.1020	0.1003	0.0985
1.3	0.0968	0.0951	0.0934	0.0918	0.0901	0.0885	0.0869	0.0853	0.0838	0.0823
1.4	0.0808	0.0793	0.0778	0.0764	0.0749	0.0735	0.0721	0.0708	0.0694	0.0681
1.5	0.0668	0.0655	0.0643	0.0630	0.0618	0.0606	0.0594	0.0582	0.0571	0.0559
1.6	0.0548	0.0537	0.0526	0.0516	0.0505	0.0495	0.0485	0.0475	0.0465	0.0455
1.7	0.0446	0.0436	0.0427	0.0418	0.0409	0.0401	0.0392	0.0384	0.0375	0.0367
1.8	0.0359	0.0351	0.0344	0.0336	0.0329	0.0322	0.0314	0.0307	0.0301	0.0294
1.9	0.0287	0.0281	0.0274	0.0268	0.0262	0.0256	0.0250	0.0244	0.0239	0.0233
2.0	0.0228	0.0222	0.0217	0.0212	0.0207	0.0202	0.0197	0.0192	0.0188	0.0183
2.1	0.0179	0.0174	0.0170	0.0166	0.0162	0.0158	0.0154	0.0150	0.0146	0.0143
2.2	0.0139	0.0136	0.0132	0.0129	0.0125	0.0122	0.0119	0.0116	0.0113	0.0110
2.3	0.0107	0.0104	0.0102	0.0099	0.0096	0.0094	0.0091	0.0089	0.0087	0.0084
2.4	0.0082	0.0080	0.0078	0.0075	0.0073	0.0071	0.0069	0.0068	0.0066	0.0064
2.5	0.0062	0.0060	0.0059	0.0057	0.0055	0.0054	0.0052	0.0051	0.0049	0.0048
2.6	0.0047	0.0045	0.0044	0.0043	0.0041	0.0040	0.0039	0.0038	0.0037	0.0036
2.7	0.0035	0.0034	0.0033	0.0032	0.0031	0.0030	0.0029	0.0028	0.0027	0.0026
2.8	0.0026	0.0025	0.0024	0.0023	0.0023	0.0022	0.0021	0.0021	0.0020	0.0019
2.9	0.0019	0.0018	0.0018	0.0017	0.0016	0.0016	0.0015	0.0015	0.0014	0.0014
3.0	0.0013	0.0013	0.0013	0.0012	0.0012	0.0011	0.0011	0.0011	0.0010	0.0010
3.1	0.0010	0.0009	0.0009	0.0009	0.0008	0.0008	0.0008	0.0008	0.0007	0.0007
3.2	0.0007	0.0007	0.0006	0.0006	0.0006	0.0006	0.0006	0.0005	0.0005	0.0005
3.3	0.0005	0.0005	0.0005	0.0004	0.0004	0.0004	0.0004	0.0004	0.0004	0.0003
3.4	0.0003	0.0003	0.0003	0.0003	0.0003	0.0003	0.0003	0.0003	0.0003	0.0002
3.5	0.0002	0.0002	0.0002	0.0002	0.0002	0.0002	0.0002	0.0002	0.0002	0.0002
3.6	0.0002	0.0002	0.0001	0.0001	0.0001	0.0001	0.0001	0.0001	0.0001	0.0001
3.7	0.0001	0.0001	0.0001	0.0001	0.0001	0.0001	0.0001	0.0001	0.0001	0.0001
3.8	0.0001	0.0001	0.0001	0.0001	0.0001	0.0001	0.0001	0.0001	0.0001	0.0001
3.9	0.0000	0.0000	0.0000	0.0000	0.0000	0.0000	0.0000	0.0000	0.0000	0.0000

TABLE C.2
Upper Percentiles of a Chi-Square Distribution

Degrees of Freedom	Upper Percentile				
	0.1	0.05	0.01	0.005	0.001
1	2.70554	3.84146	6.63489	7.87940	10.82736
2	4.60518	5.99148	9.21035	10.59653	13.81500
3	6.25139	7.81472	11.34488	12.83807	16.26596
4	7.77943	9.48773	13.27670	14.86017	18.46623
5	9.23635	11.07048	15.08632	16.74965	20.51465
6	10.64464	12.59158	16.81187	18.54751	22.45748
7	12.01703	14.06713	18.47532	20.27774	24.32130
8	13.36156	15.50731	20.09016	21.95486	26.12393
9	14.68366	16.91896	21.66605	23.58927	27.87673
10	15.98717	18.30703	23.20929	25.18805	29.58789
11	17.27501	19.67515	24.72502	26.75686	31.26351
12	18.54934	21.02606	26.21696	28.29966	32.90923
13	19.81193	22.36203	27.68818	29.81932	34.52737
14	21.06414	23.68478	29.14116	31.31943	36.12387
15	22.30712	24.99580	30.57795	32.80149	37.69777
16	23.54182	26.29622	31.99986	34.26705	39.25178
17	24.76903	27.58710	33.40872	35.71838	40.79111
18	25.98942	28.86932	34.80524	37.15639	42.31195
19	27.20356	30.14351	36.19077	38.58212	43.81936
20	28.41197	31.41042	37.56627	39.99686	45.31422
21	29.61509	32.67056	38.93223	41.40094	46.79627
22	30.81329	33.92446	40.28945	42.79566	48.26762
23	32.00689	35.17246	41.63833	44.18139	49.72764
24	33.19624	36.41503	42.97978	45.55836	51.17897
25	34.38158	37.65249	44.31401	46.92797	52.61874
26	35.56316	38.88513	45.64164	48.28978	54.05114
27	36.74123	40.11327	46.96284	49.64504	55.47508
28	37.91591	41.33715	48.27817	50.99356	56.89176
29	39.08748	42.55695	49.58783	52.33550	58.30064
30	40.25602	43.77295	50.89218	53.67187	59.70221
31	41.42175	44.98534	52.19135	55.00248	61.09799
32	42.58473	46.19424	53.48566	56.32799	62.48728
33	43.74518	47.39990	54.77545	57.64831	63.86936
34	44.90316	48.60236	56.06085	58.96371	65.24710
35	46.05877	49.80183	57.34199	60.27459	66.61917
36	47.21217	50.99848	58.61915	61.58107	67.98495
37	48.36339	52.19229	59.89256	62.88317	69.34759
38	49.51258	53.38351	61.16202	64.18123	70.70393
39	50.65978	54.57224	62.42809	65.47532	72.05504

(continued)

TABLE C.2
(continued)

Degrees of Freedom	Upper Percentile				
	0.1	0.05	0.01	0.005	0.001
40	51.80504	55.75849	63.69077	66.76605	73.40290
41	52.94850	56.94240	64.94998	68.05263	74.74412
42	54.09019	58.12403	66.20629	69.33604	76.08420
43	55.23018	59.30352	67.45929	70.61573	77.41841
44	56.36852	60.48090	68.70964	71.89234	78.74870
45	57.50529	61.65622	69.95690	73.16604	80.07755
46	58.64053	62.82961	71.20150	74.43671	81.39979
47	59.77429	64.00113	72.44317	75.70385	82.71984
48	60.90661	65.17076	73.68256	76.96892	84.03680
49	62.03753	66.33865	74.91939	78.23055	85.34987
50	63.16711	67.50481	76.15380	79.48984	86.66031
51	64.29539	68.66932	77.38601	80.74645	87.96700
52	65.42242	69.83216	78.61563	82.00062	89.27187
53	66.54818	70.99343	79.84336	83.25251	90.57257
54	67.67277	72.15321	81.06878	84.50176	91.87140
55	68.79621	73.31148	82.29198	85.74906	93.16708
56	69.91852	74.46829	83.51355	86.99398	94.46187
57	71.03970	75.62372	84.73265	88.23656	95.74998
58	72.15983	76.77778	85.95015	89.47699	97.03806
59	73.27891	77.93049	87.16583	90.71533	98.32425
60	74.39700	79.08195	88.37943	91.95181	99.60783
61	75.51409	80.23209	89.59122	93.18622	100.88685
62	76.63020	81.38098	90.80150	94.41853	102.16522
63	77.74539	82.52872	92.00989	95.64919	103.44210
64	78.85965	83.67524	93.21670	96.87794	104.71685
65	79.97299	84.82064	94.42200	98.10492	105.98766
66	81.08547	85.96494	95.62559	99.33027	107.25660
67	82.19711	87.10804	96.82768	100.55377	108.52505
68	83.30788	88.25017	98.02832	101.77574	109.79265
69	84.41787	89.39119	99.22741	102.99614	111.05540
70	85.52704	90.53126	100.42505	104.21477	112.31669
71	86.63543	91.67026	101.62144	105.43228	113.57693
72	87.74306	92.80827	102.81634	106.64732	114.83388
73	88.84994	93.94533	104.00977	107.86186	116.09165
74	89.95605	95.08146	105.20193	109.07417	117.34687
75	91.06145	96.21666	106.39285	110.28543	118.59895
76	92.16615	97.35097	107.58244	111.49537	119.85018
77	93.27017	98.48438	108.77089	112.70374	121.10075
78	94.37351	99.61696	109.95822	113.91069	122.34713
79	95.47617	100.74861	111.14403	115.11631	123.59471
80	96.57820	101.87947	112.32879	116.32093	124.83890

TABLE C.3a

Confidence Coefficients $c_{10}(a_L, a_U)$ for 90% EP Confidence Bands

a_U	a_L 0.02	0.04	0.06	0.08	0.10	0.12	0.14	0.16	0.18	0.20
0.10	2.4547	2.3049	2.1947	2.1054						
0.12	2.4907	2.3521	2.2497	2.1654	2.0933					
0.14	2.5198	2.3901	2.2942	2.2147	2.1458	2.0849				
0.16	2.5441	2.4217	2.3313	2.2561	2.1905	2.1318	2.0788			
0.18	2.5650	2.4486	2.3630	2.2917	2.2291	2.1728	2.1214	2.0742		
0.20	2.5833	2.4721	2.3906	2.3227	2.2630	2.2090	2.1594	2.1134	2.0706	
0.22	2.5997	2.4929	2.4150	2.3501	2.2930	2.2412	2.1934	2.1489	2.1071	2.0677
0.24	2.6144	2.5116	2.4368	2.3747	2.3200	2.2703	2.2242	2.1811	2.1405	2.1019
0.26	2.6278	2.5286	2.4567	2.3970	2.3445	2.2967	2.2523	2.2107	2.1712	2.1336
0.28	2.6402	2.5441	2.4748	2.4174	2.3668	2.3208	2.2781	2.2378	2.1996	2.1631
0.30	2.6517	2.5586	2.4915	2.4361	2.3874	2.3431	2.3019	2.2630	2.2260	2.1905
0.32	2.6625	2.5720	2.5071	2.4536	2.4065	2.3638	2.3240	2.2864	2.2506	2.2162
0.34	2.6727	2.5846	2.5217	2.4698	2.4244	2.3831	2.3446	2.3083	2.2736	2.2403
0.36	2.6823	2.5965	2.5354	2.4852	2.4412	2.4012	2.3640	2.3289	2.2953	2.2630
0.38	2.6915	2.6078	2.5484	2.4997	2.4570	2.4183	2.3824	2.3484	2.3159	2.2845
0.40	2.7003	2.6186	2.5608	2.5134	2.4721	2.4346	2.3997	2.3668	2.3353	2.3049
0.42	2.7088	2.6290	2.5726	2.5266	2.4865	2.4501	2.4163	2.3844	2.3539	2.3244
0.44	2.7170	2.6390	2.5840	2.5392	2.5002	2.4649	2.4321	2.4012	2.3716	2.3431
0.46	2.7249	2.6486	2.5950	2.5514	2.5134	2.4792	2.4474	2.4174	2.3887	2.3610
0.48	2.7326	2.6579	2.6056	2.5631	2.5262	2.4929	2.4620	2.4329	2.4051	2.3782
0.50	2.7402	2.6671	2.6160	2.5745	2.5386	2.5062	2.4762	2.4480	2.4210	2.3949
0.52	2.7476	2.6760	2.6261	2.5857	2.5507	2.5192	2.4900	2.4626	2.4364	2.4111
0.54	2.7548	2.6847	2.6359	2.5965	2.5625	2.5318	2.5035	2.4768	2.4514	2.4269
0.56	2.7620	2.6933	2.6456	2.6072	2.5740	2.5441	2.5166	2.4907	2.4660	2.4422
0.58	2.7691	2.7018	2.6552	2.6177	2.5853	2.5563	2.5295	2.5043	2.4804	2.4573
0.60	2.7762	2.7103	2.6647	2.6281	2.5965	2.5682	2.5422	2.5178	2.4945	2.4721
0.62	2.7833	2.7186	2.6741	2.6384	2.6076	2.5801	2.5548	2.5310	2.5084	2.4867
0.64	2.7904	2.7270	2.6835	2.6486	2.6186	2.5918	2.5672	2.5441	2.5222	2.5011
0.66	2.7975	2.7354	2.6929	2.6588	2.6296	2.6036	2.5796	2.5572	2.5359	2.5155
0.68	2.8046	2.7439	2.7023	2.6691	2.6407	2.6153	2.5920	2.5703	2.5496	2.5298
0.70	2.8119	2.7524	2.7118	2.6794	2.6517	2.6271	2.6045	2.5833	2.5633	2.5441
0.72	2.8193	2.7611	2.7214	2.6899	2.6629	2.6390	2.6170	2.5965	2.5771	2.5586
0.74	2.8269	2.7700	2.7313	2.7005	2.6743	2.6510	2.6297	2.6099	2.5911	2.5731
0.76	2.8347	2.7790	2.7413	2.7114	2.6859	2.6633	2.6427	2.6235	2.6053	2.5879
0.78	2.8428	2.7884	2.7517	2.7226	2.6979	2.6760	2.6560	2.6374	2.6198	2.6031
0.80	2.8512	2.7982	2.7624	2.7342	2.7103	2.6890	2.6697	2.6517	2.6348	2.6186
0.82	2.8601	2.8085	2.7737	2.7463	2.7232	2.7026	2.6840	2.6667	2.6504	2.6348
0.84	2.8695	2.8193	2.7856	2.7592	2.7367	2.7170	2.6990	2.6823	2.6667	2.6517
0.86	2.8797	2.8310	2.7984	2.7728	2.7513	2.7322	2.7150	2.6990	2.6840	2.6697
0.88	2.8908	2.8437	2.8123	2.7877	2.7670	2.7487	2.7322	2.7170	2.7026	2.6890
0.90	2.9032	2.8579	2.8277	2.8042	2.7844	2.7670	2.7513	2.7367	2.7232	2.7103
0.92	2.9176	2.8741	2.8454	2.8230	2.8042	2.7877	2.7728	2.7592	2.7463	2.7342
0.94	2.9348	2.8936	2.8664	2.8454	2.8277	2.8123	2.7984	2.7856	2.7737	2.7624
0.96	2.9573	2.9188	2.8936	2.8741	2.8579	2.8437	2.8310	2.8193	2.8085	2.7982
0.98	2.9919	2.9573	2.9348	2.9176	2.9032	2.8908	2.8797	2.8695	2.8601	2.8512

(continued)

TABLE C.3a
(*continued*)

a_U	0.22	0.24	0.26	0.28	0.30	0.32	0.34	0.36	0.38	0.40
0.24	2.0654									
0.26	2.0978	2.0634								
0.28	2.1280	2.0943	2.0618							
0.30	2.1563	2.1233	2.0914	2.0605						
0.32	2.1829	2.1507	2.1194	2.0890	2.0594					
0.34	2.2080	2.1766	2.1460	2.1162	2.0870	2.0585				
0.36	2.2316	2.2011	2.1713	2.1421	2.1134	2.0853	2.0577			
0.38	2.2541	2.2244	2.1953	2.1668	2.1387	2.1111	2.0839	2.0570		
0.40	2.2754	2.2466	2.2183	2.1905	2.1631	2.1360	2.1092	2.0827	2.0565	
0.42	2.2958	2.2678	2.2403	2.2132	2.1865	2.1600	2.1337	2.1076	2.0818	2.0561
0.44	2.3153	2.2881	2.2614	2.2351	2.2090	2.1831	2.1574	2.1318	2.1064	2.0810
0.46	2.3341	2.3077	2.2818	2.2561	2.2307	2.2055	2.1804	2.1554	2.1304	2.1054
0.48	2.3521	2.3265	2.3014	2.2765	2.2518	2.2272	2.2027	2.1783	2.1538	2.1293
0.50	2.3696	2.3448	2.3203	2.2962	2.2722	2.2483	2.2244	2.2006	2.1766	2.1526
0.52	2.3865	2.3625	2.3388	2.3153	2.2920	2.2688	2.2456	2.2223	2.1990	2.1755
0.54	2.4030	2.3797	2.3567	2.3339	2.3113	2.2888	2.2662	2.2436	2.2208	2.1979
0.56	2.4191	2.3965	2.3742	2.3521	2.3302	2.3083	2.2864	2.2644	2.2423	2.2200
0.58	2.4349	2.4129	2.3913	2.3699	2.3487	2.3275	2.3062	2.2848	2.2633	2.2416
0.60	2.4503	2.4291	2.4081	2.3874	2.3668	2.3463	2.3257	2.3049	2.2841	2.2630
0.62	2.4656	2.4450	2.4247	2.4046	2.3847	2.3648	2.3448	2.3248	2.3045	2.2841
0.64	2.4807	2.4607	2.4411	2.4217	2.4023	2.3831	2.3638	2.3443	2.3248	2.3049
0.66	2.4957	2.4763	2.4573	2.4385	2.4198	2.4012	2.3825	2.3638	2.3448	2.3257
0.68	2.5106	2.4919	2.4735	2.4553	2.4373	2.4192	2.4012	2.3831	2.3648	2.3463
0.70	2.5256	2.5075	2.4897	2.4721	2.4547	2.4373	2.4198	2.4023	2.3847	2.3668
0.72	2.5406	2.5231	2.5059	2.4889	2.4721	2.4553	2.4385	2.4217	2.4046	2.3874
0.74	2.5558	2.5388	2.5222	2.5059	2.4897	2.4735	2.4573	2.4411	2.4247	2.4081
0.76	2.5712	2.5548	2.5388	2.5231	2.5075	2.4919	2.4763	2.4607	2.4450	2.4291
0.78	2.5869	2.5712	2.5558	2.5406	2.5256	2.5106	2.4957	2.4807	2.4656	2.4503
0.80	2.6031	2.5879	2.5731	2.5586	2.5441	2.5298	2.5155	2.5011	2.4867	2.4721
0.82	2.6198	2.6053	2.5911	2.5771	2.5633	2.5496	2.5359	2.5222	2.5084	2.4945
0.84	2.6374	2.6235	2.6099	2.5965	2.5833	2.5703	2.5572	2.5441	2.5310	2.5178
0.86	2.6560	2.6427	2.6297	2.6170	2.6045	2.5920	2.5796	2.5672	2.5548	2.5422
0.88	2.6760	2.6633	2.6510	2.6390	2.6271	2.6153	2.6036	2.5918	2.5801	2.5682
0.90	2.6979	2.6859	2.6743	2.6629	2.6517	2.6407	2.6296	2.6186	2.6076	2.5965
0.92	2.7226	2.7114	2.7005	2.6899	2.6794	2.6691	2.6588	2.6486	2.6384	2.6281
0.94	2.7517	2.7413	2.7313	2.7214	2.7118	2.7023	2.6929	2.6835	2.6741	2.6647
0.96	2.7884	2.7790	2.7700	2.7611	2.7524	2.7439	2.7354	2.7270	2.7186	2.7103
0.98	2.8428	2.8347	2.8269	2.8193	2.8119	2.8046	2.7975	2.7904	2.7833	2.7762

TABLE C.3a
(continued)

a_U	0.42	0.44	0.46	0.48	0.50	a_L 0.52	0.54	0.56	0.58	0.60
0.44	2.0557									
0.46	2.0804	2.0555								
0.48	2.1047	2.0800	2.0553							
0.50	2.1285	2.1042	2.0798	2.0553						
0.52	2.1518	2.1280	2.1040	2.0797	2.0553					
0.54	2.1748	2.1515	2.1279	2.1040	2.0798	2.0553				
0.56	2.1974	2.1746	2.1515	2.1280	2.1042	2.0800	2.0555			
0.58	2.2197	2.1974	2.1748	2.1518	2.1285	2.1047	2.0804	2.0557		
0.60	2.2416	2.2200	2.1979	2.1755	2.1526	2.1293	2.1054	2.0810	2.0561	
0.62	2.2633	2.2423	2.2208	2.1990	2.1766	2.1538	2.1304	2.1064	2.0818	2.0565
0.64	2.2848	2.2644	2.2436	2.2223	2.2006	2.1783	2.1554	2.1318	2.1076	2.0827
0.66	2.3062	2.2864	2.2662	2.2456	2.2244	2.2027	2.1804	2.1574	2.1337	2.1092
0.68	2.3275	2.3083	2.2888	2.2688	2.2483	2.2272	2.2055	2.1831	2.1600	2.1360
0.70	2.3487	2.3302	2.3113	2.2920	2.2722	2.2518	2.2307	2.2090	2.1865	2.1631
0.72	2.3699	2.3521	2.3339	2.3153	2.2962	2.2765	2.2561	2.2351	2.2132	2.1905
0.74	2.3913	2.3742	2.3567	2.3388	2.3203	2.3014	2.2818	2.2614	2.2403	2.2183
0.76	2.4129	2.3965	2.3797	2.3625	2.3448	2.3265	2.3077	2.2881	2.2678	2.2466
0.78	2.4349	2.4191	2.4030	2.3865	2.3696	2.3521	2.3341	2.3153	2.2958	2.2754
0.80	2.4573	2.4422	2.4269	2.4111	2.3949	2.3782	2.3610	2.3431	2.3244	2.3049
0.82	2.4804	2.4660	2.4514	2.4364	2.4210	2.4051	2.3887	2.3716	2.3539	2.3353
0.84	2.5043	2.4907	2.4768	2.4626	2.4480	2.4329	2.4174	2.4012	2.3844	2.3668
0.86	2.5295	2.5166	2.5035	2.4900	2.4762	2.4620	2.4474	2.4321	2.4163	2.3997
0.88	2.5563	2.5441	2.5318	2.5192	2.5062	2.4929	2.4792	2.4649	2.4501	2.4346
0.90	2.5853	2.5740	2.5625	2.5507	2.5386	2.5262	2.5134	2.5002	2.4865	2.4721
0.92	2.6177	2.6072	2.5965	2.5857	2.5745	2.5631	2.5514	2.5392	2.5266	2.5134
0.94	2.6552	2.6456	2.6359	2.6261	2.6160	2.6056	2.5950	2.5840	2.5726	2.5608
0.96	2.7018	2.6933	2.6847	2.6760	2.6671	2.6579	2.6486	2.6390	2.6290	2.6186
0.98	2.7691	2.7620	2.7548	2.7476	2.7402	2.7326	2.7249	2.7170	2.7088	2.7003

TABLE C.3b
Confidence Coefficients $c_{05}(a_L, a_U)$ for 95% EP Confidence Bands

a_U	a_L 0.02	0.04	0.06	0.08	0.10	0.12	0.14	0.16	0.18	0.20
0.10	2.7500	2.6033	2.4874	2.3859						
0.12	2.7841	2.6506	2.5463	2.4548	2.3715					
0.14	2.8114	2.6879	2.5924	2.5090	2.4327	2.3615				
0.16	2.8341	2.7184	2.6299	2.5530	2.4827	2.4167	2.3542			
0.18	2.8535	2.7442	2.6614	2.5898	2.5245	2.4632	2.4047	2.3486		
0.20	2.8704	2.7666	2.6884	2.6213	2.5602	2.5029	2.4481	2.3953	2.3442	
0.22	2.8855	2.7862	2.7120	2.6487	2.5912	2.5374	2.4859	2.4362	2.3879	2.3407
0.24	2.8990	2.8037	2.7330	2.6729	2.6186	2.5678	2.5193	2.4724	2.4266	2.3818
0.26	2.9114	2.8196	2.7519	2.6946	2.6430	2.5949	2.5490	2.5047	2.4614	2.4188
0.28	2.9227	2.8341	2.7691	2.7143	2.6651	2.6194	2.5758	2.5338	2.4927	2.4523
0.30	2.9333	2.8475	2.7849	2.7323	2.6853	2.6417	2.6002	2.5602	2.5211	2.4827
0.32	2.9432	2.8599	2.7995	2.7490	2.7039	2.6621	2.6225	2.5844	2.5472	2.5106
0.34	2.9525	2.8716	2.8131	2.7644	2.7211	2.6811	2.6432	2.6068	2.5712	2.5363
0.36	2.9613	2.8826	2.8260	2.7789	2.7372	2.6987	2.6624	2.6275	2.5936	2.5602
0.38	2.9696	2.8930	2.8381	2.7925	2.7523	2.7153	2.6804	2.6469	2.6144	2.5825
0.40	2.9777	2.9029	2.8495	2.8055	2.7666	2.7309	2.6973	2.6651	2.6339	2.6033
0.42	2.9854	2.9124	2.8605	2.8178	2.7801	2.7456	2.7133	2.6824	2.6524	2.6230
0.44	2.9928	2.9216	2.8710	2.8295	2.7931	2.7597	2.7285	2.6987	2.6699	2.6417
0.46	3.0001	2.9304	2.8812	2.8408	2.8055	2.7732	2.7431	2.7143	2.6865	2.6594
0.48	3.0071	2.9390	2.8910	2.8517	2.8174	2.7862	2.7570	2.7293	2.7025	2.6764
0.50	3.0140	2.9473	2.9005	2.8623	2.8290	2.7987	2.7705	2.7436	2.7178	2.6926
0.52	3.0207	2.9554	2.9097	2.8726	2.8402	2.8108	2.7835	2.7575	2.7326	2.7083
0.54	3.0273	2.9634	2.9188	2.8826	2.8511	2.8226	2.7961	2.7710	2.7469	2.7234
0.56	3.0338	2.9713	2.9277	2.8924	2.8618	2.8341	2.8084	2.7841	2.7608	2.7382
0.58	3.0403	2.9790	2.9365	2.9021	2.8723	2.8454	2.8205	2.7969	2.7744	2.7525
0.60	3.0468	2.9867	2.9451	2.9116	2.8826	2.8565	2.8323	2.8095	2.7877	2.7666
0.62	3.0532	2.9944	2.9537	2.9210	2.8928	2.8674	2.8440	2.8219	2.8007	2.7803
0.64	3.0596	3.0020	2.9623	2.9304	2.9029	2.8783	2.8555	2.8341	2.8137	2.7939
0.66	3.0661	3.0096	2.9709	2.9398	2.9130	2.8891	2.8670	2.8462	2.8264	2.8074
0.68	3.0726	3.0173	2.9795	2.9492	2.9231	2.8998	2.8784	2.8583	2.8392	2.8207
0.70	3.0792	3.0251	2.9881	2.9586	2.9333	2.9107	2.8899	2.8704	2.8519	2.8341
0.72	3.0859	3.0330	2.9969	2.9682	2.9435	2.9216	2.9014	2.8826	2.8647	2.8475
0.74	3.0928	3.0411	3.0059	2.9779	2.9539	2.9326	2.9131	2.8949	2.8776	2.8610
0.76	3.0999	3.0493	3.0150	2.9878	2.9646	2.9439	2.9250	2.9074	2.8907	2.8746
0.78	3.1072	3.0579	3.0244	2.9980	2.9755	2.9554	2.9372	2.9201	2.9040	2.8886
0.80	3.1149	3.0667	3.0342	3.0086	2.9867	2.9674	2.9497	2.9333	2.9178	2.9029
0.82	3.1230	3.0761	3.0445	3.0196	2.9985	2.9798	2.9628	2.9469	2.9320	2.9178
0.84	3.1315	3.0859	3.0553	3.0312	3.0109	2.9928	2.9765	2.9613	2.9469	2.9333
0.86	3.1408	3.0965	3.0669	3.0437	3.0241	3.0067	2.9910	2.9765	2.9628	2.9497
0.88	3.1509	3.1081	3.0795	3.0572	3.0384	3.0218	3.0067	2.9928	2.9798	2.9674
0.90	3.1622	3.1209	3.0935	3.0722	3.0542	3.0384	3.0241	3.0109	2.9985	2.9867
0.92	3.1752	3.1357	3.1096	3.0892	3.0722	3.0572	3.0437	3.0312	3.0196	3.0086
0.94	3.1909	3.1534	3.1287	3.1096	3.0935	3.0795	3.0669	3.0553	3.0445	3.0342
0.96	3.2113	3.1763	3.1534	3.1357	3.1209	3.1081	3.0965	3.0859	3.0761	3.0667
0.98	3.2428	3.2113	3.1909	3.1752	3.1622	3.1509	3.1408	3.1315	3.1230	3.1149

TABLE C.3b

(continued)

a_U	0.22	0.24	0.26	0.28	a_L 0.30	0.32	0.34	0.36	0.38	0.40
0.24	2.3378									
0.26	2.3769	2.3355								
0.28	2.4123	2.3727	2.3335							
0.30	2.4447	2.4069	2.3693	2.3319						
0.32	2.4744	2.4383	2.4024	2.3664	2.3305					
0.34	2.5018	2.4673	2.4330	2.3985	2.3640	2.3293				
0.36	2.5272	2.4943	2.4614	2.4285	2.3953	2.3620	2.3284			
0.38	2.5509	2.5194	2.4880	2.4564	2.4247	2.3926	2.3603	2.3276		
0.40	2.5731	2.5430	2.5129	2.4827	2.4523	2.4215	2.3904	2.3589	2.3269	
0.42	2.5940	2.5652	2.5364	2.5074	2.4783	2.4488	2.4189	2.3885	2.3577	2.3264
0.44	2.6138	2.5862	2.5586	2.5308	2.5029	2.4746	2.4459	2.4167	2.3871	2.3568
0.46	2.6327	2.6061	2.5796	2.5530	2.5262	2.4991	2.4716	2.4436	2.4150	2.3859
0.48	2.6506	2.6251	2.5997	2.5742	2.5485	2.5225	2.4961	2.4692	2.4418	2.4138
0.50	2.6679	2.6433	2.6189	2.5944	2.5697	2.5448	2.5195	2.4937	2.4674	2.4405
0.52	2.6844	2.6609	2.6374	2.6138	2.5902	2.5662	2.5419	2.5172	2.4920	2.4661
0.54	2.7005	2.6777	2.6552	2.6325	2.6098	2.5868	2.5636	2.5398	2.5156	2.4908
0.56	2.7160	2.6941	2.6724	2.6506	2.6288	2.6068	2.5844	2.5617	2.5385	2.5147
0.58	2.7311	2.7100	2.6891	2.6682	2.6472	2.6260	2.6046	2.5828	2.5606	2.5378
0.60	2.7459	2.7256	2.7054	2.6853	2.6651	2.6448	2.6242	2.6033	2.5820	2.5602
0.62	2.7604	2.7408	2.7214	2.7020	2.6826	2.6631	2.6434	2.6233	2.6029	2.5820
0.64	2.7747	2.7558	2.7371	2.7184	2.6998	2.6811	2.6621	2.6429	2.6233	2.6033
0.66	2.7888	2.7706	2.7525	2.7346	2.7167	2.6987	2.6805	2.6621	2.6434	2.6242
0.68	2.8028	2.7852	2.7679	2.7506	2.7334	2.7161	2.6987	2.6811	2.6631	2.6448
0.70	2.8168	2.7998	2.7831	2.7666	2.7500	2.7334	2.7167	2.6998	2.6826	2.6651
0.72	2.8308	2.8145	2.7984	2.7824	2.7666	2.7506	2.7346	2.7184	2.7020	2.6853
0.74	2.8449	2.8292	2.8137	2.7984	2.7831	2.7679	2.7525	2.7371	2.7214	2.7054
0.76	2.8591	2.8440	2.8292	2.8145	2.7998	2.7852	2.7706	2.7558	2.7408	2.7256
0.78	2.8737	2.8591	2.8449	2.8308	2.8168	2.8028	2.7888	2.7747	2.7604	2.7459
0.80	2.8886	2.8746	2.8610	2.8475	2.8341	2.8207	2.8074	2.7939	2.7803	2.7666
0.82	2.9040	2.8907	2.8776	2.8647	2.8519	2.8392	2.8264	2.8137	2.8007	2.7877
0.84	2.9201	2.9074	2.8949	2.8826	2.8704	2.8583	2.8462	2.8341	2.8219	2.8095
0.86	2.9372	2.9250	2.9131	2.9014	2.8899	2.8784	2.8670	2.8555	2.8440	2.8323
0.88	2.9554	2.9439	2.9326	2.9216	2.9107	2.8998	2.8891	2.8783	2.8674	2.8565
0.90	2.9755	2.9646	2.9539	2.9435	2.9333	2.9231	2.9130	2.9029	2.8928	2.8826
0.92	2.9980	2.9878	2.9779	2.9682	2.9586	2.9492	2.9398	2.9304	2.9210	2.9116
0.94	3.0244	3.0150	3.0059	2.9969	2.9881	2.9795	2.9709	2.9623	2.9537	2.9451
0.96	3.0579	3.0493	3.0411	3.0330	3.0251	3.0173	3.0096	3.0020	2.9944	2.9867
0.98	3.1072	3.0999	3.0928	3.0859	3.0792	3.0726	3.0661	3.0596	3.0532	3.0468

(continued)

TABLE C.3b

(continued)

a_U	0.42	0.44	0.46	0.48	a_L 0.50	0.52	0.54	0.56	0.58	0.60
0.44	2.3260									
0.46	2.3561	2.3257								
0.48	2.3851	2.3556	2.3255							
0.50	2.4129	2.3845	2.3554	2.3254						
0.52	2.4396	2.4123	2.3842	2.3553	2.3254					
0.54	2.4654	2.4392	2.4121	2.3842	2.3554	2.3255				
0.56	2.4903	2.4651	2.4392	2.4123	2.3845	2.3556	2.3257			
0.58	2.5144	2.4903	2.4654	2.4396	2.4129	2.3851	2.3561	2.3260		
0.60	2.5378	2.5147	2.4908	2.4661	2.4405	2.4138	2.3859	2.3568	2.3264	
0.62	2.5606	2.5385	2.5156	2.4920	2.4674	2.4418	2.4150	2.3871	2.3577	2.3269
0.64	2.5828	2.5617	2.5398	2.5172	2.4937	2.4692	2.4436	2.4167	2.3885	2.3589
0.66	2.6046	2.5844	2.5636	2.5419	2.5195	2.4961	2.4716	2.4459	2.4189	2.3904
0.68	2.6260	2.6068	2.5868	2.5662	2.5448	2.5225	2.4991	2.4746	2.4488	2.4215
0.70	2.6472	2.6288	2.6098	2.5902	2.5697	2.5485	2.5262	2.5029	2.4783	2.4523
0.72	2.6682	2.6506	2.6325	2.6138	2.5944	2.5742	2.5530	2.5308	2.5074	2.4827
0.74	2.6891	2.6724	2.6552	2.6374	2.6189	2.5997	2.5796	2.5586	2.5364	2.5129
0.76	2.7100	2.6941	2.6777	2.6609	2.6433	2.6251	2.6061	2.5862	2.5652	2.5430
0.78	2.7311	2.7160	2.7005	2.6844	2.6679	2.6506	2.6327	2.6138	2.5940	2.5731
0.80	2.7525	2.7382	2.7234	2.7083	2.6926	2.6764	2.6594	2.6417	2.6230	2.6033
0.82	2.7744	2.7608	2.7469	2.7326	2.7178	2.7025	2.6865	2.6699	2.6524	2.6339
0.84	2.7969	2.7841	2.7710	2.7575	2.7436	2.7293	2.7143	2.6987	2.6824	2.6651
0.86	2.8205	2.8084	2.7961	2.7835	2.7705	2.7570	2.7431	2.7285	2.7133	2.6973
0.88	2.8454	2.8341	2.8226	2.8108	2.7987	2.7862	2.7732	2.7597	2.7456	2.7309
0.90	2.8723	2.8618	2.8511	2.8402	2.8290	2.8174	2.8055	2.7931	2.7801	2.7666
0.92	2.9021	2.8924	2.8826	2.8726	2.8623	2.8517	2.8408	2.8295	2.8178	2.8055
0.94	2.9365	2.9277	2.9188	2.9097	2.9005	2.8910	2.8812	2.8710	2.8605	2.8495
0.96	2.9790	2.9713	2.9634	2.9554	2.9473	2.9390	2.9304	2.9216	2.9124	2.9029
0.98	3.0403	3.0338	3.0273	3.0207	3.0140	3.0071	3.0001	2.9928	2.9854	2.9777

TABLE C.3c

Confidence Coefficients $c_{01}(a_L, a_U)$ for 99% EP Confidence Bands

a_U	a_L 0.02	0.04	0.06	0.08	0.10	0.12	0.14	0.16	0.18	0.20
0.10	3.3261	3.1910	3.0740	2.9586						
0.12	3.3563	3.2358	3.1350	3.0386	2.9408					
0.14	3.3802	3.2701	3.1805	3.0968	3.0137	2.9283				
0.16	3.3999	3.2978	3.2163	3.1418	3.0690	2.9953	2.9189			
0.18	3.4167	3.3210	3.2458	3.1780	3.1128	3.0478	2.9811	2.9117		
0.20	3.4314	3.3408	3.2706	3.2082	3.1489	3.0904	3.0311	2.9700	2.9060	
0.22	3.4443	3.3581	3.2921	3.2339	3.1793	3.1260	3.0725	3.0177	2.9610	2.9014
0.24	3.4559	3.3735	3.3109	3.2564	3.2056	3.1564	3.1075	3.0579	3.0068	2.9536
0.26	3.4665	3.3873	3.3278	3.2763	3.2287	3.1829	3.1378	3.0923	3.0458	2.9977
0.28	3.4763	3.3999	3.3430	3.2941	3.2492	3.2064	3.1643	3.1223	3.0796	3.0357
0.30	3.4853	3.4115	3.3569	3.3103	3.2678	3.2274	3.1880	3.1489	3.1094	3.0690
0.32	3.4937	3.4223	3.3698	3.3252	3.2847	3.2464	3.2094	3.1727	3.1359	3.0985
0.34	3.5017	3.4324	3.3817	3.3389	3.3002	3.2639	3.2288	3.1943	3.1598	3.1249
0.36	3.5092	3.4418	3.3929	3.3517	3.3146	3.2800	3.2467	3.2141	3.1816	3.1489
0.38	3.5163	3.4508	3.4034	3.3637	3.3281	3.2950	3.2633	3.2323	3.2016	3.1708
0.40	3.5231	3.4593	3.4133	3.3750	3.3408	3.3090	3.2787	3.2492	3.2201	3.1910
0.42	3.5297	3.4675	3.4228	3.3857	3.3527	3.3222	3.2932	3.2651	3.2374	3.2098
0.44	3.5360	3.4753	3.4319	3.3960	3.3641	3.3347	3.3069	3.2800	3.2536	3.2274
0.46	3.5422	3.4828	3.4406	3.4058	3.3750	3.3467	3.3199	3.2941	3.2689	3.2439
0.48	3.5481	3.4901	3.4490	3.4152	3.3854	3.3581	3.3323	3.3076	3.2834	3.2596
0.50	3.5540	3.4973	3.4572	3.4243	3.3955	3.3691	3.3442	3.3204	3.2973	3.2744
0.52	3.5597	3.5042	3.4651	3.4332	3.4052	3.3797	3.3557	3.3328	3.3105	3.2887
0.54	3.5653	3.5110	3.4729	3.4418	3.4147	3.3899	3.3668	3.3447	3.3233	3.3023
0.56	3.5709	3.5177	3.4805	3.4503	3.4239	3.3999	3.3776	3.3563	3.3357	3.3155
0.58	3.5763	3.5243	3.4880	3.4586	3.4329	3.4097	3.3881	3.3675	3.3477	3.3283
0.60	3.5818	3.5308	3.4954	3.4667	3.4418	3.4193	3.3984	3.3785	3.3594	3.3408
0.62	3.5873	3.5373	3.5027	3.4748	3.4506	3.4288	3.4085	3.3893	3.3709	3.3529
0.64	3.5927	3.5438	3.5100	3.4828	3.4593	3.4381	3.4185	3.3999	3.3821	3.3649
0.66	3.5982	3.5503	3.5173	3.4908	3.4680	3.4474	3.4284	3.4105	3.3933	3.3767
0.68	3.6037	3.5568	3.5247	3.4988	3.4766	3.4567	3.4382	3.4209	3.4043	3.3883
0.70	3.6093	3.5634	3.5320	3.5069	3.4853	3.4659	3.4481	3.4314	3.4154	3.3999
0.72	3.6150	3.5701	3.5395	3.5150	3.4940	3.4753	3.4580	3.4418	3.4264	3.4115
0.74	3.6208	3.5770	3.5471	3.5233	3.5029	3.4847	3.4680	3.4524	3.4375	3.4232
0.76	3.6269	3.5840	3.5549	3.5317	3.5120	3.4944	3.4782	3.4631	3.4488	3.4350
0.78	3.6331	3.5912	3.5629	3.5404	3.5212	3.5042	3.4886	3.4741	3.4602	3.4470
0.80	3.6396	3.5987	3.5712	3.5494	3.5308	3.5144	3.4993	3.4853	3.4720	3.4593
0.82	3.6464	3.6066	3.5799	3.5588	3.5408	3.5249	3.5104	3.4970	3.4842	3.4720
0.84	3.6537	3.6150	3.5891	3.5686	3.5513	3.5360	3.5221	3.5092	3.4970	3.4853
0.86	3.6615	3.6240	3.5989	3.5792	3.5626	3.5478	3.5345	3.5221	3.5104	3.4993
0.88	3.6701	3.6338	3.6096	3.5907	3.5747	3.5606	3.5478	3.5360	3.5249	3.5144
0.90	3.6796	3.6447	3.6215	3.6033	3.5881	3.5747	3.5626	3.5513	3.5408	3.5308
0.92	3.6907	3.6572	3.6350	3.6178	3.6033	3.5907	3.5792	3.5686	3.5588	3.5494
0.94	3.7040	3.6722	3.6513	3.6350	3.6215	3.6096	3.5989	3.5891	3.5799	3.5712
0.96	3.7214	3.6917	3.6722	3.6572	3.6447	3.6338	3.6240	3.6150	3.6066	3.5987
0.98	3.7481	3.7214	3.7040	3.6907	3.6796	3.6701	3.6615	3.6537	3.6464	3.6396

(continued)

TABLE C.3c
(continued)

a_U	a_L 0.22	0.24	0.26	0.28	0.30	0.32	0.34	0.36	0.38	0.40
0.24	2.8977									
0.26	2.9475	2.8946								
0.28	2.9901	2.9424	2.8920							
0.30	3.0273	2.9838	2.9381	2.8899						
0.32	3.0600	3.0201	2.9784	2.9345	2.8880					
0.34	3.0892	3.0524	3.0140	2.9738	2.9314	2.8865				
0.36	3.1156	3.0814	3.0459	3.0089	2.9700	2.9289	2.8852			
0.38	3.1396	3.1076	3.0747	3.0404	3.0045	2.9667	2.9267	2.8841		
0.40	3.1616	3.1317	3.1009	3.0690	3.0357	3.0008	2.9640	2.9249	2.8833	
0.42	3.1821	3.1539	3.1250	3.0952	3.0642	3.0319	2.9978	2.9618	2.9235	2.8826
0.44	3.2011	3.1745	3.1473	3.1193	3.0904	3.0602	3.0286	2.9953	2.9600	2.9223
0.46	3.2189	3.1937	3.1681	3.1418	3.1146	3.0864	3.0570	3.0260	2.9933	2.9586
0.48	3.2358	3.2118	3.1875	3.1627	3.1372	3.1108	3.0832	3.0544	3.0240	2.9918
0.50	3.2517	3.2290	3.2059	3.1824	3.1583	3.1335	3.1077	3.0807	3.0524	3.0225
0.52	3.2670	3.2453	3.2234	3.2011	3.1783	3.1549	3.1306	3.1053	3.0789	3.0510
0.54	3.2816	3.2608	3.2400	3.2188	3.1972	3.1751	3.1522	3.1285	3.1037	3.0777
0.56	3.2956	3.2758	3.2559	3.2358	3.2153	3.1943	3.1727	3.1504	3.1271	3.1027
0.58	3.3092	3.2903	3.2712	3.2521	3.2326	3.2127	3.1923	3.1712	3.1493	3.1264
0.60	3.3224	3.3043	3.2861	3.2678	3.2492	3.2303	3.2110	3.1910	3.1704	3.1489
0.62	3.3353	3.3179	3.3005	3.2830	3.2653	3.2474	3.2290	3.2101	3.1906	3.1704
0.64	3.3480	3.3312	3.3146	3.2978	3.2810	3.2639	3.2464	3.2286	3.2101	3.1910
0.66	3.3604	3.3443	3.3284	3.3124	3.2963	3.2800	3.2634	3.2464	3.2290	3.2110
0.68	3.3727	3.3573	3.3419	3.3267	3.3113	3.2958	3.2800	3.2639	3.2474	3.2303
0.70	3.3849	3.3701	3.3554	3.3408	3.3261	3.3113	3.2963	3.2810	3.2653	3.2492
0.72	3.3971	3.3828	3.3688	3.3548	3.3408	3.3267	3.3124	3.2978	3.2830	3.2678
0.74	3.4093	3.3956	3.3822	3.3688	3.3554	3.3419	3.3284	3.3146	3.3005	3.2861
0.76	3.4216	3.4085	3.3956	3.3828	3.3701	3.3573	3.3443	3.3312	3.3179	3.3043
0.78	3.4342	3.4216	3.4093	3.3971	3.3849	3.3727	3.3604	3.3480	3.3353	3.3224
0.80	3.4470	3.4350	3.4232	3.4115	3.3999	3.3883	3.3767	3.3649	3.3529	3.3408
0.82	3.4602	3.4488	3.4375	3.4264	3.4154	3.4043	3.3933	3.3821	3.3709	3.3594
0.84	3.4741	3.4631	3.4524	3.4418	3.4314	3.4209	3.4105	3.3999	3.3893	3.3785
0.86	3.4886	3.4782	3.4680	3.4580	3.4481	3.4382	3.4284	3.4185	3.4085	3.3984
0.88	3.5042	3.4944	3.4847	3.4753	3.4659	3.4567	3.4474	3.4381	3.4288	3.4193
0.90	3.5212	3.5120	3.5029	3.4940	3.4853	3.4766	3.4680	3.4593	3.4506	3.4418
0.92	3.5404	3.5317	3.5233	3.5150	3.5069	3.4988	3.4908	3.4828	3.4748	3.4667
0.94	3.5629	3.5549	3.5471	3.5395	3.5320	3.5247	3.5173	3.5100	3.5027	3.4954
0.96	3.5912	3.5840	3.5770	3.5701	3.5634	3.5568	3.5503	3.5438	3.5373	3.5308
0.98	3.6331	3.6269	3.6208	3.6150	3.6093	3.6037	3.5982	3.5927	3.5873	3.5818

TABLE C.3c
(continued)

a_U	0.42	0.44	0.46	0.48	0.50	a_L 0.52	0.54	0.56	0.58	0.60
0.44	2.8820									
0.46	2.9214	2.8816								
0.48	2.9575	2.9208	2.8813							
0.50	2.9908	2.9569	2.9205	2.8812						
0.52	3.0215	2.9901	2.9565	2.9203	2.8812					
0.54	3.0502	3.0211	2.9899	2.9565	2.9205	2.8813				
0.56	3.0771	3.0499	3.0211	2.9901	2.9569	2.9208	2.8816			
0.58	3.1024	3.0771	3.0502	3.0215	2.9908	2.9575	2.9214	2.8820		
0.60	3.1264	3.1027	3.0777	3.0510	3.0225	2.9918	2.9586	2.9223	2.8826	
0.62	3.1493	3.1271	3.1037	3.0789	3.0524	3.0240	2.9933	2.9600	2.9235	2.8833
0.64	3.1712	3.1504	3.1285	3.1053	3.0807	3.0544	3.0260	2.9953	2.9618	2.9249
0.66	3.1923	3.1727	3.1522	3.1306	3.1077	3.0832	3.0570	3.0286	2.9978	2.9640
0.68	3.2127	3.1943	3.1751	3.1549	3.1335	3.1108	3.0864	3.0602	3.0319	3.0008
0.70	3.2326	3.2153	3.1972	3.1783	3.1583	3.1372	3.1146	3.0904	3.0642	3.0357
0.72	3.2521	3.2358	3.2188	3.2011	3.1824	3.1627	3.1418	3.1193	3.0952	3.0690
0.74	3.2712	3.2559	3.2400	3.2234	3.2059	3.1875	3.1681	3.1473	3.1250	3.1009
0.76	3.2903	3.2758	3.2608	3.2453	3.2290	3.2118	3.1937	3.1745	3.1539	3.1317
0.78	3.3092	3.2956	3.2816	3.2670	3.2517	3.2358	3.2189	3.2011	3.1821	3.1616
0.80	3.3283	3.3155	3.3023	3.2887	3.2744	3.2596	3.2439	3.2274	3.2098	3.1910
0.82	3.3477	3.3357	3.3233	3.3105	3.2973	3.2834	3.2689	3.2536	3.2374	3.2201
0.84	3.3675	3.3563	3.3447	3.3328	3.3204	3.3076	3.2941	3.2800	3.2651	3.2492
0.86	3.3881	3.3776	3.3668	3.3557	3.3442	3.3323	3.3199	3.3069	3.2932	3.2787
0.88	3.4097	3.3999	3.3899	3.3797	3.3691	3.3581	3.3467	3.3347	3.3222	3.3090
0.90	3.4329	3.4239	3.4147	3.4052	3.3955	3.3854	3.3750	3.3641	3.3527	3.3408
0.92	3.4586	3.4503	3.4418	3.4332	3.4243	3.4152	3.4058	3.3960	3.3857	3.3750
0.94	3.4880	3.4805	3.4729	3.4651	3.4572	3.4490	3.4406	3.4319	3.4228	3.4133
0.96	3.5243	3.5177	3.5110	3.5042	3.4973	3.4901	3.4828	3.4753	3.4675	3.4593
0.98	3.5763	3.5709	3.5653	3.5597	3.5540	3.5481	3.5422	3.5360	3.5297	3.5231

TABLE C.4a

Confidence Coefficients $k_{10}(a_L, a_U)$ for 90% Hall–Wellner Confidence Bands

a_U	a_L 0.00	0.02	0.04	0.06	0.08	0.10	0.12	0.14	0.16	0.18	0.20
0.10	0.5985	0.5985	0.5979	0.5930	0.5768						
0.12	0.6509	0.6509	0.6507	0.6484	0.6405	0.6210					
0.14	0.6979	0.6979	0.6978	0.6966	0.6923	0.6819	0.6598				
0.16	0.7406	0.7406	0.7405	0.7399	0.7373	0.7310	0.7184	0.6942			
0.18	0.7796	0.7796	0.7796	0.7792	0.7776	0.7735	0.7653	0.7509	0.7249		
0.20	0.8155	0.8155	0.8155	0.8153	0.8142	0.8114	0.8058	0.7961	0.7801	0.7525	
0.22	0.8487	0.8487	0.8487	0.8485	0.8479	0.8459	0.8419	0.8349	0.8237	0.8063	0.7773
0.24	0.8795	0.8794	0.8794	0.8794	0.8789	0.8775	0.8746	0.8693	0.8611	0.8486	0.8299
0.26	0.9081	0.9081	0.9081	0.9080	0.9077	0.9067	0.9045	0.9005	0.8941	0.8847	0.8711
0.28	0.9348	0.9348	0.9348	0.9347	0.9345	0.9338	0.9320	0.9289	0.9239	0.9165	0.9060
0.30	0.9597	0.9597	0.9597	0.9597	0.9595	0.9589	0.9576	0.9551	0.9510	0.9450	0.9366
0.32	0.9829	0.9829	0.9829	0.9829	0.9828	0.9824	0.9813	0.9793	0.9760	0.9710	0.9641
0.34	1.0047	1.0047	1.0047	1.0046	1.0046	1.0042	1.0034	1.0017	0.9990	0.9948	0.9890
0.36	1.0250	1.0250	1.0250	1.0250	1.0249	1.0246	1.0239	1.0226	1.0202	1.0167	1.0118
0.38	1.0439	1.0439	1.0439	1.0439	1.0439	1.0437	1.0431	1.0419	1.0400	1.0370	1.0327
0.40	1.0616	1.0616	1.0616	1.0616	1.0616	1.0614	1.0610	1.0600	1.0583	1.0557	1.0520
0.42	1.0782	1.0782	1.0782	1.0782	1.0781	1.0780	1.0776	1.0768	1.0753	1.0730	1.0697
0.44	1.0935	1.0935	1.0935	1.0935	1.0935	1.0934	1.0931	1.0923	1.0911	1.0890	1.0862
0.46	1.1078	1.1078	1.1078	1.1078	1.1078	1.1077	1.1074	1.1068	1.1057	1.1039	1.1013
0.48	1.1211	1.1211	1.1211	1.1211	1.1211	1.1210	1.1208	1.1202	1.1192	1.1176	1.1153
0.50	1.1334	1.1334	1.1334	1.1334	1.1334	1.1333	1.1331	1.1326	1.1317	1.1303	1.1281
0.52	1.1447	1.1447	1.1447	1.1447	1.1447	1.1447	1.1445	1.1440	1.1432	1.1419	1.1400
0.54	1.1552	1.1552	1.1552	1.1552	1.1551	1.1551	1.1549	1.1545	1.1538	1.1526	1.1508
0.56	1.1647	1.1647	1.1647	1.1647	1.1647	1.1646	1.1645	1.1641	1.1635	1.1623	1.1607
0.58	1.1734	1.1734	1.1734	1.1734	1.1734	1.1733	1.1732	1.1729	1.1723	1.1712	1.1697
0.60	1.1813	1.1813	1.1813	1.1813	1.1813	1.1812	1.1811	1.1808	1.1802	1.1793	1.1778
0.62	1.1884	1.1884	1.1884	1.1884	1.1883	1.1883	1.1882	1.1879	1.1874	1.1865	1.1851
0.64	1.1947	1.1947	1.1947	1.1947	1.1947	1.1946	1.1945	1.1943	1.1938	1.1929	1.1916
0.66	1.2003	1.2003	1.2003	1.2003	1.2003	1.2002	1.2001	1.1999	1.1994	1.1986	1.1974
0.68	1.2052	1.2052	1.2052	1.2052	1.2052	1.2051	1.2050	1.2048	1.2043	1.2036	1.2024
0.70	1.2094	1.2094	1.2094	1.2094	1.2094	1.2093	1.2093	1.2090	1.2086	1.2078	1.2067
0.72	1.2130	1.2129	1.2129	1.2129	1.2129	1.2129	1.2128	1.2126	1.2122	1.2115	1.2104
0.74	1.2159	1.2159	1.2159	1.2159	1.2159	1.2159	1.2158	1.2156	1.2152	1.2145	1.2134
0.76	1.2183	1.2183	1.2183	1.2183	1.2183	1.2183	1.2182	1.2180	1.2176	1.2169	1.2159
0.78	1.2202	1.2202	1.2202	1.2202	1.2202	1.2201	1.2201	1.2199	1.2195	1.2188	1.2178
0.80	1.2216	1.2216	1.2216	1.2216	1.2216	1.2215	1.2215	1.2213	1.2209	1.2202	1.2192
0.82	1.2226	1.2226	1.2226	1.2226	1.2226	1.2225	1.2225	1.2223	1.2219	1.2212	1.2202
0.84	1.2232	1.2232	1.2232	1.2232	1.2232	1.2232	1.2231	1.2229	1.2225	1.2219	1.2209
0.86	1.2236	1.2236	1.2236	1.2236	1.2236	1.2236	1.2235	1.2233	1.2229	1.2223	1.2213
0.88	1.2238	1.2238	1.2238	1.2238	1.2238	1.2237	1.2237	1.2235	1.2231	1.2225	1.2215
0.90	1.2239	1.2238	1.2238	1.2238	1.2238	1.2238	1.2237	1.2236	1.2232	1.2225	1.2215
0.92	1.2239	1.2238	1.2238	1.2238	1.2238	1.2238	1.2238	1.2236	1.2232	1.2226	1.2216
0.94	1.2239	1.2238	1.2238	1.2238	1.2238	1.2238	1.2238	1.2236	1.2232	1.2226	1.2216
0.96	1.2239	1.2238	1.2238	1.2238	1.2238	1.2238	1.2238	1.2236	1.2232	1.2226	1.2216
0.98	1.2239	1.2238	1.2238	1.2238	1.2238	1.2238	1.2238	1.2236	1.2232	1.2226	1.2216
1.00	1.2239	1.2239	1.2239	1.2239	1.2239	1.2239	1.2238	1.2236	1.2232	1.2226	1.2216

TABLE C.4a
(continued)

a_U	0.22	0.24	0.26	0.28	0.30	a_L 0.32	0.34	0.36	0.38	0.40
0.24	0.7996									
0.26	0.8512	0.8198								
0.28	0.8913	0.8703	0.8379							
0.30	0.9251	0.9094	0.8875	0.8540						
0.32	0.9548	0.9423	0.9257	0.9028	0.8685					
0.34	0.9813	0.9710	0.9577	0.9402	0.9165	0.8812				
0.36	1.0052	0.9966	0.9855	0.9713	0.9530	0.9284	0.8924			
0.38	1.0270	1.0196	1.0102	0.9983	0.9833	0.9642	0.9389	0.9020		
0.40	1.0470	1.0406	1.0324	1.0222	1.0095	0.9938	0.9739	0.9478	0.9102	
0.42	1.0654	1.0597	1.0525	1.0436	1.0326	1.0192	1.0027	0.9821	0.9552	0.9169
0.44	1.0822	1.0772	1.0708	1.0629	1.0532	1.0415	1.0274	1.0102	0.9888	0.9613
0.46	1.0978	1.0932	1.0875	1.0804	1.0717	1.0614	1.0490	1.0342	1.0163	0.9942
0.48	1.1121	1.1079	1.1027	1.0963	1.0885	1.0792	1.0682	1.0551	1.0396	1.0210
0.50	1.1252	1.1214	1.1167	1.1108	1.1037	1.0952	1.0853	1.0735	1.0598	1.0436
0.52	1.1373	1.1338	1.1294	1.1240	1.1175	1.1097	1.1006	1.0899	1.0775	1.0631
0.54	1.1483	1.1451	1.1410	1.1360	1.1299	1.1228	1.1143	1.1046	1.0933	1.0802
0.56	1.1584	1.1554	1.1516	1.1469	1.1412	1.1345	1.1267	1.1177	1.1072	1.0952
0.58	1.1675	1.1647	1.1611	1.1567	1.1514	1.1451	1.1378	1.1293	1.1196	1.1085
0.60	1.1758	1.1731	1.1697	1.1656	1.1606	1.1546	1.1477	1.1398	1.1307	1.1203
0.62	1.1832	1.1807	1.1775	1.1735	1.1688	1.1631	1.1566	1.1490	1.1404	1.1307
0.64	1.1898	1.1874	1.1843	1.1806	1.1760	1.1707	1.1644	1.1572	1.1490	1.1398
0.66	1.1956	1.1933	1.1904	1.1868	1.1824	1.1773	1.1713	1.1644	1.1566	1.1477
0.68	1.2007	1.1985	1.1957	1.1922	1.1880	1.1830	1.1773	1.1707	1.1631	1.1546
0.70	1.2051	1.2030	1.2002	1.1969	1.1928	1.1880	1.1824	1.1760	1.1688	1.1606
0.72	1.2088	1.2067	1.2041	1.2008	1.1969	1.1922	1.1868	1.1806	1.1735	1.1656
0.74	1.2119	1.2099	1.2073	1.2041	1.2002	1.1957	1.1904	1.1843	1.1775	1.1697
0.76	1.2144	1.2124	1.2099	1.2067	1.2030	1.1985	1.1933	1.1874	1.1807	1.1731
0.78	1.2163	1.2144	1.2119	1.2088	1.2051	1.2007	1.1956	1.1898	1.1832	1.1758
0.80	1.2178	1.2159	1.2134	1.2104	1.2067	1.2024	1.1974	1.1916	1.1851	1.1778
0.82	1.2188	1.2169	1.2145	1.2115	1.2078	1.2036	1.1986	1.1929	1.1865	1.1793
0.84	1.2195	1.2176	1.2152	1.2122	1.2086	1.2043	1.1994	1.1938	1.1874	1.1802
0.86	1.2199	1.2180	1.2156	1.2126	1.2090	1.2048	1.1999	1.1943	1.1879	1.1808
0.88	1.2201	1.2182	1.2158	1.2128	1.2093	1.2050	1.2001	1.1945	1.1882	1.1811
0.90	1.2201	1.2183	1.2159	1.2129	1.2093	1.2051	1.2002	1.1946	1.1883	1.1812
0.92	1.2202	1.2183	1.2159	1.2129	1.2094	1.2052	1.2003	1.1947	1.1883	1.1813
0.94	1.2202	1.2183	1.2159	1.2129	1.2094	1.2052	1.2003	1.1947	1.1884	1.1813
0.96	1.2202	1.2183	1.2159	1.2129	1.2094	1.2052	1.2003	1.1947	1.1884	1.1813
0.98	1.2202	1.2183	1.2159	1.2129	1.2094	1.2052	1.2003	1.1947	1.1884	1.1813
1.00	1.2202	1.2183	1.2159	1.2130	1.2094	1.2052	1.2003	1.1947	1.1884	1.1813

(continued)

TABLE C.4a
(continued)

	a_L									
a_U	0.42	0.44	0.46	0.48	0.50	0.52	0.54	0.56	0.58	0.60
0.44	0.9223									
0.46	0.9660	0.9263								
0.48	0.9982	0.9693	0.9289							
0.50	1.0243	1.0009	0.9713	0.9302						
0.52	1.0462	1.0263	1.0022	0.9720	0.9302					
0.54	1.0651	1.0476	1.0270	1.0022	0.9713	0.9289				
0.56	1.0815	1.0658	1.0476	1.0263	1.0009	0.9693	0.9263			
0.58	1.0959	1.0815	1.0651	1.0462	1.0243	0.9982	0.9660	0.9223		
0.60	1.1085	1.0952	1.0802	1.0631	1.0436	1.0210	0.9942	0.9613	0.9169	
0.62	1.1196	1.1072	1.0933	1.0775	1.0598	1.0396	1.0163	0.9888	0.9552	0.9102
0.64	1.1293	1.1177	1.1046	1.0899	1.0735	1.0551	1.0342	1.0102	0.9821	0.9478
0.66	1.1378	1.1267	1.1143	1.1006	1.0853	1.0682	1.0490	1.0274	1.0027	0.9739
0.68	1.1451	1.1345	1.1228	1.1097	1.0952	1.0792	1.0614	1.0415	1.0192	0.9938
0.70	1.1514	1.1412	1.1299	1.1175	1.1037	1.0885	1.0717	1.0532	1.0326	1.0095
0.72	1.1567	1.1469	1.1360	1.1240	1.1108	1.0963	1.0804	1.0629	1.0436	1.0222
0.74	1.1611	1.1516	1.1410	1.1294	1.1167	1.1027	1.0875	1.0708	1.0525	1.0324
0.76	1.1647	1.1554	1.1451	1.1338	1.1214	1.1079	1.0932	1.0772	1.0597	1.0406
0.78	1.1675	1.1584	1.1483	1.1373	1.1252	1.1121	1.0978	1.0822	1.0654	1.0470
0.80	1.1697	1.1607	1.1508	1.1400	1.1281	1.1153	1.1013	1.0862	1.0697	1.0520
0.82	1.1712	1.1623	1.1526	1.1419	1.1303	1.1176	1.1039	1.0890	1.0730	1.0557
0.84	1.1723	1.1635	1.1538	1.1432	1.1317	1.1192	1.1057	1.0911	1.0753	1.0583
0.86	1.1729	1.1641	1.1545	1.1440	1.1326	1.1202	1.1068	1.0923	1.0768	1.0600
0.88	1.1732	1.1645	1.1549	1.1445	1.1331	1.1208	1.1074	1.0931	1.0776	1.0610
0.90	1.1733	1.1646	1.1551	1.1447	1.1333	1.1210	1.1077	1.0934	1.0780	1.0614
0.92	1.1734	1.1647	1.1551	1.1447	1.1334	1.1211	1.1078	1.0935	1.0781	1.0616
0.94	1.1734	1.1647	1.1552	1.1447	1.1334	1.1211	1.1078	1.0935	1.0782	1.0616
0.96	1.1734	1.1647	1.1552	1.1447	1.1334	1.1211	1.1078	1.0935	1.0782	1.0616
0.98	1.1734	1.1647	1.1552	1.1447	1.1334	1.1211	1.1078	1.0935	1.0782	1.0616
1.00	1.1734	1.1647	1.1552	1.1447	1.1334	1.1211	1.1078	1.0935	1.0782	1.0616

TABLE C.4b
Confidence Coefficients $k_{05}(a_L, a_U)$ for 95% Hall–Wellner Confidence Bands

a_U	0.00	0.02	0.04	0.06	0.08	0.10	0.12	0.14	0.16	0.18	0.20
0.10	0.6825	0.6825	0.6822	0.6793	0.6666						
0.12	0.7418	0.7418	0.7417	0.7405	0.7351	0.7191					
0.14	0.7948	0.7948	0.7948	0.7943	0.7916	0.7838	0.7651				
0.16	0.8428	0.8428	0.8428	0.8426	0.8412	0.8369	0.8270	0.8060			
0.18	0.8866	0.8866	0.8866	0.8865	0.8857	0.8832	0.8772	0.8655	0.8425		
0.20	0.9269	0.9268	0.9268	0.9268	0.9263	0.9247	0.9209	0.9134	0.9001	0.8753	
0.22	0.9639	0.9639	0.9639	0.9639	0.9636	0.9626	0.9600	0.9549	0.9460	0.9311	0.9049
0.24	0.9982	0.9982	0.9982	0.9982	0.9980	0.9973	0.9955	0.9919	0.9856	0.9754	0.9592
0.26	1.0299	1.0299	1.0299	1.0299	1.0298	1.0294	1.0281	1.0254	1.0208	1.0134	1.0019
0.28	1.0594	1.0594	1.0594	1.0594	1.0593	1.0590	1.0581	1.0561	1.0526	1.0470	1.0384
0.30	1.0868	1.0868	1.0868	1.0868	1.0868	1.0865	1.0859	1.0843	1.0816	1.0772	1.0706
0.32	1.1123	1.1123	1.1123	1.1123	1.1123	1.1121	1.1116	1.1104	1.1083	1.1047	1.0994
0.34	1.1360	1.1360	1.1360	1.1360	1.1360	1.1359	1.1355	1.1346	1.1328	1.1299	1.1256
0.36	1.1581	1.1581	1.1581	1.1581	1.1580	1.1580	1.1576	1.1569	1.1555	1.1531	1.1495
0.38	1.1785	1.1785	1.1785	1.1785	1.1785	1.1785	1.1782	1.1776	1.1764	1.1745	1.1714
0.40	1.1976	1.1976	1.1976	1.1976	1.1975	1.1975	1.1973	1.1968	1.1958	1.1941	1.1915
0.42	1.2152	1.2152	1.2152	1.2152	1.2152	1.2151	1.2150	1.2146	1.2137	1.2123	1.2100
0.44	1.2315	1.2315	1.2315	1.2315	1.2315	1.2314	1.2313	1.2310	1.2302	1.2290	1.2270
0.46	1.2465	1.2465	1.2465	1.2465	1.2465	1.2465	1.2464	1.2461	1.2455	1.2444	1.2426
0.48	1.2604	1.2604	1.2604	1.2604	1.2604	1.2603	1.2603	1.2600	1.2595	1.2585	1.2570
0.50	1.2731	1.2731	1.2731	1.2731	1.2731	1.2731	1.2730	1.2728	1.2723	1.2714	1.2700
0.52	1.2847	1.2847	1.2847	1.2847	1.2847	1.2847	1.2846	1.2844	1.2840	1.2832	1.2820
0.54	1.2953	1.2952	1.2952	1.2952	1.2952	1.2952	1.2952	1.2950	1.2946	1.2939	1.2928
0.56	1.3048	1.3048	1.3048	1.3048	1.3048	1.3048	1.3047	1.3046	1.3042	1.3036	1.3025
0.58	1.3134	1.3134	1.3134	1.3134	1.3134	1.3134	1.3133	1.3132	1.3129	1.3123	1.3113
0.60	1.3211	1.3211	1.3211	1.3211	1.3211	1.3211	1.3210	1.3209	1.3206	1.3201	1.3191
0.62	1.3279	1.3279	1.3279	1.3279	1.3279	1.3279	1.3278	1.3277	1.3274	1.3269	1.3261
0.64	1.3338	1.3338	1.3338	1.3338	1.3338	1.3338	1.3338	1.3337	1.3334	1.3329	1.3321
0.66	1.3390	1.3390	1.3390	1.3390	1.3390	1.3390	1.3389	1.3388	1.3386	1.3381	1.3374
0.68	1.3434	1.3434	1.3434	1.3434	1.3434	1.3434	1.3433	1.3432	1.3430	1.3426	1.3418
0.70	1.3471	1.3471	1.3471	1.3471	1.3471	1.3471	1.3470	1.3469	1.3467	1.3463	1.3456
0.72	1.3501	1.3501	1.3501	1.3501	1.3501	1.3501	1.3501	1.3500	1.3498	1.3494	1.3487
0.74	1.3525	1.3525	1.3525	1.3525	1.3525	1.3525	1.3525	1.3524	1.3522	1.3518	1.3511
0.76	1.3544	1.3544	1.3544	1.3544	1.3544	1.3544	1.3544	1.3543	1.3541	1.3537	1.3530
0.78	1.3558	1.3558	1.3558	1.3558	1.3558	1.3558	1.3558	1.3557	1.3555	1.3551	1.3544
0.80	1.3568	1.3568	1.3568	1.3568	1.3568	1.3568	1.3567	1.3567	1.3565	1.3561	1.3554
0.82	1.3574	1.3574	1.3574	1.3574	1.3574	1.3574	1.3574	1.3573	1.3571	1.3567	1.3561
0.84	1.3578	1.3578	1.3578	1.3578	1.3578	1.3578	1.3578	1.3577	1.3575	1.3571	1.3565
0.86	1.3580	1.3580	1.3580	1.3580	1.3580	1.3580	1.3580	1.3579	1.3577	1.3573	1.3567
0.88	1.3581	1.3581	1.3581	1.3581	1.3581	1.3581	1.3580	1.3580	1.3578	1.3574	1.3567
0.90	1.3581	1.3581	1.3581	1.3581	1.3581	1.3581	1.3581	1.3580	1.3578	1.3574	1.3568
0.92	1.3581	1.3581	1.3581	1.3581	1.3581	1.3581	1.3581	1.3580	1.3578	1.3574	1.3568
0.94	1.3581	1.3581	1.3581	1.3581	1.3581	1.3581	1.3581	1.3580	1.3578	1.3574	1.3568
0.96	1.3581	1.3581	1.3581	1.3581	1.3581	1.3581	1.3581	1.3580	1.3578	1.3574	1.3568
0.98	1.3581	1.3581	1.3581	1.3581	1.3581	1.3581	1.3581	1.3580	1.3578	1.3574	1.3568
1.00	1.3581	1.3581	1.3581	1.3581	1.3581	1.3581	1.3581	1.3580	1.3578	1.3574	1.3568

TABLE C.4b

(continued)

a_U	0.22	0.24	0.26	0.28	0.30	a_L 0.32	0.34	0.36	0.38	0.40
0.24	0.9315									
0.26	0.9844	0.9554								
0.28	1.0258	1.0071	0.9769							
0.30	1.0610	1.0473	1.0275	0.9962						
0.32	1.0918	1.0813	1.0666	1.0457	1.0134					
0.34	1.1194	1.1109	1.0994	1.0838	1.0619	1.0286				
0.36	1.1444	1.1374	1.1280	1.1155	1.0989	1.0761	1.0419			
0.38	1.1671	1.1612	1.1533	1.1430	1.1297	1.1122	1.0885	1.0533		
0.40	1.1878	1.1827	1.1760	1.1673	1.1562	1.1420	1.1237	1.0991	1.0631	
0.42	1.2068	1.2024	1.1966	1.1891	1.1796	1.1676	1.1526	1.1334	1.1080	1.0711
0.44	1.2242	1.2203	1.2152	1.2086	1.2003	1.1900	1.1773	1.1615	1.1414	1.1152
0.46	1.2401	1.2367	1.2321	1.2263	1.2189	1.2099	1.1988	1.1853	1.1686	1.1478
0.48	1.2547	1.2516	1.2475	1.2422	1.2357	1.2276	1.2178	1.2060	1.1916	1.1742
0.50	1.2680	1.2652	1.2615	1.2567	1.2508	1.2435	1.2347	1.2241	1.2115	1.1964
0.52	1.2801	1.2775	1.2741	1.2698	1.2643	1.2577	1.2497	1.2401	1.2288	1.2154
0.54	1.2911	1.2887	1.2856	1.2815	1.2765	1.2704	1.2630	1.2543	1.2440	1.2319
0.56	1.3010	1.2988	1.2959	1.2921	1.2874	1.2817	1.2749	1.2668	1.2574	1.2463
0.58	1.3099	1.3078	1.3051	1.3016	1.2972	1.2919	1.2855	1.2779	1.2691	1.2589
0.60	1.3178	1.3158	1.3133	1.3100	1.3058	1.3008	1.2948	1.2877	1.2794	1.2699
0.62	1.3248	1.3229	1.3205	1.3174	1.3134	1.3087	1.3030	1.2963	1.2884	1.2794
0.64	1.3309	1.3291	1.3268	1.3238	1.3201	1.3155	1.3101	1.3037	1.2963	1.2877
0.66	1.3362	1.3345	1.3323	1.3294	1.3258	1.3215	1.3162	1.3101	1.3030	1.2948
0.68	1.3407	1.3391	1.3369	1.3342	1.3307	1.3265	1.3215	1.3155	1.3087	1.3008
0.70	1.3445	1.3429	1.3408	1.3382	1.3348	1.3307	1.3258	1.3201	1.3134	1.3058
0.72	1.3476	1.3461	1.3441	1.3414	1.3382	1.3342	1.3294	1.3238	1.3174	1.3100
0.74	1.3501	1.3486	1.3466	1.3441	1.3408	1.3369	1.3323	1.3268	1.3205	1.3133
0.76	1.3520	1.3506	1.3486	1.3461	1.3429	1.3391	1.3345	1.3291	1.3229	1.3158
0.78	1.3534	1.3520	1.3501	1.3476	1.3445	1.3407	1.3362	1.3309	1.3248	1.3178
0.80	1.3544	1.3530	1.3511	1.3487	1.3456	1.3418	1.3374	1.3321	1.3261	1.3191
0.82	1.3551	1.3537	1.3518	1.3494	1.3463	1.3426	1.3381	1.3329	1.3269	1.3201
0.84	1.3555	1.3541	1.3522	1.3498	1.3467	1.3430	1.3386	1.3334	1.3274	1.3206
0.86	1.3557	1.3543	1.3524	1.3500	1.3469	1.3432	1.3388	1.3337	1.3277	1.3209
0.88	1.3558	1.3544	1.3525	1.3501	1.3470	1.3433	1.3389	1.3338	1.3278	1.3210
0.90	1.3558	1.3544	1.3525	1.3501	1.3471	1.3434	1.3390	1.3338	1.3279	1.3211
0.92	1.3558	1.3544	1.3525	1.3501	1.3471	1.3434	1.3390	1.3338	1.3279	1.3211
0.94	1.3558	1.3544	1.3525	1.3501	1.3471	1.3434	1.3390	1.3338	1.3279	1.3211
0.96	1.3558	1.3544	1.3525	1.3501	1.3471	1.3434	1.3390	1.3338	1.3279	1.3211
0.98	1.3558	1.3544	1.3525	1.3501	1.3471	1.3434	1.3390	1.3338	1.3279	1.3211
1.00	1.3558	1.3544	1.3525	1.3501	1.3471	1.3434	1.3390	1.3338	1.3279	1.3211

TABLE C.4b

(continued)

a_U	0.42	0.44	0.46	0.48	a_L 0.50	0.52	0.54	0.56	0.58	0.60
0.44	1.0775									
0.46	1.1208	1.0822								
0.48	1.1526	1.1247	1.0854							
0.50	1.1781	1.1557	1.1271	1.0869						
0.52	1.1995	1.1805	1.1573	1.1279	1.0869					
0.54	1.2177	1.2011	1.1813	1.1573	1.1271	1.0854				
0.56	1.2335	1.2185	1.2011	1.1805	1.1557	1.1247	1.0822			
0.58	1.2471	1.2335	1.2177	1.1995	1.1781	1.1526	1.1208	1.0775		
0.60	1.2589	1.2463	1.2319	1.2154	1.1964	1.1742	1.1478	1.1152	1.0711	
0.62	1.2691	1.2574	1.2440	1.2288	1.2115	1.1916	1.1686	1.1414	1.1080	1.0631
0.64	1.2779	1.2668	1.2543	1.2401	1.2241	1.2060	1.1853	1.1615	1.1334	1.0991
0.66	1.2855	1.2749	1.2630	1.2497	1.2347	1.2178	1.1988	1.1773	1.1526	1.1237
0.68	1.2919	1.2817	1.2704	1.2577	1.2435	1.2276	1.2099	1.1900	1.1676	1.1420
0.70	1.2972	1.2874	1.2765	1.2643	1.2508	1.2357	1.2189	1.2003	1.1796	1.1562
0.72	1.3016	1.2921	1.2815	1.2698	1.2567	1.2422	1.2263	1.2086	1.1891	1.1673
0.74	1.3051	1.2959	1.2856	1.2741	1.2615	1.2475	1.2321	1.2152	1.1966	1.1760
0.76	1.3078	1.2988	1.2887	1.2775	1.2652	1.2516	1.2367	1.2203	1.2024	1.1827
0.78	1.3099	1.3010	1.2911	1.2801	1.2680	1.2547	1.2401	1.2242	1.2068	1.1878
0.80	1.3113	1.3025	1.2928	1.2820	1.2700	1.2570	1.2426	1.2270	1.2100	1.1915
0.82	1.3123	1.3036	1.2939	1.2832	1.2714	1.2585	1.2444	1.2290	1.2123	1.1941
0.84	1.3129	1.3042	1.2946	1.2840	1.2723	1.2595	1.2455	1.2302	1.2137	1.1958
0.86	1.3132	1.3046	1.2950	1.2844	1.2728	1.2600	1.2461	1.2310	1.2146	1.1968
0.88	1.3133	1.3047	1.2952	1.2846	1.2730	1.2603	1.2464	1.2313	1.2150	1.1973
0.90	1.3134	1.3048	1.2952	1.2847	1.2731	1.2603	1.2465	1.2314	1.2151	1.1975
0.92	1.3134	1.3048	1.2952	1.2847	1.2731	1.2604	1.2465	1.2315	1.2152	1.1975
0.94	1.3134	1.3048	1.2952	1.2847	1.2731	1.2604	1.2465	1.2315	1.2152	1.1976
0.96	1.3134	1.3048	1.2952	1.2847	1.2731	1.2604	1.2465	1.2315	1.2152	1.1976
0.98	1.3134	1.3048	1.2952	1.2847	1.2731	1.2604	1.2465	1.2315	1.2152	1.1976
1.00	1.3134	1.3048	1.2952	1.2847	1.2731	1.2604	1.2465	1.2315	1.2152	1.1976

TABLE C.4c

Confidence Coefficients $k_{01}(a_L, a_U)$ for 99% Hall–Wellner Confidence Bands

					a_L						
a_U	0.00	0.02	0.04	0.06	0.08	0.10	0.12	0.14	0.16	0.18	0.20
0.10	0.8512	0.8512	0.8512	0.8502	0.8428						
0.12	0.9243	0.9243	0.9243	0.9240	0.9217	0.9113					
0.14	0.9895	0.9895	0.9895	0.9894	0.9886	0.9845	0.9715				
0.16	1.0483	1.0483	1.0483	1.0483	1.0479	1.0461	1.0404	1.0249			
0.18	1.1017	1.1017	1.1017	1.1017	1.1016	1.1007	1.0978	1.0903	1.0727		
0.20	1.1505	1.1505	1.1505	1.1505	1.1505	1.1500	1.1484	1.1443	1.1352	1.1157	
0.22	1.1953	1.1953	1.1953	1.1953	1.1953	1.1950	1.1941	1.1917	1.1863	1.1757	1.1544
0.24	1.2365	1.2365	1.2365	1.2365	1.2365	1.2364	1.2358	1.2343	1.2309	1.2243	1.2122
0.26	1.2745	1.2745	1.2745	1.2745	1.2745	1.2744	1.2740	1.2730	1.2708	1.2664	1.2586
0.28	1.3095	1.3095	1.3095	1.3095	1.3095	1.3095	1.3092	1.3086	1.3070	1.3039	1.2985
0.30	1.3419	1.3419	1.3419	1.3419	1.3419	1.3418	1.3417	1.3412	1.3401	1.3379	1.3340
0.32	1.3717	1.3717	1.3717	1.3717	1.3717	1.3717	1.3716	1.3713	1.3705	1.3688	1.3659
0.34	1.3993	1.3993	1.3993	1.3993	1.3993	1.3993	1.3992	1.3990	1.3984	1.3971	1.3949
0.36	1.4247	1.4247	1.4247	1.4247	1.4247	1.4247	1.4247	1.4245	1.4240	1.4231	1.4213
0.38	1.4481	1.4481	1.4481	1.4481	1.4481	1.4481	1.4481	1.4479	1.4476	1.4468	1.4454
0.40	1.4696	1.4696	1.4696	1.4696	1.4696	1.4696	1.4696	1.4695	1.4692	1.4686	1.4674
0.42	1.4893	1.4893	1.4893	1.4893	1.4893	1.4893	1.4893	1.4892	1.4890	1.4885	1.4875
0.44	1.5073	1.5073	1.5073	1.5073	1.5073	1.5073	1.5073	1.5072	1.5071	1.5066	1.5058
0.46	1.5237	1.5237	1.5237	1.5237	1.5237	1.5237	1.5237	1.5236	1.5235	1.5231	1.5225
0.48	1.5386	1.5386	1.5386	1.5386	1.5386	1.5386	1.5385	1.5385	1.5384	1.5381	1.5375
0.50	1.5520	1.5520	1.5520	1.5520	1.5520	1.5520	1.5519	1.5519	1.5518	1.5515	1.5510
0.52	1.5640	1.5640	1.5640	1.5640	1.5640	1.5640	1.5640	1.5639	1.5638	1.5636	1.5631
0.54	1.5747	1.5747	1.5747	1.5747	1.5747	1.5747	1.5747	1.5746	1.5746	1.5744	1.5739
0.56	1.5841	1.5841	1.5841	1.5841	1.5841	1.5841	1.5841	1.5841	1.5840	1.5838	1.5835
0.58	1.5924	1.5924	1.5924	1.5924	1.5924	1.5924	1.5924	1.5924	1.5923	1.5921	1.5918
0.60	1.5996	1.5996	1.5996	1.5996	1.5996	1.5996	1.5996	1.5996	1.5995	1.5993	1.5990
0.62	1.6057	1.6057	1.6057	1.6057	1.6057	1.6057	1.6057	1.6057	1.6056	1.6055	1.6052
0.64	1.6109	1.6109	1.6109	1.6109	1.6109	1.6109	1.6109	1.6109	1.6108	1.6107	1.6104
0.66	1.6152	1.6152	1.6152	1.6152	1.6152	1.6152	1.6152	1.6151	1.6151	1.6150	1.6147
0.68	1.6186	1.6186	1.6186	1.6186	1.6186	1.6186	1.6186	1.6186	1.6186	1.6184	1.6182
0.70	1.6214	1.6214	1.6214	1.6214	1.6214	1.6214	1.6214	1.6214	1.6213	1.6212	1.6209
0.72	1.6235	1.6235	1.6235	1.6235	1.6235	1.6235	1.6235	1.6235	1.6234	1.6233	1.6230
0.74	1.6250	1.6250	1.6250	1.6250	1.6250	1.6250	1.6250	1.6250	1.6249	1.6248	1.6246
0.76	1.6261	1.6261	1.6261	1.6261	1.6261	1.6261	1.6261	1.6261	1.6260	1.6259	1.6257
0.78	1.6268	1.6268	1.6268	1.6268	1.6268	1.6268	1.6268	1.6268	1.6267	1.6266	1.6264
0.80	1.6272	1.6272	1.6272	1.6272	1.6272	1.6272	1.6272	1.6272	1.6272	1.6271	1.6268
0.82	1.6275	1.6275	1.6275	1.6275	1.6275	1.6275	1.6275	1.6274	1.6274	1.6273	1.6271
0.84	1.6276	1.6276	1.6276	1.6276	1.6276	1.6276	1.6276	1.6276	1.6275	1.6274	1.6272
0.86	1.6276	1.6276	1.6276	1.6276	1.6276	1.6276	1.6276	1.6276	1.6276	1.6274	1.6272
0.88	1.6276	1.6276	1.6276	1.6276	1.6276	1.6276	1.6276	1.6276	1.6276	1.6275	1.6272
0.90	1.6276	1.6276	1.6276	1.6276	1.6276	1.6276	1.6276	1.6276	1.6276	1.6275	1.6272
0.92	1.6276	1.6276	1.6276	1.6276	1.6276	1.6276	1.6276	1.6276	1.6276	1.6275	1.6272
0.94	1.6276	1.6276	1.6276	1.6276	1.6276	1.6276	1.6276	1.6276	1.6276	1.6275	1.6272
0.96	1.6276	1.6276	1.6276	1.6276	1.6276	1.6276	1.6276	1.6276	1.6276	1.6275	1.6272
0.98	1.6276	1.6276	1.6276	1.6276	1.6276	1.6276	1.6276	1.6276	1.6276	1.6275	1.6272
1.00	1.6276	1.6276	1.6276	1.6276	1.6276	1.6276	1.6276	1.6276	1.6276	1.6275	1.6272

TABLE C.4c
(continued)

a_U	0.22	0.24	0.26	0.28	0.30	0.32	0.34	0.36	0.38	0.40
0.24	1.1893									
0.26	1.2452	1.2207								
0.28	1.2896	1.2748	1.2489							
0.30	1.3276	1.3175	1.3014	1.2742						
0.32	1.3611	1.3537	1.3425	1.3253	1.2967					
0.34	1.3912	1.3855	1.3771	1.3648	1.3464	1.3166				
0.36	1.4184	1.4139	1.4073	1.3979	1.3845	1.3650	1.3341			
0.38	1.4431	1.4395	1.4342	1.4267	1.4162	1.4018	1.3812	1.3491		
0.40	1.4655	1.4625	1.4582	1.4520	1.4436	1.4322	1.4167	1.3950	1.3619	
0.42	1.4859	1.4834	1.4797	1.4746	1.4676	1.4583	1.4459	1.4294	1.4067	1.3724
0.44	1.5045	1.5023	1.4992	1.4948	1.4888	1.4810	1.4707	1.4574	1.4399	1.4161
0.46	1.5213	1.5194	1.5167	1.5129	1.5077	1.5009	1.4922	1.4810	1.4667	1.4482
0.48	1.5365	1.5348	1.5324	1.5291	1.5245	1.5186	1.5109	1.5013	1.4892	1.4739
0.50	1.5501	1.5487	1.5465	1.5435	1.5395	1.5342	1.5274	1.5189	1.5083	1.4953
0.52	1.5623	1.5610	1.5591	1.5564	1.5527	1.5480	1.5419	1.5343	1.5249	1.5134
0.54	1.5732	1.5720	1.5703	1.5678	1.5644	1.5601	1.5545	1.5476	1.5391	1.5288
0.56	1.5828	1.5817	1.5801	1.5778	1.5747	1.5707	1.5656	1.5592	1.5514	1.5420
0.58	1.5912	1.5902	1.5887	1.5865	1.5837	1.5799	1.5751	1.5692	1.5620	1.5533
0.60	1.5984	1.5975	1.5961	1.5941	1.5914	1.5878	1.5834	1.5778	1.5710	1.5629
0.62	1.6047	1.6038	1.6024	1.6005	1.5980	1.5946	1.5904	1.5851	1.5787	1.5710
0.64	1.6099	1.6090	1.6078	1.6060	1.6035	1.6003	1.5962	1.5912	1.5851	1.5778
0.66	1.6142	1.6134	1.6122	1.6104	1.6081	1.6050	1.6011	1.5962	1.5904	1.5834
0.68	1.6177	1.6169	1.6157	1.6141	1.6118	1.6088	1.6050	1.6003	1.5946	1.5878
0.70	1.6205	1.6197	1.6186	1.6169	1.6147	1.6118	1.6081	1.6035	1.5980	1.5914
0.72	1.6226	1.6218	1.6207	1.6191	1.6169	1.6141	1.6104	1.6060	1.6005	1.5941
0.74	1.6241	1.6234	1.6223	1.6207	1.6186	1.6157	1.6122	1.6078	1.6024	1.5961
0.76	1.6252	1.6245	1.6234	1.6218	1.6197	1.6169	1.6134	1.6090	1.6038	1.5975
0.78	1.6260	1.6252	1.6241	1.6226	1.6205	1.6177	1.6142	1.6099	1.6047	1.5984
0.80	1.6264	1.6257	1.6246	1.6230	1.6209	1.6182	1.6147	1.6104	1.6052	1.5990
0.82	1.6266	1.6259	1.6248	1.6233	1.6212	1.6184	1.6150	1.6107	1.6055	1.5993
0.84	1.6267	1.6260	1.6249	1.6234	1.6213	1.6186	1.6151	1.6108	1.6056	1.5995
0.86	1.6268	1.6261	1.6250	1.6235	1.6214	1.6186	1.6151	1.6109	1.6057	1.5996
0.88	1.6268	1.6261	1.6250	1.6235	1.6214	1.6186	1.6152	1.6109	1.6057	1.5996
0.90	1.6268	1.6261	1.6250	1.6235	1.6214	1.6186	1.6152	1.6109	1.6057	1.5996
0.92	1.6268	1.6261	1.6250	1.6235	1.6214	1.6186	1.6152	1.6109	1.6057	1.5996
0.94	1.6268	1.6261	1.6250	1.6235	1.6214	1.6186	1.6152	1.6109	1.6057	1.5996
0.96	1.6268	1.6261	1.6250	1.6235	1.6214	1.6186	1.6152	1.6109	1.6057	1.5996
0.98	1.6268	1.6261	1.6250	1.6235	1.6214	1.6186	1.6152	1.6109	1.6057	1.5996
1.00	1.6268	1.6261	1.6250	1.6235	1.6214	1.6186	1.6152	1.6109	1.6057	1.5996

(continued)

TABLE C.4c

(continued)

a_U	0.42	0.44	0.46	0.48	0.50	0.52	0.54	0.56	0.58	0.60
0.44	1.3808									
0.46	1.4234	1.3870								
0.48	1.4544	1.4286	1.3912							
0.50	1.4790	1.4585	1.4316	1.3932						
0.52	1.4993	1.4821	1.4605	1.4327	1.3932					
0.54	1.5164	1.5013	1.4831	1.4605	1.4316	1.3912				
0.56	1.5308	1.5174	1.5013	1.4821	1.4585	1.4286	1.3870			
0.58	1.5430	1.5308	1.5164	1.4993	1.4790	1.4544	1.4234	1.3808		
0.60	1.5533	1.5420	1.5288	1.5134	1.4953	1.4739	1.4482	1.4161	1.3724	
0.62	1.5620	1.5514	1.5391	1.5249	1.5083	1.4892	1.4667	1.4399	1.4067	1.3619
0.64	1.5692	1.5592	1.5476	1.5343	1.5189	1.5013	1.4810	1.4574	1.4294	1.3950
0.66	1.5751	1.5656	1.5545	1.5419	1.5274	1.5109	1.4922	1.4707	1.4459	1.4167
0.68	1.5799	1.5707	1.5601	1.5480	1.5342	1.5186	1.5009	1.4810	1.4583	1.4322
0.70	1.5837	1.5747	1.5644	1.5527	1.5395	1.5245	1.5077	1.4888	1.4676	1.4436
0.72	1.5865	1.5778	1.5678	1.5564	1.5435	1.5291	1.5129	1.4948	1.4746	1.4520
0.74	1.5887	1.5801	1.5703	1.5591	1.5465	1.5324	1.5167	1.4992	1.4797	1.4582
0.76	1.5902	1.5817	1.5720	1.5610	1.5487	1.5348	1.5194	1.5023	1.4834	1.4625
0.78	1.5912	1.5828	1.5732	1.5623	1.5501	1.5365	1.5213	1.5045	1.4859	1.4655
0.80	1.5918	1.5835	1.5739	1.5631	1.5510	1.5375	1.5225	1.5058	1.4875	1.4674
0.82	1.5921	1.5838	1.5744	1.5636	1.5515	1.5381	1.5231	1.5066	1.4885	1.4686
0.84	1.5923	1.5840	1.5746	1.5638	1.5518	1.5384	1.5235	1.5071	1.4890	1.4692
0.86	1.5924	1.5841	1.5746	1.5639	1.5519	1.5385	1.5236	1.5072	1.4892	1.4695
0.88	1.5924	1.5841	1.5747	1.5640	1.5519	1.5385	1.5237	1.5073	1.4893	1.4696
0.90	1.5924	1.5841	1.5747	1.5640	1.5520	1.5386	1.5237	1.5073	1.4893	1.4696
0.92	1.5924	1.5841	1.5747	1.5640	1.5520	1.5386	1.5237	1.5073	1.4893	1.4696
0.94	1.5924	1.5841	1.5747	1.5640	1.5520	1.5386	1.5237	1.5073	1.4893	1.4696
0.96	1.5924	1.5841	1.5747	1.5640	1.5520	1.5386	1.5237	1.5073	1.4893	1.4696
0.98	1.5924	1.5841	1.5747	1.5640	1.5520	1.5386	1.5237	1.5073	1.4893	1.4696
1.00	1.5924	1.5841	1.5747	1.5640	1.5520	1.5386	1.5237	1.5073	1.4893	1.4696

TABLE C.5

Survival Function of the Supremum of the Absolute Value of a Standard Brownian Motion Process over the Range 0 to 1

$\Pr[\sup |B(t)| > x]$	x	$\Pr[\sup |B(t)| > x]$	x	$\Pr[\sup |B(t)| > x]$	x
0.01	2.8070	0.34	1.3721	0.67	0.9559
0.02	2.5758	0.35	1.3562	0.68	0.9452
0.03	2.4324	0.36	1.3406	0.69	0.9345
0.04	2.3263	0.37	1.3253	0.70	0.9238
0.05	2.2414	0.38	1.3103	0.71	0.9132
0.06	2.1701	0.39	1.2956	0.72	0.9025
0.07	2.1084	0.40	1.2812	0.73	0.8919
0.08	2.0537	0.41	1.2670	0.74	0.8812
0.09	2.0047	0.42	1.2531	0.75	0.8706
0.10	1.9600	0.43	1.2394	0.76	0.8598
0.11	1.9189	0.44	1.2259	0.77	0.8491
0.12	1.8808	0.45	1.2126	0.78	0.8383
0.13	1.8453	0.46	1.1995	0.79	0.8274
0.14	1.8119	0.47	1.1866	0.80	0.8164
0.15	1.7805	0.48	1.1739	0.81	0.8053
0.16	1.7507	0.49	1.1614	0.82	0.7941
0.17	1.7224	0.50	1.1490	0.83	0.7828
0.18	1.6954	0.51	1.1367	0.84	0.7712
0.19	1.6696	0.52	1.1246	0.85	0.7595
0.20	1.6448	0.53	1.1127	0.86	0.7475
0.21	1.6211	0.54	1.1009	0.87	0.7353
0.22	1.5982	0.55	1.0892	0.88	0.7227
0.23	1.5761	0.56	1.0776	0.89	0.7098
0.24	1.5548	0.57	1.0661	0.90	0.6964
0.25	1.5341	0.58	1.0547	0.91	0.6824
0.26	1.5141	0.59	1.0434	0.92	0.6677
0.27	1.4946	0.60	1.0322	0.93	0.6521
0.28	1.4758	0.61	1.0211	0.94	0.6355
0.29	1.4574	0.62	1.0101	0.95	0.6173
0.30	1.4395	0.63	0.9992	0.96	0.5971
0.31	1.4220	0.64	0.9883	0.97	0.5737
0.32	1.4050	0.65	0.9774	0.98	0.5450
0.33	1.3883	0.66	0.9666	0.99	0.5045

TABLE C.6
Survival Function of $W = \int_0^1 [B(t)]^2 dt$*, where $B(t)$ is a Standard Brownian Motion*

W	0.00	0.01	0.02	0.03	0.04	0.05	0.06	0.07	0.08	0.09
0.00	1.0000	1.0000	.9994	.9945	.9824	.9642	.9417	.9169	.8910	.8648
0.10	.8390	.8138	.7894	.7659	.7434	.7218	.7012	.6814	.6626	.6445
0.20	.6273	.6108	.5949	.5798	.5652	.5513	.5378	.5249	.5125	.5006
0.30	.4890	.4779	.4672	.4568	.4468	.4371	.4278	.4187	.4099	.4014
0.40	.3931	.3851	.3773	.3697	.3623	.3552	.3482	.3414	.3348	.3284
0.50	.3222	.3161	.3101	.3043	.2987	.2932	.2878	.2825	.2774	.2724
0.60	.2675	.2627	.2580	.2534	.2489	.2446	.2403	.2361	.2320	.2280
0.70	.2240	.2202	.2164	.2127	.2091	.2056	.2021	.1987	.1953	.1921
0.80	.1889	.1857	.1826	.1796	.1767	.1738	.1709	.1681	.1654	.1627
0.90	.1600	.1574	.1549	.1524	.1499	.1475	.1451	.1428	.1405	.1383
1.00	.1361	.1339	.1318	.1297	.1277	.1257	.1237	.1218	.1198	.1180
1.10	.1161	.1143	.1125	.1108	.1091	.1074	.1057	.1041	.1025	.1009
1.20	.0994	.0978	.0963	.0949	.0934	.0920	.0906	.0892	.0878	.0865
1.30	.0852	.0839	.0826	.0814	.0802	.0789	.0778	.0766	.0754	.0743
1.40	.0732	.0721	.0710	.0700	.0689	.0679	.0669	.0659	.0649	.0639
1.50	.0630	.0621	.0611	.0602	.0593	.0585	.0576	.0568	.0559	.0551
1.60	.0543	.0535	.0527	.0519	.0512	.0504	.0497	.0490	.0482	.0475
1.70	.0469	.0462	.0455	.0448	.0442	.0435	.0429	.0423	.0417	.0411
1.80	.0405	.0399	.0393	.0388	.0382	.0376	.0371	.0366	.0360	.0355
1.90	.0350	.0345	.0340	.0335	.0330	.0326	.0321	.0317	.0312	.0308
2.00	.0303	.0299	.0295	.0290	.0286	.0282	.0278	.0274	.0270	.0266
2.10	.0263	.0259	.0255	.0252	.0248	.0245	.0241	.0238	.0234	.0231
2.20	.0228	.0225	.0221	.0218	.0215	.0212	.0209	.0206	.0203	.0201
2.30	.0198	.0195	.0192	.0190	.0187	.0184	.0182	.0179	.0177	.0174
2.40	.0172	.0169	.0167	.0165	.0162	.0160	.0158	.0156	.0153	.0151
2.50	.0149	.0147	.0145	.0143	.0141	.0139	.0137	.0135	.0133	.0132
2.60	.0130	.0128	.0126	.0124	.0123	.0121	.0119	.0118	.0116	.0114
2.70	.0113	.0111	.0110	.0108	.0107	.0105	.0104	.0102	.0101	.0100
2.80	.0098	.0097	.0096	.0094	.0093	.0092	.0090	.0089	.0088	.0087
2.90	.0086	.0084	.0083	.0082	.0081	.0080	.0079	.0078	.0077	.0076
3.00	.0075	.0074	.0073	.0072	.0071	.0070	.0069	.0068	.0067	.0066
3.10	.0065	.0064	.0063	.0062	.0062	.0061	.0060	.0059	.0058	.0057
3.20	.0057	.0056	.0055	.0054	.0054	.0053	.0052	.0052	.0051	.0050
3.30	.0049	.0049	.0048	.0047	.0047	.0046	.0046	.0045	.0044	.0044
3.40	.0043	.0043	.0042	.0041	.0041	.0040	.0040	.0039	.0039	.0038
3.50	.0038	.0037	.0037	.0036	.0036	.0035	.0035	.0034	.0034	.0033
3.60	.0033	.0032	.0032	.0032	.0031	.0031	.0030	.0030	.0030	.0029
3.70	.0029	.0028	.0028	.0028	.0027	.0027	.0026	.0026	.0026	.0025
3.80	.0025	.0025	.0024	.0024	.0024	.0023	.0023	.0023	.0023	.0022
3.90	.0022	.0022	.0021	.0021	.0021	.0021	.0020	.0020	.0020	.0019

Selected upper percentage points: $W_{0.01} = 2.787$, $W_{0.025} = 2.135$, $W_{0.05} = 1.656$, $W_{0.10} = 1.196$.

TABLE C.7
Upper Percentiles of $R = \int_0^k |B^o(u)|du$, where $B^o(u)$ is a Brownian Bridge

$P(R > r)$	$k = 0.1$	$k = 0.2$	$k = 0.3$	$k = 0.4$	$k = 0.5$	$k = 0.6$	$k = 0.7$	$k = 0.8$	$k = 0.9$	$k = 1.0$
0.99	0.0003	0.0011	0.0026	0.0044	0.0068	0.0093	0.0124	0.0155	0.0184	0.0200
0.98	0.0004	0.0014	0.0032	0.0054	0.0083	0.0115	0.0151	0.0189	0.0220	0.0237
0.97	0.0004	0.0016	0.0036	0.0062	0.0094	0.0130	0.0172	0.0213	0.0249	0.0267
0.96	0.0005	0.0018	0.0040	0.0069	0.0105	0.0144	0.0189	0.0235	0.0274	0.0292
0.95	0.0005	0.0020	0.0044	0.0075	0.0114	0.0157	0.0205	0.0253	0.0295	0.0314
0.94	0.0005	0.0021	0.0047	0.0081	0.0123	0.0168	0.0220	0.0271	0.0315	0.0334
0.93	0.0006	0.0023	0.0051	0.0086	0.0131	0.0180	0.0234	0.0288	0.0334	0.0354
0.92	0.0006	0.0024	0.0054	0.0092	0.0138	0.0191	0.0248	0.0303	0.0352	0.0372
0.91	0.0007	0.0026	0.0057	0.0097	0.0147	0.0201	0.0261	0.0319	0.0369	0.0390
0.90	0.0007	0.0027	0.0060	0.0102	0.0154	0.0212	0.0273	0.0335	0.0388	0.0408
0.89	0.0007	0.0029	0.0063	0.0107	0.0162	0.0221	0.0286	0.0349	0.0404	0.0425
0.88	0.0008	0.0030	0.0066	0.0112	0.0169	0.0231	0.0299	0.0364	0.0420	0.0442
0.87	0.0008	0.0031	0.0068	0.0117	0.0176	0.0241	0.0311	0.0379	0.0437	0.0459
0.86	0.0008	0.0033	0.0071	0.0122	0.0184	0.0251	0.0322	0.0394	0.0452	0.0475
0.85	0.0009	0.0034	0.0074	0.0127	0.0191	0.0261	0.0334	0.0408	0.0468	0.0491
0.84	0.0009	0.0036	0.0077	0.0132	0.0198	0.0271	0.0346	0.0422	0.0484	0.0508
0.83	0.0010	0.0037	0.0080	0.0137	0.0205	0.0280	0.0358	0.0436	0.0500	0.0524
0.82	0.0010	0.0039	0.0083	0.0142	0.0213	0.0290	0.0370	0.0449	0.0515	0.0539
0.81	0.0010	0.0040	0.0086	0.0147	0.0220	0.0299	0.0382	0.0463	0.0532	0.0556
0.80	0.0011	0.0041	0.0089	0.0152	0.0228	0.0309	0.0394	0.0477	0.0546	0.0572
0.79	0.0011	0.0043	0.0092	0.0157	0.0236	0.0319	0.0407	0.0491	0.0562	0.0588
0.78	0.0011	0.0044	0.0096	0.0162	0.0243	0.0330	0.0419	0.0506	0.0578	0.0604
0.77	0.0012	0.0046	0.0099	0.0168	0.0250	0.0340	0.0431	0.0520	0.0594	0.0621
0.76	0.0012	0.0048	0.0102	0.0173	0.0258	0.0350	0.0443	0.0534	0.0610	0.0636
0.75	0.0013	0.0049	0.0105	0.0179	0.0266	0.0360	0.0456	0.0549	0.0626	0.0652
0.74	0.0013	0.0051	0.0109	0.0185	0.0274	0.0370	0.0468	0.0563	0.0642	0.0668
0.73	0.0014	0.0052	0.0112	0.0190	0.0283	0.0380	0.0481	0.0578	0.0657	0.0684
0.72	0.0014	0.0054	0.0116	0.0196	0.0291	0.0391	0.0494	0.0593	0.0673	0.0701
0.71	0.0014	0.0056	0.0119	0.0202	0.0299	0.0402	0.0508	0.0607	0.0690	0.0718
0.70	0.0015	0.0057	0.0123	0.0208	0.0307	0.0413	0.0521	0.0623	0.0707	0.0735
0.69	0.0015	0.0059	0.0127	0.0213	0.0316	0.0424	0.0535	0.0637	0.0723	0.0752
0.68	0.0016	0.0061	0.0131	0.0219	0.0324	0.0435	0.0549	0.0653	0.0740	0.0770
0.67	0.0016	0.0063	0.0135	0.0226	0.0333	0.0446	0.0562	0.0669	0.0757	0.0787
0.66	0.0017	0.0065	0.0138	0.0232	0.0342	0.0458	0.0576	0.0686	0.0775	0.0804
0.65	0.0017	0.0067	0.0143	0.0238	0.0351	0.0470	0.0591	0.0703	0.0793	0.0822
0.64	0.0018	0.0069	0.0147	0.0245	0.0360	0.0482	0.0605	0.0719	0.0812	0.0840
0.63	0.0018	0.0071	0.0151	0.0252	0.0370	0.0495	0.0620	0.0736	0.0831	0.0859
0.62	0.0019	0.0073	0.0155	0.0259	0.0380	0.0507	0.0636	0.0753	0.0850	0.0876
0.61	0.0019	0.0075	0.0159	0.0266	0.0390	0.0520	0.0651	0.0770	0.0869	0.0895
0.60	0.0020	0.0077	0.0164	0.0273	0.0401	0.0533	0.0668	0.0789	0.0889	0.0913
0.59	0.0021	0.0080	0.0168	0.0281	0.0412	0.0547	0.0684	0.0807	0.0909	0.0934

(continued)

TABLE C.7
(continued)

P(R > r)	k = 0.1	k = 0.2	k = 0.3	k = 0.4	k = 0.5	k = 0.6	k = 0.7	k = 0.8	k = 0.9	k = 1.0
0.58	0.0021	0.0082	0.0173	0.0288	0.0423	0.0561	0.0700	0.0825	0.0929	0.0955
0.57	0.0022	0.0084	0.0178	0.0296	0.0434	0.0575	0.0718	0.0845	0.0950	0.0974
0.56	0.0023	0.0087	0.0182	0.0304	0.0445	0.0590	0.0735	0.0865	0.0973	0.0997
0.55	0.0023	0.0089	0.0188	0.0312	0.0456	0.0605	0.0752	0.0885	0.0994	0.1018
0.54	0.0024	0.0092	0.0193	0.0321	0.0468	0.0620	0.0771	0.0905	0.1016	0.1040
0.53	0.0025	0.0094	0.0198	0.0330	0.0480	0.0636	0.0789	0.0925	0.1038	0.1063
0.52	0.0025	0.0097	0.0204	0.0338	0.0493	0.0652	0.0808	0.0947	0.1061	0.1087
0.51	0.0026	0.0100	0.0209	0.0348	0.0506	0.0668	0.0828	0.0969	0.1086	0.1111
0.50	0.0027	0.0102	0.0215	0.0357	0.0519	0.0684	0.0847	0.0991	0.1110	0.1135
0.49	0.0028	0.0105	0.0221	0.0367	0.0533	0.0702	0.0868	0.1013	0.1134	0.1159
0.48	0.0028	0.0108	0.0228	0.0377	0.0547	0.0719	0.0888	0.1035	0.1159	0.1185
0.47	0.0029	0.0111	0.0234	0.0387	0.0561	0.0737	0.0911	0.1060	0.1185	0.1210
0.46	0.0030	0.0115	0.0241	0.0398	0.0575	0.0756	0.0934	0.1084	0.1213	0.1237
0.45	0.0031	0.0118	0.0247	0.0409	0.0590	0.0776	0.0957	0.1110	0.1239	0.1263
0.44	0.0032	0.0121	0.0254	0.0421	0.0605	0.0795	0.0980	0.1136	0.1265	0.1291
0.43	0.0033	0.0125	0.0262	0.0432	0.0621	0.0816	0.1004	0.1163	0.1292	0.1320
0.42	0.0034	0.0128	0.0269	0.0444	0.0639	0.0838	0.1031	0.1191	0.1321	0.1348
0.41	0.0035	0.0132	0.0277	0.0456	0.0656	0.0860	0.1055	0.1219	0.1350	0.1380
0.40	0.0036	0.0136	0.0285	0.0468	0.0674	0.0883	0.1081	0.1248	0.1383	0.1413
0.39	0.0037	0.0140	0.0293	0.0482	0.0692	0.0907	0.1108	0.1279	0.1416	0.1445
0.38	0.0038	0.0144	0.0301	0.0496	0.0710	0.0932	0.1138	0.1310	0.1449	0.1481
0.37	0.0039	0.0148	0.0310	0.0510	0.0731	0.0958	0.1166	0.1343	0.1484	0.1516
0.36	0.0041	0.0153	0.0320	0.0525	0.0752	0.0983	0.1196	0.1378	0.1519	0.1552
0.35	0.0042	0.0158	0.0330	0.0541	0.0772	0.1012	0.1228	0.1412	0.1555	0.1587
0.34	0.0043	0.0163	0.0340	0.0556	0.0795	0.1040	0.1259	0.1446	0.1593	0.1623
0.33	0.0045	0.0168	0.0351	0.0573	0.0817	0.1068	0.1291	0.1482	0.1632	0.1663
0.32	0.0046	0.0173	0.0362	0.0590	0.0841	0.1097	0.1327	0.1523	0.1671	0.1705
0.31	0.0048	0.0179	0.0373	0.0608	0.0867	0.1129	0.1364	0.1565	0.1713	0.1746
0.30	0.0049	0.0185	0.0384	0.0627	0.0892	0.1160	0.1404	0.1606	0.1759	0.1790
0.29	0.0051	0.0191	0.0396	0.0646	0.0918	0.1195	0.1444	0.1650	0.1803	0.1837
0.28	0.0052	0.0198	0.0409	0.0666	0.0947	0.1231	0.1487	0.1698	0.1849	0.1886
0.27	0.0054	0.0204	0.0423	0.0687	0.0977	0.1267	0.1527	0.1747	0.1900	0.1934
0.26	0.0056	0.0211	0.0437	0.0709	0.1008	0.1305	0.1572	0.1796	0.1949	0.1984
0.25	0.0058	0.0218	0.0451	0.0731	0.1042	0.1346	0.1620	0.1846	0.2006	0.2037
0.24	0.0060	0.0226	0.0467	0.0756	0.1076	0.1388	0.1672	0.1900	0.2066	0.2091
0.23	0.0063	0.0234	0.0482	0.0781	0.1112	0.1434	0.1726	0.1956	0.2127	0.2150
0.22	0.0065	0.0242	0.0500	0.0808	0.1152	0.1483	0.1779	0.2015	0.2192	0.2216
0.21	0.0067	0.0251	0.0518	0.0837	0.1193	0.1532	0.1839	0.2082	0.2260	0.2285
0.20	0.0070	0.0260	0.0539	0.0868	0.1235	0.1585	0.1901	0.2146	0.2332	0.2356
0.19	0.0073	0.0271	0.0560	0.0900	0.1278	0.1645	0.1967	0.2219	0.2408	0.2427
0.18	0.0076	0.0281	0.0582	0.0936	0.1326	0.1702	0.2039	0.2295	0.2488	0.2505
0.17	0.0079	0.0293	0.0605	0.0974	0.1379	0.1764	0.2113	0.2378	0.2572	0.2589
0.16	0.0082	0.0306	0.0629	0.1013	0.1433	0.1834	0.2194	0.2462	0.2660	0.2675
0.15	0.0086	0.0318	0.0655	0.1055	0.1489	0.1909	0.2280	0.2553	0.2756	0.2776
0.14	0.0090	0.0333	0.0685	0.1104	0.1547	0.1989	0.2372	0.2652	0.2861	0.2883

TABLE C.7
(continued)

P(R > r)	k = 0.1	k = 0.2	k = 0.3	k = 0.4	k = 0.5	k = 0.6	k = 0.7	k = 0.8	k = 0.9	k = 1.0
0.13	0.0094	0.0348	0.0717	0.1155	0.1618	0.2073	0.2469	0.2759	0.2975	0.3002
0.12	0.0099	0.0365	0.0750	0.1209	0.1694	0.2169	0.2580	0.2881	0.3100	0.3122
0.11	0.0104	0.0383	0.0787	0.1272	0.1779	0.2282	0.2697	0.3015	0.3238	0.3253
0.10	0.0110	0.0404	0.0830	0.1339	0.1873	0.2403	0.2837	0.3158	0.3388	0.3406
0.09	0.0116	0.0427	0.0874	0.1414	0.1975	0.2531	0.2979	0.3323	0.3559	0.3573
0.08	0.0123	0.0454	0.0929	0.1497	0.2092	0.2679	0.3150	0.3514	0.3743	0.3774
0.07	0.0131	0.0485	0.0986	0.1589	0.2224	0.2840	0.3359	0.3723	0.3948	0.3994
0.06	0.0140	0.0520	0.1056	0.1699	0.2373	0.3038	0.3579	0.3963	0.4201	0.4239
0.05	0.0151	0.0561	0.1142	0.1823	0.2555	0.3273	0.3848	0.4259	0.4506	0.4543
0.04	0.0165	0.0612	0.1245	0.1990	0.2766	0.3567	0.4157	0.4609	0.4859	0.4941
0.03	0.0184	0.0676	0.1377	0.2206	0.3063	0.3917	0.4613	0.5074	0.5381	0.5419
0.02	0.0210	0.0768	0.1577	0.2517	0.3487	0.4473	0.5247	0.5753	0.6099	0.6106
0.01	0.0253	0.0935	0.1938	0.3098	0.4241	0.5390	0.6386	0.6848	0.7339	0.7379

D
Data on 137 Bone Marrow Transplant Patients

TABLE D.1

Data on 137 Bone Marrow Transplant Patients

g—Disease group
    1-ALL
    2-AML low-risk
    3-AML high-risk
T_1—Time (in days) to death or on study time
T_2—Disease-Free survival time (time to relapse, death or end of study)
δ_1—Death indicator
    1-Dead   0-Alive
δ_2—Relapse indicator
    1-Relapsed   0-Disease-Free
δ_3—Disease-Free survival indicator
    1-Dead or relapsed   0-Alive disease-free
T_A—Time (in days) to acute graft-versus-host disease
δ_A—Acute graft-versus-host disease indicator
      1-Developed acute graft-versus-host disease
      0-Never developed acute graft-versus-host disease
T_C—Time (in days) to chronic graft-versus-host disease
δ_C—Chronic graft-versus-host disease indicator
      1-Developed chronic graft-versus-host disease
      0-Never developed chronic graft-versus-host disease
T_P—Time (in days) to return of platelets to normal levels
δ_P—Platelet recovery indicator
      1-Platelets returned to normal levels
      0-Platelets never returned to normal levels
Z_1—Patient age in years
Z_2—Donor age in years
Z_3—Patient sex
    1-Male   0-Female
Z_4—Donor Sex
    1-Male   0-Female
Z_5—Patient CMV status
    1-CMV positive   0-CMV negative
Z_6—Donor CMV status
    1-CMV positive   0-CMV negative
Z_7—Waiting time to transplant in days
Z_8—FAB
    1-FAB Grade 4 Or 5 and AML   0-Otherwise
Z_9—Hospital
    1-The Ohio State University   2-Alfred
    3-St. Vincent   4-Hahnemann
Z_{10}—MTX used as a graft-versus-host-prophylactic
    1-Yes   0-No

TABLE D.1
(continued)

g	T_1	T_2	δ_1	δ_2	δ_3	T_A	δ_A	T_C	δ_C	T_P	δ_P	Z_1	Z_2	Z_3	Z_4	Z_5	Z_6	Z_7	Z_8	Z_9	Z_{10}
1	2081	2081	0	0	0	67	1	121	1	13	1	26	33	1	0	1	1	98	0	1	0
1	1602	1602	0	0	0	1602	0	139	1	18	1	21	37	1	1	0	0	1720	0	1	0
1	1496	1496	0	0	0	1496	0	307	1	12	1	26	35	1	1	1	0	127	0	1	0
1	1462	1462	0	0	0	70	1	95	1	13	1	17	21	0	1	0	0	168	0	1	0
1	1433	1433	0	0	0	1433	0	236	1	12	1	32	36	1	1	1	1	93	0	1	0
1	1377	1377	0	0	0	1377	0	123	1	12	1	22	31	1	1	1	1	2187	0	1	0
1	1330	1330	0	0	0	1330	0	96	1	17	1	20	17	1	0	1	1	1006	0	1	0
1	996	996	0	0	0	72	1	121	1	12	1	22	24	1	0	0	0	1319	0	1	0
1	226	226	0	0	0	226	0	226	0	10	1	18	21	0	1	0	0	208	0	1	0
1	1199	1199	0	0	0	1199	0	91	1	29	1	24	40	1	1	0	1	174	0	3	1
1	1111	1111	0	0	0	1111	0	1111	0	22	1	19	28	1	1	0	1	236	0	3	1
1	530	530	0	0	0	38	1	84	1	34	1	17	28	1	1	0	0	151	0	3	1
1	1182	1182	0	0	0	1182	0	112	1	22	1	24	23	0	0	0	1	203	0	2	1
1	1167	1167	0	0	0	39	1	487	1	1167	0	27	22	0	1	1	1	191	0	2	1
1	418	418	1	0	1	418	0	220	1	21	1	18	14	1	1	0	0	110	0	1	0
1	417	383	1	1	1	417	0	417	0	16	1	15	20	1	1	0	0	824	0	1	0
1	276	276	1	0	1	276	0	81	1	21	1	18	5	0	0	0	0	146	0	1	0
1	156	104	1	1	1	28	1	156	0	20	1	20	33	1	1	0	1	85	0	1	0
1	781	609	1	1	1	781	0	781	0	26	1	27	27	1	0	1	1	187	0	1	0
1	172	172	1	0	1	22	1	172	0	37	1	40	37	0	0	0	1	129	0	1	0
1	487	487	1	0	1	487	0	76	1	22	1	22	20	1	1	0	0	128	0	1	0
1	716	662	1	1	1	716	0	716	0	17	1	28	32	1	1	0	0	84	0	1	0
1	194	194	1	0	1	194	0	94	1	25	1	26	32	0	1	0	0	329	0	1	0
1	371	230	1	1	1	371	0	184	1	9	1	39	31	0	1	0	1	147	0	1	0
1	526	526	1	0	1	526	0	121	1	11	1	15	20	1	1	0	0	943	0	1	0
1	122	122	1	0	1	88	1	122	0	13	1	20	26	1	0	0	1	2616	0	1	0
1	1279	129	1	1	1	1279	0	1279	0	22	1	17	20	0	0	0	0	937	0	3	1
1	110	74	1	1	1	110	0	110	0	49	1	28	25	1	0	1	0	303	0	3	1
1	243	122	1	1	1	243	0	243	0	23	1	37	38	0	1	1	1	170	0	3	1
1	86	86	1	0	1	86	0	86	0	86	0	17	26	1	0	1	0	239	0	3	1
1	466	466	1	0	1	466	0	119	1	100	1	15	18	1	1	0	0	508	0	3	1
1	262	192	1	1	1	10	1	84	1	59	1	29	32	1	1	1	0	74	0	3	1
1	162	109	1	1	1	162	0	162	0	40	1	36	43	1	1	1	0	393	0	2	1
1	262	55	1	1	1	262	0	262	0	24	1	23	16	0	1	1	1	331	0	2	1
1	1	1	1	0	1	1	0	1	0	1	0	42	48	1	1	0	0	196	0	2	1
1	107	107	1	0	1	107	0	107	0	107	0	30	19	1	1	1	1	178	0	2	1
1	269	110	1	1	1	269	0	120	1	27	1	29	20	0	1	1	1	361	0	2	1
1	350	332	0	1	1	350	0	350	0	33	1	22	20	1	0	0	0	834	0	2	1
2	2569	2569	0	0	0	2569	0	2569	0	21	1	19	13	1	1	1	0	270	1	1	0
2	2506	2506	0	0	0	2506	0	2506	0	17	1	31	34	1	1	0	0	60	0	1	0
2	2409	2409	0	0	0	2409	0	2409	0	16	1	35	31	1	1	1	1	120	0	1	0
2	2218	2218	0	0	0	2218	0	2218	0	11	1	16	16	1	1	1	0	60	1	1	0
2	1857	1857	0	0	0	1857	0	260	1	15	1	29	35	0	0	1	0	90	0	1	0
2	1829	1829	0	0	0	1829	0	1829	0	19	1	19	18	1	1	1	0	210	0	1	0
2	1562	1562	0	0	0	1562	0	1562	0	18	1	26	30	1	1	1	1	90	0	1	0
2	1470	1470	0	0	0	1470	0	180	1	14	1	27	34	1	1	0	1	240	0	1	0

(continued)

TABLE D.1
(continued)

g	T_1	T_2	δ_1	δ_2	δ_3	T_A	δ_A	T_C	δ_C	T_P	δ_P	Z_1	Z_2	Z_3	Z_4	Z_5	Z_6	Z_7	Z_8	Z_9	Z_{10}
2	1363	1363	0	0	0	1363	0	200	1	12	1	13	24	1	1	1	0	90	0	1	0
2	1030	1030	0	0	0	1030	0	210	1	14	1	25	29	0	0	0	0	210	0	1	0
2	860	860	0	0	0	860	0	860	0	15	1	25	31	0	1	0	1	180	0	1	0
2	1258	1258	0	0	0	1258	0	120	1	66	1	30	16	0	1	1	0	180	0	2	1
2	2246	2246	0	0	0	52	1	380	1	15	1	45	39	0	0	0	0	105	0	4	0
2	1870	1870	0	0	0	1870	0	230	1	16	1	33	30	0	0	1	1	225	0	4	0
2	1799	1799	0	0	0	1799	0	140	1	12	1	32	23	1	0	0	0	120	0	4	0
2	1709	1709	0	0	0	20	1	348	1	19	1	23	28	0	1	1	0	90	1	4	0
2	1674	1674	0	0	0	1674	0	1674	0	24	1	37	34	1	1	0	0	60	1	4	0
2	1568	1568	0	0	0	1568	0	1568	0	14	1	15	19	1	0	0	0	90	0	4	0
2	1527	1527	0	0	0	1527	0	1527	0	13	1	22	12	0	1	0	1	450	1	4	0
2	1324	1324	0	0	0	25	1	1324	0	15	1	46	31	1	1	1	1	75	0	4	0
2	957	957	0	0	0	957	0	957	0	69	1	18	17	1	1	0	0	90	0	4	0
2	932	932	0	0	0	29	1	932	0	7	1	27	30	0	0	0	0	60	1	4	0
2	847	847	0	0	0	847	0	847	0	16	1	28	29	1	1	0	0	75	0	4	0
2	848	848	0	0	0	848	0	155	1	16	1	23	26	1	1	0	0	180	0	4	0
2	1850	1850	0	0	0	1850	0	1850	0	9	1	37	36	0	0	0	1	180	0	3	1
2	1843	1843	0	0	0	1843	0	1843	0	19	1	34	32	0	0	1	1	270	0	3	1
2	1535	1535	0	0	0	1535	0	1535	0	21	1	35	32	0	1	0	0	180	1	3	1
2	1447	1447	0	0	0	1447	0	220	1	24	1	33	28	0	1	1	1	150	0	3	1
2	1384	1384	0	0	0	1384	0	200	1	19	1	21	18	0	0	0	0	120	0	3	1
2	414	414	1	0	1	414	0	414	0	27	1	21	15	1	1	0	1	120	1	1	0
2	2204	2204	1	0	1	2204	0	2204	0	12	1	25	19	0	0	0	1	60	0	1	0
2	1063	1063	1	0	1	1063	0	240	1	16	1	50	38	1	0	1	0	270	1	1	0
2	481	481	1	0	1	30	1	120	1	24	1	35	36	1	0	1	1	90	1	1	0
2	105	105	1	0	1	21	1	105	0	15	1	37	34	1	0	1	1	120	0	1	0
2	641	641	1	0	1	641	0	641	0	11	1	26	24	1	1	0	0	90	0	1	0
2	390	390	1	0	1	390	0	390	0	11	1	50	48	1	1	0	0	120	0	1	0
2	288	288	1	0	1	18	1	100	1	288	0	45	43	1	1	1	1	90	0	1	0
2	522	421	1	1	1	25	1	140	1	20	1	28	30	1	1	0	1	90	1	1	0
2	79	79	1	0	1	16	1	79	0	79	0	43	43	0	0	0	0	90	0	1	0
2	1156	748	1	1	1	1156	0	180	1	18	1	14	19	1	0	0	0	60	0	1	0
2	583	486	1	1	1	583	0	583	0	11	1	17	14	0	1	0	0	120	0	1	0
2	48	48	1	0	1	48	0	48	0	14	1	32	33	0	1	1	0	150	1	1	0
2	431	272	1	1	1	431	0	431	0	12	1	30	23	0	1	1	0	120	1	1	0
2	1074	1074	1	0	1	1074	0	120	1	19	1	30	32	1	1	1	0	150	1	1	0
2	393	381	1	1	1	393	0	100	1	16	1	33	28	0	0	0	0	120	1	1	0
2	10	10	1	0	1	10	0	10	0	10	0	34	54	1	0	1	1	240	0	2	1
2	53	53	1	0	1	53	0	53	0	53	0	33	41	0	1	1	1	180	0	2	1
2	80	80	1	0	1	10	1	80	0	80	0	30	35	0	0	0	1	150	0	2	1
2	35	35	1	0	1	35	0	35	0	35	0	23	25	0	1	1	1	150	0	2	1
2	1499	248	0	1	1	1499	0	1499	0	9	1	35	18	1	1	0	1	30	0	4	0
2	704	704	1	0	1	36	1	155	1	18	1	29	21	0	1	1	0	105	0	4	0
2	653	211	1	1	1	653	0	653	0	23	1	23	16	1	0	0	0	90	1	4	0
2	222	219	1	1	1	222	0	123	1	52	1	28	30	1	1	1	1	120	1	3	1
2	1356	606	0	1	1	1356	0	1356	0	14	1	33	22	1	1	1	0	210	1	3	1

TABLE D.1

(continued)

g	T_1	T_2	δ_1	δ_2	δ_3	T_A	δ_A	T_C	δ_C	T_P	δ_P	Z_1	Z_2	Z_3	Z_4	Z_5	Z_6	Z_7	Z_8	Z_9	Z_{10}
3	2640	2640	0	0	0	2640	0	2640	0	22	1	18	23	1	1	0	0	750	0	1	0
3	2430	2430	0	0	0	2430	0	2430	0	14	1	29	26	1	1	0	1	24	0	1	0
3	2252	2252	0	0	0	2252	0	150	1	17	1	35	31	1	0	0	0	120	0	1	0
3	2140	2140	0	0	0	2140	0	220	1	18	1	27	17	1	1	1	1	210	0	1	0
3	2133	2133	0	0	0	2133	0	250	1	17	1	36	39	0	1	0	0	240	0	1	0
3	1238	1238	0	0	0	1238	0	250	1	18	1	24	28	1	0	1	1	240	0	1	0
3	1631	1631	0	0	0	1631	0	150	1	40	1	27	21	1	0	1	0	690	1	2	1
3	2024	2024	0	0	0	2024	0	180	1	16	1	35	41	0	1	0	0	105	1	4	0
3	1345	1345	0	0	0	32	1	360	1	14	1	50	36	1	1	1	1	120	0	4	0
3	1136	1136	0	0	0	1136	0	140	1	15	1	47	27	1	0	1	0	900	0	3	1
3	845	845	0	0	0	845	0	845	0	20	1	40	39	0	0	1	1	210	1	3	1
3	491	422	1	1	1	491	0	180	1	491	0	22	21	0	0	0	0	210	1	1	0
3	162	162	1	0	1	162	0	162	0	13	1	22	23	1	0	0	1	300	0	1	0
3	1298	84	1	1	1	1298	0	1298	0	1298	0	8	2	0	0	1	0	105	1	1	0
3	121	100	1	1	1	28	1	121	0	65	1	39	48	1	1	1	1	210	1	1	0
3	2	2	1	0	1	2	0	2	0	2	0	20	19	1	1	0	0	75	1	1	0
3	62	47	1	1	1	62	0	62	0	11	1	27	25	1	1	0	0	90	1	1	0
3	265	242	1	1	1	265	0	210	1	14	1	32	32	1	0	0	0	180	1	1	0
3	547	456	1	1	1	547	0	130	1	24	1	31	28	1	0	1	1	630	1	1	0
3	341	268	1	1	1	21	1	100	1	17	1	20	23	0	1	1	1	180	1	1	0
3	318	318	1	0	1	318	0	140	1	12	1	35	40	0	1	1	1	300	0	1	0
3	195	32	1	1	1	195	0	195	0	16	1	36	39	1	1	0	0	90	1	1	0
3	469	467	1	1	1	469	0	90	1	20	1	35	33	0	0	1	0	120	0	1	0
3	93	47	1	1	1	93	0	93	0	28	1	7	2	1	1	0	0	135	1	1	0
3	515	390	1	1	1	515	0	515	0	31	1	23	25	1	1	1	0	210	1	1	0
3	183	183	1	0	1	183	0	130	1	21	1	11	7	0	1	0	0	120	1	1	0
3	105	105	1	0	1	105	0	105	0	105	0	14	18	1	0	0	0	150	1	1	0
3	128	115	1	1	1	128	0	128	0	12	1	37	35	0	0	1	1	270	0	1	0
3	164	164	1	0	1	164	0	164	0	164	0	19	32	0	0	0	1	285	1	1	0
3	129	93	1	1	1	129	0	129	0	51	1	37	34	0	1	1	0	240	1	1	0
3	122	120	1	1	1	122	0	122	0	12	1	25	29	0	1	1	1	510	1	1	0
3	80	80	1	0	1	21	1	80	0	0	1	35	28	1	0	0	0	780	1	1	0
3	677	677	1	0	1	677	0	150	1	8	1	15	14	1	1	1	0	150	1	1	0
3	73	64	1	1	1	73	0	73	0	38	1	45	42	0	1	1	0	180	1	2	1
3	168	168	1	0	1	168	0	200	1	48	1	32	43	0	1	1	1	150	1	2	1
3	74	74	1	0	1	29	1	74	0	24	1	41	29	0	1	1	1	750	0	2	1
3	16	16	1	0	1	16	0	16	0	16	0	27	36	0	0	1	0	180	0	4	0
3	248	157	1	1	1	248	0	100	1	52	1	33	39	0	0	1	1	180	1	4	0
3	732	625	1	1	1	732	0	732	0	18	1	39	43	0	1	1	1	150	1	4	0
3	105	48	1	1	1	105	0	105	0	30	1	17	14	0	1	0	0	210	1	4	0
3	392	273	1	1	1	392	0	122	1	24	1	43	50	1	1	1	0	240	0	3	1
3	63	63	1	0	1	38	1	63	0	16	1	44	37	1	1	0	0	360	1	3	1
3	97	76	1	1	1	97	0	97	0	97	0	48	56	1	1	1	1	330	0	3	1
3	153	113	1	1	1	153	0	153	0	59	1	31	25	0	1	1	1	240	0	3	1
3	363	363	1	0	1	363	0	363	0	19	1	52	48	1	1	1	0	180	0	3	1

E
Selected Solutions
to Exercises

Solutions to Chapter 2

2.1 (a) 1000
 (b) 693.15
 (c) 0.1353

2.3 (a) 50 days: 0.2205
       100 days: 0.0909
       150 days: 0.0516
 (b) 21.5 days
 (c) The inflection point is at 13.572
 (d) 52.1 days

2.5 (a) Mean = 210.2985 days
       Median = 23.9747 days
 (b) 100 days: 0.2466
       200 days: 0.1544
       300 days: 0.1127

2.7 (a) 0.3027
 (b) 0.4303
 (c) 15 months

2.9   (a)

	Treatment A	Treatment B
1 year	0.5030	0.4042
2 years	0.3673	0.2779
5 years	0.2127	0.1475

(b)

	Treatment A	Treatment B
1 year	0.5019	0.4397
2 years	0.4161	0.3569
5 years	0.3106	0.2598

2.11   (a) $f(x) = 0,$ $x < \phi$

$\quad\quad\quad \lambda\alpha(x - \phi)^{(\alpha-1)} \exp\{-\lambda(x - \phi)^{\alpha}\}, \quad x \geq \phi$

$\quad\quad h(x) = 0,$ $x < \phi$

$\quad\quad\quad \lambda\alpha(x - \phi)^{(\alpha-1)},$ $x \geq \phi$

(b) Mean lifetime = 233.3333

Median lifetime = 192.4196

2.13   (a) $S(x) = 1,$ $x < 1$

$\quad\quad\quad 1 - p,$ $1 \leq x < 2$

$$\vdots$$

$\quad\quad\quad (1 - p)^i, \quad i \leq x < i + 1, i = 1, 2, \ldots$

(b) $h(x) = p, \quad x = 1, 2, \ldots$

$\quad\quad\quad 0,$ elsewhere

For both the exponential and the geometric distributions the hazard rate is constant and is equal to the reciprocal of the mean.

2.15   (a) $S(x) = \exp[-x(\alpha + \beta x/2)]$

(b) $f(x) = (\alpha + \beta x) \exp[-x(\alpha + \beta x/2)]$

2.17   (a) $E(X) = 10$

(b) $h(x) = 2/(x + 10)$

(c) $S(x) = 100/(x + 10)^2$

2.19   (a) $S_X(x) = 1 - x, 0 < x < 1$

(b) $CI(x) = x - .75x^2 + .5x^3 - .25x^4, 0 < x < 1$

Solutions to Chapter 3

3.1   (a) Generalized Type I right censoring

(b) Generalized Type I right censoring

3.3   (a) Left censoring at 42 days

(b) Type I right censoring at 140 days

(c) Interval censoring in 84–91 days

(d) Random right censoring at 37 days

(e) $L \propto [1 - S(42)]S(140)[S(84) - S(91)]S(37)$

3.5   From (3.5.1) the likelihood has the form

$$L \propto \left[\frac{\alpha\lambda(0.5)^{\alpha-1}}{(1 + \lambda(0.5)^\alpha)^2}\right] \left[\frac{\alpha\lambda(1)^{\alpha-1}}{(1 + \lambda)^2}\right] \left[\frac{\alpha\lambda(0.75)^{\alpha-1}}{(1 + \lambda(0.75)^\alpha)^2}\right]$$

$$\left[1 - \frac{1}{1 + \lambda(0.25)^\alpha}\right] \left[1 - \frac{1}{1 + \lambda(1.25)^\alpha}\right]$$

3.7   (a) First 4 observations are interval-censored

Last 4 observations are Type I right-censored

(b) $L \propto [\exp(-\lambda 55^\alpha) - \exp(-\lambda 56^\alpha)][\exp(-\lambda 58^\alpha) - \exp(-\lambda 59^\alpha)]$

$[\exp(-\lambda 52^\alpha) - \exp(-\lambda 53^\alpha)][\exp(-\lambda 59^\alpha) - \exp(-\lambda 60^\alpha)]$

$[\exp(-\lambda 60^\alpha)]^4$

Solutions to Chapter 4

4.1   (a) $\hat{S}(12) = 0.9038$, $SE(\hat{S}(12)) = 0.0409$, $\hat{S}(60) = 0.6538$, $SE(\hat{S}(60)) = 0.0660$

(b) $\hat{H}(60) = 0.4178$ $SE(\hat{H}(60)) = 0.0992$
$\exp\{-\hat{H}(60)\} = 0.6585$

(c) $(0.5245, 0.7832)$

(d) $(0.5083, 0.7659)$

(e) $(0.5205, 0.7759)$

(f, g)

Time	EP Band (f)	Hall–Wellner Band (g)
36–41	(0.5060–0.8200)	(0.5014–0.8221)
41–51	(0.4866–0.8045)	(0.4855–0.8050)
51–65	(0.4674–0.7887)	(0.4692–0.7878)
65–67	(0.4281–0.7724)	(0.4520–0.7699)
67–70	(0.4089–0.7558)	(0.4345–0.7520)
70–72	(0.3900–0.7389)	(0.4169–0.7340)

(h) $\hat{\mu} = 146.6$ Confidence interval: $(92.4, 200.9)$

(i) Median $= 93$ Confidence interval: $(67–157)$

4.3   Solution to parts a and c

Time	Part a		Part b	
	Estimated Survival	Standard Error	Estimated Survival	Standard Error
0–22	1.0000	0.0000	1.0000	0.0000
22–27	0.9600	0.0392	0.9600	0.0392
27–50	0.9200	0.0543	0.9200	0.0543
50–68	0.8800	0.0650	0.8800	0.0650
68–99	0.8400	0.0733	0.8400	0.0733
99–101	0.8400	0.0733	0.8400	0.0733
101–108	0.8400	0.0733	0.8000	0.0800
108–121	0.8400	0.0733	0.8000	0.0800
121–131	0.8400	0.0733	0.8000	0.0800
131–134	0.8400	0.0840	0.8000	0.0800
134–136	0.7906	0.0922	0.7600	0.0854
136–139	0.7412	0.0984	0.7200	0.0898
139–144	0.6918	0.1030	0.6800	0.0933
144–186	0.6424	0.1030	0.6400	0.0960
186–191	0.6424	0.1030	0.6400	0.0960
191–198	0.6424	0.1030	0.6000	0.0980
198–203	0.6424	0.1030	0.6000	0.0980
203–210	0.5840	0.1090	0.5600	0.0993
210–217	0.5256	0.1126	0.5200	0.0999
217–224	0.5256	0.1126	0.5200	0.0999
224–231	0.5256	0.1126	0.4800	0.0999
231–248	0.5256	0.1126	0.4800	0.0999
248–256	0.4505	0.1190	0.4400	0.0993
256–290	0.4505	0.1190	0.4000	0.0980
290–306	0.4505	0.1190	0.4000	0.0980
306–308	0.4505	0.1190	0.4000	0.0980
308–320	0.4505	0.1190	0.3600	0.0960
320–363	0.4505	0.1190	0.3600	0.0960
363–410	0.2252	0.1700	0.3200	0.0933
410–441	0.2252	0.1700	0.3200	0.0933
441–482	0.2252	0.1700	0.2800	0.0898
482–511	0.2252	0.1700	0.2400	0.0854
511–559	0.2252	0.1700	0.2000	0.0800
559–561	0.2252	0.1700	0.1600	0.0733
561–580	0.2252	0.1700	0.1200	0.0650
580–683	0.2252	0.1700	0.0800	0.0543
683–724	0.2252	0.1700	0.0400	0.0392
724 infinity	0.2252	0.1700	0.0000	0.0000

(b) For $t > 363$, estimate $S(t)$ by $\exp\{-0.004t\}$

(d) $\hat{u} = 312.3$, $SE = 70.9$ days

(e) $\hat{\mu} = 294.6$, $SE = 42.7$

(f) $\bar{x} = 294.6$, $s = 213.4$, $SE = 42.7$

4.5 (a)

Time in Days	Estimated Survival	Standard Error
0–40	1.0000	0.0000
40–45	0.9828	0.0171
45–106	0.9655	0.0240
106–121	0.9480	0.0293
121–229	0.9297	0.0339
229–344	0.9115	0.0378
344–864	0.8912	0.0421
864–929	0.8672	0.0473
929–943	0.8383	0.0539
943–1016	0.8093	0.0592
1,016–1,196	0.7782	0.0646
1,196–2,171	0.7458	0.0696
2,171–2,276	0.6884	0.0846
2,276–2,650	0.6258	0.0974
>2,650	0.5364	0.1176

(b) (0.8087, 0.9737)

(c) (0.7731, 0.9497)

(d) (0.7960, 0.9497)

4.7

Time	Y	Left Truncated $S(t \mid \text{Alive at } 60)$	$S(t \mid \text{Alive at } 65)$	Y	No Truncation $S(t \mid \text{Alive at } 60)$	$S(t \mid \text{Alive at } 65)$
58	2	1.0000	1.0000	30	1.0000	1.0000
59	3	1.0000	1.0000	30	1.0000	1.0000
60	5	0.8000	1.0000	30	0.9667	1.0000
61	6	0.8000	1.0000	30	0.9667	1.0000
62	9	0.7111	1.0000	29	0.9333	1.0000
63	10	0.6400	1.0000	28	0.9000	1.0000
64	10	0.6400	1.0000	28	0.9000	1.0000
65	10	0.5120	0.8000	27	0.8333	0.9259
66	10	0.4608	0.7200	25	0.8000	0.8889
67	12	0.4608	0.7200	25	0.8000	0.8889
68	13	0.3899	0.6092	24	0.7333	0.8148
69	14	0.3342	0.5222	22	0.6667	0.7407
70	13	0.2828	0.4419	18	0.5926	0.6584
71	12	0.2357	0.3682	16	0.5185	0.5761
72	12	0.1964	0.3068	14	0.4444	0.4938
73	11	0.1785	0.2790	11	0.4040	0.4489
74	9	0.1587	0.2480	9	0.3591	0.3991
76	7	0.1360	0.2125	7	0.3078	0.3420
77	5	0.1088	0.1700	5	0.2463	0.2736
78	4	0.1088	0.1700	4	0.2463	0.2736
79	3	0.1088	0.1700	3	0.2463	0.2736
80	1	0.1088	0.1700	1	0.2463	0.2736

4.9   Parts a,b

Time	Thymic Lymphoma		Reticulum Cell Sarcoma		Other Causes		Overall Survival:
	CI	KME	CI	KME	CI	KME	KME
200	0.0633	0.0641	0.0000	0.0000	0.0127	0.0127	0.9241
300	0.2025	0.2061	0.0000	0.0000	0.0380	0.0433	0.7595
400	0.2278	0.2326	0.0000	0.0000	0.0506	0.0598	0.7215
500	0.2785	0.2880	0.0127	0.0182	0.0633	0.0766	0.6456
600	0.3038	0.3159	0.0253	0.0386	0.0759	0.0954	0.5949
700	0.3165	0.3314	0.1139	0.2082	0.1899	0.2827	0.3797
800	0.3418	0.3877	0.1646	0.3358	0.2658	0.4397	0.2278
900	0.3418	0.3877	0.1772	0.3773	0.4051	0.8008	0.0759
1,000	0.3418	0.3877	0.1899	0.5849	0.4430	0.9004	0.0253

(c) The Kaplan–Meier estimator (KME) for thymic lymphoma estimates the probability of having died from this cause in a hypothetical world where no other cause of death is possible.

(d) At 500 days we have 0.3014, at 800 days we have 0.6000. These are estimates of the conditional probability of dying from thymic lymphoma among survivors who have not died from one of the other two causes of death.

Solutions to Chapter 5

5.1

Age in Years	Estimated Survival Function
0–14	1.000
14–15	0.984
15–16	0.961
16–17	0.882
17–18	0.780
18–19	0.741
19–20	0.717
20–21	0.702
21–22	0.693
22–23	0.667
23–26	0.644
>26	0.515

5.3

Time	Estimated Survival Function
0–9	1.000
9–24	0.889
24–36	0.694
36–42	0.580
42–60	0.435
>60	0.000

5.5

T_i	X_i	R_i	d_i	Y_i	$P[X \leq x_i \mid X \leq 42]$
2	30	12	1	10	0.9000
4	27	15	1	13	0.8308
7	25	17	1	14	0.7714
14	19	23	1	17	0.7261
20	18	24	1	16	0.6807
18	17	25	1	16	0.6381
8	16	26	2	16	0.5584
13	16	26			
17	15	27	3	14	0.4387
26	15	27			
20	15	27			
15	13	29	2	11	0.3589
23	13	29			
5	12	30	1	9	0.3191
16	11	31	1	8	0.2792
15	9	33	1	8	0.2443
11	8	34	3	7	0.1396
6	8	34			
33	8	34			
4	7	35	1	5	0.1117
8	6	36	3	5	0.0447
35	6	36			
10	6	36			
36	4	38	2	2	0.0000
25	4	38			

5.7

1	2	3	4	5	6	7	8	9	10	11
Time	Y'_j	W_j	Y_j	d_j	$\hat{S}(a_j)$	$\hat{f}(a_{mj})$	$\hat{h}(a_{mj})$	$SE(\hat{S})$	$SE(\hat{f}(a_{mj}))$	$SE(\hat{h}(a_{mj}))$
45–50	1571	29	1556.5	29	1.0000	0.0022	0.0022	0	0.0005	0.0005
50–55	1525	60	1495.0	60	0.9891	0.0048	0.0049	0.0026	0.0008	0.0008
55–60	1429	83	1387.5	83	0.9653	0.0085	0.0091	0.0047	0.0011	0.0012
60–65	1284	441	1063.5	441	0.9221	0.0132	0.0148	0.0070	0.0015	0.0017
65–70	767	439	547.5	439	0.8562	0.0156	0.0192	0.0098	0.0021	0.0027
70–75	278	262	147.0	262	0.7780	0.0031	0.0126	0.0138	0.0031	0.0042
75–80	7	7	3.5	7	0.7304	0	0	0.0201		

5.9

t	$\hat{S}(t)$
0–46	1.000
46–49	0.877
49–54	0.754
54–61	0.631
61–62	0.508
62–64	0.385
64–68	0.308
68–120	0.231
120–150	0.154
150–160	0.077
> 160	0.000

Solutions to Chapter 6

6.1   (a) $\hat{h} = 0.0369, SE = 0.0144$

(b) $\hat{h} = 0.0261, SE = 0.0142$

(c) $\hat{h} = 0.0258, SE = 0.0154$

(d) Uniform Kernel $\hat{h} = 0.0223, SE = 0.0100$

Epanechnikov Kernel $\hat{h} = 0.0263, SE = 0.0122$

Biweight Kernel $\hat{h} = 0.0293, SE = 0.0142$

6.3   (a) $\hat{h} = 0.0161$ for Surgical Placement, $\hat{h} = 0.0229$ for Percutaneous Placement

(b) $\hat{h} = 0.0326$ for Surgical Placement, $\hat{h} = 0.0014$ for Percutaneous Placement

6.5   (a, b)

time	$B(t)$	Se$[B(t)]$
28	182	182
32	382	270
49	604	350
84	854	430
357	1140	516

(c)

time	Nelson-Aalen	$\Theta(t)$	$A(t)$
28	0.091	0.014	0.077
32	0.191	0.016	0.175
49	0.302	0.023	0.279
84	0.427	0.036	0.392
357	0.570	0.122	0.448
933	0.570	0.279	0.290
1078	0.570	0.312	0.257
1183	0.570	0.332	0.238
1560	0.570	0.383	0.187
2114	0.570	0.433	0.137
2144	0.570	0.435	0.135

6.7

Time	Estimate Using Dirichlet Prior	Estimate Using Beta Prior	Kaplan-Mejer Estimate	Prior
20	0.868	0.852	1.000	0.670
40	0.580	0.634	0.667	0.449
60	0.454	0.448	0.556	0.301

Solutions to Chapter 7

7.1 Log rank $\chi^2 = 6.03$, p-value $= 0.03$

7.3 (a) Log rank $\chi^2 = 3.793$, p-value $= 0.052$
(b) Gehan $\chi^2 = 2.864$, p-value $= 0.09$
(c) Tarone–Ware $\chi^2 = 3.150$, p-value $= 0.08$

7.5 Log rank $\chi^2 = 22.763$, p-value < 0.0001

7.7 (a) Log rank $\chi^2 = 33.380$, p-value $=< 0.0001$
(b) Untreated vs. Radiation
   Log rank $\chi^2 = 11.412$, p-value $= 0.0007$
Untreated vs. Radiation + BPA
   Log rank $\chi^2 = 21.671$, p-value < 0.0001
Radiation vs. Radiation + BPA
   Log rank $\chi^2 = 10.148$, p-value $= 0.0014$

The Bonferroni multiple comparison procedure would test at the $0.05/3 = 0.0167$ level. All pairwise comparisons are significant.
(c) Log rank trend test $\chi^2 = 30.051$, p-value < 0.0001

7.9   (a) Log rank $\chi^2 = 4.736$, p-value $= 0.19$
Gehan $\chi^2 = 3.037$, p-value $= 0.39$

(b) For males,
Log rank $\chi^2 = 0.097$, p-value $= 0.76$
Gehan $\chi^2 = 0.366$, p-value $= 0.55$
For females,
Log rank $\chi^2 = 4.847$, p-value $= 0.03$
Gehan $\chi^2 = 2.518$, p-value $= 0.11$

To test the hypothesis that blacks have a higher mortality rate than whites, after adjusting by stratification for sex, we get

$$Z = 1.064, \quad p\text{-value} = 0.30$$

7.11   Stratified $\chi^2 = 23.25$, p-value < 0.0001

Prior to 1975 log rank $\chi^2 = 12.00$, p-value $= 0.007$

1975 or later log rank $\chi^2 = 11.59$, p-value $= 0.009$

7.13   (a) Log rank $\chi^2 = 5.4943$, p-value $= 0.0195$

(b) $Q = 2.34$, p-value $= 0.04$

(c) $Q_1 = 1.10$, p-value $= 0.27$

(d) $W_{KM} = 116.22$, $Z = 1.90$, p-value $= 0.06$

7.15   Let treatment 1, 2, and 3 = ALL, AML-Low, and AML-High, respectively. Then

$Z_{12}(365) = -0.83$, two-sided p-value $= 0.41$,

$Z_{13}(365) = 0.94$, two-sided p-value $= 0.35$,

$Z_{23}(365) = 1.88$, two-sided p-value $= 0.06$.

Solutions to Chapter 8

8.1   (a) $Z_1 = 1$ if HOD Allo patient, 0 otherwise
$Z_2 = 1$ if NHL Auto patient, 0 otherwise
$Z_3 = 1$ if HOD Auto patient, 0 otherwise

(b) Let
$Z_1 = 1$ if Auto patient, 0 otherwise
$Z_2 = 1$ if HOD patient, 0 otherwise
$Z_3 = Z_1 X Z_2$

(c) $\beta_1 = 1.5, \beta_2 = 2, \beta_3 = -3$

8.3   (a) Score test using Breslow method gives a p-value of 0.098.

(b) $b = -0.461$, $se(b) = 0.281$, $RR = 0.63$, 95% CI is $(0.36, 1.09)$

(c) p-value $= 0.106$

(d) p-value $= 0.100$

8.5   (a)

Testing Global Null Hypothesis

Test	Chi-Square	DF	p-value
Likelihood Ratio	7.89	3	0.048
Score	11.08	3	0.011
Wald	9.26	3	0.026

ANOVA Table

Variable	DF	Parameter Estimate	Standard Error	Chi-Square	p-value
HOD Allo	1	1.830	0.675	7.34	0.007
NHL Auto	1	0.664	0.564	1.38	0.24
HOD Auto	1	0.154	0.589	0.07	0.79

(b) The global tests are the same as in 8.5(a) above.

ANOVA Table

Variable	DF	Parameter Estimate	Standard Error	Chi-Square	p-value
Auto	1	0.664	0.564	1.38	0.24
HOD	1	1.830	0.675	7.34	0.007
Auto & HOD	1	-2.340	0.852	7.55	0.006

Likelihood ratio p-value $= 0.007$ and Wald p-value $= 0.006$. Thus we conclude that there is a significant interaction between disease type and transplant type.

(c) Either model gives a relative risk for an NHL Auto transplant patient to an NHL Allo transplant patient of 1.94 with a 95% confidence interval of $(0.64, 5.87)$.

(d) Comparing Allo patients gives a p-value of 0.007. Comparing Auto patients gives a p-value of 0.31.

(e) The test statistic is 8.50, which has a chi-square distribution with 2 degrees of freedom and the p-value is 0.014.

Note that the inferential conclusions in this entire problem do not depend upon the coding scheme employed.

8.7 (a) For the routine bathing care, the cut point is 25% of total surface area burned. For the chlorhexidine gluconate method, the cut point is 22% of total surface area burned.

(b) For routine bathing care method, $Q = .8080$, p-value > 0.30, $RR = 1.59$.

For chlorhexidine gluconate method, $Q = 1.3404$, p-value $= 0.055$, $RR = 2.31$.

(c) For routine bathing care method, $b = 0.007$, $SE = 0.009$, p-value $= 0.44$, $RR = 1.01$.

For chlorhexidine gluconate method, $b = 0.007$, $SE = 0.012$, p-value $= 0.54$, and $RR = 1.01$.

8.9 (a) Let $Z_1 = $ type of disinfectant, $Z_4 = $ % of surface area burned, p-value $= 0.056$.

(b) p-value $= 0.077$.

(c) Tests of hypothesis that the times to staphylococcus infection are the same for the two disinfectant groups adjusting for each of the listed factors in a separate model are shown below.

Tests for Z_1 adjusted for gender, p-value $= 0.040$; race, p-value $= 0.045$; area burned, p-value $= 0.077$; type of burn, p-value $= 0.045$.

(d) The final model along with the parameter estimates is given below. Although we have used Wald tests in this exercise, similar conclusions are obtained if the likelihood ratio statistic is used throughout.

Variable	DF	Parameter Estimate	Standard Error	Chi-Square	p-value	Hazard Ratio
Disinfectant	1	−0.601	0.298	4.07	0.044	0.55
Type of Burn*	3			7.94	0.047	
Scald	1	1.557	1.087	2.05	0.15	4.7
Electric	1	2.151	1.086	3.92	0.048	8.6
Flame	1	0.999	1.016	0.97	0.33	2.7
Race	1	2.269	1.025	4.90	0.027	9.7

*Chemical burn is referent group.

The local Wald test of the primary hypothesis of no difference between the times to staphylococcus infection for the two disinfectant groups has a p-value of 0.044, which suggests that the times to staphylococcus infection are different for the two disinfectant groups after adjustment for the type of the burn and race of the patient. Note that the test was insignificant when no cofounders were adjusted for in part a.

8.11   (a)

Test	Chi-Square	DF	p-value
Likelihood Ratio	16.58	1	<.001
Score	15.04	1	<.001
Wald	13.62	1	<.001

Variable	DF	Parameter Estimate	Standard Error	Chi-Square	p-value	Hazard Ratio
Z	1	−1.095	0.297	13.62	<.001	0.33

(b) Tests for Z adjusted for: mother's age p-value $<$ .001; urban p-value $<$ .001; alcohol p-value $<$ .001; smoking status p-value $<$ .001; region p-value $<$ .001; birthweight p-value $<$ .001; poverty p-value $<$ .001; race p-value $<$ .001; siblings p-value $<$ .001.

(c) The final model along with the parameter estimates is given below.

Variable	DF	Parameter Estimate	Standard Error	Chi-Square	p-value	Hazard Ratio
Z	1	−0.882	0.303	8.49	0.004	0.4
Smoking status*	2			9.39	0.009	
< 1 pack/day	1	0.751	0.256	8.60	0.003	2.1
> 1 pack/day	1	0.632	0.349	3.28	0.070	1.9
siblings	1	0.387	0.124	9.77	0.002	1.5
mother's age	1	−0.121	0.050	5.88	0.015	0.9

*Referent group is nonsmokers.

8.13   At 20 days for a patient with 25% of the total body area burned, 95% confidence intervals for the survival functions based on the log transformation (4.3.2), for the two bathing solutions, are $(0.52, 0.75)$ and $(0.67, 0.85)$, respectively.

Solutions to Chapter 9

9.1   Using $g(t) = \log(t)$ the Wald p-value is 0.47. No evidence of a departure from proportional hazards.

9.3   (a) Relative Risk = 0.90, Confidence Interval = $(0.58, 1.39)$

(b) Using $g(t) = \log(t)$ the Wald p-value is 0.009.

(c) The best cut point is at 254 days.

(d) Up to 254 days the relative risk of Chemo only to Chemo+Radiation is 0.24 95% Confidence Interval $(0.10, 0.56)$. Among 254-day survivors the relative risk is 1.89 $(104, 3.44)$.

9.5 (a)

Variable	DF	Parameter Estimate	Standard Error	Chi Square	Pr > Chi Square
Stage II	1	0.112	0.464	0.06	0.81
Stage III	1	0.619	0.356	3.03	0.082
Stage IV	1	1.697	0.443	14.86	<0.001
Age	1	0.017	0.015	1.30	0.25

(b) From a model with common covariate values in each strata $-2 \text{ LOG } L = 323.869$ is greater than a model with different covariates values in each strata $-2 \text{ LOG } L = 320.806$, Chi Square $= 3.06$, $p = .55$.

(c) Chi Square $= 2.82$, $p = 0.59$

9.7 Step 1: Waiting time: df $= 1$, $p = 0.075$, FAB Class: df $= 1$, $p = 0.883$.

MTX: df $= 1$, $p = 0.091$. Sex: df $= 3$, $p = 0.752$. CMV: df $= 3$, $p = 0.049$.

Age: df $= 3$, $p = 0.17$.

$\Rightarrow$ Add CMV.

Step 2: Waiting time: $p = 0.26$. FAB Class: $p = 0.67$.

MTX: $p = 0.062$. Sex: $p = 0.46$. Age: $p = 0.29$.

Final Model:

Variable	DF	Parameter Estimate	Standard Error	Chi-Square	Pr > Chi Square	Hazard Ratio
AML low-risk ($Z1$)	1	0.343	0.658	0.27	0.60	1.41
AML high-risk ($Z2$)	1	2.136	0.884	5.84	0.016	8.47
Donor CMV positive ($Z9$)	1	1.764	0.778	5.14	0.023	5.83
Patient CMV positive ($Z10$)	1	0.401	0.955	0.18	0.67	1.49
Both CMV positive ($Z11$)	1	-2.530	1.346	3.53	0.060	0.08

Solutions to Chapter 10

10.1 (a) Let $Z = 1$ if Allo, 0 if Auto, and model $h(t \mid Z) = \beta_0(t) + \beta_1(t)$. Estimation restricted to 1,345 days.

Time	$B_1(t)$	Standard Error
0–2	0	0
2–4	0.0625	0.0625
4–28	0.1292	0.0914
28–30	0.2006	0.1160
30–32	0.1636	0.1218
32–36	0.2405	0.1440
36–41	0.2020	0.1491
41–42	0.1620	0.1543
42–49	0.1204	0.1599
49–52	0.2037	0.1803
52–53	0.1602	0.1855
53–57	0.1148	0.1909
57–62	0.0671	0.1968
62–63	0.0171	0.2030
63–72	−0.0355	0.2098
72–77	0.0554	0.2286
77–79	0.1554	0.2495
79–81	0.2665	0.2731
81–84	0.1554	0.2842
84–108	0.2804	0.3105
108–132	0.2179	0.3167
132–140	0.1512	0.3237
140–252	0.0798	0.3314
252–357	−0.0111	0.3437
357–524	0.1318	0.3722
524–1345	−0.0682	0.4225

(b) $U = 3.37$, $V(U) = 117.74$, chi-square $= 0.96$, $p = 0.7560$.

(c) Inference is restricted to 0–79 days.

	Disease		Type of Transplant		Interaction	
time	$B_1(t)$	SE	$B_2(t)$	SE	$B_3(t)$	SE
0–2	0.0000	0.0000	0.0000	0.0000	0.0000	0.0000
2–4	0.0000	0.0000	0.2000	0.2000	−0.2000	0.2000
4–28	0.0000	0.0000	0.4500	0.3000	−0.4500	0.3202
28–30	0.0000	0.0000	0.4500	0.3202	−0.3591	0.3328
30–32	−0.0667	0.0667	0.3833	0.3202	−0.2924	0.3394
32–36	−0.0667	0.0067	0.3833	0.3270	−0.1924	0.3538
36–41	−0.1381	0.0977	0.3119	0.3270	−0.1210	0.3610
41–42	−0.2150	0.1244	0.2350	0.3347	−0.0441	0.3691
42–49	−0.1317	0.1497	0.2350	0.3435	−0.1274	0.3784
49–52	−0.1317	0.1497	0.2350	0.3435	−0.1063	0.3944
52–53	−0.2150	0.1713	0.1516	0.3435	0.0670	0.4031
53–57	−0.1241	0.1940	0.1516	0.3534	−0.0239	0.4132
57–62	−0.0241	0.1282	0.1516	0.3534	−0.1239	0.4251
62–63	−0.1150	0.2364	0.0607	0.3649	−0.0330	0.4347
63–72	−0.0039	0.2612	0.0607	0.3649	−0.1441	0.4487
72–77	−0.0039	0.2612	0.3941	0.4943	−0.4774	0.5590
77–79	−0.0039	0.2612	0.8941	0.7031	−0.9744	0.7500
79	−0.0039	0.2612	1.8941	1.2224	−1.9774	1.2500

Analysis of Variance (Using number at risk as weights)

Effect	Chi-Square	df	p-value
Disease	0.6587	1	0.4170
Type of transplant	0.0113	1	0.9153
Interaction	6.6123	1	0.0101

10.3 (a) $b = -0.00023$, SE $= 0.00064$, $\chi^2 = 0.129$, $p = 0.7193$.

(b) Effect	α	Standard Error	Chi-Square	df	p-value
Type of Transplant	0.0012	0.0010	1.38	1	0.2390
Disease	0.0170	0.0097	3.09	1	0.0786
Interaction	−0.0182	0.0098	3.45	1	0.0632

Solutions to Chapter 11

11.1 (a) From the Martingale residual plot a quadratic or a threshold model (see section 8.6) is suggested.

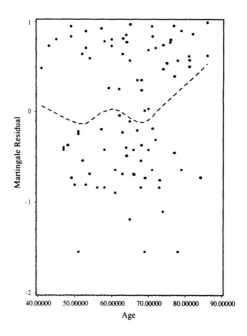

(b) The covariate year of transplant seems to enter the model as a linear term. The plot suggests its regression coefficient is not significantly different from zero.

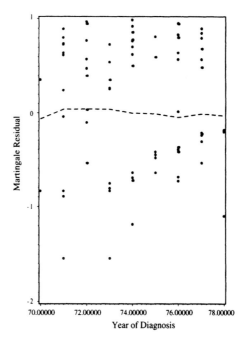

(c) The model seems to fit the data well.

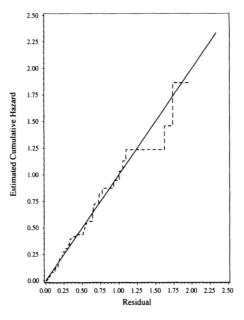

11.3   All four plots seem to suggest that proportional hazards is suspect for Stage IV as compared to Stage I.

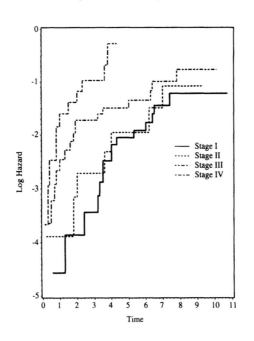

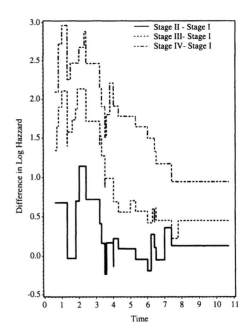

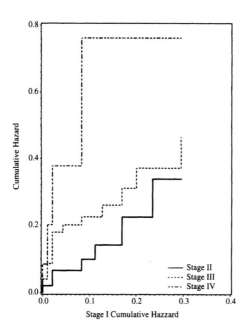 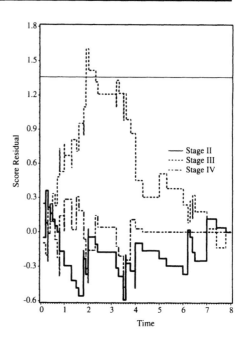

11.5 (a) The deviance residual suggests that there are a number of data points for which the model does not fit well. These are points with a deviance residual greater than 2. The worst three points are observations 529, 527, 526, which all have a risk score of −0.248. These patients, who the model predicts are good-risk patients, die very early at 2, 3, and 7 days after transplant. Most of the "outliers" are of this nature, patients who died too soon after transplant. A possible remedy is to add additional covariates which could help to explain these early deaths.

(b) For race these are observations 444, 435, 434, and 433. All are the four black males with the shortest survival time. For gender these are observations 532, 529, 527, and 526, the four white females who die the soonest after transplant. For the interaction term these are observations 812, 809, 807, and 806, the four black females with the shortest survival time.

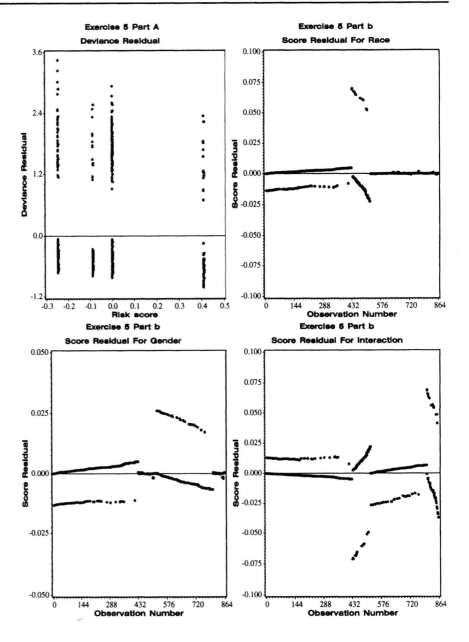

Solutions to Chapter 12

12.1   (a) For aneuploid group, $\hat{\lambda} = 0.016$, $\hat{\alpha} = 0.832$ with standard errors 0.010 and 0.128, respectively, and for diploid group, $\hat{\lambda} = 0.036$, $\hat{\alpha} = 0.775$ with standard errors 0.023 and 0.136, respectively.

(b) For aneuploid group, L.R. p-value is 0.21 and Wald p-value is 0.27; for diploid group, L.R. p-value is 0.12 and Wald p-value is 0.19.

(c) MLE of median of aneuploid and diploid groups, respectively, are 91.8 and 45.8. SE of median of aneuploid and diploid groups, respectively, are 19.8 and 17.6.

(d) LR and Wald p-values are 0.059 and 0.057, respectively. Estimate of RR is 0.58 and the 95% confidence interval is $(0.34, 1.01)$. Estimate of the acceleration factor is 0.51 and the 95% confidence interval is $(0.26, 1.02)$. This means that the median lifetime for diploid patients is between 0.26 and 1.02 times that of aneuploid patients with 95% confidence.

12.3   (a) LR global test p-value $= 0.004$.

ANOVA Table for $\hat{\mu}$, $\hat{\sigma}$, and $\hat{\gamma}_i$

Variable	df	Parameter Estimate	Standard Error	Chi-Square	p-Value
Intercept ($\hat{\mu}$)	1	7.831	0.753		
Scale ($\hat{\sigma}$)	1	1.653	0.277		
Auto ($\hat{\gamma}_i$)	1	-2.039	0.930	4.81	0.028
Hod ($\hat{\gamma}_2$)	1	-4.198	1.067	15.48	<0.001
Auto by hod ($\hat{\gamma}_3$)	1	5.358	1.377	15.14	<0.001

ANOVA Table for $\hat{\lambda}$, $\hat{\alpha}$, and $\hat{\beta}_i$

Variable	df	Parameter Estimate	Standard Error	Chi-Square	p-Value
Intercept ($\hat{\lambda}$)	1	0.009	0.007		
Scale ($\hat{\alpha}$)	1	0.605	0.101		
Auto ($\hat{\beta}_1$)	1	1.233	0.574	4.61	0.032
Hod ($\hat{\beta}_2$)	1	2.539	0.699	13.2	<0.001
Auto by hod ($\hat{\beta}_3$)	1	-3.241	0.878	13.6	<0.001

(b) p-value < 0.001.

(c) $RR = 3.4$ and 95% confidence interval for RR is $(1.1, 10.6)$.

(d) p-value < 0.001 and 0.17, respectively.

(e) Using contrast matrix $C = \begin{pmatrix} 1 & 0 & 0 \\ 1 & 0 & 1 \end{pmatrix}$, p-value < 0.001.

12.5   (a) For aneuploid group, $\hat{\lambda} = 0.009$, $\hat{\alpha} = 1.048$ with standard errors 0.007 and 0.163, respectively, and for diploid group, $\hat{\lambda} = 0.022$, $\hat{\alpha} = 1.035$ with standard errors 0.016 and 0.181, respectively.

(b) p-value > 0.6 for both aneuploid and diploid groups.

(c) MLE of median of aneuploid and diploid groups, respectively, are 87.2 and 39.6. SE of median of aneuploid and diploid groups, respectively, are 21.2 and 12.9.

(d) LR and Wald p-values are both 0.051. Estimate of relative odds is 0.44 and the 95% confidence interval is $(0.19, 1.01)$. Estimate of the acceleration factor is 0.45 and the 95% confidence interval is $(0.21, 1.002)$.

12.7   (a) For aneuploid group $\hat{\mu} = 4.46$, $\hat{\sigma} = 1.72$, and var-cov matrix for $\hat{\mu}$ and $\hat{\sigma}$ is

$$\begin{matrix} 0.07 & 0.02 \\ 0.02 & 0.06 \end{matrix}$$

For diploid group $\hat{\mu} = 3.64$, $\hat{\sigma} = 1.634$ and var-cov matrix for $\hat{\mu}$ and $\hat{\sigma}$ is

$$\begin{matrix} 0.10 & 0.01 \\ 0.01 & 0.07 \end{matrix}$$

(b) For aneuploid group $\hat{\mu} = 112.21$, $\hat{\sigma} = 96.71$ and var-cov matrix for $\hat{\mu}$ and $\hat{\sigma}$ is

$$\begin{matrix} 233.22 & 61.47 \\ 61.47 & 169.44 \end{matrix}$$

For diploid group $\hat{\mu} = 71.32$, $\hat{\sigma} = 70.67$ and var-cov matrix for $\hat{\mu}$ and $\hat{\sigma}$ is

$$\begin{matrix} 194.27 & 19.89 \\ 19.89 & 121.03 \end{matrix}$$

(c) For aneuploid group $\hat{\mu} = 4.75$, $\hat{\sigma} = 1.44$, $\hat{\theta} = 0.53$ and var-cov matrix for $\hat{\mu}$, $\hat{\sigma}$, and $\hat{\theta}$ is

$$\begin{matrix} 0.15 & -0.09 & 0.19 \\ -0.09 & 0.15 & -0.20 \\ 0.19 & -0.20 & 0.38 \end{matrix}$$

For diploid group $\hat{\mu} = 3.86$, $\hat{\sigma} = 1.55$, $\hat{\theta} = 0.32$ and var-cov matrix for $\hat{\mu}$, $\hat{\sigma}$, and $\hat{\theta}$ is

$$\begin{matrix} 0.40 & -0.14 & 0.45 \\ -0.14 & 0.13 & -0.20 \\ 0.45 & -0.20 & 0.65 \end{matrix}$$

(d) For aneuploid and diploid group the Wald p-values for testing $\hat{\theta} = 0$ is 0.39 and 0.69, respectively. Cannot reject log normal fit for both groups.

(e) For aneuploid and diploid group the Wald p-values for testing $\hat{\theta} = 1$ are 0.75 and 0.70, respectively. Cannot reject Weibull fit for both groups.

(f)

		Aneuploid	*Diploid*
Exponential	Log Likelihood	−77.14	−48.59
	AIC	156.28	99.18
Weibull	Log Likelihood	−76.36	−47.40
	AIC	156.72	98.80
Log logistic	Log likelihood	−76.09	−47.51
	AIC	156.18	99.02
Log normal	Log Likelihood	−76.42	−47.15
	AIC	156.84	98.30
Generalized gamma	Log Likelihood	−76.08	−47.08
	AIC	158.16	100.16

For aneuploid group log logistic is slightly better fit and for diploid group lognormal is slightly better fit.

12.9   The hazard plots, all of which should be linear, suggest that any of the models would be reasonable.

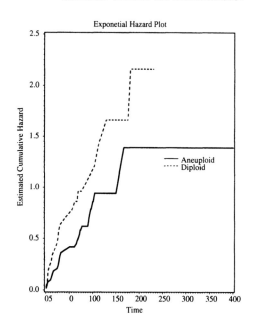

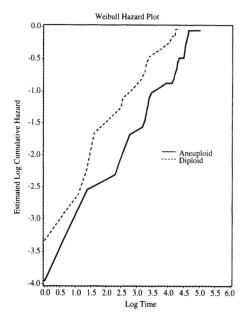

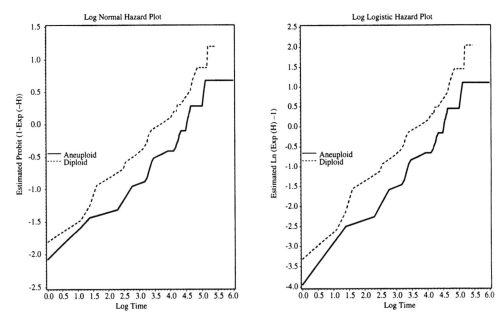

12.11 The plot appears to be roughly linear with a slope of about 3.5. The slope is a crude estimate of the acceleration function.

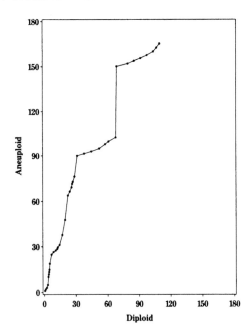

12.13   The model seems to fit well to the Weibull but is suspect for the log logistic model. Both Deviance residual plots suggest the models do not fit well to early events.

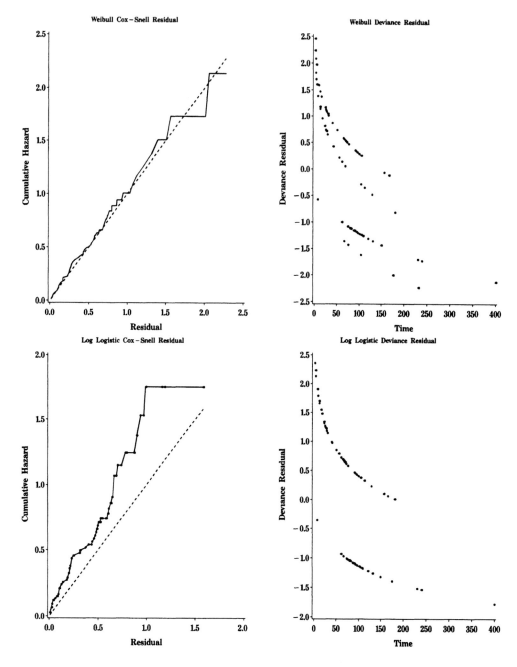

Solutions to Chapter 13

13.1 $T = 14.8$, $V = 107.5$, $Z = 1.4$, $p = 0.1539$. No evidence of random effect.

13.3   (a) Standard Cox model $b = -1.035$, SE $(b) = 0.44$, $p = 0.0187$.

       (b) Gamma Frailty Model $b = -1.305$, SE $(b) = 0.528$, $p = 0.0133$.

           Estimate of $\theta = 0.713$, SE $= 0.622$, Wald p-value of test of $\theta = 0 = 0.2517$, likelihood ratio p-value $= 0.1286$.

13.5   (a) See 13.1.

       (b) Adjusted SE $= 0.3852$, test statistic $= -1.035/0.3852$, $p = .0072$.

Bibliography

Aalen, O. O. Statistical Inference for a Family of Counting Processes. Ph.D. Dissertation, University of California, Berkeley, 1975.

Aalen, O. O. Nonparametric Estimation of Partial Transition Probabilities in Multiple Decrement Models. *Annals of Statistics* 6 (1978a): 534–545.

Aalen, O. O. Nonparametric Inference for a Family of Counting Processes. *Annals of Statistics* 6 (1978b): 701–726.

Aalen, O. O. A Model for Non-Parametric Regression Analysis of Counting Processes. In *Lecture Notes on Mathematical Statistics and Probability*, 2, W. Klonecki, A. Kozek, and J. Rosiski, eds. New York: Springer-Verlag, 1980, pp. 1–25.

Aalen, O. O. Heterogeneity in Survival Analysis. *Statistics in Medicine* (1988): 1121–1137.

Aalen, O. O. A Linear Regression Model for the Analysis of Lifetimes. *Statistics in Medicine* 8 (1989): 907–925.

Aalen, O. O. Modeling Heterogeneity in Survival Analysis by the Compound Poisson Distribution. *Annals of Applied Probability* 2 (1992): 951–972.

Aalen, O. O. Further Results on the Nonparametric Linear Regression Model in Survival Analysis. *Statistics in Medicine* 12 (1993): 1569–1588.

Aalen, O. O. and Johansen, S. An Empirical Transition Matrix for Nonhomogeneous Markov Chains Based on Censored Observations. *Scandinavian Journal of Statistics* 5 (1978): 141–150.

Akaike, H. Information Theory and an Extension of the Maximum Likelihood Principle. In *2nd International Symposium of Information Theory and Control*, E. B. N. Petrov and F. Csaki, eds. Akademia Kiado, Budapest, pp. 267–281.

American Joint Committee for Staging and End-Result Reporting. *Manual for Staging of Cancer*, 1972.

Andersen, P. K. Testing Goodness of Fit of Cox's Regression and Life Model. *Biometrics* 38 (1982): 67–77. Correction: 40 (1984): 1217.

Andersen, P. K., Borgan, Ø., Gill, R. D., and Keiding, N. Linear Nonparametric Tests for Comparison of Counting Processes, with Application to Censored Survival Data (with Discussion). *International Statistical Review*, 50 (1982): 219–258. Amendment: 52 (1984): 225.

Andersen, P. K., Borgan, Ø., Gill, R. D., and Keiding, N. Censoring, Truncation and Filtering in Statistical Models Based on Counting Processes. *Contemporary Mathematics* 80 (1988): 19–60.

Andersen, P. K., Borgan, Ø., Gill, R. D. and Keiding, N. *Statistical Models Based on Counting Processes.* New York: Springer-Verlag, 1993.

Andersen, P. K., Bentzon, M. W., and Klein, J. P. Estimating the Survival Function in the Proportional Hazards Regression Model: A Study of the Small Sample Size Properties. *Scandinavian Journal of Statistics* 23 (1996): 1–12.

Andersen, P. K., Klein, J. P., Knudsen, K. M., and Tabanera-Palacios, R. Estimation of the Variance in a Cox's Regression Model with Shared Frailties. *Biometrics* 53 (1997): 1475–1484.

Andersen, P. K. and Væth, M. Simple Parametric and Nonparametric Models for Excess and Relative Mortality. *Biometrics* 45 (1989): 523–535.

Arjas, E. A Graphical Method for Assessing Goodness of Fit in Cox's Proportional Hazards Model. *Journal of the American Statistical Association* 83 (1988) 204–212.

Arjas, E. A. and Haara, P. A Note on the Exponentiality of Total Hazards Before Failure. *Journal of Multivariate Analysis* 26 (1988): 207–218.

Avalos, B. R., Klein, J. L., Kapoor, N., Tutschka, P. J., Klein, J. P., and Copelan, E. A. Preparation for Marrow Transplantation in Hodgkin's and Non-Hodgkin's Lymphoma Using Bu/Cy. *Bone Marrow Transplantation* 13 (1993): 133–138.

Barlow, R. E. and Campo, R. Total Time on Test Processes and Application to Failure Data Analysis. In *Reliability and Fault Tree Analysis*, R. E. Barlow, J. Fussell, and N. D. Singpurwalla, eds. SIAM, Philadelphia, (1975) pp. 451–481.

Batchelor, J. R. and Hackett, M. HLA Matching in the Treatment of Burned Patients with Skin Allographs. *Lancet* 2 (1970): 581–583.

Beadle, G. F., Come, S., Henderson, C., Silver, B., and Hellman, S. A. H. The Effect of Adjuvant Chemotherapy on the Cosmetic Results after Primary Radiation Treatment for Early Stage Breast Cancer. *International Journal of Radiation Oncology, Biology and Physics* 10 (1984a): 2131–2137.

Beadle, G. F., Harris, J. R., Silver, B., Botnick, L., and Hellman, S. A. H. Cosmetic Results Following Primary Radiation Therapy for Early Breast Cancer. *Cancer* 54 (1984b): 2911–2918.

Berman, S. M. Note on Extreme Values, Competing Risks and Semi-Markov Processes. *The Annals of Mathematical Statistics* 34 (1963): 1104–1106.

Berretoni, J. N. Practical Applications of the Weibull Distribution. *Industrial Quality Control* 21 (1964): 71–79.

Beyer, W. H. *CRC Handbook of Tables for Probability and Statistics.* Boca Raton, Florida: CRC Press, 1968.

Bie, O., Borgan, Ø., and Liestøl, K. Confidence Intervals and Confidence Bands for the Cumulative Hazard Rate Function and Their Small Sample Properties. *Scandinavian Journal of Statistics* 14 (1987): 221–233.

Billingsley, P. *Convergence of Probability Measures.* New York: John Wiley and Sons, 1968.

Borgan, Ø. and Liestøl, K. A Note on Confidence Intervals and Bands for the Survival Curve Based on Transformations. *Scandinavian Journal of Statistics* 17 (1990): 35–41.

Breslow, N. E. A Generalized Kruskal–Wallis Test for Comparing K Samples Subject to Unequal Patterns of Censorship. *Biometrika* 57 (1970): 579–594.

Breslow, N. E. Covariance Analysis of Censored Survival Data. *Biometrics* 30 (1974): 89–99.

Breslow, N. E. Analysis of Survival Data under the Proportional Hazards Model. *International Statistics Review* 43 (1975): 45–58.

Brookmeyer, R. and Crowley, J. J. A Confidence Interval for the Median Survival Time. *Biometrics* 38 (1982a): 29–41.

Brookmeyer, R. and Crowley, J. J. A K-Sample Median Test for Censored Data. *Journal of the American Statistical Association* 77 (1982b): 433–440.

Brown, J. B. W., Hollander, M., and Korwar, R. M. Nonparametric Tests of Independence for Censored Data, with Applications to Heart Transplant Studies. In *Reliability and Biometry: Statistical Analysis of Lifelength*, F. Proschan and R. J. Serfling, eds. Philadelphia: SIAM, 1974, pp. 327–354.

Buckley, J. D. Additive and Multiplicative Models for Relative Survival Rates. *Biometrics* 40 (1984): 51–62.

Chiang, C. L. *Introduction to Stochastic Processes in Biostatistics.* New York: John Wiley and Sons, 1968.

Chiang, C. L. *The Lifetable and its Applications.* Malabar, Florida: Krieger, 1984.

Christensen, E., Schlichting, P., Andersen, P. K., Fauerholdt, L., Schou, G., Pedersen, B. V., Juhl, E., Poulsen, H., and Tygstrup, N. Updating Prognosis and Therapeutic Effect Evaluation in Cirrhosis Using Cox's Multiple Regression Model for Time-Dependent Variables. *Scandinavian Journal of Gastroenterology* 21 (1986): 163–174.

Chung, C. F. Formulae for Probabilities Associated with Wiener and Brownian Bridge Processes. Technical Report 79, Laboratory for Research in Statistics and Probability, Ottawa, Canada: Carleton University, 1986.

Clausius, R. Ueber Die Mittlere Lange Der Wege. *Ann. Phys. Lpzg* 105 (1858): 239–58.

Clayton, D. G. A Model for Association in Bivariate Life Tables and its Application in Epidemiological Studies of Familial Tendency in Chronic Disease Incidence. *Biometrika* 65 (1978): 141–151.

Clayton, D. G. and Cuzick, J. Multivariate Generalizations of the Proportional Hazards Model (with Discussion). *Journal of the Royal Statistical Society A* 148 (1985): 82–117.

Cleveland, W. S. Robust Locally Weighted Regression and Smoothing Scatter Plots. *Journal of the American Statistical Association,* 74 (1979): 829–836.

Collett, D. *Modeling Survival Data in Medical Research.* New York: Chapman and Hall, 1994.

Commenges, D. and Andersen, P. K. Score Test of Homogeneity for Survival Data. *Lifetime Data Analysis* 1 (1995): 145–160.

Contal, C. and O'Quigley, J. An Application of Change Point Methods in Studying the Effect of Age on Survival in Breast Cancer. *Computational Statistics and Data Analysis* 30 (1999): 253–270.

Copelan, E. A., Biggs, J. C., Thompson, J. M., Crilley, P., Szer, J., Klein, J. P., Kapoor, N., Avalos, B. R., Cunningham, I., Atkinson, K., Downs, K., Harmon, G. S., Daly, M. B., Brodsky, I., Bulova, S. I., and Tutschka, P. J. Treatment for Acute Myelocytic Leukemia with Allogeneic Bone Marrow Transplantation Following Preparation with Bu/Cy. *Blood* 78 (1991): 838–843.

Cornfield, J. A. and Detre, K. Bayesian Life Table Analysis. *Journal of the Royal Statistical Society Series B* 39 (1977): 86–94.

Costigan, T. M. and Klein. J. P. Multivariate Survival Analysis Based On Frailty Models, A. P. Basu, ed. *Advances In Reliability,* New York: North Holland, pp. 43–58.

Cox, D. R. The Analysis of Exponentially Distributed Lifetimes with Two Types of Failure. *The Journal of the Royal Statistical Society* B21 (1959): 411–421.

Cox, D. R. *Renewal Theory.* London: Methune, 1962.

Cox, D. R. Regression Models and Life Tables (with Discussion). *Journal of the Royal Statistical Society B* 34 (1972): 187–220.

Cox, D. R. and Hinkley, D. V. *Theoretical Statistics.* New York: Chapman and Hall, 1974.

Cox, D. R. and Oakes, D. *Analysis of Survival Data.* New York: Chapman and Hall, 1984.

Cox, D. R. and Snell, E. J. A General Definition of Residuals (with Discussion). *Journal of the Royal Statistical Society B* 30 (1968): 248–275.

Crowley, J. J. and Storer, B. E. Comment On 'A Reanalysis of the Stanford Heart Transplant Data', by M. Aitkin, N. Laird, and B. Francis. *Journal of the American Statistical Association,* 78 (1983): 277–281.

Cutler, S. J. and Ederer, F. Maximum Utilization of the Life Table Method in Analyzing Survival. *Journal of Chronic Diseases* 8 (1958): 699–712.

Dabrowska, D. M., Doksum, K. A., and Song, J. K. Graphical Comparison of Cumulative Hazards for Two Populations. *Biometrika* 76 (1989): 763–773.

David, H. A. *Order Statistics.* New York: John Wiley and Sons, 1981.

David, H. A., and Moeschberger, M. L. *The Theory of Competing Risks.* London: Charles Griffin, 1978.

Davis, D. J. An Analysis of Some Failure Data. *Journal of the American Statistical Association* 47 (1952): 113–150.

Doll, R. The Age Distribution of Cancer: Implications for Models of Carcinogens. *Journal of the Royal Statistical Society, Series A* 134 (1971): 133–66.

Efron, B. The Two Sample Problem with Censored Data. In *Proceedings of the Fifth Berkeley Symposium On Mathematical Statistics and Probability.* New York: Prentice-Hall, (1967): 4, 831-853.

Efron, B. The Efficiency of Cox's Likelihood Function for Censored Data. *Journal of the American Statistical Association* 72 (1977): 557–565.

Elandt–Johnson, R. C. and Johnson, N. L. *Survival Models and Data Analysis.* New York: John Wiley and Sons, 1980.

Epstein, B. The Exponential Distribution and Its Role in Life Testing. *Industrial Quality Control* 15 (1958): 2–7.

Epstein, B. and Sobel, M. Some Theorems Relevant to Life Testing from an Exponential Distribution. *Annals of Mathematical Statistics* 25 (1954): 373–81.

Escobar, L. A. and Meeker, W. Q. Assessing Influence in Regression Analysis with Censored Data. *Biometrics* 48 (1992): 507–528.

Feigl, P. and Zelen, M. Estimation of Exponential Survival Probabilities with Concomitant Information. *Biometrics* 21 (1965): 826–838.

Feinleib, M. A Method of Analyzing Log Normally Distributed Survival Data with Incomplete Follow-Up. *Journal of the American Statistical Association,* 55 (1960): 534–545.

Ferguson, T. S. A Bayesian Analysis of Some Nonparametric Problems. *Annals of Statistics* 1 (1973): 209–230.

Ferguson, T. S. and Phadia, E. G. Bayesian Nonparametric Estimation Based on Censored Data. *Annals of Statistics* 7 (1979): 163–186.

Finkelstein, D. M. A Proportional Hazards Model for Interval-Censored Failure Time Data. *Biometrics* 42 (1986): 845–854.

Finkelstein, D. M. and Wolfe, R. A. A Semiparametric Model for Regression Analysis of Interval-Censored Failure Time Data. *Biometrics* 41 (1985): 933–945.

Fleming, T. R. and Harrington, D. P. A Class of Hypothesis Tests for One and Two Samples of Censored Survival Data. *Communications In Statistics* 10 (1981): 763–794.

Fleming, T. R. and Harrington, D. P. *Counting Processes and Survival Analysis.* New York: John Wiley and Sons, 1991.

Fleming, T. R., Harrington, D. P., and O'Sullivan, M. Supremum Versions of the Log-Rank and Generalized Wilcoxon Statistics. *Journal of the American Statistical Association* 82 (1987): 312–320.

Fleming, T. R., O'Fallon, J. R., O'Brien, P. C., and Harrington, D. P. Modified Kolmogorov-Smirnov Test Procedures with Application to Arbitrarily Right Censored Data. *Biometrics* 36 (1980): 607–626.

Freireich, E. J., Gehan, E., Frei, E., Schroeder, L. R., Wolman, I. J., Anbari, R., Burgert, E. O., Mills, S. D., Pinkel, D., Selawry, O. S., Moon, J. H., Gendel, B. R., Spurr, C. L., Storrs, R., Haurani, F., Hoogstraten, B., and Lee, S. The Effect of 6-Mercaptopurine on the Duration of Steroid-Induced Remissions in Acute Leukemia: A Model for Evaluation of Other Potentially Useful Therapy. *Blood* 21 (1963): 699–716.

Galambos, J. Exponential Distribution. In *Encyclopedia of Statistical Science*, N. L. Johnson and S. Kotz, eds. New York: John Wiley and Sons, Vol. 2, pp. 582–587.

Galambos, J. and Kotz, S. *Characterizations of Probability Distributions. Lecture Notes in Mathematics* 675. Heidelberg: Springer-Verlag, 1978.

Gasser, T. and Müller, H. G. Kernel Estimation of Regression Functions. In *Smoothing Techniques for Curve Estimation, Lecture Notes in Mathematics* 757. Berlin: Springer-Verlag, 1979, pp. 23–68.

Gastrointestinal Tumor Study Group. A Comparison of Combination and Combined Modality Therapy for Locally Advanced Gastric Carcinoma. *Cancer* 49 (1982): 1771–1777.

Gatsonis, C., Hsieh, H. K., and Korway, R. Simple Nonparametric Tests for a Known Standard Survival Based on Censored Data. *Communications in Statistics—Theory and Methods* 14 (1985): 2137–2162.

Gehan, E. A. A Generalized Wilcoxon Test for Comparing Arbitrarily Singly Censored Samples. *Biometrika* 52 (1965): 203–223.

Gehan, E. A. and Siddiqui, M. M. Simple Regression Methods for Survival Time Studies. *Journal of the American Statistical Association* 68 (1973): 848–856.

Gelfand, A. E. and Smith, A. F. M. Sampling-Based Approaches to Calculating Marginal Densities. *Journal of the American Statistical Association* 85 (1990): 398–409.

Gill, R. D. Censoring and Stochastic Integrals. *Mathematical Centre Tracts*. Amsterdam: Mathematisch Centrum, 1980, 124.

Gill, R. D. Discussion of the Paper by D. Clayton and J. Cuzick. *Journal of the Royal Statistical Society A* 148 (1985): 108–109.

Gill, R. D. and Schumacher, M. A Simple Test of the Proportional Hazards Assumption. *Biometrika* 74 (1987): 289–300.

Gomez, G., Julia, O., and Utzet, F. Survival Analysis for Left Censored Data. In *Survival Analysis: State of the Art*, J. P. Klein and P. Goel, eds. Boston: Kluwer Academic Publishers, 1992, pp. 269–288.

Gompertz, B. On the Nature of the Function Expressive of the Law of Human Mortality and on the New Mode of Determining the Value of Life Contingencies. *Philosophical Transactions of the Royal Society of London* 115 (1825): 513–585.

Gooley, T. A., Leisenring, W., Crowley, J., and Storer, B. Estimation of Failure Probabilities in the Presence of Competing Risks: New Representations of Old Estimators, *Statistics in Medicine* 18 (1999): 695–706.

Greenwood, M. The Natural Duration of Cancer. In *Reports On Public Health and Medical Subjects* 33. London: His Majesty's Stationery Office, 1926, pp. 1–26.

Gross, S. and Huber–Carol, C. Regression Analysis for Discrete and Continuous Truncated and Eventually Censored Data. In *Survival Analysis: State of the Art*, J. P. Klein and P. Goel, eds. Boston: Kluwer Academic Publishers, pp. 289–308.

Guerts, J. H. L. On the Small Sample Performance of Efron's and Gill's Version of the Product Limit Estimator Under Proportional Hazards. *Biometrics* 43 (1987): 683–692.

Gumbel, E. J. *Statistics of Extremes*. New York: Columbia University Press, 1958.

Hall, W. J. and Wellner, J. A. Confidence Bands for a Survival Curve from Censored Data. *Biometrika* 67 (1980): 133–143.

Hamburg, B. A., Kraemer, H. C., and Jahnke, W. A Hierarchy of Drug Use in Adolescence Behavioral and Attitudinal Correlates of Substantial Drug Use. *American Journal of Psychiatry* 132 (1975): 1155–1163.

Harrington, D. P. and Fleming, T. R. A Class of Rank Test Procedures for Censored Survival Data. *Biometrika* 69 (1982): 133–143.

Heckman, J. J. and Honore, B. E. The Identifiability of the Competing Risks Model, *Biometrika* 76 (1989): 325–330.

Hjort, N. L. Nonparametric Bayes Estimators Based on Beta Processes in Models for Life History Data. *Annals of Statistics* 18 (1990): 1259–1294.

Hjort, N. L. Semiparametric Estimation of Parametric Hazard Rates. In *Survival Analysis: State of the Art*, J. P. Klein and P. Goel, eds. Boston: Kluwer Academic Publishers, 1992, pp. 211–236.

Hoel, D. G. and Walburg, H. E. Survival Analysis of Survival Experiments, *Journal of the National Cancer Institute* 49 (1972): 361–372.

Horner, R. D. Age at Onset of Alzheimer's Disease: Clue to the Relative Importance of Etiologic Factors? *American Journal of Epidemiology*, 126 (1987): 409–414.

Hougaard, P. A Class of Multivariate Failure Time Distributions. *Biometrika* 73 (1986a): 671–678.

Hougaard, P. Survival Models for Heterogeneous Populations Derived from Stable Distributions. *Biometrika* 73 (1986b): 387–396.

Howell, A. *A SAS Macro for the Additive Regression Hazards Model*. Master's Thesis, Medical College of Wisconsin, Milwaukee, Wisconsin, 1996.

Huffer, F. W. and McKeague, I. W. Weighted Least Squares Estimation for Aalen's Additive Risk Model. *Journal of the American Statistical Association* 86 (1991): 114–129.

Hyde, J. Testing Survival under Right Censoring and Left Truncation. *Biometrika* 64 (1977): 225–230.

Hyde, J. Survival Analysis with Incomplete Observations. In *Biostatistics Casebook*, R. G. Miller, B. Efron, B. W. Brown, and L. E. Moses, eds. New York: John Wiley and Sons, 1980, pp. 31–46.

Ichida, J. M., Wassell, J. T., Keller, M. D., and Ayers, L. W. Evaluation of Protocol Change in Burn-Care Management Using the Cox Proportional Hazards Model with Time-Dependent Covariates. *Statistics in Medicine* 12 (1993): 301–310.

Izenman, A. J. Recent Developments in Nonparametric Density Estimation. *Journal of the American Statistical Association* 86 (1991): 205–224.

Jesperson, N. C. B. *Discretizing a Continuous Covariate in the Cox Regression Model*, Research Report 86/2, Statistical Research Unit, University of Copenhagen, 1986.

Johansen, S. An Extension of Cox's Regression Model. *International Statistical Review* 51 (1983): 258–262.

Johnson, N. L. and Kotz, S. *Distributions in Statistics: Continuous Multivariate Distributions.* New York: John Wiley and Sons, 1970.

Johnson, W. and Christensen, R. Bayesian Nonparametric Survival Analysis for Grouped Data. *Canadian Journal of Statistics* 14, (1986): 307–314.

Kalbfleisch, J. D. and Prentice, R. L. Marginal Likelihoods Based on Cox's Regression and Life Model. *Biometrika* 60 (1973): 267–278.

Kalbfleisch, J. D. and Prentice, R. L. *The Statistical Analysis of Failure Time Data.* New York: John Wiley and Sons, 1980.

Kao, J. H. K. A Graphical Estimation of Mixed Weibull Parameters in Life-Testing Electron Tubes. *Technometrics* 1 (1959): 389–407.

Kaplan, E. L. and Meier, P. Nonparametric Estimation from Incomplete Observations. *Journal of the American Statistical Association* 53 (1958): 457–481.

Kardaun, O. Statistical Analysis of Male Larynx-Cancer Patients—A Case Study. *Statistical Nederlandica* 37 (1983): 103–126.

Kay, R. Goodness-of-Fit Methods for the Proportional Hazards Model: A Review. *Reviews of Epidemiology Santé Publications* 32 (1984): 185–198.

Keiding, N. Statistical Inference in the Lexis Diagram. *Philosophical Transactions of the Royal Society of London A* 332 (1990): 487–509.

Keiding, N. Independent Delayed Entry. In *Survival Analysis: State of the Art*, J. P. Klein and P. Goel, eds. Boston: Kluwer Academic Publishers, 1992, pp. 309–326.

Keiding, N. and Gill, R. D. Random Truncation Models and Markov Processes. *Annals of Statistics* 18 (1990): 582–602.

Kellerer, A. M. and Chmelevsky, D. Small-Sample Properties of Censored-Data Rank Tests. *Biometrics* 39 (1983): 675–682.

Klein, J. P. Small-Sample Moments of Some Estimators of the Variance of the Kaplan–Meier and Nelson–Aalen Estimators. *Scandinavian Journal of Statistics* 18 (1991): 333–340.

Klein, J. P. Semiparametric Estimation of Random Effects Using the Cox Model Based on the EM Algorithm. *Biometrics* 48 (1992): 795–806.

Klein, J. P. and Moeschberger, M. L. The Asymptotic Bias of the Product-Limit Estimator Under Dependent Competing Risks. *Indian Journal of Productivity, Reliability and Quality Control* 9 (1984): 1–7.

Klein, J. P. and Moeschberger, M. L. Bounds on Net Survival Probabilities for Dependent Competing Risks. *Biometrics* 44 (1988): 529–538.

Klein, J. P., Keiding, N., and Copelan, E. A. Plotting Summary Predictions in Multistate Survival Models: Probability of Relapse and Death in Remission for Bone Marrow Transplant Patients. *Statistics in Medicine* 12 (1994): 2315–2332.

Klein, J. P., Keiding, N., and Kreiner, S. Graphical Models for Panel Studies, Illustrated on Data from the Framingham Heart Study. *Statistics in Medicine* 14 (1995): 1265–1290.

Klein, J. P., Moeschberger, M. L., Li, Y. H. and Wang, S.T. Estimating Random Effects in the Framingham Heart Study. In *Survival Analysis: State of the Art*, J. P. Klein and P. Goel, eds. Boston: Kluwer Academic Publishers, 1992, pp. 99–120.

Koziol, J. A. A Two-Sample Cramér–Von Mises Test for Randomly Censored Data. *Biometrical Journal* 20 (1978): 603–608.

Kuo, L. and Smith, A. F. M. Bayesian Computations in Survival Models via the Gibbs Sampler. In *Survival Analysis: State of the Art*, J. P. Klein and P. Goel, eds. Boston: Kluwer Academic Publishers, 1992, pp. 11–24.

Lagakos, S. W. The Graphical Evaluation of Explanatory Variables in Proportional Hazards Regression Models. *Biometrika* 68 (1981): 93–98.

Lagakos, S. W., Barraj, L. M., and Degruttola, V. Nonparametric Analysis of Truncated Survival Data, with Application to AIDS. *Biometrika* 75 (1988): 515–523.

Lai, T. L. and Ying, Z. Estimating a Distribution Function with Truncated and Censored Data. *Annals of Statistics* 19 (1991): 417–442.

Latta, R. B. A Monte Carlo Study of Some Two-Sample Rank Tests with Censored Data. *Journal of the American Statistical Association* 76 (1981): 713–719.

Lee, E. W., Wei, L. J., and Amato, D. A. Cox-Type Regression Analysis for Large Numbers of Small Groups of Correlated Failure Time Observations. In *Survival Analysis: State of the Art*, J. P. Klein and P. Goel, eds. Boston: Kluwer Academic Publishers, 1992, pp. 237–248.

Lee, L. and Thompson, W. A., Jr. Results on Failure Time and Pattern for the Series System. In *Reliability and Biometry: Statistical Analysis of Lifelength*, F. Proshan and R. J. Serfling, eds. Philadelphia: SIAM, 1974, pp. 291-302.

Lee, P. N. and O'Neill, J. A. The Effect Both of Time and Dose Applied on Tumor Incidence Rate in Benzopyrene Skin Painting Experiments. *British Journal of Cancer* 25 (1971): 759–70.

Lee, S. and Klein, J. P. Bivariate Models with a Random Environmental Factor. *Indian Journal of Productivity, Reliability and Quality Control* 13 (1988): 1–18.

Li, Y., Klein, J. P., and Moeschberger, M. L. Effects of Model Misspecification in Estimating Covariate Effects in Survival Analysis for a Small Sample Size. *Computational Statistics and Data Analysis* 22 (1996): 177–192.

Liang, K. -Y., Self, S. G., and Chang, Y. -C. Modeling Marginal Hazards in Multivariate Failure-Time Data. *Journal of Royal Statistical Society* B 55; 441–463, 1993.

Lieblein, J. and Zelen, M. Statistical Investigation of the Fatigue Life of Deep-Groove Ball Bearings. *Journal of Research, National Bureau of Standards* 57 (1956): 273–316.

Lin, D. Y. MULCOX2: A General Program for the Cox Regression Analysis of Multiple Failure Time Data. *Computers in Biomedicine* 40 (1993): 279–293.

Lin, D. Y. and Wei, L. J. Robust Inference for the Cox Proportional Hazards Model. *Journal of the American Statistical Association* 84 (1989): 1074–1078.

Lin, D. Y. and Ying, Z. Semiparametric Analysis of the Additive Risk Model. *Biometrika* 81 (1994): 61–71.

Lin, D. Y. and Ying, Z. Semiparametric Analysis of General Additive-Multiplicative Hazard Models for Counting Processes. *Annals of Statistics* 23 (1995): 1712–1734.

Lin, D. Y. and Ying, Z. Additive Regression Models for Survival Data. In *Proceedings of the First Seattle Symposium in Biostatistics: Survival Analysis*, D. Y. Lin and T. R. Fleming, eds. New York: Springer, 1997, pp. 185–198.

Lindley, D. V. and Singpurwalla N. A. Multivariate Distributions for the Reliability of a System of Components Sharing a Common Environment. *Journal of Applied Probability* 23 (1986): 418–431.

Makeham, W.M. On the Law of Mortality and the Construction of Annuity Tables. *Journal of the Institute of Actuaries*, 8 (1860): 301–310.

Mantel, N., Bohidar, N. R., and Ciminera, J. L. Mantel–Haenszel Analysis of Litter-Matched Time-To-Response Data, with Modifications for Recovery of Interlitter Information. *Cancer Research* 37 (1977): 3863–3868.

Marquardt, D. An Algorithm for Least-Squares Estimation of Nonlinear Parameters. *SIAM Journal of Applied Mathematics* 11 (1963): 431–441.

Matthews, D. E. and Farewell, V. T. On Testing for a Constant Hazard Against a Change-Point Alternative (Corr. v41, 1103). *Biometrics* 38 (1982): 463–468.

McCarthy, D. J., Harman, J. G., Grassanovich, J. L., Qian, C., and Klein, J. P. Combination Drug Therapy of Seropositive Rheumatoid Arthritis. *The Journal of Rheumatology* 22 (1995): 1636–1645.

McCullagh, P. and Nelder, J. A. *Generalized Linear Models*, 2nd Ed. London: Chapman and Hall, 1989.

McGilchrist, C. A. and Aisbett, C. W. Regression with Frailty in Survival Analysis. *Biometrics* 47 (1991): 461–466.

McKeague, I. W. Asymptotic Theory for Weighted Least-Squares Estimators in Aalen's Additive Risk Model. *Contemporary Mathematics* 80 (1988): 139–152.

Miller, R. G. and Siegmund, D. Maximally Selected Chi-Square Statistics. *Biometrics* 38 (1982): 1011–1016.

Moeschberger, M. L and Klein, J. P. A Comparison of Several Methods of Estimating the Survival Function When There is Extreme Right Censoring. *Biometrics* 41 (1985): 253–259.

Morsing, T. *Competing Risks in Cross-Over Designs*. Technical Report, Department of Mathematics, Chalmers University of Technology, Goteborg, 1994.

Nahman, N. S., Middendorf, D. F., Bay, W. H., McElligott, R., Powell, S., and Anderson, J. Modification of the Percutaneous Approach to Peritoneal Dialysis Catheter Placement Under Peritoneoscopic Visualization: Clinical Results in 78 Patients. *Journal of The American Society of Nephrology* 3 (1992): 103–107.

Nair, V. N. Confidence Bands for Survival Functions with Censored Data: A Comparative Study. *Technometrics* 14 (1984): 265–275.

National Longitudinal Survey of Youth. *NLS Handbook.* Center for Human Resource Research. The Ohio State University, Columbus, Ohio, 1995.

Nelson, W. Theory and Applications of Hazard Plotting for Censored Failure Data. *Technometrics* 14 (1972): 945–965.

Nelson, W. *Applied Life Data Analysis.* New York: John Wiley and Sons, 1982.

Nielsen, G. G., Gill, R. D., Andersen, P. K., and Sørensen, T. I. A. A Counting Process Approach to Maximum Likelihood Estimation in Frailty Models. *Scandinavian Journal of Statistics* 19 (1992): 25–43.

Odell, P. M., Anderson, K. M., and D'Agostino, R. B. Maximum Likelihood Estimation for Interval-Censored Data Using a Weibull-Based Accelerated Failure Time Model. *Biometrics* 48 (1992): 951–959.

Peace, K. E. and Flora, R. E. Size and Power Assessment of Tests of Hypotheses on Survival Parameters. *Journal of the American Statistical Association* 73 (1978): 129–132.

Pepe, M. S. Inference for Events with Dependent Risks in Multiple Endpoint Studies. *Journal of the American Statistical Association* 86 (1991): 770–778.

Pepe, M. S. and Fleming, T. R. Weighted Kaplan–Meier Statistics: A Class of Distance Tests for Censored Survival Data. *Biometrics* 45 (1989): 497–507.

Pepe, M. S. and Fleming, T. R. Weighted Kaplan–Meier Statistics: Large Sample and Optimality Considerations. *Journal of the Royal Statistical Society B* 53 (1991): 341–352.

Pepe, M. S. and Mori, M. Kaplan-Meier, Marginal or Conditional Probability Curves in Summarizing Competing Risks Failure Time Data? *Statistics in Medicine* 12 (1993): 737–751.

Pepe, M. S., Longton, G. Pettinger, Mori, M., Fisher, L. D., and Storb, R. Summarizing Data on Survival, Relapse, and Chronic Graft-Versus-Host Disease After Bone Marrow Transplantation: Motivation for and Description of New Methods. *British Journal of Haematology* 83 (1993): 602–607.

Peterson, A. V., Jr. Bounds for a Joint Distribution Function with Fixed Sub-Distribution Functions: Application to Competing Risks. *Proceedings of the National Academy of Sciences* 73 (1976): 11–13.

Peto, R. and Lee, P. N. Weibull Distributions for Continuous-Carcinogenesis Experiments. *Biometrics,* 29 (1973): 457–470.

Peto, R. and Peto, J. Asymptotically Efficient Rank Invariant Test Procedures (with Discussion). *Journal of the Royal Statistical Society A* 135 (1972): 185–206.

Peto, R. and Pike, M. C. Conservatism of the Approximation $\Sigma(0 - E)^2/E$ in the Log Rank Test for Survival Data or Tumor Incidence Data. *Biometrics* 29 (1973): 579–584.

Pike, M. C. A Method of Analysis of a Certain Class of Experiments in Carcinogenesis. *Biometrics* 22 (1966): 142–161.

Prentice, R. L. and Marek, P. A Qualitative Discrepancy Between Censored Data Rank Tests. *Biometrics* 35 (1979): 861–867.

Qian, C. Time-Dependent Covariates in a General Survival Model with Any Finite Number of Intermediate and Final Events. Unpublished Doctoral Dissertation, The Ohio State University, Columbus, Ohio, 1995.

Ramlau–Hansen, H. The Choice of a Kernel Function in the Graduation of Counting Process Intensities. *Scandinavian Actuarial Journal* (1983a): 165–182.

Ramlau–Hansen, H. Smoothing Counting Process Intensities by Means of Kernel Functions. *Annals of Statistics* 11 (1983b): 453–466.

Rosen, P. and Rammler, B. The Laws Governing the Fineness of Powdered Coal. *Journal of Inst. Fuels* 6 (1933): 29–36.

Sacher, G. A. On the Statistical Nature of Mortality with Special References to Chronic Radiation Mortality. *Radiation* 67 (1956): 250–257.

Schoenfeld, D. Partial Residuals for the Proportional Hazards Regression Model. *Biometrika* 69 (1982): 239–241.

Schumacher, M. Two-Sample Tests of Cramér–Von Mises and Kolmogorov–Smirnov Type for Randomly Censored Data. *International Statistical Review* 52 (1984): 263–281.

Sedmak, D. D., Meineke, T. A., Knechtges, D. S., and Anderson, J. Prognostic Significance of Cytokeratin-Positive Breast Cancer Metastases. *Modern Pathology* 2 (1989): 516–520.

Sheps, M. C. Characteristics of a Ratio Used to Estimate Failure Rates: Occurrences Per Person Year of Exposure. *Biometrics* 22 (1966): 310–321.

Sickle-Santanello, B. J., Farrar, W. B., Keyhani-Rofagha, S., Klein, J. P., Pearl, D., Laufman, H., Dobson, J., and O'Toole, R. V. A Reproducible System of Flow Cytometric DNA Analysis of Paraffin Embedded Solid Tumors: Technical Improvements and Statistical Analysis. *Cytometry* 9 (1988): 594–599.

Slud, E. V. Nonparametric Identifiability of Marginal Survival Distributions in the Presence of Dependent Competing Risks and a Prognostic Covariate. In *Survival Analysis: State of the Art*, J. P. Klein and P. Goel, eds. Boston: Kluwer Academic Publishers, 1992, pp. 355–368.

Slud, E. V. and Rubinstein, L. V. Dependent Competing Risks and Summary Survival Curves. *Biometrika* 70 (1983): 643–649.

Smith, R. M. and Bain, L. J. An Exponential Power Life-Testing Distribution. *Communications in Statistics-Theory and Methods* 4 (1975): 469–481.

Stablein, D. M. and Koutrouvelis, I. A. A Two-Sample Test Sensitive to Crossing Hazards in Uncensored and Singly Censored Data. *Biometrics* 41 (1985): 643–652.

Storer, B. E. and Crowley, J. J. A Diagnostic for Cox Regression and General Conditional Likelihoods. *Journal of the American Statistical Association* 80 (1985): 139–147.

Susarla, V. and Van Ryzin, J. Nonparametric Bayesian Estimation of Survival Curves from Incomplete Observations. *Journal of the American Statistical Association* 61 (1976): 897–902.

Tarone, R. E. and Ware, J. H. On Distribution-Free Tests for Equality for Survival Distributions. *Biometrika* 64 (1977): 156–160.

Therneau, T. M., Grambsch, P. M., and Fleming, T. R. Martingale-Based Residuals for Survival Models. *Biometrika* 77 (1990): 147–160.

Thisted, R. A. *Elements of Statistical Computing.* New York: Chapman and Hall, 1988.

Thompson, W. A., Jr. On the Treatment of Grouped Observations in Life Studies. *Biometrics* 33 (1977): 463–470.

Thomsen, B. L. A Note on the Modeling of Continuous Covariates in Cox's Regression Model. Research Report 88/5, Statistical Research Unit, University of Copenhagen, 1988.

Tsai, W. -Y. Testing The Assumption of Independence of Truncation Time and Failure Time. *Biometrika* 77 (1990): 169–177.

Tsiatis, A. A Nonidentifiability Aspect of the Problem of Competing Risks. *Proceedings of the National Academy of Sciences* 72 (1975): 20–22.

Tsuang, M. T. and Woolson, R. F. Mortality in Patients with Schizophrenia, Mania and Depression. *British Journal of Psychiatry*, 130 (1977): 162–166.

Turnbull, B. W. Nonparametric Estimation of a Survivorship Function with Doubly Censored Data. *Journal of the American Statistical Association* 69 (1974): 169–173.

Turnbull, B. W. The Empirical Distribution Function with Arbitrarily Grouped, Censored and Truncated Data. *Journal of the Royal Statistical Society B* 38 (1976): 290–295.

Turnbull, B. W. and Weiss, L. A Likelihood Ratio Statistic for Testing Goodness of Fit with Randomly Censored Data. *Biometrics* 34 (1978): 367–375.

U.S. Department of Health and Human Services. Vital Statistics of the United States, 1959.

U.S. Department of Health and Human Services. Vital Statistics of the United States, 1990.

Wagner, S. S. and Altmann, S. A. What Time Do the Baboons Come Down from the Trees? (An Estimation Problem). *Biometrics* 29 (1973): 623–635.

Wang, S. T., Klein, J. P., and Moeschberger, M. L. Semiparametric Estimation of Covariate Effects Using the Positive Stable Frailty Model. *Applied Stochastic Models and Data Analysis* 11 (1995): 121–133.

Ware, J. H. and DeMets, D. L. Reanalysis of Some Baboon Descent Data. *Biometrics* 32 (1976): 459–463.

Wei, L. J., Lin, D. Y., and Weissfeld, L. Regression Analysis of Multivariate Incomplete Failure Time Data by Modeling Marginal Distributions. *Journal of the American Statistical Association* 84 (1989): 1065–1073.

Weibull, W. A Statistical Theory of the Strength of Materials. *Ingeniors Vetenskaps Akakemien Handlingar* 151 (1939): 293–297.

Weibull, W. A Statistical Distribution of Wide Applicability. *Journal of Applied Mechanics* 18 (1951): 293–297.

Weissfeld, L. A. and Schneider, H. Influence Diagnostics for the Weibull Model to Fit to Censored Data. *Statistics and Probability Letters* 9 (1990): 67–73.

Wellner, J. A. A Heavy Censoring Limit Theorem for the Product Limit Estimator. *Annals of Statistics* 13 (1985): 150–162.

Woolson, R. F. Rank Tests and a One-Sample Log Rank Test for Comparing Observed Survival Data to a Standard Population. *Biometrics* 37 (1981): 687–696.

Wu, J.-T. Statistical Methods for Discretizing a Continuous Covariate in a Censored Data Regression Model. Ph.D. Dissertation, The Medical College of Wisconsin, 2001.

Zheng, M. and Klein, J. P. A Self-Consistent Estimator of Marginal Survival Functions Based on Dependent Competing Risk Data and an Assumed Copula. *Communications in Statistics-Theory and Methods* A23 (1994): 2299–2311.

Zheng, M. and Klein, J. P. Estimates of Marginal Survival for Dependent Competing Risks Based on an Assumed Copula. *Biometrika* 82 (1995): 127–138.

Author Index

Subject Index

CPSIA information can be obtained at www.ICGtesting.com
Printed in the USA
LVOW030009170112

264181LV00001B/2/P

9 781441 929853